REVIEWS in MINERALOGY   Volume 35

GEOMICROBIOLOGY:

INTERACTIONS BETWEEN MICROBES AND MINERALS

Edited by

Jillian F. Banfield
University of Wisconsin–Madison

Kenneth H. Nealson
University of Wisconsin–Milwaukee

COVER: Bacterium (DOE SMCC BO693) attached to the surface of etched bytownite plagioclase via strands of extracellular poly-saccharide. High-pressure cryofixation and freeze-fracture sample preparation preserved the true structure of the highly hydrated polymer. Field emission gun high-resolution scanning electron micrograph: 1.5 kV accelerating voltage. The cell is one micrometer in length.

Series Editor: Paul H. Ribbe
Department of Geological Sciences
Virginia Polytechnic Institute & State University
Blacksburg, Virginia 24061 U.S.A.

Mineralogical Society of America
Washington, D.C.

MINERALOGICAL SOCIETY OF AMERICA

REVIEWS IN MINERALOGY

(Formerly: SHORT COURSE NOTES)

ISSN 0275-0279

Volume 35

GEOMICROBIOLOGY:
INTERACTIONS BETWEEN MICROBES AND MINERALS

ISBN 0-939950-45-6

ADDITIONAL COPIES of this volume as well as those listed on the following page may be obtained at moderate cost from:

THE MINERALOGICAL SOCIETY OF AMERICA
1015 EIGHTEENTH STREET, NW, SUITE 601
WASHINGTON, DC 20036 U.S.A.

List of volumes currently available in the
Reviews in Mineralogy series

Vol.	Year	Pages	Editor(s)	Title
1-7	*out*	*of*	*print*	
8	1981	398	A.C. Lasaga R.J. Kirkpatrick	KINETICS OF GEOCHEMICAL PROCESSES
9A	1981	372	D.R. Veblen	AMPHIBOLES AND OTHER HYDROUS PYRIBOLES— MINERALOGY
9B	1982	390	D.R. Veblen, P.H. Ribbe	AMPHIBOLES: PETROLOGY AND EXPERIMENTAL PHASE RELATIONS
10	1982	397	J.M. Ferry	CHARACTERIZATION OF METAMORPHISM THROUGH MINERAL EQUILIBRIA
11	1983	394	R.J. Reeder	CARBONATES: MINERALOGY AND CHEMISTRY
12	1983	644	E. Roedder	FLUID INCLUSIONS (Monograph)
13	1984	584	S.W. Bailey	MICAS
14	1985	428	S.W. Kieffer A. Navrotsky	MICROSCOPIC TO MACROSCOPIC: ATOMIC ENVIRONMENTS TO MINERAL THERMODYNAMICS
15	1990	406	M.B. Boisen, Jr. G.V. Gibbs	MATHEMATICAL CRYSTALLOGRAPHY (Revised)
16	1986	570	J.W. Valley H.P. Taylor, Jr. J.R. O'Neil	STABLE ISOTOPES IN HIGH TEMPERATURE GEOLOGICAL PROCESSES
17	1987	500	H.P. Eugster I.S.E. Carmichael	THERMODYNAMIC MODELLING OF GEOLOGICAL MATERIALS: MINERALS, FLUIDS, MELTS
18	1988	698	F.C. Hawthorne	SPECTROSCOPIC METHODS IN MINERALOGY AND GEOLOGY
19	1988	698	S.W. Bailey	HYDROUS PHYLLOSILICATES (EXCLUSIVE OF MICAS)
20	1989	369	D.L. Bish, J.E. Post	MODERN POWDER DIFFRACTION
21	1989	348	B.R. Lipin G.A. McKay	GEOCHEMISTRY AND MINERALOGY OF RARE EARTH ELEMENTS
22	1990	406	D.M. Kerrick	THE Al_2SiO_5 POLYMORPHS (Monograph)
23	1990	603	M.F. Hochella, Jr. A.F. White	MINERAL-WATER INTERFACE GEOCHEMISTRY
24	1990	314	J. Nicholls J.K. Russell	MODERN METHODS OF IGNEOUS PETROLOGY— UNDERSTANDING MAGMATIC PROCESSES
25	1991	509	D.H. Lindsley	OXIDE MINERALS: PETROLOGIC AND MAGNETIC SIGNIFICANCE
26	1991	847	D.M. Kerrick	CONTACT METAMORPHISM
27	1992	508	P.R. Buseck	MINERALS AND REACTIONS AT THE ATOMIC SCALE: TRANSMISSION ELECTRON MICROSCOPY
28	1993	584	G.D. Guthrie B.T. Mossman	HEALTH EFFECTS OF MINERAL DUSTS
29	1994	606	P.J. Heaney, C.T. Prewitt, G.V. Gibbs	SILICA: PHYSICAL BEHAVIOR, GEOCHEMISTRY, AND MATERIALS APPLICATIONS
30	1994	517	M.R. Carroll J.R. Holloway	VOLATILES IN MAGMAS
31	1995	583	A.F. White S.L. Brantley	CHEMICAL WEATHERING RATES OF SILICATE MINERALS
32	1995	616	J. Stebbins P.F. McMillan D.B.Dingwell	STRUCTURE, DYNAMICS AND PROPERTIES OF SILICATE MELTS
33	1996	862	E.S. Grew L.M. Anovitz	BORON: MINERALOGY, PETROLOGY AND GEOCHEMISTRY
34	1996	438	P.C. Lichtner C.I. Steefel E.H. Oelkers	REACTIVE TRANSPORT IN POROUS MEDIA

GEOMICROBIOLOGY

INTERACTIONS BETWEEN MICROBES AND MINERALS

FOREWORD

"Microorganisms cause mineral precipitation and dissolution and control the distribution of elements in diverse environments at and below the surface of the Earth. Conversely, mineralogical and geochemical factors exert important controls on microbial evolution and the structure of microbial communities." This was the rationale for the Short Course on Geomicrobiology presented by the Mineralogical Society of America on October 18 and 19, 1997, at the Alta Peruvian Lodge in Alta, Utah. The conveners, Jillian Banfield of the University of Wisconsin–Madison and Kenneth Nealson of the University of Wisconsin–Milwaukee, also served as editors of this volume. Remarkably, Jill did most of her Promethean work by *e*-mail from her sabbatical home at the University of Tokyo.

This is Volume 35 in the *Reviews in Mineralogy* series. Many volumes of the *RiM* series, begun in 1974, are still available from the Mineralogical Society. See the opposite page.

Paul H. Ribbe, Series Editor
Blacksburg, Virginia
September 1, 1997

TABLE OF CONTENTS

MICROBIAL DIVERSITY IN MODERN SUBSURFACE, OCEAN, SURFACE ENVIRONMENTS

Chapter 3   J.F. Banfield & R.J. Hamers

PROCESSES AT MINERALS AND SURFACES WITH RELEVANCE TO MICROORGANISMS AND PREBIOTIC SYNTHESIS

Chapter 4   B.J. Little, P.A. Wagner & Z. Lewandowski

SPATIAL RELATIONSHIPS BETWEEN
BACTERIA AND MINERAL SURFACES

Chapter 5   D. Fortin, F.G. Ferris & T.J. Beveridge

SURFACE-MEDIATED MINERAL DEVELOPMENT
BY BACTERIA

Chapter 6   **D.A. Bazylinksi & B.M. Moskowitz**

MICROBIAL BIOMINERALIZATION OF MAGNETIC IRON MINERALS:
MICROBIOLOGY, MAGNETISM
AND ENVIRONMENTAL SIGNIFICANCE

Chapter 7   B.M. Tebo, W.C. Ghiorse,
                                      L.G. van Waasbergen, P.L. Siering, & R. Caspi

BACTERIALLY-MEDIATED MINERAL FORMATION:
INSIGHTS INTO MANGANESE(II) OXIDATION FROM MOLECULAR
GENETIC AND BIOCHEMICAL STUDIES

Chapter 8   E.W. de Vrind-de Jong & J.P.M. de Vrind

ALGAL DEPOSITION OF CARBONATES AND SILICATES

Chapter 9   **A.T. Stone**

REACTIONS OF EXTRACELLULAR ORGANIC LIGANDS WITH DISSOLVED METAL IONS AND MINERAL SURFACES

Chapter 10   S. Silver

THE BACTERIAL VIEW OF THE PERIODIC TABLE: SPECIFIC FUNCTIONS FOR ALL ELEMENTS

Chapter 11   D.K. Nordstrom & G. Southam

GEOMICROBIOLOGY OF SULFIDE MINERAL OXIDATION

Chapter 12   **W.W. Barker, S.A. Welch & J.F. Banfield**

BIOGEOCHEMICAL WEATHERING OF SILICATE MINERALS

Chapter 13   **D.J. DesMarais**

LONG-TERM EVOLUTION OF THE
BIOGEOCHEMICAL CARBON CYCLE

PREFACE

H. Catherine W. Skinner

Department of Geology and Geophysics
Yale University
P. O. Box 208109
New Haven, Connecticut 06520 U.S.A.

Minerals have been known and honored since humans realized their essential contributions to the "terra firma" and stone tools thrust our species on the path of cultural evolution. Microbes are the oldest living creatures, probably inhabiting at least a few salubrious environments on the earth as early as 3.8 billion years ago. At this moment in history we are only beginning to appreciate the intimate juxtaposition and interdependence of minerals and microbes. We have been nudged into this position by the realization that our earth is finite, and the recognition of many global environmental problems that minerals and microbes contribute to, both positively and negatively. In addition, our globe may not be the only site in the solar system where 'life' arose, or may persist.

What all of these concerns enunciate is that we as scientists only dimly comprehend our own dynamic "terrestrial halls." This short course and volume have been generated with great enthusiasm for grasping as much as possible of the whole panorama of possibilities that involve both the inorganic and biologic realms.

Over 3600 mineral species have been defined and their relationships to each other and the environments in which they form have been documented. This vast data base, collected over the past several hundred years and constantly added to and upgraded, is a monument to the research efforts of many geoscientists focused on the inorganic realm. Much of this data has come from investigators intrigued by the novelty, beauty, and versatility of minerals, direct expressions of the chemistry and physics of geologic processes. We are now adding a new dimension to questions of mineral formation, dissolution, and distribution: what were, are, and will be the contributions of microbes to these basic components of the environment.

Microbes have also been known for hundreds of years. However, their small size (0.5 to 5 μm in diameter) and the difficulties associated with identifying a species unless it was grown in the laboratory (cultured), precluded thorough analysis. The advent of molecular biology has only recently made it possible to evaluate microbial evolutionary relatedness (phylogeny) and physiological diversity. These techniques are now being applied to study of microbial populations in natural environments.

It is becoming very clear that the surface of Earth is populated by far more species of microbes than there are types of minerals. We are now exploring every portion of the globe and finding the relationships under the rubric "geomicrobiology." The ocean deeps are characterized by a diversity of microorganisms, including those associated with manganese nodules. The profusion and concentration of minerals created at ocean ridges and vents matches the variety of microorganisms, large animals, and plants there. The snowy tops of mountain ranges and glaciers of Antarctica harbor not just ice but whole bacterial communities whose cellular types and activities need elucidation. The equatorial jungles and the deserts, with their enormous diversity of ecological niches, further challenge us.

The diversity of geographic, geologic, and biologic environments, including some contributed by humans (e.g. mines, air-conditioning equipment), can now also be explored in detail. Modern studies use protocols developed to preserve or measure *in situ* chemical and physical characteristics. Electron microscopes allow direct characterization of mineral and biological morphology and internal structures. Spectroscopic techniques permit complimentary chemical analysis, including determination of oxidation states, with very high spatial resolution. Other studies quantitatively measure isotopic abudances. These data serve to distinguish biologically mediated, or biologically controlled formation of the mineral from an abiotic process and mechanism.

Each ecological niche requires accurate characterization of the mineralogic and biologic entities in order for us to begin to understand the range of dynamic relationships. We can pose many questions. Is the mineral only a substrate, or is its occurrence and stability impacted by microbiologic activity and metabolic requirements? Which minerals are of microbiological rather than inorganic origin and what are the mechanisms by which organisms dictate the morphology and structure of the solid phase formed? How do organic metabolic products bind metals and change their form and distribution, with implications for metal toxicity and geochemical cycles? How do inorganic reactions such as mineral dissolution and precipitation impact microbial populations through control of their physical and chemical environments? Clearly, new and excitingly research areas exist for all varieties of scientists. Although published by the Mineralogical Society of America, the authors of this volume include microbiologists, molecular biologists, biochemists, biophysicists, bioengineers as well as biomineralogists. Here, they bring together their respective expertise and perspectives to provide disciplinary and interdisciplinary background needed to define and further explore the topic of geomicrobiology.

The volume is organized so as to first introduce the nature, diversity, and metabolic impact of microorganisms and the types of solid phases they interact with. This is followed by a discussion of processes that occur at cell surfaces, interfaces between microbes and minerals, and within cells, and the resulting mineral precipitation, dissolution, and changes in aqueous geochemistry. The volume concludes with a discussion of the carbon cycle over geologic time. In detail:

Nealson and Stahl acquaint us with the basic properties of prokaryotes, including their size and structure. They define the types and ranges of microorganisms and their metabolisms and describe their impacts on some important biogeochemical cycles.

Barns and Nierzwicki-Bauer document the phylogenetic relationships and evolution of microorganisms, begging some fundamental questions that might be now just beyond our grasp: What was the 'last common ancestor'? The physiology, biochemistry and ecology of hyperthermophilic, and the many diverse geologically important microbial species from the lithosphere and hydrosphere, as well as some of the techniques employed, are presented.

Banfield and Hamers describe and integrate the processes acting on minerals and at surfaces relevant to microorganisms, examining the factors that control mineralogy, mineral forms, and the stability of phases. Surface properties and reaction rates for dissolution, precipitation, and growth of important classes of minerals are discussed. The possible role of mineral surfaces in formation of prebiotic molecules needed to explain the origin of life is examined.

Little, Wagner and Lewandowski describe biofilms, an essential interface between microbes and minerals. They demonstrate that these membranes, with their unique morphological and structural attributes, are sites where much activity related to dissolution and/or formation of minerals takes place. Biology makes it possible to move molecules and elements against a gradient. Many questions regarding the transfer of elements from minerals to microbes at this important heterogeneous interface remain.

Fortin, Ferris and Beveridge review surface-mediated mineral development by bacteria. Fresh or oceanic waters, anaerobic or aerobic environments provide discretely different ecologies, bacterial entities, and resulting mineralogies. It is obvious from this presentation that investigators have just scratched the surface of microbial mineralization processes.

Bazlinski and Moskowitz review the magnetic biominerals and provide insights into the environmental and biological significance of these few tens of nanometer-sized mineral products. The magnetosome chemistry and biochemistry is probably the best understood of any biologically precipitated mineral. Their formation and unique properties underscore the roles these biomaterials play in the rock magnetic record and in geochemical cycles.

Tebo, Ghiorse, van Waasbergen, Siering and Caspi contribute data on the roles of Mn-minerals and Mn(II) oxidation in geologic environments. Their chapter encompasses molecular genetic and biochemical investigations. Manganese oxides and oxyhydroxides are notoriously difficult to identify and the crystal chemistry of these phases is a research effort on its own. The prospect of learning how microbes utilize the multiple oxidation states of Mn (2+, 3+ and 4+) as a source of energy sharpens the motivation for interdisciplinary study. Manganese is also known as a cofactor in the production and activation of the enzymes that digest large biomolecules that must be the source of the smaller molecular species and ultimately the building blocks of C, N, O, H required by all species. How have the mechanisms identified in the bacterial systems been transferred up the phylogenetic tree to plants and humans? This is an expanding and intriguing area for further investigation.

DeVrind-de Jong and de Vrind address silicate and carbonate deposition by algae (eukaryotic photosynthetic microorganisms). This chapter documents the mechanisms of biomineralization of diatoms and coccoliths. These abundant aquatic organisms are responsible for huge volumes of siliceous sediments and calcium carbonate deposits world wide. The implications of algal biomineralization for climatic variation throughout much of the Earth's history may be quite significant.

Stone leads us though a quantitative approach to evaluating reactions between organic molecules and cations. He considers available extracellular organic ligands and the roles these play in uptake of metals. He documents the basic chemical speciation and complexation for several elements, making metal to metal comparisons. Remaining challenges involve coordinating the organic and inorganic results of biologic activity.

Following the discussion of biomineralization and interactions between organic compounds and cations, Silver discusses the strategies microorganisms have evolved to deal with toxic metal concentrations in solution. Beyond the fundamental biological significance, this has important implications for understanding microbial populations in contaminated environments. The impact on the geochemical form (speciation) and distribution of elements is also discussed.

Nordstrom and Southam summarize sulfide mineral oxidation and dissolution kinetics and devote considerable effort to describing the specific contributions of microorganisms, mostly bacteria. Despite the vast amount of accumulated information, many unanswered questions remain.

Barker, Welch and Banfield address weathering of silicate minerals. This topic encompasses not only mineralogy but geomorphology, microbiology, and geochemistry. The necessary interdisciplinary mode of these investigations is highlighted by discussion of the role(s) of bacterial nutrition, groundwater chemistry, and biochemistry. There are obvious implications for hazardous waste storage, a currently daunting and politicized topic that requires predictions over thousands to millions of years.

Finally, Des Marais treats the long term evolution of the carbon cycle, adopting a biogeochemical view. He discusses the sources, sinks and the transfer of the element over geologic time. Consideration of such a basic series of questions relating to the partitioning of carbon necessitate interdisciplinary crossovers. It is a fitting conclusion to a dialogue in progress.

Chapter 1

MICROORGANISMS AND BIOGEOCHEMICAL CYCLES: WHAT CAN WE LEARN FROM LAYERED MICROBIAL COMMUNITIES?

Kenneth H. Nealson

Center for Great Lakes Studies
University of Wisconsin-Milwaukee
Milwaukee, Wisconsin 53204 U.S.A.

David A. Stahl

Department of Civil Engineering
Northwestern University
Evanston, Illinois 60208 U.S.A.

INTRODUCTION

A discussion of some of the properties of prokaryotes, dealing with basic issues like size, structure, and metabolic diversity is followed by a discussion of some emerging techniques and some new findings in microbial ecology that have relevance to the role(s) that the prokaryotes play in biogeochemical cycling. Only the two prokaryotic domains of life [Archaea and Bacteria] are dealt with here. These are illustrated diagramatically In Figure 1, where the three domains of life, as defined by sequences of 16S ribosomal RNA sequences (Olson et al. 1986, 1993; Pace et al. 1986, 1993; Stahl and Amann 1991, Woese 1987), are shown, and compared to the more classical view of the five kingdoms (Margulis 1981) derived from considerations of anatomical structures. Such molecular phylogenetic trees (see Barns and Nierswicki-Bauer, this volume, for more detailed presentations), suggest that the major genetic diversity on the planet resides among single cell organisms. For example, in terms of evolutionary distance, humans are only slightly removed from the fungi or the plants, while the distance between seemingly similar bacteria is quite impressively large. This view is consistent with what we know of evolution of the biota, given that the planet is believed to have been inhabited by prokaryotes for more than 3.5 billion years. There is a growing sense that eukaryotes (those organisms with chromosomes, nuclei, nuclear membranes, and many visible exterior structures) arose from within the Archaea and, therefore, are more recent inhabitants. The prokaryotes have remained small and simple throughout their evolutionary history; their diversity is expressed in terms of physiology and metabolism, while that of the larger eukaryotes is expressed in terms of structures and behavior.

In terms of the abilities of microbes to drive biogeochemical cycles, it is useful to begin with a general treatment of the major chemical species with which microbes interact to change oxidation state. In this case, one can examine a profile of a typical stratified environment, like a fjord (Fig. 2a) or a sediment (Fig. 2b) and see a progression of thermodynamically predictable redox changes with depth. This begins with oxygen depletion, and typically ends with the reduction of sulfate (in marine environments) to produce sulfide, or the reduction of CO_2 (in freshwater environments) to produce methane. Oxygen solubility in water limits the level of oxygen to the order of 300 to 400 μM (depending on water temperature), so that if organic matter is present that can be aerobically

0275-0279/97/0035-0001$05.00

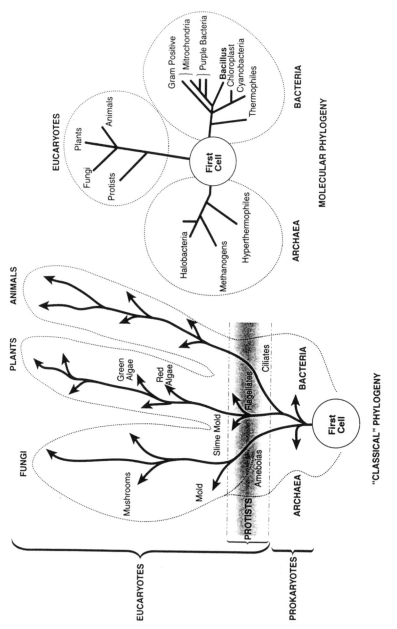

Figure 1. The domains of life, as defined by 16S ribosomal RNA sequence analysis, and compared to the classical 5 kingdoms of life defined by morphology and behavior. On the left the 5-kingdom phylogeny is shown, which presents the tree of life and diversity with the eukaryotic kingdoms as the dominant and diverse types. On the right is the view obtained from sequence analysis of the 16S ribosomal RNA genes. Here many of the functional groups that are discussed in this review are included, so that they can be placed in the context of the review article. The relative distances between groups show the phylogenetic separations as estimated from molecular sequence analysis.

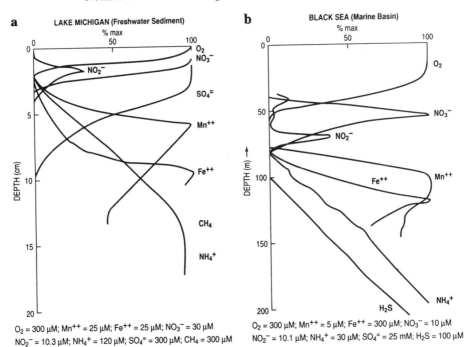

$O_2 = 300 \mu M; Mn^{++} = 25 \mu M; Fe^{++} = 25 \mu M; NO_3^- = 30 \mu M$
$NO_2^- = 10.3 \mu M; NH_4^+ = 120 \mu M; SO_4^= = 300 \mu M; CH_4 = 300 \mu M$

$O_2 = 300 \mu M; Mn^{++} = 5 \mu M; Fe^{++} = 300 \mu M; NO_3^- = 10 \mu M$
$NO_2^- = 10.1 \mu M; NH_4^+ = 30 \mu M; SO_4^= = 25 mM; H_2S = 100 \mu M$

Figure 2. Vertical profiles in typical freshwater (Lake Michigan) and marine (Black Sea) systems. These profiles are meant to act as general guidelines of what might be expected upon analysis of porewater components from marine or freshwater sediments. The upper regions are oxic, and thus consistent with eukaryotic life, while lower portions are anoxic, and primarily the domain of prokaryotes. The depth of oxygen depletion will be a function of the amount of organic carbon that reaches the sediment. The primary difference between the freshwater and marine sediments relates to the amount of sulfate in the latter, and the resulting dominance of the sulfur cycle, while in the freshwater sediments, methane formation is the terminal step, dominating carbon metabolism at depth.

respired, it is virtually a universal property of undisturbed sediments that they become anoxic with depth. After oxygen is depleted, a series of rather stable horizontal gradients are set up within the sediments, in which various electron acceptors are then consumed, usually in the order of decreasing redox potentials (Fig. 3). The gradients are a function of organic input, microbial metabolic abilities, and the geochemistry of the environment (marine vs. freshwater, mineral content, etc.).This is an uncommon abiotic phenomenon and a common biological one, as microbes are excellent catalysts that lead to rapid consumption of reactants at rates greater than diffusion. Similar layered communities are found throughout the world, ranging from scales of many meters in the Black Sea (Fig. 2a) to centimeters or millimeters in lake sediments (Fig. 2b), to micrometers in algal mats and biofilms (see Fig. 2 in Little and Lewandowski, this volume). A major difference between stratified water columns and the more common stratified sediments and soils, is the abundance of minerals (clays, carbonates, silicates, metals oxides, etc.). Minerals can be both reactants with and/or products of microbial metabolism, and undoubtedly impact the microbial ecology and metabolism of their environments, both structurally and functionally.

The primary inhabitants of these stratified environments are the simpler, smaller prokaryotes. As opposed to eukaryotes, the prokaryotes have few modes of behavior other than growth and division: their small size and rigid cell walls preclude their being predators in the classical sense. Diversity is expressed in terms of metabolism rather than structure,

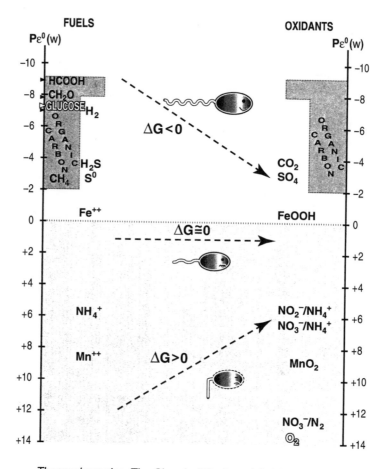

Thermodynamics: The Chemical Fuels and Oxidants of Life

Figure 3. Electron donors and acceptors of life. This cartoon is designed to show the various redox couples known to be utilized by living organisms. On the left are the fuels or energy sources (organic and inorganic), while on the right are oxidants used to burn these fuels. If the arrow between energy source and oxidant has a negative slope, then, kinetic properties allowing, an organism should be able to harvest this energy. Prokaryotes have exploited most of these energetic niches, while eukaryotes are confined to just a few organic compounds as fuels, and to molecular oxygen as oxidant.

and prokaryotes have optimized their biochemistry for the uptake and utilization of a wide variety of nutrients. In sedimentary environments, prokaryotes so efficiently exploit the energy sources that they outcompete the larger, metabolically more restricted eukaryotes. Many aspects of prokaryotic metabolic diversity will be dealt with breifly here; for more detailed treatments of metabolism and organisms, one is referred to The Prokaryotes (Balows et al. 1991).

Metabolic optimization includes versatility. Eukaryotes, for the most part, are limited to energy sources that can be converted into glucose or breakdown products of glucose, like pyruvate, and to oxygen as the only electron acceptor used to burn their metabolic fuel through respiration. In contrast, prokaryotes use a wide variety of different energy sources, exploit many different alternative electron acceptors, or "oxygen substitutes" for respiration in the absence of molecular oxygen, making them extremely versatile with regard to energy. Some microbes are quite versatile themselves, while others specialize with remarkable efficiency in their own niches, and it is common to find these microbial specialists in intricate metabolic symbioses with other specialists (Schink 1991). This contrasts with the eukaryotic domain, where metabolic versatility is rare, but morphological (and resulting behavioral) versatility is the rule, resulting in complex food chains, predator-prey relationships, behavioral symbioses, and many other phenomena.

PROPERTIES OF PROKARYOTES

This section begins with introductory material that we hope will allow the uninitiated to understand the basic language (jargon) that appears throughout this volume. It may be remedial for some readers and, we hope, valuable for others. A diagram of a Gram-negative bacterial cell is shown in Figure 4, along with a transmission electron microcope (TEM) image of the Gram-negative bacterium, *Shewanella putrefaciens*. While the diagram shows a capsule or eps (extracellular polysaccharide) layer, electron micrographs rarely see this layer due to the dehydration procedures used in TEM preparation. The bacteria are operationally divided into two groups, Gram-positives and Gram-negatives, based on the ability to take up a stain called the Gram's stain, and which is indicative of a basic differences in the cell wall structure. The Gram-negative types have a very thin rigid cell wall, with an inner membrane and an outer membrane, making the structure quite complex. The Gram-positive types, on the other hand, have an inner plasma membrane and a very thick rigid cell wall, and lack the second outer membrane. These structural differences render the two groups quite different with regard to antibiotic sensitivity, outer wall charge and hydrophobicity, and other general properties. Figure 5 shows diagramatically some of the components found in bacterial cells, and their relative sizes. All bacteria possess a phospholipid membrane of approximately 7 nm in thickness. This water insoluble lipid bilayer separates the inside of the cell from the outside, and is the site on which transport proteins, electron transport systems, ion channels, etc. are located. These components are operationally similar for all groups of bacteria. The DNA is the informational molecule containing all the information (in the form of triplet codons) needed for bacterial metabolism, growth and replication. This information is commonly contained on one major chromosome, a circular piece of double stranded DNA. Often other smaller pieces of DNA, called plasmids, are present containing DNA that is non-essential but easily transferred between bacterial cells. Such plasmids often carry information for functions like antibiotic resistance, toxic metal resistance and other such functions. Three types of RNA are present: (1) messenger RNA (mRNA), which copies one strand of DNA to create a message(in triplet code) for the synthesis of proteins; (2) transfer RNAs (tRNAs), which are small RNA molecules that carry amino acids (a given amino acid for each triplet) to the ribosomes for protein synthesis; and (3) ribosomal RNA (rRNA) which forms the bulk of the structure of the ribosome, which is the "machine" on which proteins are synthesized. The ribosomes are composed of rRNA and ribosomal proteins, are common to all living organisms on Earth, and are present at the level of several thousand per cell in actively growing bacteria. Mutations that lead to changes in ribosome structure are often lethal to bacteria, so these structures evolve quite slowly—this is one of the reasons that rRNA was chosen as a molecule for comparison of bacterial evolution. Indeed, the 16S rRNA has proven to be a valuable molecule for the study of bacterial taxonomy and phylogeny, as discussed below.

Size

Given that the major role of microbes is metabolism, and that the uptake of "food" and the elimination of waste products are expected to be surface limited processes, it is probably no accident that they have, for the most part, remained small over evolutionary

Gram-negative

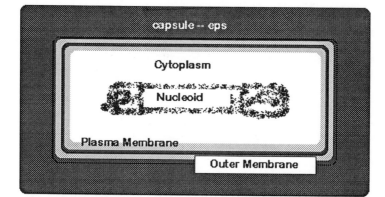

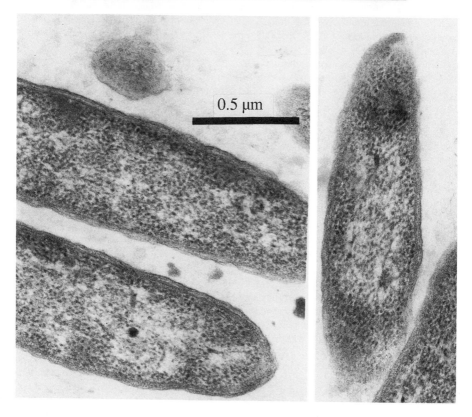

Figure 4. Bacterial structure. The simple structure of the prokaryotic cell is demonstrated diagramatically (top), in which the outer capsule or eps [extracellular polysaccharide) is shown as the outer component. Since this is meant to be a gram negative cell, it has an outer lipololysaccharide membrane enclosing a thin rigid cell wall (not labelled), which is again around an inner plamsa membrane. Gram positive cells have no outer membrane, and have a much thicker rigid cell wall layer. The TEM images shown below the diagram show the nucleoid layer in the center and many black dots (ribosomes), and the layers of the cell wall structure can be seen.

THE SIZE AND FUNCTION OF "THINGS"

"THINGS"	FUNCTION(S)	SIZE	DEPICTION
Plasma Membrane	Permeability barrier, site of electron transport system	7 nm across surrounds cell	
ATP (adenosine tri phosphate)	Energy currency of living cells	2 X 4 nm	
DNA (deoxyribonucleic acid)	Carrier of genetic code	3 nm in diameter > 1,000's nm	
t-RNA (transfer RNA)	Carrier of amino acids to site of protein synthesis	7 X 10 nm	
m-RNA (messenger RNA)	Carrier of genetic message to ribosomes for protein synthesis	2 nm in diameter > 100's nm in length	
ribosome (r-RNA plus r-proteins)	Site ("machine") for protein synthesis	21 X 25 nm	
Enzymes (Catalytic Proteins)	A. Glycolytic Enzymes (glucose --> pyruvate)	10 enzymes needed Each about 8 nm in diameter	
	B. Pyruvate Dehydrogenase Complex (Pyruvate --> Acetyl CoA	Multi-enzyme complex (20 nm in diameter)	

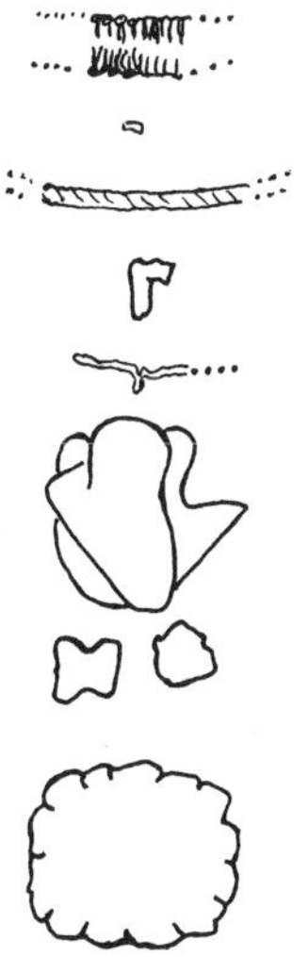

Figure 5. The size and function of "things" This diagram is meant to show, in a proportional way, the sizes of some cell components, and to briefly list the functions of each. The plasma membrane, which encloses the cytoplasm of the cell contains the energy donating molecules like ATP, the genetic material (DNA) the information-translating molecules (various RNAs), and the protein catalysts (enzymes) that perform the cellular metabolic work.

time, thus maximizing their surface to volume ratios. Small bacterial cells, with their cell size of 0.5 to a few micrometers in diameter, have S/V values 100 to 1000 times higher than typical eukaryotic cells, which may range from 20 micrometers to millimeters in diameter (Nealson 1982), making it difficult for eukaryotes to energetically compete. This must also be kept in mind in consideration of terms like "biomass." Clearly, if the S/V value of bacterial biomass is 100 times that of its eukaryotic neighbors, then even when bacteria represent only a few weight percent of the total biomass, they have a potential reactivity and environmental impact equal to that of the total. In many sediments, the bacterial biomass probably accounts for 90 to 99% of the biomass, thus making the prokaryotes by far the dominant group in terms of metabolic potential. Clearly, it is an advantage for prokaryotes to remain small—given 3.5 billion years, if it were an advantage to get larger, they would have done it! In marked contrast, the eukaryotes tend to get larger in response to bahavioral advantages such as predator-prey, etc.

However, there must also be lower size limits to metabolicaly effective life; the small size of bacteria may put limits on the chemistry that is possible. For example, the intracellular volume of a bacteria that is 0.5 µm is sufficiently small that at a pH value of 7.5, there are only a few free protons per cell, and as bacteria get even smaller, problems are encountered in terms of liquid phase chemistry. The number of molecules of solutes in cells smaller than 100 nm may in fact be prohibitively small (Nealson 1997a,b), and given that concentrations of various compounds (substrates, solutes, etc.) range from micromolar to several millimolar, it would seem that when cells reach the size of 20 nm or less, life, as we know it, is impossible, and even for 50-nm size cells, it is extremely difficult. In a 20-nm sphere, for example, even at concentrations of 1 mM, there is not sufficient volume for even one molecule of a given solvent! At the level of substrate concentrations typical of many enzyme reactions (1 to 10 μM), at cell sizes <0.1 μm, the number of molecules per cell approaches only 1. Thus, the trend to become small as a biological advantage must be limited by chemical reality as a function of available volume. Given the need for enzyme catalysts, DNA, and ribosomes, the volume available to substrates and solvents may be in fact substantially less, especially in the smaller cells, where these components will occupy a larger percentage of the small volume.

Structure

Two special items are noted in terms of bacterial structure: (1) simple cellular architecture, and (2) rigid cell walls. While prokaryotic cells are very sophisticated and highly regulated metabolic machines, they are structurally simple in comparison to eukaryotes. They have a single piece of double-stranded DNA ("chromosome"), with no nucleus or nuclear membranes. They have almost no intracellular organelles, and no tissues or organs characteristic of eukaryotes: few complex parts that can fail. For these reasons they are often found in extreme environments; conditions that preclude the growth or survival of the structurally complex (fragile) eukaryotes.

The rigid structure of the bacterial cell wall imposes certain restrictions on metabolism of the cell and the way in which bacteria interact with their environment. A common mode of nutrient uptake available to eukaryotic cells is that of phagocytosis (surrounding food particles by an extended cell membrane) and food engulfment; it is one property that lends advantage to being large and developing predatory behavior. Bacteria do not have this luxury—their rigid cell wall precludes the phagocytotic mode, and probably in response, they have advanced two modes of nutrient modification and uptake to a state of near perfection: (1) the use of extracellular enzymes to convert large polymeric molecules into smaller oligomers and monomers, and (2) the use of specific transport systems to move nutrients against concentration gradients into the cytoplasm. There are many variations on these themes among the different groups of bacteria, but the generalization that extracellular enzymes and specific transport systems predominate in bacteria must always be remembered when considering their potential ecological role(s).

Metabolic versatility

While eukaryotic versatility is expressed in terms of structures (and behaviors that are possible because of these structures), prokaryotic versatility tends to reside in cell metabolism. While all prokaryotes share a similar mechanism energy conservation, namely the generation of an electrochemical gradient or electron motive force (emf), and the use of this emf for the generation of ATP, they are decidedly different from the eukaryotes with regard to the variety and types of substrates that can be utilized to generate reduced biological intermediates, and the array of oxidants to which these reduced molecules can be coupled to generate the emf. Of all the oxidants and reductants shown on Figure 3, the eukaryotes use only a few organic molecules (glucose, pyruvate, etc.) as fuels, and only O_2 as the oxidant.

The diversity of prokaryotic metabolism requires a separate vocabulary for the description of the metabolic groups (Table 1). The vocabulary is based on the energy source (light or chemical) used by the bacteria, so that they are called either phototrophs or chemotrophs, and among the chemotrophs are included those that use organic carbon (chemoorganotrophs) or inorganic energy sources (chemolithotrophs). If an organism uses organic carbon as its source of carbon, it is referred to as a heterotroph, while if it fixes all its carbon from CO_2, it is an autotroph. Combinations of these terms thus surface as the versatility of a given organism is revealed, and the situation can be semantically complex if an organism is capable of many types of metabolism; for example, some organisms, called mixotrophs can grow both autotrophically and heterotrophically. However, Table 1 does

Table 1. Metabolic patterns of Prokaryotes and Eukaryotes.

PROCESS	DETAILS OF PROCESS	PROK	EUK
Autotrophy[e]			
Photo-	Oxygenic	+	+
	Anoxygenic		
	Electron Donor – H_2	+	-
	H_2S, S^o	+	-
	Organic C	+	-
	Fe^{2+}	+	
Litho-[c]	Energy Source		
	H_2	+	-
	HS^-, S^o, $S_2O_3^{2-}$		
	NH_4^+, NO_3^-		
	Fe^{2+}		
	Mn^{2+}		
Heterotrophy[b]			
	Fermentation	+	-[d]
	Aerobic Respiration	+	+
	Anaerobic Respiration		
	NO_3^-, NO_2^-	+	-
	S^o, $S_2O_3^{2-}$, SO_3^{2-}, SO_4^{2-}	+	-
	Fe^{3+}, Mn^{4+}	+	-
	CO_2	+	-
	fumarate (organic C)	+	-

[a] Autotrophy refers to the ability to grow with CO_2 as the source of carbon
[b] Heterotrophy refers to the need to to have organic carbon for growth
[c] Lithotrophy refers to the ability to use inorganic compounds as a source of
   energy for growth. Usually these energy sources are oxidized by oxygen,
   but it is also possible to use some of the other electron acceptors shown
   in the Anaerobic Respiration section below.
[d] Most eukaryotic cells will produce lactic acid if oxygen stressed, but among the
   eukaryotes, only a few yeasts can grow by fermentation.
[e] These marks indicate that major new knowledge has been gained about these
   metabolic groups since the Viking expedition.

serve to point out that bacterial groups are known that utilize many of the environmental redox pairs. Such bacteria reside in a remarkable array of environments where energy would normally flow, and using enzyme catalysts to speed up otherwise slow reactions, making use of a large number of energy sources and oxidants that are not available to eukaryotes. It seems likely that, if chemical kinetics are sufficiently slow to allow bacteria to compete, then almost any redox couple that yields energy will be exploited.

Mechanisms of energy conservation and patterns of metabolism

One of the revelations of the past 25 years in biology was the elucidation of a central mechanism of metabolism that allowed microbial energy conservation to be understood as a unified feature. The so-called chemiosmotic theory (see Gottschalk 1994), proposed that it should be possible to transform chemical energy of a variety of forms into an electrochemical potential across a membrane: a so-called proton motive force (pmf). This pmf is then used to drive the synthesis of ATP via membrane-bound enzymes called ATPases, which utilized the energy in the electrochemical gradient to drive the synthesis of high energy phosphate bonds that could then be used by cells for many purposes. The chemiosmotic model is unifying in the sense that all living organisms utilize it, or some variaton of it, to synthesize ATP and drive cellular functions. It allows, in principle, any electron potential between electron donor and acceptor to be used to pump protons to the exterior of the cell, thus establishing the pmf that can be utilized to synthesize ATP. The energy source can be organic carbon, inorganic substrates, or light—it really makes no difference as long as the cell possesses the enzymatic machinery to harvest the energy and transfer it to the membrane bound electron transport chain. The difference between the versatile prokaryotes and the limited eukaryotes is that most eukaryotes have very few choices of reductants, and can utilize only one oxidant, molecular oxygen. It may be instructive in considering the biogeochemical cycling discussed in the next section, to keep this mechanism in mind, remembering that almost all of the reactants and products that will be discussed are either oxidants or reductants that fit nicely into the model proposed by the chemiosmotic scheme. A schematic diagram of the chemiosmotic mechanism of energy conservation is presented in Figure 6.

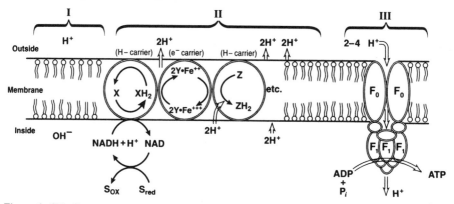

Figure 6. This diagram presents many of the basics of the chemiosmotic theory—in essence, how living organisms on Earth harvest chemical (redox) energy from the environment and conserve it as biologically useful energy (ATP). Three basic features are shown: (1) A semipermeable membrane, which is impermeable to charged molecules and can thus be used to separate charges. Once a charge separation is achieved, energy can be harvested. (2) A vectorial electron transport chain, in which H-carriers and e-carriers alternate in the flow of reducing power from substrate to oxidant. As electrons flow toward the oxidant, protons are pumped to the outside of the membrane, creating the electrochemical gradient (a protonmotive force, pmf, consisting of a combination of pH and charge gradient). (3) An enzyme to convert the pmf into useful cellular energy. In this case, the enzyme shown is the membrane bound ATPase, which allows protons to flow back into the cell through pores in the membrane it creates, and during this flow, uses the energy to synthesize ATP from ADP + Pi. In addition to the ATPase, other systems are present that are activated by the pmf, such as transport systems to bring in, or excrete substrates, as well as the flagellar "motor" which causes the bacterial flagella to rotate, giving the organism its motility.

THE PROCESSES

Profiles like those shown in Figure 2 are the result of a series of redox processes and ultimately driven by the input of organic carbon to the sediments. Commonly measured components in the study of stratified environments are shown in the center column of Table 2, while the processes and organisms that generate them are in the left column, and those that consume them are in the right column. It is these processes that will be discussed in this section, followed by a discussion of the contributing organisms. In the example of Table 2, we deal with a rather simple situation, out of the photic zone, with minimal mixing. For more complex systems, the situation gets appropriately more difficult to dissect in terms of microbial populations and chemical interactions. The following discussions proceed downward into the sediment, discussing each component listed in Table 2 and shown as profiles in Figure 2. While such profiles can be used to identify general zones of metabolic processes, some, like fermentation, for which there are no easily measured or distinctive

Table 2. Commonly measured pore water components: Biological sources, sinks, processes, and organisms involved.

Source (Process)/Organism(s)	Component	Sink (Process)/Organism(s)
Oxygenic Photosynthesis /Cyanobacteria and Algae	Oxygen	Organic Carbon Oxidation/Aerobic Heterotrophs; Inorganic (H_2, H_2S, etc.) Oxidation/Aerobic Chemolithotrophs
Ammonia Oxidation/Nitrifying Bacteria	Nitrate (NO_3^-)	Organic Carbon Oxidation/Denitrifiers; Inorganic (H_2,H_2S, Fe^{++}) Oxidation/Anaerobic Chemolithotrophs
Ammonification/Protein Degrading Bacteria	Ammonia (NH_4^+)	Aerobic Oxidation/Nitrifiers
Manganese Reduction/Manganese Reducers	Manganese (Mn^{++})	Manganese oxidation/Manganese oxidizers
Iron Reduction/Fe Reducers	Iron (Fe^{++})	Iron Oxidation/Aerobic Fe and Anaerobic (NO_3^-) Fe oxidizers;/Photosynthetic Fe oxidizers
Sulfur (H_2S,$S°$, $S_2O_3^{2-}$) oxidation/Aerobic	Sulfate (SO_4^{2-})	Heterotrophic sulfate reduction/Sulfate Reducting Bacteria; Sulfate Assimilaton/Most bacteria
Sulfur ($S°$,$S_2O_3^{2-}$,SO_4^{2-}) Reduction/Sulfate, Sulfur, and Thiosulfate Reducers	Hydrogen Sulfide (H_2S)	Aerobic Sulfide Oxidation/Aerobic Sulfide Oxidizers; Anaerobic Sufide Oxidatition/Anaerobic Sulfide Oxidizers; Photosynthetic Sulfide Oxidation/Sulfur Photosynthetic Bacteria
Methanogenesis/Methanogens	Methane (CH_4)	Aerobic Methane Oxidation/Methanotrophs; Anaerobic Mehtane Oxidation/Unknown Consortium
Fermentation/Fermentative bacteria; Proton Reduction/Syntrophic Bacteria	Hydrogen (H_2)	Aerobic H Oxidation/Hydrogen Chemolithotrophs; Anaerobic H Oxidation/Many Bacteria; Methanogenesis/Methanogens; Acetogenesis/Acetogens
Fermentation/Fermentative bacteria; Respiration/Many Bacteria	Carbon Dioxide (CO_2)	Autotrophy/Chemo or Photoautotrophs; Methanogenesis/Methanogens; Acetogenesis/Acetogens

chemical markers like oxygen, sulfide, methane, or Mn(II), are much more difficult to spatially fix in the sediment environment.

Aerobic respiration

Organic matter that reaches the sediments is aerobically respired (to CO_2 and H_2O) until consumption exceeds the amount of oxygen that can be delivered to the site by diffusion. In sediments overlain by deep water, much of the organic matter is respired during transit to the sediments, leaving very little carbon to be further oxidized. The result is that deep sea surface sediments are usually oxidized, and oxygen can remain at high levels for many centimeters downward. However, in shallower, more carbon-rich sediments, it is usual to see oxygen depletion within millimeters or centimeters of the sediment surface.

Nitrification and denitrification

Proceeding downward past the oxygen depletion zone, it is usual to see a zone in which nitrate concentration increases. This occurs at low concentrations of oxygen, where ammonia diffusing upwards from below is converted into nitrate via a process called nitrification. Nitrate then decreases due to denitrification, which occurs as a result of the oxidation of organic carbon by nitrate, with the concomitant production of CO_2. The magnitudes of nitrification and denitrification are not easy to measure because they occur in spatially adjacent samples, and because, for both, the product of one process is the reactant of the other. Thus, with small concentrations of nitrate, the effect on the nitrogen and carbon cycles can be substantial if the cycling rate is large. The use of molecular tools to directly study the distribution and activity of contributing microorganisms should provide important insights into the magnitude of these elusive rates (below).

Methane oxidation

Although some anaerobic methane oxidation occurs, the major oxidant for methane is molecular oxygen, and the profiles of methane observed in sediments usually show methane depletion at the intersection of the oxygen minimum.

Manganese and iron oxidation

As soluble Mn(II) and Fe(II) diffuse upwards from reduced sediments, they are deposited as insoluble metal oxide layers or crusts in the presence of low levels of oxygen. The layer of oxidized manganese, commonly in the form of MnO_2 typically overlays a layer of oxidized iron, which is more rapidly oxidized in the presence of low levels of oxygen, and precipitates just below the MnO_2. Such ferromanganese layers are common in sediments. The oxidation of both Mn(II) and Fe(II) are thermodynamically favored, but the kinetics of the two processes are substantially different at neutral pH values common to most sediments. Mn(II) is kinetically stable (Stumm and Morgan 1981) and usually requires biological catalysis, while Fe(II) oxidation is very rapid at neutral pH, and biological catalysis is assumed to be unnecessary.

Sulfur oxidation

Reduced sulfur species such as sulfide, thiosulfate or polysulfide, are produced as a result of organic carbon oxidation in deep sediments, and as they diffuse upwards, they are oxidized. These processes are chemically complex, and poorly quantified in sediments. As sulfide diffuses upwards, it is oxidized by Fe(III), Mn(IV), and oxygen, with the latter two reactions being quite rapid. Each oxidant generates different sulfur intermediates, that can interact with other compounds, making the system sufficiently complex as to defy most efforts to quantify the separate parts. In most systems the oxidizing potential of the sediment is such that sulfide is consumed within the sediment, either by oxygen itself, or by other oxidants, such as nitrate or metals.

Manganese and iron reduction

Below the nitrate reduction zone, are the zones of metal reduction, where organic carbon is oxidized by manganese and iron oxides, resulting in increased levels of pore water $Mn(II)$ and $Fe(II)$. Because $Fe(II)$ is a good reductant of MnO_2 (Myers and Nealson 1988b), $Mn(II)$ increases, followed by the appearance of $Fe(II)$. In freshwater sediments the profiles are often clearly defined, while in marine sediments, iron can be difficult to follow, in part because of the active sulfur cycle in marine systems, in which upward diffusion of sulfide tends to remove $Fe(II)$ as iron sulfide (pyrite). Manganese and iron reduction can be either biological or abiological. Manganese is easily reduced by organic compounds (Stone et al. 1994) as well as several inorganics, such as sulfide (Burdige and Nealson 1986) or $Fe(II)$ (Myers and Nealson 1988). Iron can also be reduced by sulfide (Stumm and Morgan 1981) or organics, but it is considerably more resistant to chemical reduction, and some reports maintain that all iron reduction in nature is due to biological catalysis (Lovley et al. 1991).

Sulfate reduction

Sulfate reduction is well characterized in sediments, while thiosulfate and sulfur reduction are much less well quantified. With the exception of reduction by very high temperatures, such as those found in the hydrothermal waters, sulfate is stable unless reduced biologically—probably no chemical reduction of sulfate occurs in sediment systems. Once below the zone of metal reduction, the next major reduced species to appear in the pore waters is sulfide, and this is generally attributed to sulfate-reducing bacteria. The importance of other sulfur intermediates, such as thiosulfate or elemental sulfur (polysulfide) remains to be elucidated, probably because of the complexity of sulfur chemistry.

The production of sulfide (and the generation of other reduced sulfur species) as a result of sulfate reduction is probably the major biogeochemical difference between freshwater and marine systems. In freshwater, sulfate lies in the range of 100 to 250 mM, while in marine systems it is 25 µM or more. Sulfate is thus the dominant electron acceptor in marine sediments, dwarfing even oxygen. In contrast, in freshwater systems, the sulfur cycle is less dominant than in the marine systems. The profiles in Figure 1 point out that in marine sediments, oxidants are consumed by reduced sulfur species diffusing upward, and that these reduced species are produced due to the oxidation of organic carbon in the sediments.

Methanogenesis

As with sulfate reduction, methanogenesis is a process that occurs only as a result of biological catalysis at temperatures and conditions common to most sediments. As sulfate reduction to sulfide dominates marine sediments, so does CO_2 reduction to methane dominate freshwater ones. Methane appears in porewaters just below the oxic/anoxic interface, and is the major indicator of organic carbon turnover in fresh water sediment systems.

THE ORGANISMS

Microbes accumulate at redox interfaces because of the energy to be harvested, and should thus be expected to vary in type and number as the input of energy to the given sediment varies in quantity and quality. The organisms can be discussed in terms of the processes observed in the sediments and discussed above, but this approach can be unsatisfying because many organisms are facultative, capable of crossing the boundaries between the processes. Thus to speak of denitrifiers as a group may miss an important point that many denitrifiers also use oxygen, and many use other electron acceptors and/or are fermentative bacteria. Given the variety of energy conserving mechanisms outlined above, metabolic plasticity is not particularly surprising. Many bacteria enjoy a wide versatility, with sulfate reducers being known that will utilize electron acceptors up to the potential of nitrate, and aerobes that can utilize sulfite and elemental sulfur. This versatility also emphasizes the point that simply identifying a given organism will not necessarily give information as to what it is actually doing; it is only the first step in defining the microbial ecology.

Aerobic heterotrophs

A wide array of bacteria possess the ability to degrade organic matter at the expense of molecular oxygen as an electron acceptor. Only a few of these are obligately aerobic, and even for these, there may be alternative modes of survival or growth under anoxic conditions. For example, some of the groups long considered to be aerobes, such as *Pseudomonas* and *Bacillus* species have quite a number of species that do well under anoxic conditions, either through some form of fermentation, or via the use of alternate electron aceptors like nitrate. It is safe to say that there will not be a shortage of aerobic heterotrophs as long as there is oxygen and organic carbon, and that these efficient organisms, given a high enough flux of oxygen, will leave little organic carbon undegraded. It is common for a single aerobe to completely degrade its carbon source to CO_2, unlike many obligate anaerobes that only partially oxidize their substrates.

Chemolithotrophs

In marked contrast to the aerobic heterotrophs, the chemolithotrophs (organisms that utilize inorganic energy sources) are restricted to a few recognized groups of organisms, and tend to be specialists (Shively and Barton 1991). An exception is the group of H_2-utilizing bacteria (Table 3), which are widespread, both environmentally and biologically.

Hydrogen-oxidizing bacteria. The most widespread of the chemolithotrophs are the H_2-oxidizing bacteria, which have the ability to utilize H_2 as an energy source via the enzyme hydrogenase, coupling this to the reduction of some electron acceptor. A wide range of organisms are known to utilize H_2 as an energy source, ranging from aerobes like *Alcaligenes* and strains of *Pseudomonas*, to facultative organisms like *Shewanella*, which couples hydrogen oxidation to the reduction of a variety of different electron acceptors, to sulfate reducing bacteria like *Desulfovibrio* spp., and finally, to acetogens and methanogens. Some of these organisms are autotrophs, growing with CO_2 as the sole source of carbon, while others, like *S. putrefaciens* simply use the hydrogen as a source of energy, growing heterotrophically on organic carbon.

Sulfur-oxidizing bacteria. Organisms that oxidize sulfur compounds lithotrophically are usually quite specialized. Once the commitment is made to this metabolism, other modes of metabolism are not common. The substrate of the sulfur oxidizers is usually thought of as hydrogen sulfide, although many sulfur oxidizers will also oxidize elemental sulfur and/or thiosulfate as well. Given the rapid kinetics of sulfide oxidation by molecular oxygen, sulfide oxidizing organisms are in a continuous struggle with chemical oxidation. They are often thus found at interfaces, where anoxic waters are mixing slowly with oxic waters above them. At such boundaries, sulfur oxidizers position themselves between the two reactants and take advantage of natural gradients (that they also help to maintain), thus harvesting abundant energy.

Table 3. Hydrogen-utilizing bacteria.

I. AEROBES

Many aerobic bacteria use hydrogen as an energy source, and some use it as the sole energy source with CO_2 as the carbon source: e.g. growing lithoautotrophically on hydrogen.

Organism Type	Example	Comments
Chemolithoautotroph	*Alcaligenes eutrophus*	autotrophic H_2 oxidizer
Chemolithoheterotroph	*Shewanella putrefaciens*	heterotrophic H_2 oxidizer

II. ANAEROBES

Under anaerobic conditions, one sees many other types of metabolism of hydrogen, commonly utilizing this energy source to reduce a wide array of different electron acceptors, thus driving the metabolism an anaerobic ecosystems, and generating many different products in the process. Some examples are:

Organism Type	Example (genus)	Electron Acceptor	Major Product[a]
Methanogen	*Methanococcus*	CO_2	CH_4
Acetogen	*Acetobacterium*	CO_2	CH_3COOH
Sulfate Reducer	*Desulfovibrio*	SO_4^{2-}	H_2S
Sulfur Reducer	*Desulfuromonas*	S°	H_2S
Iron Reducer	*Shewanella*	Fe^{3+}	Fe^{2+}
Manganese Reducer	*Shewanella*	Mn^{4+}	Mn^{2+}

[a] It should be noted that the products produced alter the environment by acting as substrates for other organisms and setting up the possibility for biogeochemical cycles.

Another approach to utilizing sulfide in an oxic world is seen with the sulfide oxidizing symbionts of hydrothermal vents eukaryotes. In some of these systems, the eukaryotic hosts have developed sulfide binding proteins that harvest sulfide from the vent waters, and transport the sulfide (and oxygen) to the bacteria in their symbiotic organelle (e.g. the trophosome of *Riftia* species of vent tube worms) where the bacteria can deal with each substrate separately (Jannasch 1995).

The sulfur oxidizers include those that are restricted to low pH environments (acidophiles) and the neutral pH types. The acidophiles are commonly isolated from acid mine drainage environments, where sulfur rich coals or ores are exposed to oxygen, and these bacteria efficiently convert it to sulfuric acid, thus creating a very exclusive environment of pH 4 or less, where they thrive, and other organisms are excluded. The neutral pH sulfur oxidizers include several *Thiobacillus* species as well as an array of structurally complex sulfur bacteria in the groups *Beggiatoa*, *Thioploca*, *Thiothrix* and others. These bacteria have been traditionally difficult to culture, but are often abundant in both marine and freshwater sediments. In some cases, such as *Beggiatoa*, the bacteria are clearly capable of growth on sulfide. In others, the involvement with sulfide is suspected because of the environment from which they are isolated, and often, from the presence of either intra-or extracellular sulfur granules associated with the cells. Fossing et al (1995) reported *Thioploca* cells with vesicles in which nitrate is stored for later utilization, so that the organisms could accumulate nitrate (up to 500 µM!) in oxic environments, then migrate downward as much as 10 cm, where they could oxidize sulfide at the expense of the stored nitrate. Similar observations have been made for vacuolate marine *Beggiatoa* species (McHatton et al. 1996). If such microbial "SCUBA divers" are common, the delicate profiles presented in Figure 1 could be missing major metabolic microenvironments.

Despite the difficulties in culturing the sulfur bacteria, many have now been identified using 16S rRNA probes, so their abundance and distribution, as well as some aspects of their diversity will undoubtedly be elucidated in the near future.

Iron-oxidizing bacteria. As with the sulfur oxidizers, the iron oxidizers are in competition with chemical oxidation -- at neutral pH, the oxidation of ferrous to ferric iron is extremely rapid, so it is difficult to compete except in gradient-like environments. At low pHs as occur in acid mines, where Fe(II) is abundant, some species of *Thiobacillus (T. ferrooxidans)* grow very well using ferrous iron as their only source of energy. In neutral pH sediments, other iron oxidizers may compete. For most neutral pH iron oxidizers like *Gallionella* species, the proof of iron oxidation as a mode of energy conservation has been difficult. In the laboratory, it is hard to duplicate environments where reduced iron is produced and made available to aerobic bacteria without the interference from abiotic oxidation.

Nitrifying bacteria and methanotrophs. The nitrifying bacteria consist of two groups, those that oxidize ammonia to nitrite, and those that oxidize nitrite to nitrate. These organisms are ubiquitous in sedimentary environments where ammonia diffuses upwards from below, and where oxygen is available for the reaction. They share many properties, such as complex internal membranes and location at the oxic/anoxic interface, with the methane oxidizing bacteria (C1 organisms that are not considered chemolithotrophs), which are commonly referred to as the methanotrophs.

Methanogens. All known methanogenic bacteria are found in the domain of the Archaea. The reactions required to produce methane are unique to this group, and the organisms, in genera such as *Methanobacter, Methanococcus, Methanobacterium*, etc. are widely distributed in anaerobic niches throughout the world, including many thermophilic environments. Some of these bacteria use hydrogen and CO_2 as their substrates for methane formation, while others utilize acetate.

Acetogens. Bacteria collectively referred to as acetogens, and consisting of several genera (*Clostridia, Acetobacterium, Acetogenium*, and others), have the capacity to utilize hydrogen as the reductant to fix CO_2 into acetate, which they then utilize for growth. Since these substrates are also used by methanogens, these physiological groups are often thought to compete. However, acetogens have additional fermentative capacities, and this may account in part for their co-existence with methanogens in the environment.

Fermentative bacteria

Fermentative bacteria consist of a wide array of different metabolic types that are specialists in the disproportionation of organic carbon. The carbon is taken into the cell, usually split into smaller molecules, and part of it oxidized while other parts are reduced. This results in the excretion of both oxidized and reduced fermentation end-products (H_2, CO_2, acetate, lactate, etc.) almost all of which are important in the metabolism of other organisms in the anaerobic food chain.

Nitrate-respiring bacteria

A great number of bacteria can catalyze the reduction of nitrate to N_2 gas in the process called dentrification, and although nitrate is often at low concentrations in the environment, these organisms are ubiquitous, and play a role in the cycling of carbon and nitrogen in sedimentary systems. When oxygen becomes limiting, nitrate is typically the next major biological electron acceptor utilized. There is a great variation among organisms that accomplish nitrate reduction, with some reducing the nitrate all the way to ammonia, and other bacteria reducing nitrate even in the presence of molecular oxygen.

Metal-respiring bacteria

For many years, metal profiles like those shown in Figure 1 were regarded as the result of chemical reduction, primarily because the substrates involved were solids (Mn and Fe oxides) and thought to be non-accessible to direct bacterial reduction. This view has changed dramatically in the past few years, with the discovery of several groups of bacteria that grow via the dissimilatory reduction of iron and/or manganese (see reviews by Lovley 1993, and Nealson and Saffarini 1994) resulting in the oxidation of organic matter and the reduction of Fe(II) or Mn(II). The organisms so far identified consist of both facultative anaerobes in the genus *Shewanella* (DiChristina et al. 1988, Myers and Nealson 1988a), and obligate anaerobes in organisms closely related to the *Geobacter* group (Lovley and Phillips 1988, Lonergan et al. 1996).

Sulfur- and sulfate-reducing bacteria

Much has been written about bacteria that live by the dissimilatory reduction of sulfur compounds, and just a few groups are mentioned here. Among the sulfur reducers are both facultative anaerobes like *S. putrefaciens* (Moser and Nealson 1996) as well as many anaerobic bacteria, in groups such as *Wollinella, Desulfuromonas*, and others (see Balows et al. 1991). The sulfate reducers, on the other hand, are all obligate anaerobes, with the commonly encountered genera known as *Desulfobacter, Desulfovibrio*, and others.

Proton reducers (syntrophic bacteria)

One of the major modes of bacterial symbiosis involves the exchange of hydrogen between different organisms (Schink 1991), and some of the bacteria commonly involved occur in anaerobic environments, where they eke out an existence by oxidizing fermentation end products like acetate, propionate or fatty acids, and producing H_2 gas. Such reactions are energetically feasible only when the hydrogen is maintained at very low concentrations, so that such bacteria (in the genera *Syntrophobacter, Syntrophomonas, Syntrophus*, etc.) must live in co-culture with H_2-utilizing bacteria such as methanogens, acetogens, and sulfate reducers.

NEW METHODS AND APPROACHES

Given the state of sediment microbiology today, it is inevitable that one area of future work will involve dissecting the diversity and interactions of the microbial populations. This will certainly involve the use of new molecular approaches, which have revolutionized our way of thinking about the phylogeny and diversity of microbes. The power of these tools, as outlined below, allows us to answer questions like, 'Who's there?' and 'Is there anything new here?' even without culturing organisms: such approaches have revealed an abundance of new organisms,even suggesting what types they are—information that necessitates the study of new isolates with unique metabolic abilities—this point is demonstrated with a discussion of some new organisms involved in the biogeochemical cycling of metals. In addition, two emerging areas which will require a particularly interdisciplinary approach are discussed. These include the interactions between microbes and minerals, and the study of ecosystems where energy is delivered as a result of geological processes. The particular topics chosen for discussion are meant to focus on areas that might be less well known to the casual reader, and to act as points of departure for further reading and thought.

Molecular approaches to the microbiology of sediments

One of the most exciting recent happenings in microbial ecology has been the introduction of the tools of molecular biology to the field (Amann et al. 1992, 1995; Olsen

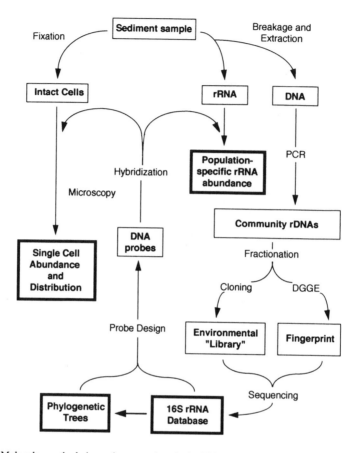

Figure 7. Molecular methods in environmental analysis. This flow diagram shows one approach that is now being used to exploit molecular (rRNA) methods for the study of microbial ecosystems. The first step is the extraction of nucleic acids, with both DNA and RNA being extracted. Next, the rRNA is interrogated with probes of various specificity, and the quantitative response to these probes is used to estimate the types and amounts of the various groups of microbial rRNA's that were extracted. This gives a view of the population in a rather quick way, but is limited in the sense that if organisms are present for which probes are not available, they will not be scored. The next step is the amplification (by PCR) of the rRNA genes, followed by cloning and sequencing random clones. An alternative to cloing is to use of a fractionation method called Denaturing Gradient Gel Electrophoresis (DGGE) (Muyzer et al., 1993, 1995). This provides for the electrophoretic separation of DNA fragments of differing sequence (derived from different populations) on acrylamide gels. The gel-isolated DNAs can be sequenced directly or following a cloning step. This now allows one to determine the major components of the population that have been amplified, and even to place them in a taxonomic context, based on rRNA sequence analysis. Finally, hybridization probes can be made and labelled with fluorescent dyes, and these fluorescent probes used to go back to the environmental samples and look for the presence of specific organisms, using flurorescence microscopy (flurorescence *in situ* hybridization or FISH). The hybridization methods allow one not only to ask where specific types of organisms reside, but provide an internal check on whether the major groups obtained by this approach are also the major groups seen in the environmental samples.

et al. 1986, Pace et al. 1986, 1993; Stahl 1986, Stahl and Amann 1991, Stahl et al. 1988). Molecular approaches (Fig. 7) allow one to use informational molecules like ribosomal RNA (rRNA) for the direct characterization of environmental populations, inferring the amounts of total and specific microbial biomass, and even locating within a given environment, certain organisms of interest—all this without the need for culturing the microbe(s) of interest.

Ribosomal RNA-based approaches. The ribosome is composed of ribosomal RNA and specific ribosomal proteins, and is the highly complex 3-D subcellular structure on which protein synthesis occurs (Lake 1985). Although rRNA is single stranded, it has substantial internal secondary structure and is thus stable and easily isolated. It can be separated into different size groups (5S, 16S, and 23S), of which the 16S has most been extensively studied by sequence comparison. Analysis of 16S rRNA sequences led to the realization (Woese 1987) that all organisms are related to one another and that sequence comparisons could be used to infer a universal "tree of life" (phylogeny). The sequences of more than 5000 different 16S rRNAs have been used to generate molecular phylogenies of the type shown in Figure 1 (Maidak et al. 1994), and in more detail by Barnes and Nierswicki-Bauer (this volume). Once such phylogenies are established on the basis of cultured organisms, they can be used to place any new sequence obtained from the environment into this context, so that one can, in principle, identify an organism on the basis of its nucleotide sequence, never having to culture that organism.

Some parts of the ribosome have evolved so slowly, they are identical in all living organisms (universally conserved). Thus it is possible to synthesize an oligonucleotide (DNA probe) that is complementary to such a sequence and will anneal with (hybridize to) any known 16S rRNA. Such sites are also used as universal priming sites to initiate DNA synthesis for PCR (polymerase chain reaction) amplification, as discussed below. Other areas of the 16S rRNA are less well conserved, and these more variable regions can be used for designing hybridization probes that are domain-specific, group specific, or even species specific (Devereaux et al. 1989, 1992, 1994; Kohring et al. 1994, Pace et al. 1993; Table 4). Thus, the specific rRNA sequence information provides not only phylogenetic information, but allows, at several levels of resolution, the placement of rRNA sequences of microbes to functional groups and taxa (Johnson 1984, Stackebrandt and Liesack 1992, Stahl and Amann 1991, Woese 1987). It is this latter property that is now being exploited for molecular ecological analyses. Through the analysis of rRNA, or the DNA coding for rRNA, it is thus possible to infer a great deal about the populations of microbes in the environment being studied. Such approaches have been very successful in many different ecosystems (Amann et al. 1992, Barns et al. 1994, Giovannoni et al 1990, Liesack and Stackebrandt 1992, Pace et al. 1986, Risatti et al. 1994, Stahl et al. 1988, Teske et al. 1996, Ward et al. 1992, Weller and Ward 1989), although they have only recently been applied to the study of sediments (MacGregor et al. 1996, Devereux et al. 1994, 1996; Rochelle et al. 1994) .

While many molecular approaches can be taken to characterize an environment, two that involve 16S rRNA will be discussed here in order to demonstrate some of the power of the techniques. Both are outlined on Figure 7. First, one can extract the rRNA directly, and interrogate this RNA with specific probes. The probes are labelled with ^{32}P or some other appropriate label that can be quantified, and the RNA is hybridized to each probe separately, and the radioactivity used to estimate the relative levels of that population (Fig. 8). Thus, universal probes may be used to estimate total rRNA (a proxy of total biomass), and more specific probes used to ask whether more specific groups of bacteria are present (e.g. domain- (Archaea-) specific) probes; group-specific probes (e.g. for sulfate-reducing bacteria; Devereaux et al. 1987, 1992), or even species-specific probes (Table 4). In the absence of culturing organisms, one can thus get a molecular snapshot of the major metabolic groups as represented in the rRNA that is extracted.

Table 4. Selected oligionucleotide hybridization probes.

Probe Name	Target Group	Probe or Primer Sequence 5' to 3'	Original Reference	Old Name
All Organisms				
S-*-Univ-1390-a-A-18	All Organisms	GACGGGCGGTGTGTACAA	Zheng et al. 1996	
Domains				
S-D-Arch-0344-a-A-20	Archaea	TCGCGCCTGCTGCICCCCGT	Raskin et al., 1994	ARC344
S-D-Arch-0915-a-A-20	Archaea	GTGCTCCCCCGCCAATTCCT	Stahl and Amann, 1991	ARC915
S-D-Bact-0338-a-A-18	Bacteria	GCTGCCTCCCGTAGGAGT	Amann et al., 1990a	EUB338
S-D-Euca-0502-a-A-16	Eucarya	ACCAGACTTGCCCTCC	Amann et al., 1990a	EUK516
Methanogens				
S-O-Mcoc-1109-a-A-20	Methanococcales	GCAACATAGGGCACGGGTCT	Raskin et al., 1994	MC1109
S-F-Mbac-0310-a-A-22	Methanobacteriaceae	CTTGTCTCAGGTTCCATCTCCG	Raskin et al., 1994	MB310
S-F-Mbac-1174-a-A-22	Methanobacteriaceae	TACCGTCGTCCACTCCTTCCTC	Raskin et al., 1994	MB1174
S-O-Mmic-1200-a-A-21	Methanomicrobiales	CGGATAATTCGGGGCATGCTG	Raskin et al., 1994	MG1200
S-O-Msar-0860-a-A-21	Methanosarcinales	GGCTCGCTTCACGGCTTCCCT	Raskin et al., 1994	MSMX860
S-F-Msar-1414-a-A-21	Methanosarcinaceae	CTCACCCATACCTCACTCGGG	Raskin et al., 1994	MS1414
S-G-Msar-0821-a-A-24	*Methanosarcina*	CGCCATGCCTGACACCTAGCGAGC	Raskin et al., 1994	MS821
S-G-Msae-0825-a-A-23	*Methanosaeta*	TCGCACCGTGGCCGACACCTAGC	Raskin et al., 1994	MX825
"Cold Water" Crenarchaeota				
S-*-Cren-0667-a-A-15	Uncultured Crenarchaeota	CCGAGTACCGTCTAC	DeLong ett al. 1994	
Sulfate-Reducing Bacteria				
S-*-Srb-0385-a-A-18	Delta Proteobacteria	CGGCGTCGCTGCGTCAGG	Amann et al. 1990	SRB385
S-F-Dsv-0687-a-A-16	Desulfovibrionaceae and some metal reducers	TACGGATTTCACTCCT	Devereux et al., 1992	SRB687R
S-G-Dsb-0804-a-A-18	Desulfobacter group	CAACGTTTACTGCGTGGA	Devereux et al., 1992	SRB804R
S-G-Dsbm-0221-a-A-20	*Desulfobacterium*	TGCGCGGACTCATCTTCAAA	Devereux et al., 1992	SRB221R
S-G-Dsb-0129-a-A-18	*Desulfobacter*	CAGGCTTGAAGGCAGATT	Devereux et al., 1992	SRB129R
S-*-Dcoc-0814-a-A-18	Desulfococcus group	ACCTAGTGATCAACGTTT	Devereux et al., 1992	SRB814R
S-G-Dsbb-0660-a-A-20	*Desulfobulbus*	GAATTCCACTTTCCCCTCTG	Devereux et al., 1992	SRB660R

Probe and primer names have been standardized as follows (Alm ,et al. 1996): S or L for Large or Small subunit rRNA as the target; letter(s) designating the taxonomic level targeted; letters designating the target group of the oligonucleotide probe or primer; nucleotide position (*E. coli* numbering) in the target where the 3' end of the probe or primer binds; letter designating the version of the probe or primer (a for version 1, b for version 2, etc.); A or A for Antisense or Sense direction; number indicating the length in nucleotides of the probe or primer.

The oligomeric DNA hybridization probes can also be used for microscopic observation of the individual cells comprising a probe-target population. The environmental specimen is first chemically "fixed" to stabilize the cells and then treated in a way to allow short DNA probes to pass through the cell wall and hybridize with the ribosomal RNAs. Since actively growing bacteria contain thousands of ribosomes, such probes, when labelled with fluorescent dyes and hybridized to environmental samples, can be used to physically locate the cells by fluorescence *in situ* hybridization (FISH) (Fig. 9). This method provides a very high resolution analysis of population spatial distribution and also has the advantage that only living (rapidly growing) cells contain enough ribosomes to give a signal, so that dead cells will not be scored.

Second, one can extract DNA from the samples, and amplify the genes for 16S rRNA from these extracts. The amplification is made possible because of the universally conserved areas within the rRNA molecule, which can be used as primers for sequencing

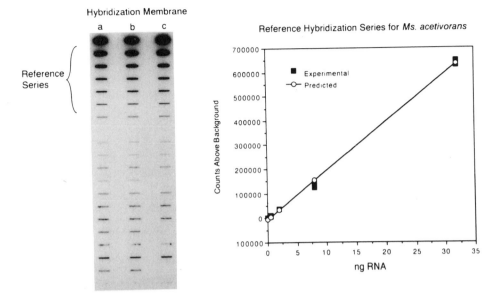

Hybridization Membrane

Reference Hybridization Series for *Ms. acetivorans*

Figure 8. Total RNA samples extracted from different depths of Lake Michigan sediment were applied to a nylon hybridization membrane using as blotting device for gridding (replicate and triplicate applications, columns 1-3)(left) and hybridized with a ^{32}P-labelled DNA probe for the genus *Methanosarcina*. A reference series of known amounts of RNA extracted from a cultured species (*Methanosarcina acetivorans*) was applied to the upper region of the membrane, providing the basis to relate the amount of probe bound (hybridization intensity) to RNA abundance in the sediment (right). In this example, the amount of ^{32}P bound to an RNA application position was measured using a phosphorimaging system.

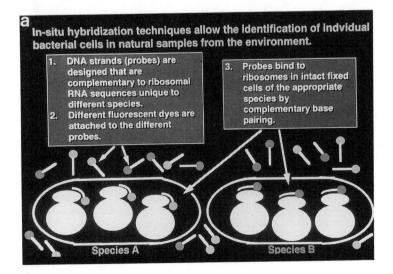

Figure 9. Fluorescent *in situ* hybridization. The early events of colonization of a glass surface by bacteria in an anaerobic sulfidogenic reactor system. The sample (glass coupon) was hybridized with a general probe (S-*-Srb-0385-1-A-18) for many Gram-negative sulfate-reducing bacteria (Table 4). (a) Methodological schematic. Figure 9b,c on next page.

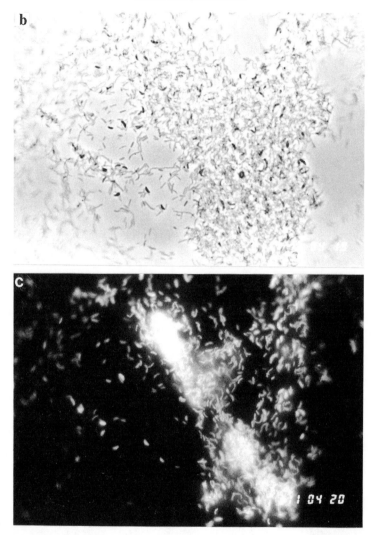

Figure 9, continued. (b) Phase contrast and (c) fluorescence micrographs of the same field. Cells that hybridize to this probe are conspicuous during early surface colonization (biofilm formation).

reactions and for PCR (polymerase chain reaction) amplification. Once these molecules are amplified, they can be cloned, random clones sequenced, and these sequences identified as appropriate metabolic groups or taxa, as described above. By this approach, one can get a more specific snapshot of the taxonomic and phylogentic groups of bacteria within a given environment.

An example of this combined approach to study Lake Michigan sediments is shown in Figures 10 and 11. The domain probe for the Archaea revealed an unexpected distribution in the sediment (Fig. 10). Archaea were most abundant in the oxic zone. Since the only Archaea known to inhabit this environment are methanogens, and they would be restricted to the anoxic region, this distribution suggested some kind of novelty. This was sub-

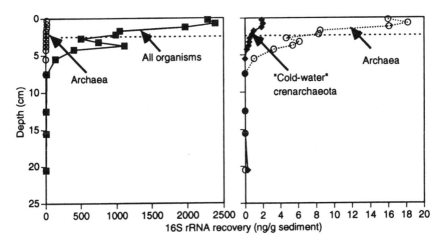

Figure 10. Depth distribution of total biomass (all organisms probe), Archaea, and "cold water" Crenarchaeota in Lake Michigan sediment. DNA probes (Table 4) were hybridized to sediment-extracted RNA as depicted in Figure 8. The dashed line shows the approximate depth of oxygen depletion.

sequently demonstrated by domain-selective PCR amplification, cloning, and sequencing (Fig. 11). Representatives of a new lineage of Archaea within the kingdom Crenarchaeota were identified and their localization within the oxic zone verified by hybridization (Fig. 10). Other members of this novel lineage were first identified (by sequence) in marine plankton and more recently in soils and sediments (DeLong et al 1994, Hershberger et al. 1996, Jurgens et al. 1997, Schleper et al. 1997, Bintrim et al. 1997). However, since none have yet to be recovered in pure culture, their physiology is unknown. As we continue to explore this environment, we anticipate that this will be the more common observation—sediment chemistry is controlled by microorganisms unknown to biology.

Isolation of new organisms—metal active microbes

While the techniques outlined above are of tremendous value for determining populations, it is probably safe to say that the environment will never be understood if the organisms are not grown and studied in culture. That is, just knowing who is there is not enough, one must examine the metabolic plasticity and regulation of the population in order to understand how the ecosystem might function under different conditions. Thus the isolation of new organisms remains an important and guiding part of the field. One difference, of course, is that on the basis of the molecular data obtained above, it is now posible to suspect the presence of types of organisms that may not have been previously suspected, and to design enrichment cultures and growth media accordingly.

For example, some of the new findings obtained with metal-active microbes illustrate that the "art" of isolation and characterization of new organisms represents an active area with tremendous potential for further development. The examples include iron oxidizing and iron reducing bacteria, new metabolic groups unknown until recently.

Iron-oxidizing bacteria. In the past decade, two new groups of iron oxidizing bacteria have been isolated, and while the product of both these bacteria is oxidized [ferric] iron, the metabolism of the two groups that produce the Fe(III) is decidedly different. The first, described by Ehrenreich and Widdell (1994), are photosynthetic bacteria, which use the energy of sunlight to drive the fixation of CO_2 to organic carbon $(CH_2O)n$ under anaerobic conditions. The electron donor for this reaction is ferrous iron, so that the

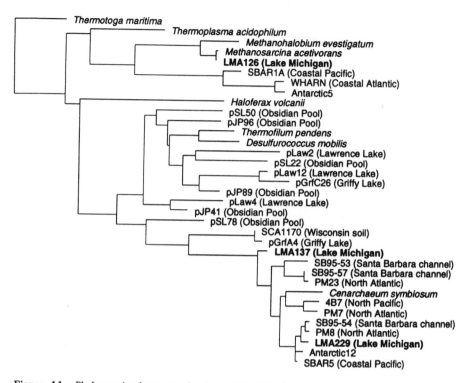

Figure 11. Phylogenetic placement of archaeal 16S rRNA sequences recovered from Lake Michigan sediment by domain-specific PCR amplification. Sequences from the study discussed in the text are named LMA###. Other sequences affiliated with these novel Crenarchaeota have been recovered from marine plankton, soil, and other sediments.

product that accumulates during the reaction is ferric iron in a solid form like ferrihydrite $[Fe(OH)_3]$.

$$Fe(II) + CO_2 + H_2O \xrightarrow{\text{sunlight}} Fe(III) + (CH_2O)n$$

The second group, described by Straub et al (1996) are chemolithotrophic bacteria that utilize ferrous iron as the source of energy, while using nitrate as the oxidant for anaerobic growth. These bacteria use the energy of the $Fe(II)/NO_3$-couple to supply energy for carbon fixation and resulting autotrophic growth.

The iron-reducing bacteria. As discussed above, until a few years ago, it was not accepted that bacteria could use solid substrates like iron or manganese oxides for dissimilatory metabolism. This changed in 1988 , with the report of two organisms *Shewanella putrefaciens* and *Geobacter metalloreducens* capable of living anaerobically by coupling their heterotrophic growth to the reduction of iron or manganese oxides. Such bacteria were capable of catalyzing the general reaction shown below:

$$Fe(III) + \text{organic carbon} \longrightarrow Fe(II) + CO_2 \text{ (or other oxidized organic C)}$$

Anaerobic biogeochemical cycling of iron. With the knowledge that such organisms exist, it is possible to "construct" a hypothetical biogeochemiscal cycle of iron, driven by light, and cycling organic carbon, as shown in Figure 12. This anaerobic system involves no volatile components as redox members.

Hypothetical Iron Cycle

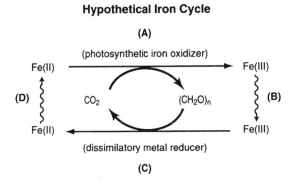

Figure 12. A hypothetical biogeochemical cycle of iron and carbon, catalyzed by two recently isolated groups of prokaryotes, as described in the text. It is meant only to stimulate thought and to emphasize that many different kinds of elemental cycles are, in principle, possible.

Minerals and microbes

Sediments are dominated by solid phase minerals of many different kinds (carbonates, silicates, clays, metal oxides, metal sulfides, etc.) and these minerals are often the reactants and/or products of microbial reactions. They also serve as solid substrates for microbial colonization, irrespective of whether they are directly involved with microbial metabolism. This general area represents one of the most unknown and uninvestigated areas of microbial ecology, and is one of central importance to the ultimate understanding of the dynamics of sediments. For example, the magnetic mineral magnetite can be formed (Bell et al. 1987) or dissolved (Kostka and Nealson 1995) by iron reducing bacteria, depending on the conditions. Similarly, although it is well known that microbially reduced iron can end up as any one of a variety of mineral end-products (iron sulfide, phosphate, carbonate, etc.), very little is known of any specific roles that microbes play in the formation and/or dissolution of most minerals. In terms of undertanding past sedimentary deposits, such knowledge would be extremely valuable, but it represents an area in which there is a paucity of understanding.

New microscopic techniques

The molecular studies outlined above, as well as the study of mineral-microbe interactions have both benefitted from recent advances in microscopy, which will undoubtedly continue, along with advances in image processing, to enhance our ability to do environmental microscopy of sediments. In particular, the confocal laser scanning microscope has allowed one to image microbes on surfaces, and to reconstruct their three dimensional environments. Also, the environmental scanning electron microscope (Little et al. 1991; Fig. 13) allows one to examine microbes under ambient atmospheric conditions, obviating the need for vacuum preparations and gold coating of the sample. These two techniques, in combination with fluorescent probes of the type discussed above, will have major impacts on our appreciation of sediment microbiology in the next decade.

New microbial ecosystems

Many new microbial ecosystems undoubtedly await description, and as the diversity of microbial consortia and populations are unravelled, the description of these systems may represent one of the major challenges of the future. The delicate symbioses into which microbes are entered may well explain why many remain uncultured—we may simply need to understand that many organisms are dependent on syntrophs to either supply or remove a given nutrient or product before growth can occur.

Apart from these speculations, some ecosystems have emerged in the recent past that suggest that the interaction between the geology of the Earth and the microbial ecosystems that reside there might be much more intricate that had been imagined a few decades ago. The two examples I choose relate to the generation of reduced energy sources by geological processes, and the potential development of major bacterial populations in response to these energy sources.

Hydrothermal energy inputs. When subsurface seawater is entrained, exposed to very high temperatures, and redirected to the ocean via the mid-oceanic ridges, major changes in its chemistry occur due to high temperature catalysis (see Jannasch 1995). These so-called

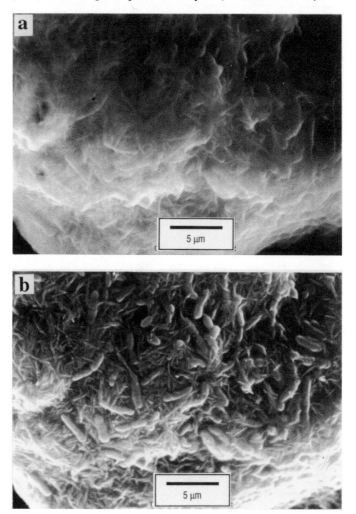

Figure 13. Bacteria on the surface of a metal oxide. This figure shows a metal oxide MnO_2 in a late stage of dissolution. The surface is covered by a layer of bacteria that are growing with $Mn(IV)$ as the electron acceptor. The bacteria produce a polysaccharide layer that obscures them from view when the samples are wet, but upon drying, they become easily visible. (a) Image of hydrated sample of MnO_2 with bacteria imaged with the environmental scanning electron microscope (ESEM. (b) ESEM image of the same field after dehydration showing abundant bacteria on the surface of the metal oxide.

hydrothermal vent environments have some very exciting characteristics concerning microbiology, and many of these are directly relevant to deep sediments on Earth. Under hydrothermal conditions, reduction of sulfur, sulfate and CO_2 are all thermodynamically favored processes. In fact, not only are H_2, H_2S and CH_4 predicted under these conditions, but so are a series of organic compounds, like fatty acids, alcohols and ketones (Shock 1996). Thus, under defined hydrothermal conditions, it is possible to use the geothermal energy of the Earth to create both inorganic (chemolithotrophic) and organic (heterotrophic) energy sources that should be easily exploited by thermophilic microbes. That this occurs on the contemporary Earth is elegantly shown by the rich populations of organisms around the deep sea hydrothermal vents—whether it could occur in other deep sedimentary environments with hydrothermal input is an interesting queston.

Hydrogen driven, rock-based ecosystems. A recent report by Stevens and McKinley (1995) raises the possibility of other types of ecosystems, also possibly isolated from the surface environment, and also driven ultimately by geochemical energy rather than sunlight. This is the report of hydrogen production as a result of the interaction anaerobic water with basaltic rocks. These reactions have been verified in the laboratory using rock/water systems, and are consistent with the presence of rich populations of anaerobic microorganisms in the deep subsurface—populations that exist entirely as a result of the input of hydrogen gas and the availability of CO_2 from the carbonate rocks.

SUMMARY

One might note that nearly everything discussed in the final section of the review involves more than one discipline. The spatial and temporal scales involved often require disciplines ranging from geology and geochemistry at one end, to microbiology and molecular biology at the other. It is now possible to define the microbial populations that can not be cultured. It is also readily accepted that microbes exploit many energy sources and oxidants not known a few years ago, and that some environments exist where energy is provided through geological processes rather than sunlight. Despite these advances in knowledge, it must be admitted that our understanding of the ways that sedimentary microbes interact with each other and with their mineral rich environment is only now beginning to emerge.

In terms of microbes and minerals, what we learn from the study of stratified communities is that many of the elements that form and dissolve minerals are the products and/or reactants of the myriad of microbes that accumulate at these fronts, dramatically affecting the mineral content and quality through ways that remain mostly the object of speculation.

ACKNOWLEDGMENTS

We thank Liz Alm, Barbara MacGregor, Bill Caplan, and Jennifer Becker for their contributions to this manuscript. This work was in part supported by NSF grant DEB-9408243 and ONR grant N00014-94-1-1171 to DAS. KHN also thanks NASA (Exobiology Program) for support (#144DG40).

REFERENCES

Amann RI, Binder BJ, Olson RJ, Chisholm SW, Devereux R, Stahl DA (1990) Combination of 16S rRNA-targeted oligonucleotide probes with flow cytometry for analyzing mixed microbial populations. Appl Environ Microbiol 56:1919-1925

Amann RI, Stromley J, Devereux R, Key R, Stahl DA (1992) Molecular and microscopic identification of sulfate-reducing bacteria in multispecies biofilms. Appl Environ Microbiol 58:614-623

Amann RI, Ludwig W, Schleifer K-H (1995) Phylogenetic identification and *in situ* detection of individual microbial cells without cultivation. Microbiol Rev 59:143-169

Balows A, Trueper HG, Dworkin M, Harder W, Schleifer K-H (1991) The Prokaryotes, 2nd ed. Berlin: Springer-Verlag

Barns SM, Fundyga RE, Jeffries MW, Pace NR (1994) Remarkable archaeal diversity detected in a Yellowstone National Park hot spring environment. Proc Natl Acad Sci 91:1609-1613

Bell PE, Mills AL, Herman JS (1987) Biogeochemical conditions favoring magnetite formation during anaerobic iron reduction. Appl Environ Microbiol 53:2610-2616

Bintrim SB, Donohue TJ, Handelsman J, Roberts GP, Goodman R (1997) Molecular phylogeny of Archaea from soil. Proc Natl Acad Sci 94:277-282

Burdige DJ, Nealson KH (1986) Chemical and microbiological studies of sulfide mediated manganese reduction. Geomicrobiol J 4:361-387

DeLong EF, Wu KY, Prézelin BB, Jovine RVM (1994) High abundance of archaea in Antarctic marine picoplankton. Nature 371:695-697

Devereux R, Delaney M, Widdel F, Stahl DA (1989) Natural relationships among sulfate-reducing eubacteria. J Bacteriol 171:6689-6695

Devereux R, Kane MD, Winfrey J, Stahl DA (1992) Genus- and group-specific hybridization probes for determinative and environmental studies of sulfate-reducing bacteria. System Appl Microbiol 15:601-609

Devereux R, Mundfrom GW (1994) A phylogenetic tree of 16S rRNA sequences form sulfate-reducing bacteria in a sandy marine sediment. Appl Environ Microbiol 60:3437-39

Devereux R, Winfrey MR, Winfrey J, Stahl DA (1996) Depth profiles of sulfate-reducing bacterial ribosomal RNA and mercury methylation in an estuarine sediment. FEMS Microbiol Ecol 20:23-31

DiChristina TJ, Arnold RG, Lidstrom ME, Hoffmann MR (1988) Dissimilative Fe(III) reduction by the marine eubacterium Alteromonas putrefaciens 200. Water Sci Technol 20:69-79

Ehrenreich A, Widdell F (1994) Anaerobic oxidation of ferrous iron by purple bacteria, a new type of phototrophic metabolism. Appl Environ Microbiol 60:4517-4526

Fossing H, Gallardo VA, Jorgensen BB, Huettel M, Nielsen LP et al. (1995) Concentration and transport of nitrate by the mat-forming sulfur bacterium Thioploca. Nature 374:713-715

Giovannoni SJ, Britschgi TB, Moyer CL, Field KG (1990) Genetic diversity in sargasso sea bacterioplankton. Nature 345:60-63

Gottschalk G (1994) Microbial Metabolism, 2nd ed. Berlin: Springer-Verlag.

Hershberger KL, Barns SM, Reysenbach AL, Dawson SC, Pace NR (1996) Wide diversity of Crenarchaeota. Nature 384:420

Jannasch HW (1995) Microbial interactions with hydrothermal fluids. In: Seafloor Hydrothermal Systems: Physical, Chemical, Biological and Geological Interactions. Geophys Monograph 91:273-296

Johnson JL (1984) Nucleic acids in Bacterial Classification. In: NR Krieg, J Holt (eds) p 8-11, Bergey's Manual of Systematic Bacteriology. Baltimore, Maryland: Williams & Wilkins

Jurgens G, Lindström K, Saano A (1997) Novel group within the kingdom Crenarchaeota from boreal forest soil. Appl Environ Microbiol 63:803-805

Kohring L, Ringelberg D, Devereux R, Stahl DA, Mittelman MW, White DC (1994) Comparison of phylogenetic relationships based on phospholipid fatty acid profiles and rRNA sequence similarities among dissimilatory sulfate-reducing bacteria. FEMS Microbiol Lett 119:303-308

Kostka JE, Nealson KH (1995) Dissolution and reduction of magnetite by bacteria. Env Sci Technol 29:2535-2540

Lake J (1985) Evolving ribosome structure: domains in archaebacteria, eubacteria, eocytes and eukaryotes. Ann Rev Biochem 54:507-530

Liesack W, Stackebrandt E (1992) Occurrence of novel groups of the domain Bacteria as revealed by analysis of genetic material isolated from an Australian terrestrial environment. J Bacteriol 174:5072-5078

Little BL, Wagner P, Ray R, Pope R, Sheetz R (1992) Biolfilms: an ESEM evaluation of artifacts introduced during SEM preparation. J Industr Microbiol 8:213-222

Lonergan DJ, Jenter HI, Coates JD, Phillips EJP, Schmidt TM, Lovley DR (1996) Phylogenetic analysis of dissimilatory Fe(III)-reducing bacteria. J Bacteriol 178:2402-2408

Lovley DR, Phillips EF (1988) Novel mode of microbial energy metabolism: Organic carbon oxidation coupled to dissimilatory reduction of iron or manganese. Appl Environ Microbiol 51:683-689

Lovley DR, Phillips EJ, Lonergan DJ (1991) Enzymatic versus nonenzymatic mechanisms for Fe(III) reduction in aquatic sediments. Environ Sci Technol 25:1062-1067

Lovley DR (1993) Dissimilatory Fe and Mn reduction. Microbiol Rev 55:259-287

MacGregor BJ, Moser DP, Nealson KH, Stahl DA (1996) Crenarchaeota in Lake Michigan sediments. Appl Environ Microbiol 63:1178-1181

McHatton SC, Barry JP, Jannasch HJ, Nelson DC (1996) High nitrate concentrations in vacuolate, autotrophic marine Beggiatoa spp. Appl Environ Microbiol 62:954-958

Maidak BL, Larsen N, McCaughey MJ, Overbeek R, Olsen GJ, Fogel K, Blandy J, Woese CR (1994) The ribosomal database project. Nucl Acids Res 22:3485-3487

Moser DP, Nealson KH (1996) Growth of the facultative anaerobe Shewanella putrefaciens by elemental sulfur reduction. Appl Environ Microbiol 62:2100-2105

Muyzer G, DeWall EC, Uitterlinden AG (1993) Profiling of complex microbial populations by denaturing gradient gel electrophoresis analysis of polymerase chain reaction-amplified genes coding for 16S rRNA. Appl Environ Microbiol 59:695-700

Muyzer G, Hottenträger S, Teske A, Wawer C (1995) Denaturing gradient gel electrophoresis of PCR-amplified 16S rDNA - A new molecular approach to analyse the genetic diversity of mixed microbial communities. In: ADL Akkermans, JD van Elsas, FJ de Bruijn (eds) Molecular Microbial Ecology Manual. Dordrecht, The Netherlands: Kluwer Press

Myers CR, Nealson KH (1988) Bacterial manganese reduction and growth with manganese oxide as the sole electron acceptor. Science 240:1319-1321

Myers CR, Nealson KH (1988) Microbial reduction of Mn oxides: interactions with iron and sulfur. Geochim Cosmochim Acta 52:2727-2732

Nealson KH (1982) Bacterial ecology of the deep sea. In: Ernst WG, Morin JG (eds) The Environment of the Deep Sea, Vol II, p 179-200 Englewood Cliffs, NJ: Prentice Hall

Nealson KH, Saffarini DA (1994) Iron and manganese in anaerobic respiration: environmental significance, physiology, and regulation. Ann Rev Microbiol 48:311-43

Olsen GJ, Lane DJ, Giovannoni SJ, Pace NR, Stahl DA (1986) Microbial ecology and evolution: A ribosomal RNA approach. Ann Rev Microbiol 40:337-365

Olsen GJ, Woese CR, Overbeek R (1994) The winds of (evolutionary) change: Breathing new life into microbiology. J Bacteriol 176(1):1-6

Pace NR, Stahl DA, Lane DJ, Olsen GJ (1986) The analysis of natural microbial populations by ribosomal RNA sequences, p 1-55. In: KC Marshall (ed) Advances in Microbial Ecology. New York: Plenum Press

Pace NR, Angert ER, DeLong EF, Schmidt TM, Wickham GS (1993) New perspective on the natural microbial world. In: RH Baltz, GD Hegeman (eds) Industrial Microorganisms: Basic and Applied Molecular Genetics, p 77-84 Washington, DC: Am Soc Microbiol

Raskin L, Stromley JM, Rittmann BE, Stahl DA (1994) Group-specific 16S rRNA hybridization probes to describe natural communities of methanogens. Appl Environ Microbiol 60:1232-1240

Risatti JB, Capman WC, Stahl DA (1994) Community structure of a microbial mat: The phylogenetic dimension. Proc Natl Acad Sci 91:10173-10177

Rochelle PA, Cragg BA, Fry JC, Parkes RJ, Weightman AJ (1994) Effect of sample handling on estimation of bacterial diversity in marine sediments by 16S rRNA gene sequence analysis. FEMS Microbiol Ecol 15:215-226

Schleper C, Holben W, Klenk H-P (1997) Recovery of Crenarchaeotal ribosomal DNA sequences from freshwater-lake sediments. Appl Environ Microbiol 63:321-323

Schmidt TM, DeLong EF, Pace NR (1991) Analysis of a marine picoplankton community by 16S rRNA gene cloning and sequencing. J Bacteriol 173:4371-4378

Shively JM, Barton LL (1991) Variations in Autotrophic Life. New York: Academic Press

Shock E (1996) High temperature life without photosynthesis as a model for Mars. J Geophys Res (in press)

Stackebrandt E, Liesack W (1993) Nucleic acids and classification. In: M Goodfellow, AG O'Donnell (eds) Handbook of New Bacterial Systematics, p 151-194, London: Academic Press

Schink B (1991) Syntrophism among prokaryotes. In: A Balows, HG Trueper, M Dworkin, W Harder, K-H Schleifer (eds) The Prokaryotes, p 300-312 Berlin: Springer-Verlag

Stahl DA (1986) Evolution, ecology and diagnosis: Unity in variety. Bio/Technology 4:623-628

Stahl DA, Flesher B, Mansfield H, Montgomery L (1988) Use of phylogenetically based hybridization probes for studies of ruminal microbial ecology. Appl Environ Microbiol 54:1079-84

Stahl DA, Amann R (1991) Development and application of nucleic acid probes in bacterial systematics. In: E Stackebrandt, M Goodfellow (eds) Sequencing and Hybridization Techniques in Bacterial Systematics, p 205-248 Chichester, UK: John Wiley & Sons

Stevens TO, McKinley JP (1995) Lithotrophic microbial ecosystems in deep basalt aquifers. Science 270:450-453

Stone AT, Godtfredsen KL, Deng B (1994) Sources and reactivity of reductants encountered in aquatic environments. In: Chemistry of Aquatic Systems. G Bidoglio, W Stumm (eds) Dordrecht, The Netherlands: Kluwer

Straub KL, Benz M, Schink B, Widdel F (1996) Anaerobic, nitrate-dependent microbial oxidation of ferrous iron. Appl Environ Microbiol 62:1458-1460

Stumm W, Morgan JJ (1981) Aquatic Chemistry, 2nd ed. New York: Wiley

Teske A. Wawer C, Muyzer G, Ramsing NB (1996) Distribution of sulfate-reducing bacteria in a stratified fjord (Mariager Fjord, Denmark) as evaluated by most-probable number counts and denaturing gradient gel electrophoresis of PCR-amplified DNA fragments. Appl Environ Microbiol 62:1405-1415

Ward DM, Bateson MM, Weller R, Ruff-Roberts AL (1992) Ribosomal RNA analysis of microorganisms as they occur in nature. In: KC Marshall (ed) Advances in Microbial Ecology, p 219-286 New York, Plenum Press

Weller R, Ward DM (1989) Selective recovery of 16S rRNA sequences from natural microbial communities in the form of cDNA. Appl Environ Microbiol 55:1818-1822

Woese CR (1987) Bacterial evolution. Microbiol Rev 51:221-271

Zheng D, Alm EW, Stahl DA, Raskin L (1996) Characterization of universal small subunit rRNA hybridization probes for quantitative molecular microbial ecology studies. Appl Environ Microbiol 62:4504-4513

Chapter 2

MICROBIAL DIVERSITY IN OCEAN, SURFACE AND SUBSURFACE ENVIRONMENTS

Susan M. Barns

Environmental Molecular Biology
M888, Life Sciences Division
Los Alamos National Laboratory
Los Alamos, New Mexico 87545 U.S.A.

Sandra A. Nierzwicki-Bauer

Department of Biology MRC 306
Rensselaer Polytechnic Institute
Troy, New York 12181 U.S.A.

INTRODUCTION

This chapter will provide an introduction to the evolution and diversity of geoloically important microorganisms. It will begin with a brief consideration of the origin of life and the evolutionary history of microorganisms, and then present an overview of the diversity of geomicrobial agents and thier activities. On a practical note, we will next introduce some of the methods used most commonly for the study of such microbes, both in the laboratory and *in situ* in the environment. Finally, we will present some recent "case studies" in which newly developed techniques have been used to study organisms in unusual environments and have revealed unexpected diversity and novelty in the composition of the microbial community.

PHYLOGENETIC INTRODUCTION TO THE DIVERISTY AND EVOLUTION OF MICROORGANISMS

Molecular phylogeny of the microbial world

The earth's biosphere has been dominated by microbes since its inception, so that understanding the history of the planet requires an understanding of the evolution of microorganisms. Microbes, and all life, were born out of the geochemical processes of the nascent earth, and have evolved with those processes, changing and being changed by them. Charting the course of microbial evolution, therefore, allows us to understand the history not just of genetic lineages, but of the metabolic processes that formed the biosphere and transform the atmosphere, lithosphere and hydrosphere. It is impossible to fully explore the biology or geology of the earth without consideration of microbial diversity and evolution.

Unfortunately, determining the evolutionary history of microorganisms poses some special problems, and as a result, is a very recent endeavor. Most of these problems arise from the small size and simple shapes of microbes. The evolutionary relationships, or *phylogenies*, of macroscopic organisms have generally been inferred by comparison of morphological characters between extant species, and between extant and fossil organisms. This approach has been quite successful in determining both taxonomy and evolutionary history for plants and animals. Microbes, on the other hand, being fragile cells of a limited

0275-0279/97/0035-0002$05.00

variety of shapes, generally do not leave an informative fossil record, so that it is nearly impossible to derive an evolutionary history for most extant species from preserved ancestors. Microbiologists therefore have turned to the examination of physiological traits to reconstruct evolutionary relationships for microbes. These properties were also used as a basis for microbial taxonomy, but because of the ephemeral nature of many biochemical traits, neither a predictive taxonomic system nor a meaningful evolutionary history was determinable. By the 1960s, many microbiologists abandoned the quest, lamenting that :

> ... the general course of evolution [for bacteria] will probably never be known, and there is simply not enough objective evidence to base their classification on phylogenetic grounds... For these and other reasons, most modern taxonomists have explicitly abandoned the phylogenetic approach (Stanier et al. 1976).

As a result, evolutionary trees of the time tended to show "higher" organisms, such as plants, animals and fungi, resolved into well-ordered kingdoms, while microbes (protists and "monera" [bacteria]) were generally lumped into one or two homogenous groups at the base (Margulis and Schwartz 1982).

Biologists since the 1950s had recognized the value of using comparative analysis of molecular sequences to determine the evolutionary history of organisms (Zuckerkandl and Pauling 1965). The basis of this approach, termed *molecular phylogeny*, is to compare sequences of proteins and nucleic acids (DNA and RNA) having a single origin (*homologous* sequences) from different species. By determining how similar the sequences of a particular molecule are from two different organisms, one can infer an evolutionary history for that molecule and hence for the organisms from which it was obtained. Two closely related species will have accumulated few changes in the sequence since they arose from a common ancestor; their sequences should be very similar. Sequences from two more distantly related organisms will have accumulated many differences. By counting the number of differences between the sequences of a protein or nucleic acid from various species we then have a measure of the evolutionary distance between those species. We can transform these distances into line segments and use them to draw a map (or *phylogenetic tree*) of evolutionary relationships between organisms, such as that shown in Figure 1.

It was not until the pioneering work of Carl Woese in the 1970s that this approach was applied to determine a phylogeny for microbes. Woese and his colleagues chose the sequence of ribosomal RNA (rRNA) for their analyses, for a number of reasons (Fox et al. 1980, Woese 1987, Woese and Fox 1977). Small subunit rRNA (SSU rRNA) is the most widely used of the rRNAs, as it is small enough (611 to 2023 nucleotides; Maidak et al. 1997), average bacterial rRNA ~1500 nucleotides) to facilitate laboratory manipulations, but large enough to carry sufficient information for analysis, and SSU rRNA is the most widely used phylogenetic marker for microbial studies today. A large (>10,000 sequences) database now exists for comparison (Maidak et al. 1997), and is growing at the rate of hundreds of sequences per year.

"Universal" phylogenetic trees based on rRNA (and other) sequences show a picture of life's diversity (Fig. 1) that is radically different from what had been proposed based on morphology. Most remarkable is the observation that all life can be grouped into three primary lines of descent, or *domains*; Eucarya, Archaea and Bacteria (formerly termed eukaryotes, archaebacteria and eubacteria; Woese et al. 1990). [A similar picture is obtained from analysis of other highly conserved molecular sequences (Brown and Doolittle 1995, Doolittle and Brown 1994) and whole genome sequences (Bult et al. 1996), although trees based on some types of sequences disagree (Baldauf et al. 1996, Forterre 1997, Golding and Gupta 1995).] Microbial species occur in all three domains; Archaea and Bacteria are comprised solely of microbial species, while Eucarya include microbes such as protists and

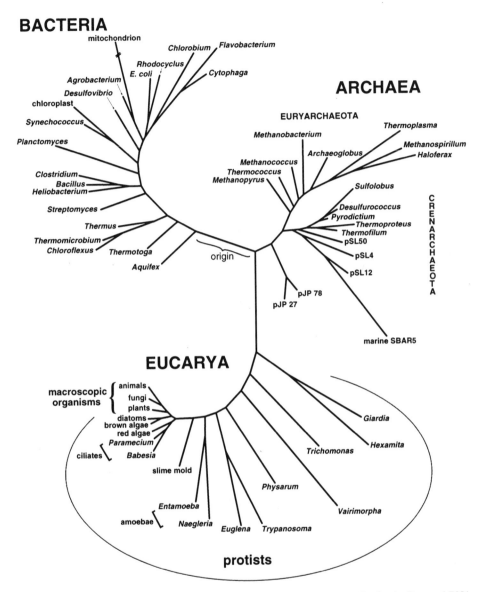

Figure 1. Diagrammatic "Universal" phylogenetic tree of life, based on small-subunit ribosomal RNA sequences. Based on analyses of Barns et al. (1996b), Olsen et al. (1994), and Sogin (1994).

algae, as well as "higher" organisms—plants, animals and fungi. Also striking is the observation that the evolutionary diversity of macroscopic life (plants, animals and fungi) is very limited compared to that of microbes (all other lineages). This is reflected in the enormous variety of metabolic capablities of microorganisms (as will be described later in this chapter) relative to that of "higher" organisms. Indeed, rRNA sequence analysis has also been used to confirm that much of eukaryotic metabolism, such as photosynthesis and respiration, is actually of bacterial origin (see below). Such analyses make clear that the vast majority of evolutionary, genetic and metabolic diversity lies in the microbial world.

By many criteria, all organisms appear to have descended from a single "last common ancestor" (the nature of which will be discussed below). However, phylogenetic trees from single gene sequences, like those based on rRNA, cannot indicate where position of that ancestor, the "root" of the tree, lies. Several recent analyes have compared sequences of genes which duplicated in that ancestor, and have shown that the position of the root is proabably along the bacterial line (Baldauf et al. 1996, Brown and Doolittle 1995, Gogarten et al. 1989, Iwabe et al. 1989; also see Forterre 1997). This means that the Archaea and Eucarya shared a common ancestor for awhile before branching off into separate domains, and this common heritage is reflected in the sequences of many genes (Bult et al. 1996).

Unfortunately, it is difficult to calibrate sequence-based trees to geologic time. This is because rates of sequence change differ between lineages, and even vary *within* lineages across time. Where a fossil record exists, it is possible to infer the time of some nodes, such as the divergence of the "higher" Eucarya (including plants, animals and fungi) about 1 billion years ago (Doolittle et al. 1996, Sogin 1994). For some microbial groups it is possible to estimate a minimum age, such as ca. 2 billion years for cyanobacteria (Schopf 1994) and 2.1 billion years for Eucarya (Han and Runnegar 1992). Difficulties arise when attempting to look further back in time, for instance to the times of diversification of the domains and the last common ancestor, and some highly controversial estimates have been made based on sequence analyses (Doolittle et al. 1996, Gogarten et al. 1996, Hasegawa and Fitch 1996).

The origin of life and the Last Common Ancestor

These recent advances in understanding the evolutionary history of microorganisms have also shed light on the origin of life and the nature of the last common ancestor of all organisms. Our ability to map evolutionary relationships between extant species and infer the "root" of the universal tree has provided new perspectives on how life may have arisen. It is now possible to infer some aspects of the nature of the last ancestor based on the properties which are common to those organisms whose lineages branch closest to it, deeply in the tree. Examining these extant species may help us sort through many of the possible scenarios for the origin of life.

Cellular life appears to have arisen quickly, within the first billion years of the Earth's history (Chang 1994). Carbon isotope analysis of apatite specimens from Greenland suggests the presence of organisms as far back as 3.85 billion years (Ga) (Mojzsis et al. 1996). Fossilized microbial mats have been identified in 3.55 Ga old sedimentary rocks, and recognizable 3.4 Ga old filamentous microbial fossils have been found (Schopf 1994) . Most of the Earth's surface was probably covered by oceans at 3.8 Ga, and the early atmosphere was probably dominated by CO_2, H_2O and N_2, lacking significant amounts of oxygen (Lowe 1994).

There are two main current hypotheses concerning the steps leading to the first living cell. (It has also been proposed that life may have been imported to Earth from other planets on comets or other interplantary debris [the "panspermia" hypothesis; Crick and Orgel 1973, Davies 1996]).These have been recently reviewed (Lazcano and Miller 1996). The older of the two, sometimes referred to as the Oparin-Haldane theory (Oparin 1938, see also Lazcano and Miller 1996), proposes that life arose in a dilute "organic soup" of sugars, amino acids, nucleotides and other molecules abiotically synthesized in the early ocean or perhaps brought in on comets (Oro 1994). These organic building blocks then self-assembled, perhaps templated on the surface of clays (Cairns-Smith et al. 1992), and polymerized into carbohydrates, proteis, RNA and perhaps DNA. Eventually, lipids aggregated into vesicles, surrounding the polymers to produce encapsulated cells.

While the Oparin-Haldane scenario postulates that the earliest cells were *heterotrophic*, obtaining carbon and energy from pre-made organic molecules available in the surrounding "soup," a more recent hypothesis proposes an *autotrophic* origin, in which life arose by reactions of inorganic substrates (Wachtershauser 1990). This type of metabolism requires only simple inorganic substrates for energy generation, all of which were likely to have been abundant on the early Earth. In this view, the first synthetic processes assembled inorganic building blocks, such as CO_2, phosphate and ammonia, on positively-charged mineral surfaces, such as pyrite. The energy and electrons for this process could have been provided by the oxidation of hydrogen sulfide:

$$FeS + H_2S \rightarrow FeS_2 + 2H^+ + 2e^-$$

(Wachtershauser 1994). This reaction produces pyrite, on the surface of which negatively-charged molecules such as CO_2 could collect and possibly polymerize, allowing production of organic carbon compounds (*carbon fixation*). As in the Oparin-Haldane scenario, lipid vesicles then surrounded these reactions to produce cellular life. Eventually, the cell-evolved enzymes which replaced the pyrite crystal as the site of polymer synthesis.

While experimental evidence and arguments for both theories have accumulated rapidly in recent years, it is still too early to decide on either (or rule out many other possibilities). However, examination of extant organisms, in light of what we currently know about microbial phylogeny, favors an autotrophic origin. All of the archaeal and bacterial lineages which branch closest to the root in the universal tree of life are those of autotrophic organisms. These include members of the bacterial genera *Aquifex* and *Thermotoga*, and many archaeal genera such as *Methanopyrus*, *Thermococcus*, *Methanococcus* and *Thermofilum*, shown in Figure 1. All of these organisms can obtain energy for growth from inorganic substrates such as H_2, S^0 H_2S, and many can also fix CO_2 to organic carbon (Stetter 1996b). While the mechanisms by which these metabolic reactions occur in modern organisms are relatively complex, the presence of autotrophic capability in all of the most ancient lineages of life indicates its antiquity. In addition, comparison of metabolic genes found in the recently completed sequences of archaeal and bacterial genomes indicate that many autotrophic pathways have an ancient evolutionary origin (Bult et al. 1996).

It is also clear from examination of the properties common to all extant organisms that the last common ancestor was a fairly sophisticated creature. It already contained a mechanism for synthesis of DNA, although the diversity of systems found in organisms today suggests that it may have been a rudimentary one. RNA and protein synthesis appears to have been more highly evolved, as extant species have many elements of these systems in common. Many elements of metabolism, including enzymes for autotrophic carbon fixation, are shared between all organisms, indicating that they were probably present in the last common ancestor (Olsen and Woese 1996).

Intriguingly, the available phylogenetic evidence also points to a high-temperature origin of life. Hyperthermophilic microorganisms (organisms with optimum growth temperatures over 80°C) constitute all of the archaeal and bacterial lineages which branch closest to the root in the universal tree of life (Stetter 1996b, Woese 1987). It seems likely, therefore, that thermophily was an ancestral characteristic of Bacteria and Archaea. (To date, no extremely thermophilic Eucarya have been discovered. The deepest branches within that domain are populated by protists, but none of these lineages lie very close to the root.) There is also some geological evidence that the Earth's surface at the time of the origin of life may have been quite warm (80° to 100°C), due to "greenhouse" trapping of planetary heat by high pressures of atmospheric CO_2 (Pace 1991, Schwartzman et al. 1993). The high concentrations of reduced chemicals present in hydrothermal vent fluids have prompted the suggestion that life may have arisen in ancient sea-floor vents (Shock 1996).

It is important to recognize, however, that the nature of the first "life" and of the "last common ancestor" may be different (Lazcano and Miller 1996). For instance, it has been proposed that life may have arisen in a low temperature, "organic soup" setting, diversified to include a few species which adapted to hydrothermal environments, then been nearly extinguished by bolide impacts (Sleep et al. 1989). Such impacts would have heated the earth to the point where only heat-adapted organisms could survive, leading to a "bottleneck" through which a hyperthermophilic last common ancestor would emerge. Based on the current universal tree, this organism would presumably have been an autotroph as well. Unfortunatly, the phylogenetic approach does not allow us to look beyond the last common ancestor and address this alternative possibility.

Beyond the Last Common Ancestor—diversification of the domains

The diversity of the microbial life we see today is the result of genetic mutation and environmental adaptation over ~3.5 billion years. Evolution has taken a different course in each of the three domains to produce the variety of species which inhabit the current biosphere. This section will outline the major events and trends in evolution and diversification within the three domains. More extensive discussions have been presented by Woese (1987), Kandler (1994) and Sogin (1994).

Eucarya. The nature of the first eukaryotic cell is unknown, and examination of genetic structure and sequence data suggests that it may have arisen from a fusion of early eukaryotic, bacterial and archaeal genomes (Golding and Gupta 1995, Sogin 1994). The earliest branching eukaroytic lineages are all populated by heterotrophic species lacking mitochondria. It has been postulated, therefore, that the proto-eukaryotic cell was an anaerobic heterotroph, probably feeding on the organic compounds produced by Bacteria and Archaea (Kandler 1994). Since the origin of the domain is as old as that of the Bacteria and Archaea, and the deepest eukaryotic branches are far from the inferred "root" of the tree (Fig. 1), the question of the true phenotype of the earliest species is unresolved. Given how little we know of the diversity of microbial species, early-branching, autotrophic Eucarya may exist and await discovery (Barns et al. 1996a).

Most of the history of eukaryotic diversification is dominated by the emergence of protist lineages. Protist species are mainly single-celled organisms, but also include macroscopic red and brown algae. They are enormously diverse in phenotype and include both photosynthetic and non-photosynthetic organisms of a dazzling array of shapes and habitats. Microbial protists do not group into a cohesive evolutionary assemblage. Rather, as can be seen in Figure 1, protists occupy a succession of primary branchings from the eukaryotic line of descent, and are united mainly by their type of cellular organization (Sogin 1994). The first three protist lineages (represented in Fig. 1 by *Trichomonas*, *Giardia/Hexamita* and *Vairimorpha/Encephalitozooan*) lack mitochondria, and this is believed to be the primordial condition of Eucarya (*Entamoeba* also lacks mitochondria, and may branch more deeply than is suggested by Fig. 1; Hasegawa et al. 1993). Molecular phylogenetic analysis has been used to identify the origin of mitochondria from within the bacterial domain (see Fig. 1; Yang et al. 1985) probably as a result of an endosymbiotic event (Margulis 1970). Similarly, it is clear that chloroplasts of protists, algae and plants arose from endosymbiotic cyanobacteria at a somewhat later time (Zablen et al. 1975). Thus, both oxygen utilization and photosynthesis in Eucarya are of bacterial origin. Parallel studies of phylogeny, physiology and ultrastructure are currently helping to better define the diversity and relationships between protist groups (Sogin 1994).

The five "crown" kingdoms, plants (including green algae), animals, fungi, and the unicellular alveolates (including ciliates and dinoflagellates) and stramenopiles (diatoms, brown algae, chrysophytes) appear to have radiated roughly simultaneously, ~1 Ga ago.

The triggering event for this diversification is unknown, but may have included a sudden increase in atmospheric oxygen, a cataclysmic event or invention of new genetic mechanisms (Sogin 1994). Unfortunatly, this rapid radiation has largely obscured the branching order of these groups, so that it is difficult to define the orgins of each kingdom. It is likely that the two unicellular kingdoms branched off prior to the divergence of plants, animals and fungi. Several criteria suggest that animals and fungi shared a common ancestor after their divergence from plants (Wainright et al. 1993).

Archaea. As previously discussed, the ancestral archaeal cell was probably a thermophilic autotroph. Early in the evolution of the Archaea, a fundamental split appears to have taken place, producing two discrete lineages, the kingdoms Crenarchaeota and Euryarchaeota (see Fig. 1; Woese et al. 1990). Members of the Crenarchaeota which have been grown in laboratory culture all share a themophilic phenotype, and most are anaerobic and utilize sulfur in some aspect of energy metabolism. [Recent molecular sequenced-based analyses, however, indicate the existance of previously unknown, aerobic, low-temperature Crenarchaeota; see end of this chapter.] The Euryarchaeota, on the other hand, display a wide diversity of phenotypes. Methane generation for energy production seems to have been invented in this kingdom, and several methanogenic groups are scattered throughout the Euryarchaeota (*Methanopyrus, Methanococcus, Methanobacterium, Methanospirillum* in Fig. 1). Species with tolerance to and requirement for high salt concentrations for growth (*Haloferax*) evidently evolved from within a methanogenic lineage (Woese 1993). Thermophilic, sulfur-utilizing lineages are also present (*Thermococcus, Thermoplasma, Archaeoglobus*). The earliest lineages in both kingdoms are anaerobic organisms; oxygen dependance appears to have arisen later (*Sulfolobus, Haloferax*).

Recent analysis of two sequences (pJP27 and pJP78 in Fig. 1) recovered from a hot spring sediment suggest the existance of a third archaeal kingdom, tentatively named "Korarchaeota" (Barns et al. 1996b). One of these organisms has recently been grown in anaerobic mixed culture at 85°C (Burggraf et al. 1997a), however nothing else is known of its physiology. The branching position of this group, close to the root of the universal tree, suggests that these organisms may hold clues to the nature of the origin of life, and further supports the thermophilic character of the last common ancestor.

Bacteria. The bacterial domain displays a dazzling array of primary lineages, and an equally diverse assortment of physiologies. Well-studied bacteria group into about 12 distinct lineages, or kingdoms (Olsen et al. 1994, Woese et al. 1990). [However, molecular sequence-based analyses of naturally occurring microbial species is rapidly increasing this number.] A rigid cell wall composed of murein was a primordial invention, and may have helped bacteria colonize many different habitats (Kandler 1994). The earliest branching lineages (*Aquifex, Thermotoga, Thermus/Thermomicrobium*) are all inhabited by thermophilic species, supporting a high-temperature origin for the domain. Beyond these early lineages, a rapid radiation of kingdoms may have occurred, as the branching order between these groups is ambiguous (Van de Peer et al. 1994, Woese 1987).

The distribution of metabolic types across the bacterial tree is not always consistent with phylogenetic groupings of species. This can be due to gene loss, convergent evolution and/or gene transfer between species over the course of evolution. As a result, geologically important properties such as photosynthesis, nitrogen fixation, manganese reduction and oxidation, iron oxidation and reduction, sulfur oxidation and reduction, and hydrogen oxidation are distributed among several distinct kingdoms within the domain. Conversely, physiologies such as nitrification and methanotrophy are restricted to distinct kingdoms. Needless to say, this scattered arrangement of conspicuous physiological properties has led to conflicting classifications in many instances.

Photosynthesis was probably an early invention, as it is found distributed among at least five kingdoms, each with distinct photosynthetic pigment types (Pierson 1993, Woese 1987). Plant-like, oxygen-evolving photosynthesis appears to have evolved only once, however, within the cyanobacteria/chloroplast lineage. This allowed species to exploit an almost unlimited supply of water as a source of electrons, and probably precipitated an explosion of cyanobacterial colonization. It is this invention which first made oxygen abundant on Earth, and allowed the diversification of aerobic organisms, such as animals and fungi. Oxygen production also allowed accumulation of an ozone layer, protecting surface organisms from the lethal effects of UV radiation.

GEOMICROBIOLOGICAL AGENTS IN DEPTH

Eucarya: fungi, lichens, algae and protozoa

Eukaryotic microorganisms that can play a role in certain geological processes include protozoa, algae, fungi, and lichens. A very brief description of each of these groups follows.

As a group, the protozoa are characterized by obtaining their nutrition mainly by ingestion. Thus, protozoans are important as predators. Certain of the protozoa inhabit fresh water and marine environments as well as being abundant in soils. The amoebas and their relatives are all unicellular protozoa. Other protozoa have shells that become important components of marine or fresh water sediments. In some instances the shells are made of silicates which are recycled when the organisms die and in other instances shells composed of organic material and calcium carbonate contribute to sediments. A number of protozoans as well as some eukaryotic algae lay down $CaCO_3$ as cell-support structures. Other protozoans are known to be able to deposit iron on their cells. Protozoans may also play a significant role related to the geomicrobiology of manganese because some of them can grow on ferromanganese nodules. Since they feed on bacterial cells, it is likely that they significantly impact the populations of Mn(II)-oxidizing bacteria.

The algae consist mainly of photosynthetic organisms, although some phyla also include heterotrophs. These organisms are relatively simple aquatic organisms that have chlorophyll a, as well as other accessory pigments that are variable. Included in this group are members of *Dinoflagellata, Chrysophyta, Bacillariophyta, Euglenophyta, Chlorophyta, Phaeophyta*, and *Rodophyta*. The different groups of algae vary in their abundance in either marine or fresh water habitats. In addition to these common habitats, members of *Chlorophyta* live in diverse environments including damp soil, the surface of snow or as symbionts. Among the dinoflagellates are both photosynthetic and heterotrophic forms. In a number of eukaryotic algae, calcium carbonate is laid down either as calcite or aragonite. The amount of carbon incorporated into carbonate as a result of algal photosynthesis may be a significant portion of the total carbon assimilated. In diatoms which have high silicon content, silicon seems to play a metabolic role in the synthesis of chlorophyll and the synthesis of DNA, in addition to its structural function.

Fungi are heterotrophs that acquire their nutrients by absorption and are mostly multicellular. More specifically, they absorb small organic molecules from the surrounding environment. In the environment, fungi play important functions as scavengers and decomposers. As decomposers fungi are able to break down complex molecules such as polysaccharides and proteins to carbon dioxide, nitrate and other simple inorganic compounds. Fungi are found in a wide variety of environments including terrestrial and aquatic habitats. Fungi are extremely important geomicrobial agents in that some of them are able to take up silicon, precipitate $CaCO_3$, carry out nitrification, reduce arsenic, perform mercury methylation, reduce ferric iron, reduce and oxidize manganese, etc.

Lichens are symbiotic associations of photosynthetic algae or cyanobacteria, and fungi. The fungal component is most commonly an ascomycete, and the photosynthetic partner is usually a unicellular or filamentous green alga (chlorophytes) or a cyanobacterium. Lichens are able to survive in many inhospitable habitats including bare rocks, arctic tundra and deserts. Lichens have been shown to form organic compounds that can dissolve rock minerals and thereby cause weathering.

Archaea

Introduction. The archaeal domain is comprised of organisms of enormous physiological and ecological diversity. Within it are species having the highest known growth temperatures, which thrive in boiling, sulfurous pools and in hydrothermal vents on the sea floor. Others flourish in saturated brines of the Dead Sea and Great Salt Lakes; some species are nearly square in shape. Yet others can exist only in environments devoid of oxygen, eating hydrogen and breathing carbon dioxide. In fact, it has only been very recently that microbiologists have detected Archaea which appear to live in relatively "normal" environments, in cool oxygenated sea water and terrestrial soils (see below). To date, none of these new, "atypical" Archaea have been cultivated, and their relatively aberrant physiologies remain a mystery.

Across this diversity, however, is unity. Archaea have much basic biochemistry in common with Bacteria and Eucarya, but are distinguished from these groups by several features. As discussed above, rRNA sequence information was first used to show the coherence of the archaeal domain, and its distiction from the domains Bacteria and Eucarya. Archaeal cell membranes are unique in being composed of lipids which lack fatty acids, having instead hydrocarbons called isoprenes. These membranes also have ether linkages between lipids and glycerol molecules, while Bacteria and Eucarya both have ester linkages (Zillig 1991).

Archaea exhibit a wide range of metabolic capabilities and adaptations to their environments. The ability to obtain energy and carbon for growth from inorganic sources is widespread and takes several forms. A few species are capable of light-driven energy metabolism (phototrophy), while many others exist by gaining energy from oxidation of H_2, S^0, Fe, pyrite and other chemical sources (lithotrophy). Archaea utilize many different oxidants as electron sinks, including O_2 (aerobic respiration) and S^0, NO_3^-, Fe^{3+}, SO_4^{2-}, CO_2 and various organic compounds (anaerobic respiration), and some species can grow using both strategies. Within the Archaea are both primary producers and consumers of organic carbon. Many species can obtain carbon for growth by fixation of CO_2 (autotrophy), and use of exogenous organic carbon for carbon and energy metabolism is also common.

A variety of unusual adaptations allows Archaea to utilize resources in some rather extreme (from our perspective) environments. *Hyperthermophiles,* Archaea with optimum growth temperatures above 80°C, require thermostabilization of cellular components to withstand existance in hot springs and hydrothermal vents. In these organisms, DNA, RNA, proteins and other molecules are stabilized against denaturation by a variety of mechanisms, and generally require high temperatures for proper function. The *extreme halophiles* require high concentrations of salt for growth, and evidently dominate microbial communities in saline environments. Halophilic Archaea have evolved specialized mechanisms to allow energy generation in the presence of high ionic concentrations, and cell wall components and proteins in these species require high cation (Na^+, K^+) concentrations for stability. The third group, the *methanogens*, contains a variety of organisms united by an anaerobic physiology which generates methane as an end-product of energy metabolism. These organisms are restricted to anoxic habitats, where they commonly live in cooperation with other species with which they exchange nutrients.

Although many archaeal habitats may seem unusual, they are in fact widespread around the planet, making Archaea important players in global chemistry and ecology. For instance, terrestrial hot springs exist on every continent, and hydrothermal vent systems extend for thousands of kilometers along the mid-ocean ridges. Hydrothermal fluids often contain energy-rich chemical components obtained when geothermally heated water reacts with crustal minerals. Hyperthermophilic Archaea readily exploit these solutes for growth (Shock 1996), allowing the formation of highly productive biological communities (Lutz et al. 1994). There is even mounting evidence for a hot subsurface environment which may harbor enormous biomass (Cragg et al. 1995, Gold 1992, Parkes et al. 1994). Anoxic environments are even more common, and include freshwater and marine sediments, soils and animal and insect digestive tracts. Again, there is evidence for extensive microbial life in subsurface anoxic environments (Parkes et al. 1994, Stevens and McKinley 1995). Finally, molecular-sequence based analyses of uncultivated Archaea in low-temperature, oxic ocean and terrestrial environments have indicated that members of the domain are probably abundant in these habitats, too (see end of this chapter for detailed discussion; Bintrim et al. 1997, DeLong 1992, Fuhrman et al. 1992, Hershberger et al. 1996). As such environments cover most of the surface of the earth, these novel Archaea may be among the most numerous and ecologically important microbes in the biosphere.

Hyperthermophiles.   The archaeal domain contains organisms with the highest growth temperatures known, the *hyperthermophiles*. These are defined as organisms which grow optimally at temperatures above 80°C, and generally cannot grow below 60°C (Segerer et al. 1993). Many species flourish above 100°C; the highest known growth temperature is 113° (Stetter 1995). Most of these species utilize sulfur in some way and are most commonly isolated from active volcanic habitats, such as heated soils, hot springs and hydrothermal vents. The majority of hyperthermophilic species known are members of the kingdom Crenarchaeota; indeed all Crenarchaeota cultivated to date are thermophiles or hyperthermophiles (but see below for evidence of low-temperature Crenarchaeota). These include the genera *Sulfolobus*, *Metallosphaera*, *Pyrodictium*, *Desulfurococcus*, *Staphylothermus*, *Thermofilum*, *Thermoproteus*, *Pyrobaculum*, and a few others (Burggraf et al. 1997b). The euryarchaeal genera *Thermococcus*, *Archaeoglobus* and *Thermoplasma* share many features with Crenarchaeota. Several hyperthermophilic methanogens are also known-- they will be discussed separately. A number of review articles on hyperthermophiles have been published recently (Blochl et al. 1995, Burggraf et al. 1997b, Segerer et al. 1993, Stetter 1995, Stetter 1996a).

Physiology.   As a group, the hyperthermophiles are quite metabollically diverse, and there is often much nutritional flexibility within species. Several energy-yielding metabolic reactions and the genera which carry them out are given in Table 1. As can be seen, sulfur is central to many of these reactions, either as reductant or oxidant. Many genera can gain energy using H_2 or S^0 as electron donor and S^0, SO_4^{2-}, CO_2, NO_3^- or O_2 as acceptor (Schonheit and Schafer 1995). Many of these same species can live organotrophically, oxidizing organic compounds and utilizing S^0 or O_2 as electron acceptors. Remarkably, a few genera such as *Acidianus* can oxidize or reduce S^0 depending on growth conditions (Segerer et al. 1993). The only genus known to reduce sulfate to sulfide is *Archaeoglobus*, which also produces small amounts of methane (Achenbach-Richter et al. 1987). Members of the genus *Sulfolobus* can both oxidize Fe^{2+} to Fe^{3+} aerobically, and reduce Fe^{3+} to Fe^{2+} anaerobically (Madigan et al. 1997). Most S^0 oxidizing species can also use Fe^{2+} and molybdate as electron acceptors (Segerer et al. 1993), and a strain of *Metallosphaera* which is capable of oxidizing S^0, pyrite, spalerite, chalcopyrite or H_2 was recently isolated from a uranium mine (Fuchs et al. 1996).

Table 1. Major energy-yielding reactions of hyperthermophilic Archaea.
Adapted from Madigan et al. (1997) and Stetter (1995 1996b).

Nutritional mode	Energy-yielding reaction	Example genera
Chemo-lithotrophic	$H_2 + S^0 \rightarrow H_2S$	*Pyrodictium, Thermoproteus, Pyrobaculum, Acidianus*
	$2H_2 + O_2 \rightarrow 2H_2O$	*Sulfolobus, Pyrobaculum, Acidianus*
	$2S^0 + 3O_2 + 2H_2O \rightarrow 2H_2SO_4$	*Sulfolobus, Acidianus, Metallosphaera*
	$4H_2S + 2O_2 \rightarrow 4H_2SO_4$	*Sulfolobus, Acidianus, Metallosphaera*
	$H_2 + NO_3^- \rightarrow NO_2^- + H_2O$	*Pyrodictium, Pyrobaculum, Thermoproteus*
Chemo-organo-trophic	Organic cpd $+ S^0 \rightarrow H_2S + CO_2$	*Desulfurococcus, Thermoproteus, Thermofilum, Thermococcus*
	Organic cpd $+ O_2 \rightarrow H_2O + CO_2$	*Sulfolobus*
	Organic cpd $\rightarrow CO_2 + H_2$	*Pyrococcus*
	Organic cpd $\rightarrow CO_2 +$ fatty acids	*Staphylothermus*

Volcanic environments are generally low in oxygen and rich in reductants (Stetter 1995), and most hyperthermophiles accordingly are anaerobes. Although several remarkably acidophilic species are known, such as *Sulfolobus acidocaldarius*, *Metallosphaera sedula* and *Acidianus infernus* which have growth optima in the pH 1-3 range (Ingledew 1990, Segerer et al. 1993, Stetter 1996a), most other species are neutrophiles. Few hyperthermophiles can grow below 60°C but, surprisingly, many will survive for years at room or refrigerated temperatures, even in the presence of oxygen (Stetter 1995). This property may allow for their wide dispersal between thermal habitats around the globe (Stetter 1996b).

Ecology. Hyperthermophiles have been isolated from a variety of terrestrial and marine high-temperature environments. These areas are formed cheifly by volcanic or tectonic activity in which water is heated and rises convectively, carrying many dissolved minerals (Edwards 1990). Terrestrial habitats include sulfur-rich hot springs, mud pots and heated soils, often occurring together in areas called sulfataras. The pH of hot springs exhibits a bimodal distribution, most either highly acidic (pH 0-3) or nearly neutral (pH 5-8), depending on local geochemistry (Brock 1978). Acid present in these systems is largely H_2SO_4 of biological origin, the product of S^0 and H_2S oxidation by hyperthermophiles (Madigan et al. 1997). At the surface of solfataric soils, S^0 often accumulates due to oxidation of H_2S, and the enviroment is generally acidic. Ferric iron is often present. Below the surface, the environment becomes anaerobic and reducing, producing ferrous iron and nearly neutral pH. Sulfate (30 mM) is present in both regimes (Stetter 1996b). Genera isolated from terrestrial environments include *Sulfolobus, Acidianus, Metallosphaera, Desulfurococcus, Thermofilum, Thermoproteus,* and *Thermoplasma*.

Marine environments may be shallow or deep-sea hydrothermal outflows. These include the spectacular "black smoker" vents along the mid-ocean ridges, which spew mineral-laden 400°C fluids and support thriving (e.g. 10^8 cells/g) archaeal communities in the surrounding rock chimneys built up by mineral precipitation (Stetter 1996b). More easily studied, if less impressive, archaeal habitats also exist in shallow marine sediments heated by volcanic activity. The pH of these environments is generally slightly acidic to alkaline (pH 5-8.5). Volcanic emissions often contain CO_2, H_2S, CO, H_2, CH_4 and N_2 (Stetter 1996b), in addition to NaCl and sulfate from seawater. Genera isolated from marine habitats include *Acidianus, Staphylothermus, Pyrodictium, Thermococcus, Pyrococcus,* and *Archaeoglobus* (Segerer et al. 1993). Submarine volcanic eruptions are probably a major mode of dispersal of hyperthermophilic (and other) organisms (Stetter 1996b).

Hyperthermophiles have also been isolated from man-made environments, including power-plant outflows and self-heated coal refuse piles (Stetter 1995). Several strains are being evaluated for use in metal bioleaching and coal desulfurization processes (Clark and Norris 1996, Karavaiko and Lobyreva 1994). Recently, evidence has accumulated for the presence of *Thermococcus, Pyrococcus* and *Archaeoglobus* strains in deep, high temperature oil reservoirs. These organisms exhibited growth on crude oil as a sole carbon and energy source, producing H_2S, and thus may be responsible for crude oil souring (Grassia et al. 1996, Stetter et al. 1993).

As is suggested by their varied metabolic abilities, hyperthermophiles often exist in diverse communities, maintaining complex trophic interactions with their environment and each other. The ability of many species to fix CO_2 by way of inorganic redox reactions utilizing readily available H_2, S^0 and O_2 means that primary production in these communities can be carried out independantly of solar energy input. This allows active communities to develop in environments which are too hot (thermal environments) or too dark (deep-sea hydrothermal vents) to allow photosynthesis. Metabolic flexibility in many species allows them also to derive nutrients and energy from organic materials made available by primary producers. In some cases, these interactions evidently have become obligatory. For instance, in the laboratory, *Thermofilum* species can only be grown in coculture with *Thermoproteus*, or with extracts from *Thermoproteus* in the medium. This suggests that these organisms may closely associate in nature (Madigan et al. 1997). Recent molecular evidence suggests that species diversity in some environments greatly exceeds that known from cultivation studies (see below; Barns et al. 1996b). It is clear that we have only begun to understand the complexity of hyperthermophilic communities.

Halophiles. Extremely halophilic organisms, species which not only tolerate but require high salt concentrations for growth, are found mainly within the Archaea. Most require 1.5 to 4 M NaCl for optimal growth, but many can also grow slowly at salinities of 5.5 M, the limit of saturation for NaCl. Molecular sequence-based techniques (see below) have even detected the presence of halophilic species in saltern ponds at salt concentrations above the saturation point (Benlloch et al. 1996). Genera of extreme halophiles include *Halobacterium, Halorubrum, Haloferax, Haloarcula, Halococcus and Natronobacterium,* and *Natronococcus,* and are classified in the familiy Halobacteriaceae (Tindall 1992). Phylogenetic analysis of 16S rRNA sequences from these organisms indicate that they are most closely related to the methanogen lineage (Burggraf et al. 1991a), and some methanogens are also halophilic. However, the halophiles have diverged considerably and display a number of unusual physiological adaptations to rather extreme environmental conditions.

Physiology. All cultivated species of extremely halophilic Archaea are chemoorganotrophs, mostly utilizing amino acids or organic acids derived from the decay of Bacteria and Eucarya in the environment, for nutrients. Most species grow best aerobically, while some can survive by sugar fermentation and anaerobic respiration of nitrate or fumerate. Under anaerobic, nutrient-limited conditions some *Halobacterium* species are capable of light-driven production of ATP. This utilizes a unique light harvesting protein called bacteriorhodopsin which mediates light-driven pumping of H^+ ions into the cell (Madigan et al. 1997). Some halophiles are able to fix CO_2, while others may be capable of using H_2 as a reductant (Tindall 1992), indicating that there is considerable nutritional flexibility in the group.

Extreme halophiles have several adaptations that allow them to thrive under (and require) very high salt concentrations in their environment. For instance, the cell wall of *Haloabacterium* sps. contains many negatively charged (acidic) amino acids which require high Na^+ concentrations for stabilization. Dilution of Na^+ results in cell lysis. Within the cell, most proteins are also acidic, but require high K^+ levels. This is acheived by using the bacteriorhodopsin-mediated, light-driven accumulation of protons to pump Na^+ out of the cell, allowing uptake of K^+ ions. A second light-driven system utilizes halorhodopsin to pump Cl^- ions into the cell to act as anions for the K^+. High intracellular K^+ concentrations also prevent the cell from dehydrating due to osmotic imbalance. Because of these adaptations, *Halobacterium salinarum* requires an external Na^+ concentration of at least 3.0M, and maintains an internal K^+ concentration of 4.0 M (Gilmour 1990).

Ecology. Extreme halophiles have been isolated mostly from salty habitats, such as inland salt lakes like the Dead Sea and Great Salt Lake, and commercial salterns. Such environments occur where evaporation rates of saline water are high, relative to replenishment (Gilmour 1990). The reddish-purple coloration often seen in these environments is due to pigment production (including bacteriorhodopsin and halor-hodopsin) by halophiles. While all of these environments are high in salt concentration, they can vary greatly in ionic composition, depending on the surrounding minerology. For instance, Dead Sea brines are slightly acidic and higher in $MgCl_2$ than in NaCl, while those of the Great Salt Lake are slightly alkaline, contain sulfate and are dominated by NaCl (Gilmour 1990). Soda lakes, such as those of the Rift Valley, Kenya, can be highly alkaline (pH 10-12) due to the presence of carbonate minerals and low concentrations of Mg^{2+} and Ca^{2+} in the surrounding rocks (Tindall 1992). Species of the genera *Halobacterium* and *Halococcus* seem to predominate in the slightly acidic and slightly alkaline environments, while those of *Natronobacter* and *Natronococcus* dominate in soda lake communities (Gilmour 1990). Intriguingly, recent molecular sequence-based analysis of Archaea in a marine saltern indicated increasing species diversity with increasing salinity (Benlloch et al. 1996).

In these environments, halophilic Archaea are supported largely by organic material from primary production by bacterial and algal species present. Photosynthetic bacteria, such as *Ectothiorhodospira* and *Chromatium*, and algae, such as *Dunaliella*, provide fixed carbon. A number of other aerobic and anaerobic archaeal and bacterial species have also been isolated from these environments, indicating the presence of complex communities. Halophiles produce a suite of enzymes useful for degradation of proteins, lipids and nucleic acids, allowing them to harvest these nutrients from the decay of other organisms. As noted above, some species are also able to live autotrophically, fixing carbon from CO_2, and possibly utilizing molecular hydrogen for reductant (Tindall 1992).

The presence of halophilic organisms in saline environments can influence mineralogical processes. As outlined above, cells are constantly altering interior and exterior ionic concentrations by pumping Na^+, K^+, Cl^- and other ions across membranes. Bacteriorhodopsin, halorhodopsin and carotenoids color microbial blooms a deep reddish purple, which in turn traps solar heat, raising the surrounding temperature and increasing evaporation rates. Organisms also alter the structure of precipitated salt crystals. These processes likely occurred in ancient salt lakes as well, and halophiles can sometimes be cultured from within salt crystals (Tindall 1992).

Methanogens. All known organisms which form methane (CH_4) as the primary product of their energy metabolism group within the Archaea and are termed *methanogens.* As will be described below, these strictly anaerobic organisms occupy an important niche in anoxic environments and play a major role in global carbon cycling. They form complex, interdependent relationships with many disparate organisms including bacteria, protozoa and cows, and many creatures in between. All species characterized to date belong to the Euryarchaeota and comprise 17 genera clustering into seven groups, based on SSU rRNA sequence analysis (Madigan et al. 1997). Many aspects of methanogen ecology and physiology have been extensively studied (for review, see Ferry 1993).

Physiology. Methanogenic Archaea use simple organic and inorganic carbon compounds as energy sources, converting them stoichiometrically to methane. These include CO_2, formate, acetate, di- and trimethylamine, dimethylsulfide and several alcohols (Madigan et al. 1997). Almost all known genera can use H_2 as an electron donor, and most can gain energy from metabolism of CO_2 and H_2 (Whitman et al. 1991):

$$CO_2 + 4 H_2 \rightarrow CH_4 + 2 H_2O$$

In some species, formate and some alcohols can be used as electron donors instead of H_2 (Madigan et al. 1997).

When growing on C-1 compounds (other than CO_2), many species disproportionate the substrate, using some molecules as electron donors and some as acceptors:

$$4 CH_3OH \rightarrow 3 CH_4 + CO_2 + 2 H_2O$$

This occurs commonly in marine sediments, where C-1 compounds are abundant. In anoxic fresh water sediments where acetate is available, a few methanogen species are able grow on this substrate alone, oxidizing the carboxyl carbon to CO_2 and reducing the methyl carbon to methane:

$$CH_3COO^- + H_2O \rightarrow CH_4 + HCO_3^-$$

Essentially, these reactions represent a type of anaerobic respiration in which CO_2 or a methyl carbon is reduced (Whitman et al. 1991). The pathway of methanogenesis involves seven steps, and has been extensively investigated (Ferry 1993, Rouviere and Wolfe 1988).

In many species, the pathways for energy production (methanogenesis) and carbon assimilation are closely linked, and share many intermediates (Madigan et al. 1997). About half of the known species can grow autotrophically, producing organic carbon from CO_2. Other species obtain organic carbon exogenously from acetate (Whitman et al. 1991). Nitrogen and sulfur are obtained from organic and inorganic sources, and a few species can fix molecular nitrogen. In addition, methanogens have requirements for several metals, such as nickel, iron and cobalt, which are used in trace amounts for the catalytic activities of many enzymes (Madigan et al. 1997).

All methanogens are extreme anaerobes, and many are killed quickly by exposure to oxygen. This is probably due, in part, to the oxygen sensitivity of many of the enzymes involved in methanogenesis and other cellular processes. In addition, several species require reduced sulfur for growth, and are unable to use oxidized sulfur (Whitman et al. 1991).

Ecology. In the global carbon cycle, atmospheric carbon (CO_2) is fixed to an organic form by photosynthesis and other processes. This organic carbon is eventually degraded to CO_2 and methane, mostly by microbial activities. After CO_2, methane is the most abundant carbon-containing ("greenhouse") gas in the atmosphere, released from the surface at a rate of about 5×10^{14} g per year. Most of this methane is biologically produced, and its concentration in the atmosphere is increasing approximately 1% annually (Whitman et al. 1991). In addition, a portion of atomospheric CO_2 is produced as a result of oxidation of methane by methanotrophic bacteria (Madigan et al. 1997). Clearly, methanogenesis is of major importance both to global climate and to the carbon cycle.

Methanogens are abundant in anoxic habitats such as anaerobic freshwater and marine sediments, marshes, swamps, soils, and animal and insect gastrointestinal tracts. Oxygen depletion in these environments is acheived largely through the rapid respiration of other organisms relative to the rate of O_2 influx. In laboratory culture, most methanogens are intolerant to even small amounts of oxygen. It is likely, then, that they maintain intimate associations with O_2-respiring organisms in their natural environments. Methanogens can also be found in generally aerated environments which contain anoxic microenvironments, such as the interior of soil particles. These species are also restricted to environments low in NO_3^-, Fe^{3+} and SO_4^{2-}. Where these oxidants are present, other organisms outcompete methanogens for available energy sources (Whitman et al. 1991).

Due to their requirement for H_2 and a limited number of simple carbon compounds for growth, methanogens depend on the activities of other organisms to produce these substrates from complex organic matter. In some cases, these nutrients are supplied by bacterial species termed *syntrophs*. Syntrophic microbes live in cooperation with other species to which they are nutritionally linked. In the case of methanogens, syntrophic species are often those which can ferment long chain organic acids, excreting H_2 and C-1 or C-2 compounds which methanogens can metabolize. This fermentation, however, is inhibited by even very low concentrations of H_2 and formate in the environment. Methanogens oblige these species by actively consuming H_2 and formate, keeping concentrations vanishingly small. This process is termed *interspecies hydrogen transfer*, and generally involves intimate associations between methanogens and syntrophic species (Thiele and Zeikus 1988). It is thought to be the foundation for the obligate symbioses between methanogen species and many protozoa of ruminant, freshwater and marine environments (Goosen et al. 1988, Tokura et al. 1997, Wagener et al. 1990).

Several methanogen species are found in hydrothermal environments, such as hot springs, solfataras and deep-sea vents. In such environments, H_2 is produced geo-thermally, and syntrophic associations are unneccessary (Whitman et al. 1991). Among these is the genus *Methanopyrus*, which has been isolated from sediments and chimneys of deep-sea "black smoker" vents, and is capable of growth to 110°C (Stetter 1996b). *Methanopyrus* is the archaeal isolate which branches most deeply in the tree of life (Burggraf et al. 1991b). Intriguingly, this organism obtains all of its energy from H_2 and CO_2 alone, suggesting that this form of autotrophy may be a primitive one.

Molecular sequence-based analyses of oxygenated marine environments had recently revealed a new group of organisms whose closest known relatives are methanogens (see below; [DeLong 1992, Fuhrman et al. 1993]). Since none of these organisms have been cultured, it is not yet known how metabolically similar they are to known methanogens. It is possible that they may be true anaerobic methanogens, living in close association with aerobic organisms which provide locally anoxic conditions. Alternatively, they may be as different from known methanogens as the halophiles are. Whatever their physiology, they appear to be abundant in the oceans, and may therefore be important in global ecology.

Bacteria: abundance and diversity

There are numerous examples of members of the domain Bacteria that serve as geological agents. These microorganisms belong to diverse phylogenetic groups that can be placed into broad physiological groups. Chemolithotrophs (which include members of both Bacteria and Archaea) obtain their energy from the oxidation of inorganic compounds and assimilate carbon as CO_2, HCO_3^-, or CO_3^{2-}. Photolithotrophs (Bacteria only) obtain their energy from sunlight and use as a source of carbon CO_2, HCO_3^-, or CO_3^{2-}. Some of these organisms are anoxygenic while others are oxygenic. Mixotrophs (including both Bacteria and Archaea members) derive energy simultaneously from oxidation of reduced carbon compounds and oxidizable inorganic compounds. They may obtain their energy totally from the oxidation of an inorganic compound but their carbon from organic compounds. Photoheterotrophs (mostly Bacteria members) obtain their energy from sunlight and carbon by assimilating organic compounds. Heterotrophs include both members of Bacteria and Archaea, obtain energy from the oxidation of organic compounds, and assimilate organic compounds as their main carbon source.

Biogeochemical cycling by bacteria: Introduction. Many bacteria play important roles in biogeochemical cycling. Examples of important physiological groups include, but are not limited to, iron, sulfur, and manganese -oxidizing and -reducing bacteria, as well as the nitrifying and denitrifying bacteria. A brief description of these physiological groups is provided below. For information regarding other important physiological groups and in some instances more detailed information regarding these groups, the reader is referred to other chapters in this volume.

Iron-oxidizing and -reducing organisms. Iron is the most abundant element in the Earth as a whole, and is the fourth most abundant element in the Earth's crust. It is found in a number of minerals in rocks, soils and sediments. Both oxidative and reductive reactions of iron brought about by microorganisms play important roles in the iron cycle. Microorganisms can affect the mobility of iron as well as its accumulation. Iron can serve as an energy source for some bacteria. The most widely studied acidophilic iron-oxidizing bacterium is *Thiobacillus ferrooxidans,* however, other acidophilic bacteria such as *Leptospirillum ferrooxidans* and *Metallogenium* sp. can also oxidize ferrous iron enzymatically. At neutral pH, *Gallionella* (which can grow autotrophically and mixotrophically) is also capable of enzymatic iron oxidation. Other Bacteria that have also been associated with iron oxidation include sheathed species such as *Sphaerotilus*, *Leptothrix* sp., *Crenothrix polyspora*, *Clonothrix* sp. and *Lieskeella bifida* and some encapsulated organisms such as the *Siderocapsaceae* group. However, many of these are likely to be iron-depositing bacteria that bind preoxidized iron at their cell surface (see Ehrlich 1995).

Ferric iron may be microbiologically reduced to ferrous iron. As is true for iron oxidation, iron reduction may be enzymatic or nonenzymatic. There are a number of bacteria for which iron (III) serves as a terminal electron acceptor. Examples of organisms in which ferric iron reduction accompanies fermentation, include a variety of *Bacillus* species (including *B. polymyxa, circulans, hypermegas*, etc.), *Pseudomonas* spp., *Vibrio* sp., and anaerobes such as *Clostridium* spp., *Bacteroides hypermegas*, *Desulfovibrio desulfuricans*, and *Desulfotomaculum nigrificans*. Additionally, there are strictly anaerobic and facultative bacteria that carry out heterotrophic ferric iron respiration. Examples include *Geobacter* sp., various sulfate-reducing bacteria, *Shewanella* sp. and *Pseudomonas* sp.

Sulfur-oxidizing and -reducing bacteria. Sulfur, which occurs in organic and inorganic forms in nature, is one of the more common elements in the biosphere. Different organisms may assimilate sulfur in an organic or inorganic form. The oxidation and reduction reactions involving sulfur and sulfur compounds are especially important geomicrobiologically. Microorganisms play an important role in mineralizing organic sulfur compounds in both soil and aqueous environments. Inorganic sulfur may exist in various oxidation states in nature, most commonly as sulfide (-2), elemental sulfur (0), and sulfate (+6). Different members of the domain Bacteria (and Archaea) play an important role in the interconversion of these oxidation states. Among the microorganisms that oxidize reduced forms of sulfur are chemolithotrophs, anoxygenic and oxygenic photolithotrophs, mixotrophs, and heterotrophs. Most chemolithotrophs and mixotrophs use oxygen as the oxidant, but a few chemolithotrophs can substitute nitrate or ferric iron when oxygen is absent. The *Thiobacillaceae* group, which includes obligate and facultative autotrophs as well as mixotrophs, is one of the most widespread terrestrial groups of aerobic sulfur-oxidizing bacteria. Another group of hydrogen-sulfide oxiders of importance in fresh water and marine environments are the *Beggiatoaceae*. Two examples of bacteria that are facultatively anaerobic sulfur oxidizers are *Thiobacillus denitrificans* and *Thermothrix thiopara*. The strictly anaerobic sulfur oxidizers include the photosynthetic purple *(Chromatiaceae)* and green bacteria *(Chlorobiaceae)*.

Most of the known sulfate-reducing bacteria belong to the domain Bacteria, and are gram-negative cells. Morphologically, sulfate-reducing bacteria are a diverse group, including cocci, sarcinae, rods, vibrios, spirilla and filaments. Many of the sulfate-reducers are able to use H_2 as an energy source. Most require an organic carbon source, but a few can grow autotrophically on hydrogen. Frequently isolated sulfate-reducing bacteria include those belonging to the *Desulfovibrio*, *Desulfomonas*, *Desulfotomaculum*, and *Desulfobacterium* genera. Additionally, some *Clostridia* spp. have been found to be strong sulfite reducers. Furthermore, in the Black Sea, *Shewanella putrefaciens* is able to reduce partially oxidized sulfur compounds and appears to play a significant role in the sulfur cycle (see Ehrlich 1995).

Manganese-oxidizing and -reducing bacteria. Manganese is found as a component in more than 100 naturally occurring minerals, with major accumulations occurring in the form of oxides, carbonates, and silicates. Manganese-oxidizing and -reducing bacteria play an important role in the immobilization and mobilization of manganese in soil, fresh water and marine environments. The manganese-oxidizing bacteria often contribute to the accumulation of manganese oxides on and in sediments, while the manganese-reducing bacteria may mobilize manganese into the water column or environment.

There are a number of manganese-oxidizing bacteria that exist which are phylogenetically unrelated. However, all of the manganese-oxidizing bacteria appear to be aerobic members of the domain Bacteria. As a group, the manganese-oxidizing bacteria include both gram-negative and gram-positive organisms, sporeformers and non-sporeformers, sheathed and appendaged bacteria. In almost all instances, growth in the presence of manganese is either mixotrophic or heterotrophic. The most notable examples of manganese-oxidizing bacteria include: *Arthrobacter* spp., *Leptothrix* spp., *Hyphomicrobium* spp., *Oceanospirillum*, *Vibrio* and *Metallogenium*. Manganese-oxidizing bacteria have been detected in very diverse environments including rock surfaces, soils, fresh water lakes and streams, and ocean water and sediments (see Ehrlich 1995).

As is the case with the manganese-oxidizing bacteria, a number of different phylogenetic groups of bacteria have been found to reduce manganese either enzymatically or nonenzymatically. The manganese-reducing bacteria include aerobes, and strict and facultative anaerobes. They encompass both gram-positive and gram-negative forms and include *Bacillus* spp., *Geobacter* sp., *Shewanella*, and *Acinetobacter* sp. Manganese-reducing bacteria have been found in diverse environments, including soil, fresh water and marine habitats.

Nitrifying and denitrifying bacteria. In fresh and ocean waters, soils, and sediments, nitrogen exists in both inorganic and organic forms. The geomicrobially important inorganic forms include ammonia and ammonium ion, nitrite, nitrate, and gaseous oxides of nitrogen. Geomicrobially important organic nitrogen compounds include humic and fulvic acids, proteins, peptides, amino acids, purines, amides, etc. Nitrification is a mobilization process of particular importance in soils, which results in the transfer of inorganic fixed forms of nitrogen from surface soils to subsurface groundwater reservoirs.

Nitrifying bacteria play a key role in the conversion of ammonia into an anionic nitrogen species. The majority of nitrifying bacteria are aerobic autotrophs that are members of the domain Bacteria. There are two steps in the nitrification process, the oxidation of ammonia to nitrite and the formation of nitrate from nitrite. These reactions are carried out by different groups of nitrifying bacteria. Representative bacteria that oxidize ammonia to nitrite include *Nitrosomonas*, *Nitrosospira* and *Nitrocystis* species. Those that oxidize nitrite to nitrate include members of the *Nitrobacter*, *Nitrospina* and *Nitrococcus* genera. Nitrifying bacteria are found in diverse environments including soil, fresh water and seawater. In soils, *Nitrosomonas* is the dominant bacterial genus involved in the oxidation of ammonia to nitrite, while *Nitrobacter* is the dominant genus in the oxidation of nitrite to nitrate (Ehrlich 1995).

The conversion of fixed forms of nitrogen to molecular nitrogen by denitrifying bacteria is another important process in biogeochemical cycling of nitrogen. Denitrification occurs when nitrate ions serve as terminal electron acceptors in anaerobic respiration, with the reduction of nitrate leading to the production of gaseous forms of nitrogen (including nitric oxide, nitrous oxide, and molecular nitrogen). Only a limited number of bacteria are capable of denitrification. Those with this capability include members of genera *Pseudomonas*, *Moraxella*, *Spirillum*, *Thiobacillus*, and *Bacillus* (Atlas 1984). All of the nitrate respiratory processes have been found to occur in terrestrial, fresh water and marine environments.

Bacteria in the lithosphere and hydrosphere: Introduction. The litho-sphere and hydrosphere support the existence of diverse groups of microorganisms. In these environments microorganisms are critical agents in biogeochemical cycling. The majority of our knowledge regarding the diversity, abundance, and activities of microorganisms in different environmental settings has come from traditional culture-based approaches. However, in recent years a variety of studies on bacterial community structure and diversity have been carried out using molecular approaches. A majority of these studies have utilized 16S rRNA phylogenetic analyses to survey bacterial species present without the need for culturing. It is of particular interest that regardless of the specific environment under study, the following two results have usually been obtained: (1) sequences from non-cultured organisms in environmental samples rarely exhibit an exact match to any of the 16S rDNA sequences in existing databases, suggesting that most of the bacteria in the environment are novel species and, (2) similar groups and clusters of bacteria are often recovered in independent examinations of the same or similar habitat. Results of both culture based and molecular studies (of culturable and non-culturable organisms) support the idea that certain groups of phylogenetically related organisms are adapted to particular habitat types.

In the sections that follow, general information is provided regarding the lithosphere and hydrosphere as microbial habitats. Additionally, information that has been obtained from culture based studies and selected examples from more recent molecular studies regarding the abundance and diversity of bacteria in these environments is provided.

Bacteria in the lithosphere. The major habitat of terrestrial microorganisms is soil. In soils containing varying amounts of clay, silt and sand particles, diverse phylogenetic and physiological groups of microorganisms are found. The number of microorganisms in soil habitats typically range from 10^6 to 10^9 bacteria per gram, which generally is higher than that found in fresh water or marine habitats. Within the soil column, microorganisms are unevenly distributed. Usually the organically rich surface layers contain greater numbers of bacteria than the underlying mineral soils or subsurface sediments. Based on morphological characteristics, there are gram-positive rods and cocci, gram-negative rods and spirals, sheathed bacteria, stalked bacteria, mycelial bacteria, budding bacteria, etc. Physiological types include a variety of bacteria involved with nitrogen cycling, sulfur cycling, iron cycling, manganese cycling, etc. Many different bacterial genera are commonly found in soil, however *Arthrobacter*, *Bacillus* and *Actinomycetes* species usually comprise more than 5% of the culturable types. While sporeforming rods are frequently encountered when culturing soil in the laboratory, they are not that abundant *in situ*. However, of greater importance than numerical abundance is the activity level of different organisms and their biochemical significance in the soil. As an example, while bacteria involved with nitrogen cycling may be lower in total numbers than other bacteria, they are vitally important for microbial and plant communities.

Molecular studies of bacterial diversity and abundance in sediments and soils. Studies in which microbial diversity in soils has been examined using the culture-independent approach of direct DNA extraction from soil samples, followed by amplification and sequence analysis of SSU rDNA, are still few in number. In the early 1990s (Liesack and Stackebrandt 1992, Stackebrandt et al. 1993), microbial diversity using this approach was examined in subtropical soil from Queensland Australia. Analysis of thirty sequences from this soil revealed primarily alpha Proteobacteria, with some planctomycetes and distant relatives of the planctomycetes. In a different study, 17 SSU rDNA clones from a soil sample collected from a soybean field in Japan were analyzed and found to contain a diverse group of Proteobacteria, green sulfur bacteria, an archeaon, a high G+C content gram-positive strain and several bacteria from novel groups within the domain Bacteria (Ueda et al. 1995). Finally, Borneman et al. (1996) estimated microbial diversity in a pasture soil from southern Wisconsin by analysis of 124 clones, of which 98.4% were bacterial. Within the Bacteria, three kingdoms were highly represented: the Proteobacteria (16.1%), the *Cytophaga-Flexibacter-Bacteroides* group (21.8%) and the low G+C gram-positive group (21.8%). Some kingdoms, such as the *Thermotogales*, the green nonsulfur group, the *Fuso*Bacteria, and the *Spirochaetes*, were absent. Additionally, a large number of the sequences (39.4%) were distributed among several clades that are not among the major taxa in the existing database.

In another study, bacterial diversity was examined in sediments of Carolina Bay, which has acidic water and moderate dissolved organic carbon concentrations (Wise et al. 1997). Clones were found that related to taxa found previously in DNA extracted from Australian soils and other habitats (Liesack and Stackebrandt 1992, Stackebrandt et al. 1993). Sequences that were consistently found were associated with five major groups: the Proteobacteria (including members of the alpha, beta, and delta subdivisions), the *Acidobacterium* subdivision of the *Fibrobacter* division, the *Verrucomicrobium* subdivision of the *Planctomyces* division, the low G+C gram-positive bacteria and the green nonsulfur bacteria (Wise et al. 1997).

Many of the taxonomic groups represented from clones in the aforementioned studies also have been commonly found in culture-based studies of soil microbial diversity. However, there are some interesting differences. For example, the high G+C gram-positive group can comprise as much as 92% of the cultured isolates from soil. Additionally, dominant microorganisms that are found in soil by culture methods are of the genera

Arthrobacter (5-60%), *Bacillus* (7-67%), *Pseudomonas* (3-15%), *Agrobacterium* (1-20%), *Alcaligenes* (1-20%), *Flavobacterium* (1-20%) and the order *Actinomycetales* (5-20%) (Atlas 1984). Of these groups, only *Bacillus* (19%) and *Actinomycetales* (0.8%) were found in the study of Borneman et al. (1996).

Bacteria in the hydrosphere. The hydrosphere is mainly marine, with oceans occupying approximately 70% of the Earth's surface. Most of the ocean floor is covered by sediments that may consist of sand, silt, and/or clays. Marine sediments contain large numbers of microbes as well as other organisms. While viable bacteria have been recovered from significant depths (35 m below the sediment surface), their numbers usually decrease with increasing depth in the sediment column. Additionally, the metabolic activity bacteria of deep-sea sediments has been shown to be at least 50 times lower than that of microorganisms in shallow waters or in sediments at shallow depths (Jannasch et al. 1971, Wirsen and Jannasch 1974). The distribution of microorganisms in open oceans is not uniform geographically or throughout the water column. There are a number of factors that affect this distribution, including available energy sources, carbon, nitrogen and phosphorus levels, temperature, hydrostatic pressure and salinity. Nevertheless, most of the culturable marine bacteria are gram-negative and motile, with members of *Pseudomonas* and *Vibrio* species often being predominant.

Fresh water is found on land in lakes, ponds, swamps and streams above ground, and in aquifers below ground. Lake water varies in composition from low to high salt concentrations, and from low to high organic content. Temperature, pH, nutrient concentrations, and oxygen levels are also important parameters that influence the occurrence of specific microorganisms in fresh water habitats. In lakes, the metabolic activities of chemolithotrophic bacteria are important in the cycling of nitrogen, sulfur, and iron. Members of the genera *Nitrosomonas*, *Nitrobacter* and *Thiobacillus* are particularly important for the cycling of nitrogen and sulfur.

Lake bottoms vary in composition ranging from sediments overlying bedrock to sands, clays, and muds. The sediments are a major habitat for microorganisms which are usually different species from those found in the over-lying waters. Culturable bacteria range from 10^2 to 10^5 per ml of lake water and in the order of 10^6 per gram of lake sediment. Culturable bacteria from lakes predominantly consist of gram-negative rods, although gram-positive sporeforming rods and actinomycetes have also been isolated (especially from sediments). Within sediments, anaerobic microorganisms are extremely important in the cycling of carbon, nitrogen and sulfur (see Ehrlich 1995).

Water that collects below the land surface in soil, sediment and permeable rock strata, is called groundwater. Groundwaters are derived from surface waters that seep into the ground and accumulate above impervious rock strata as aquifers. Studies have demonstrated the occurrence of as many as 10^6 culturable groundwater microorganisms per gram of sediment from the vadose zone or shallow aquifers. Gram-positive microorganisms are usually more abundant than gram-negative ones (Wilson et al. 1983). In one study, as many as 10^4 to 10^6 colony-forming heterotrophic bacteria per gram were obtained from sediments from deep vadose zones and aquifers from the Atlantic Coastal Plain (Fredrickson et al. 1991). Diverse physiological groups including autotrophs, such as sulfur oxidizers, nitrifiers and methanogens, as well as heterotrophs, including sulfate-reducers and denitrifiers, have been obtained from shallow and deep aquifer material.

Molecular studies of bacterial diversity and abundance in the hydrosphere: Seawater. There are abundant heterotrophic marine bacterioplankton species that have only recently been studied using analyses of small subunit rRNA gene clones. In general, cell counts estimated from cultures are orders of magnitude lower than those estimated from DAPI or other direct count methods. Thus, direct analyses of DNA extracted from environmental organisms, without culturing, have been used to complement studies of culture based identifications of organisms.

In one such study, the small-subunit rDNAs of 127 isolates obtained from a water sample off the Oregon coast and 58 bacterial rDNAs cloned directly from the DNA of the same water samples were characterized (Suzuki et al. 1997). Sequencing of the isolates' rDNAs revealed the presence of the following phylogenetic groups (listed in order of abundance): members of the gamma and alpha subdivisions of the Proteobacteria, of the *Cytophaga-Flavobacter-Bacteriodes* line of decent, and the high- and low- G+C gram-positive phyla (Suzuki et al. 1997). The most common members of the gamma *Proteo*-Bacteria were similar or related to *Azospirillum* sp., the *Pseudomonas* subgroup, the *Colwellia* assemblage and the *Alteromonas* group. The most abundant alpha Proteobacteria were members of the *Sphingomonas* and *Roseobacter* groups. The other abundant species were members of the *Vesiculatum*, *Cytophaga lytica* subgroups, *Arthrobacter* and *Bacillus* groups. The direct environmental rDNA clones were, in order of abundance, members of the gamma, alpha and beta subdivisions of the Proteobacteria and members of the high-G+C gram-positive phylum. Groups that were most frequently represented were members of the SAR86 cluster, organisms closely related to sulfur-oxidizing symbionts, *Pseudomonas* spp., SAR11 cluster, SAR16 cluster, or SAR83 cluster within the *Rhodobacter* group. There was little overlap between the partial SSU rDNA sequences from the isolate and environmental clones (Suzuki et al. 1997). These results indicate that many of the most abundant bacterioplankton species are not readily culturable by standard methods. The possibility that biases introduced by molecular techniques caused these differences seem unlikely because investigations of bacterio-plankton diversity by different molecular techniques have revealed the same phylogenetic lineages regardless of the specific water sample examined (Furhman et al. 1993, Giovannoni et al. 1990, Schmidt et al. 1991). In fact, independent studies have shown many similar novel taxa occurring in samples that are geographically separated, such as the Atlantic Ocean, the Pacific Ocean and the Antarctic sea. Additionally, many of the heterotrophic bacterioplankton that are culturable are strains for which few similar sequences are currently present in existing databases. Unfortunately, these results are less informative regarding the biogeochemical impact of bacterioplankton, since metabolic pathways of microorganisms belonging to phylogenetically cohesive groups may be quite diverse.

Fresh water. Fresh water aquatic bacteria play crucial roles in decomposition, food chains and biogeochemical cycling, yet they have received little attention and remain largely uncharacterized, in part due to difficulties in obtaining pure cultures that are representative of natural populations. In a recent study, bacterial communities of seven lakes in the southwest region of the Adirondacks were sampled for rRNA gene sequence analysis. Results for 109 sequences, the first report of an extensive rDNA sequence dataset from fresh water lakes (Hiorns et al. 1997), revealed that phylogenetically diverse bacteria occur in the lakes studied, and differ from many of the bacteria characterized from oceanic (Mullins et al. 1995) and subterranean waters (Pedersen 1996) and soils (Ueda et al. 1995, Borneman, et al. 1996). Many of the 16S rRNA sequences were in the Proteobacteria kingdom (alpha-subdivision 19%; beta-subdivision, 31%; and gamma subdivision, 9%), the *Cytophaga-Bacteriodes-Flavo*Bacteria group (15%) and the High G+C gram-positive group (18%). A small number of clone sequences were phylogenetically associated with *Verrucomicrobium spinosum*, and a single sequence was associated with *Acidobacterium capsulatum*. However, few of the sequence types obtained were closely related to those of characterized species. No representation of the *Planctomycetales* was found, which is somewhat surprising given the widespread occurrence of these organisms in fresh water (Ward et al. 1995). The relative abundances of the groups of sequences differed among the lakes examined, suggesting that bacterial population structure varies, and that it may be possible to relate aquatic bacterial community structure to water chemistry.

A reoccurring theme in environmental studies using rRNA techniques is the discovery of lineages that are well-separated from those of described microorganisms. For instance, two lineages that were found in the Adirondack fresh water lake clones of Hiorns et al. (1997) appear to be related closely to environmental rDNA sequences from soil of Queensland, Australia (Liesack and Stackebrandt 1992, Stackebrandt et al. 1993). Additionally, it appears that the prosthecate fresh water bacteria members of *Verrucomicrobiales* and some strains of *Prosthecobacter* are found in many types of environments that are geographically widespread.

Subterranean groundwater ecosystems. Bacterial diversity has been studied in diverse subterranean ecosystems including the following: the Stripa research mine and the Aspo Hard Rock Laboratory, Sweden, the Bangombe fossil nuclear reactor in Gabon, Africa, the alkaline groundwater of Maqarin in Jordan, and bentonite-sand buffer material from Pinawa, Canada (see Pedersen 1996). From these different environments over 500 totally or partially sequenced 16S rRNA genes have been obtained from over 50 independent samples. More than 155 identifiable species sequences have been found, scattered over 11 branches of the phylogenetic tree for the domain Bacteria. Representatives include those of the alpha, beta, gamma, delta, and epsilon groups of the Proteobacteria, Gram-positive bacteria, as well as five unidentified clusters (Pedersen 1996).

The 16S rRNA gene diversity for attached bacteria in groundwater from the Stripa research mine revealed that specific beta-group Proteobacteria predominate along with a gamma Proteobacterial species (Ekendahl et al. 1994). In contrast, attached and unattached bacteria from 10 boreholes in the Aspo HRL area, were found to be distributed over all of the aforementioned branches (except one of the unidentified branches). Sequences grouping with the Gram-positives predominated among the attached bacteria, while organisms branching with the gamma Proteobacteria were predominant among the unattached bacteria (Pedersen et al. 1996b). The distribution of the 16S rRNA genes was related to the different types of groundwater studied (at least 4 types existed). A number of the sequences obtained could be identified at the genus level as belonging to *Acinetobacter*, *Bacillus*, *Desulfovibrio* or *Thiomicrospira*. Additionally, the 16S rRNA genes from 20 selected isolates were closely related to the sulfate reducers, *Desulfo-microbium baculatum* or *Desulfovibrio* sp., the iron reducer *Shewanella putrefaciens*, or distantly related to the Gram-positive genus *Eubacterium* (Pedersen et al. 1996b).

The natural nuclear reactor analogue groundwater in Bangombe was inhabited by representatives of the beta, gamma, and delta Proteobacteria groups, gram-positive bacteria and an unidentified cluster. These groups were unevenly distributed over the different boreholes. The deepest borehole only harbored beta Proteobacteria, while all groups were detected in the three other boreholes (Pedersen et al. 1996a). In the bentonite-sand buffer material around the Canadian test canister, gamma Proteobacteria were dominant with a few gram-positive sequences (Stroes-Gascoyne et al. 1995). There was only one genus that occurred at all five sites, which was *Acinetobacter*. Acinetobacters are ubiquitous organisms that may constitute as much as 0.001% of the total aerobic population of soil and water (Tower 1992). Thus, finding this genera at all sites was not unexpected.

TECHNIQUES FOR STUDYING
GEOLOGICALLY IMPORTANT MICROORGANISMS

Culture-based techniques

It has been frequently cited that in soil microbiology far less than 10% of the total bacterial population has been culturable (Alexander 1977). Nevertheless, the approach of culturing organisms from environmental samples followed by isolation and subsequent identification has been important in learning more about geologically important microorganisms. Isolation is achieved by the physical separation of microorganisms. The success of the isolation method includes the maintenance and the growth of a pure culture. There exists a wide variety of identification methods that are being applied to bacterial taxonomy, which include morphological and physiological characteristics, chemical analysis of bacterial components (e.g. lipids, polysaccharides, proteins, etc.), as well as the analysis of nucleic acids extracted from environmental isolates. However, continued development of additional methods for the determination of microbial diversity remains critical in furthering our understanding of microbial ecology.

The classical method that has been used to study the diversity of bacterial communities is to inoculate different solid and liquid media with environmental samples and then incubate at various temperatures and gas compositions. Once bacterial growth has occurred it is then possible to enumerate and classify the organisms according to their characteristics (e.g. morphology, physiology, etc.). Viable bacterial population densities or colony-forming units (CFU) are commonly determined by plating samples on laboratory media. However, viable counts for the enumeration of bacteria depend strongly on the ability to develop growth media that can support the growth of all present bacteria. Because there is not a medium that satisfies these needs, viable counts almost always underestimate the number of bacteria present in a sample. Aerobic heterotrophic plate counts have been used in conjunction with environmental samples by initially mixing the sample in a buffered solution and then plating the diluted sample on appropriate media. Culturable spores have been enumerated from environmental samples by growth following heat treatment. Another traditional approach to bacterial enumeration of organisms is most-probable-number (MPN) analysis (see Atlas 1984).

There exists both general and selective media for the growth of microorganisms. Examples of general media that have been frequently used include: peptone tryptone yeast glucose (PTYG) extract, tryptic soy agar (TSA), plate count broth, R2A, potato dextrose agar, etc. Frequently the medium is diluted to 1% or 10% concentration (low nutrient) for the isolation of environmental organisms. The rationale for this has been that some environmental microorganisms may be oligotrophs and could therefore be inhibited if high concentrations of nutrients are used. Examples of selective culture media for environmental microorganisms include: Thio Citrate Biole Salts Agar (for *Vibrio* spp., and *Pseudomonas* spp.), Hektoen Enteric Agar (for *Salmonella* spp., *Shigella* spp., and *Aeromonas* spp.) M-Endo agar (for *Enterobacteriaceae*), actinomycete agar (for actinomycetes), etc.

The enrichment culture technique has also proven valuable for isolating specific groups of microorganisms. This technique involves using culture media and incubation conditions designed to favor the growth of particular microorganisms. The technique models many natural situations in which the growth of a particular microbial population is favored by the chemical composition of the system and the specific environmental conditions. Following growth of the initial enrichment culture, repeated subculturing of single colonies from solid to liquid media can be used to obtain a pure culture. Often, long incubation times are required for the growth of environmental microorganisms. Incubation periods of 3 to 8 weeks are common, and in one study incubation times up to two months were used (Kieft et al. 1995).

Identification and physiological studies

While many organisms can be identified on the basis of morphological features, bacteria having similar morphologies are often difficult to identify this way. Thus, numerous techniques for bacterial identification based on other criteria, such as physiological and biochemical tests, have been developed. A number of commercially available kits for such purposes now exist (e.g. API test strips, Biolog, Miniteck, Micro-ID systems, etc.). In a variety of studies, the distribution of organisms from environmental samples have been described by comparing characteristics of recovered isolates based on colony and cellular morphology; physiological profiles provided by API-rapid-NFT strips and/or Biolog microtiter plates (Bone and Balkwill 1988, Kolbel-Boelke et al. 1988); and antibiotic and metal resistances (Amy et al. 1992, Haldeman and Amy 1993). General profiles obtained using these methods have proven valuable in describing organisms from different environmental samples. However, specific organism identification has proven a more difficult task (Balkwill 1989, Amy et al. 1992). Below, the API and Biolog tests which have been used to aid in the identification of environmental isolates are briefly described.

API tests. Diversity of many types of bacterial isolates can be assessed by examination of certain physiological traits. One method for doing this is use of the API NFT testing system (BioMerieux-Vitek, Inc. Hazelwood Missouri), which was originally designed for identification of nonfermentative, gram-negative bacteria of clinical importance. This system includes 21 different tests for nine specific enzymatic capabilities and the ability to aerobically utilize 12 different compounds as sole sources of carbon. There is also the API 20E testing system, designed for members of the family *Entero*Bacteria*ceae* and other gram-negative bacteria. The API testing method, which is relatively simple and rapid, compares test results with a computerized database containing results from known bacteria. In a number of studies, the API system has been used for phenotypic identification in the discrimination of environmental bacterial species. However, in most cases it has been determined that phenotypic tests alone do not provide an accurate assessment of the structure and composition of heterotrophic bacterial communities (see Jimenez 1990). Thus, while API strips are extremely useful for the identification of clinical isolates, at the present time due to a lack of extensive information in the database, they are only of limited usefulness for environmental bacteria (Brown and Leff 1996). Nevertheless, as more environmental bacteria are collected and their profiles added to the database, this identification method may become more important.

Biolog tests. The Automated Microbial Identification System of Biolog Inc., was originally developed for the rapid identification of pure cultures of microorganisms, especially clinically important strains, on the basis of their substrate utilization profiles. In 1991, Garland and Mills showed that direct incubations of environmental samples in Biolog microplates with 95 different substrates produce utilization patterns. Additionally, it has been suggested that the Biolog system should be suitable for identifying and characterizing a wide variety of environmental bacteria (and yeasts). In Biolog's latest release (3.50), 824 bacterial species belonging to 120 genera are included, encompassing a considerable number of the most important culturable soil and water bacteria. Most recently the Biolog system was evaluated for its usefulness in studying the diversity of bacterial communities (Wunsche and Babel 1996).

Studies have been conducted to compare the value of using phenotypic traits versus rRNA sequence determination for the identification of bacteria. Boivin-Jahns et al. (1995)

isolated 74 bacterial strains from a mine gallery and found that misidentification of bacteria was much less common using rRNA sequence analysis than the more traditional methods including morphology, gram staining, enzyme activities, and the utilization of different substrates as sole carbon and energy sources. This result was not unexpected given the power of molecular methodologies for phylogenetic analyses.

Analysis of microbial populations without cultivation

Introduction. Microbiologists suspected from the beginning that cultivation-based methods for the study of naturally occurring microorganisms might exhibit biases. Due to the difficulty of accurately reproducing conditions of natural environments in the laboratory, it was clear to microbial ecologists like Winogradsky that "The [natural] microflora is not understood either qualitatively or quantitatively" (Ward et al. 1992) when studied by such methods. When compared with the total number of cells that can be visualized microscopically in natural samples, the quantity recovered in culture is very small (0.001 to 15%, depending on environment; Amann 1995). Elective enrichment techniques tend to recover those species which flourish on laboratory media and outcompete other organisms, and even cultivation on agar plates is selective for species which can grow under relatively simple conditions. In addition, many cells apparently exist in a "viable but non-culturable" state in the environment, and cannot be revived in the laboratory (Amann et al. 1995).

In recent years, culture-independent methods of characterizing naturally-occurring microbial populations have been developed. These include direct recovery and analysis of cellular components, such as DNA or RNA, from organisms in natural samples and *in situ* analysis of populations and their metabolic activities. While such techniques are not without biases of their own, we believe they are providing a clearer picture of the natural microbial world than previously was possible. Molecular sequence-based surveys of environmental samples have already astonished us with the enormous diversity of the natural microbial world, detecting previously unknown phyla and even kingdoms of organisms in almost every instance. In parallel, development of methods for locating and quantifying organisms *in situ* in their environments, and for studying their physiological activities, has begun to reveal the complex structure and activities of microbial communities. With such tools in hand, we are rapidly gaining insight on the diversity of microorganisms and their roles in natural processes.

Molecular sequence-based phylogenetic analysis of microbial populations: molecular microbial ecology. Survey of constituent species. In the mid-1980s, Norman Pace and colleagues had the insight that it was possible to identify organisms in naturally-occurring microbial populations without first cultivating them (Olsen et al. 1986, Pace et al. 1986). The techniques they, and many other laboratories since that time, developed revolve around the use of ribosomal RNA (rRNA), or the genes that encode it, extracted directly from cells in environmental samples. Comparison of these sequences with rRNAs from cultivated species allows determination of evolutionary (phylogenetic) relationships between the unknown organisms and known ones. In addition, these sequences can be used as "molecular signatures" for organisms, whose use will be described below.

While several different approaches to obtaining and analyzing rRNA sequences from organisms in environmental samples have been developed (reviewed recently by Amann et al. 1995 and Ward et al. 1992), the one most widely used at this time is outlined below. In this, microbial cells in environmental samples are lysed, and cellular DNA (including genes encoding rRNAs) is chemically extracted and purified (More et al. 1994, Sayler and Layton 1990, Tsai and Olson 1991, Zhou et al. 1996). Small subunit rRNA genes (rDNAs) are then selectively amplified from the bulk DNA via the polymerase chain reaction (PCR; Devereux and Willis 1995, Saiki et al. 1988)) utilizing primers specific for rDNA sequences (Edwards et al. 1989, Weisburg et al. 1991, Wilson et al. 1990). This produces a mixture of rDNA copies from the organisms of the starting population. This mixture is separated into individual copies by ligation into a plasmid vector (a circular piece of DNA into which one can insert pieces of DNA) and transformed into *E. coli* host cells. Because each cell only takes up one plasmid containing one rDNA insert, this effectively creates cells lines, each of which contains a clone of an rDNA gene from an organism in the starting population. [For more details on this procedure, see Sambrook et al. (1989)]. These plasmid clones are then sequenced, and the sequences compared with each other and with the SSU rRNA sequences of cultured organisms. Phylogenetic comparative analysis of these sequences (see above) can then be used to give an estimate of the genetic diversity of organisms in the community.

At this time, it is generally problematic to infer physiological properties or cultivation conditions of most organisms based on their phylogenetic relationship to other organisms. Unless an environmental rDNA sequence is found to be very closely related to that of a cultivated species, it is difficult to say much about the activities of the unknown organism. This is because most environmentally important characteristics of organisms, what they eat and excrete, often are not well conserved across broad phylogenetic groups. Exceptions to this include the cyanobacteria, which are apparently all oxygenic photosynthetic organisms, and archeael methanogens and halophiles (see above).

In spite of this shortcoming, however, this approach to the analysis of natural populations is proving to be a powerful one for several reasons. Microbial ecologists can now hope to attain a truly exhaustive census of the species present in natural environments, a goal unreachable by cultivation-based methods. In almost every case where even a small number of environmental rDNA clones has been analysed, previously unexpected species diversity has been revealed (for examples, see Barns et al. 1996b, DeLong 1992, Giovannoni et al. 1990, Liesack and Stackebrandt 1992, Ward et al. 1990, Ward et al. 1992). Indeed, it is rare for a sequence recovered from a natural sample to match that of any cultivated species available in the sequence database. In several cases, this has caused microbiologists to rethink assumptions about the properties of formerly "well-known" groups of organisms (Angert et al. 1993, DeLong 1992, Liesack and Stackebrandt 1992). Analyses of environmental sequences have begun to reshape the universal phylogenetic tree (Barns et al. 1996b, DeLong 1992), and given us additional perspective on the nature of the last common ancestor of all life (Barns et al. 1996b, Reysenbach et al. 1994). With the knowledge of "who's there?" in an environment, microbial communities are no longer "black boxes", and researchers can undertake informed, directed studies of the activities of the constituent organisms, using the techniques described below.

rRNA sequence-based in situ *methods.* Ribosomal RNA sequences also can be used as "molecular signatures" to identify and study organisms in their natural environments. The variability in rate of sequence change in different portions of the SSU rRNA molecule allows one to select signature regions specific for individual species (using hypervariable portions of the sequence) or whole kingdoms (using more conserved sequence regions) or all of life (using highly conserved, "universal" rRNA sequence regions). Indeed, it is generally possible to find regions suitable for detection of any level of phylogenetic grouping by comparison of aligned sequences in the large rRNA database now available. Once identified, these signatures are exploited by design of generally short (12 to 25 nucleotide) DNA oligonucleotide primers or probes of complimentary sequence, whose use will be outlined below. Such probes can be quite specific, as descrimination has been acheived based on a single nucleotide difference between target and non-target sequences (Amann et al. 1990b). Methods for design and testing of primers and probes have been described (Amann et al. 1995, Stahl and Amann 1991, Zheng et al. 1996), and a useful program for comparison of probe sequences against the database of rRNA sequences is available (CHECK_PROBE of the Ribosomal Database Project; Maidak et al. 1994). An electronic database of primers and probes and their properties has recently been established (The Oligonucleotide Probe Database; Alm et al. 1996).

PCR-based analyses. Species- or group-specific PCR primers can be used to indicate the presence of DNA of target microorganism(s) in environments. Several techniques have been developed which exploit this idea to detect and monitor organisms of interest, and to compare populations between samples or through time. Generally, this involves PCR amplification of bulk DNA extracted from environmental samples, using one or two primers specific for the rDNA sequence of the target organism(s). If a known concentration of an internal-standard target DNA is included in the reaction, it is possible to roughly quantitate the number of rRNA genes in the environmental DNA, and thereby infer the number of organisms present (Lee et al. 1996, Leser et al. 1995). Through the use of special denaturing gradient gel electrophoresis (DGGE) conditions, it is possible to separate individual sequence types present in a mixed PCR product. This provides a "fingerprint" for the community present which can be used to compare species composition across time and environments. It has been used to survey populations and reveal rare sequence types, as well as to monitor laboratory enrichment cultures for the growth of particular species (Ferris et al. 1996, Muyzer et al. 1993, Santegoeds et al. 1996). A somewhat lower-resolution approach to analysing the constituents of mixed population PCR products involves "restriction fragment length polymorphism" (RFLP) analysis. Restriction enzymes cut DNA molecules at specific places, based on sequence, and produce fragments of different lengths, depending on the places where the DNA was cut. This technique has been used to screen rDNA clone libraries (Hershberger et al. 1996, Moyer et al. 1994), analyze diversity and structure in natural microbial populations (Britschgi and Giovannoni 1991, Martinez-Murcia et al. 1995, Moyer et al. 1994), and compare populations across a salinity gradient (Martinez-Murcia et al. 1995). Use of RFLP patterns for phylogenetic analyis has also been proposed (Moyer et al. 1996).

DNA probe hybridization analyses. DNA oligonucleotide probes, complimentary to signature sequences, can be labelled isotopically or fluorescently and used to detect the rRNA of target species. In natural population analyses, isotopically-labelled probes are used to hybridize to bulk community rRNA bound to filter membranes. The amount of rRNA present from a particular species is then determined by quantitation of bound radioactivity. When divided by the amont of radioactively-labelled universal probe bound to the same rRNA sample, the relative proportion of rRNA contributed by the particular species can be calculated. Since rRNA content of cells is correlated with growth rate, cell number cannot be directly obtained from this measurement. However, it is possible to infer the relative physiological contribution of the target species to the activity of the community (Amann et al. 1995, DeLong et al. 1994, Giovannoni et al. 1990).

Probes can also be hybridized to rRNA present in intact microbial cells, a technique generally known as whole cell hybridization. Fluorescently-labelled probes are most widely utilized, as they are easy to use and can be readily visualized by fluorescence microscopy (Amann et al. 1990b, Amann 1995, Amann et al. 1995, DeLong et al. 1989, Ward et al. 1992). In this technique, organisms are fixed *in situ* in environmental samples by treatment with formaldehyde, which stabilizes cellular structure, after which the cell membranes are permeabilized. Samples are then bathed in an aqueous solution containing oligonucleotide probes at a temperature at which the probes should only bind to fully complimentary target sequences. Excess probe is washed away, and the bound probe-containing cells visualized with an epifluorescence microscope. Confocal laser scanning microscopy and digital image analysis techniques have been used to great effect for signal enhancement and quantitative fluorescence measurements (Leser et al. 1995, Poulsen et al. 1993, Wagner et al. 1994), as well as spatial analysis of target species within samples (Amann et al. 1996, Moller et al. 1996). Difficulties with this technique occur due to cell impermeability to probes (Braun-Howland et al. 1992), poor rRNA target sequence accessibility, and low ribosome content and small cell sizes commonly encountered in enviromental samples (Amann 1995, Ward et al. 1992).

In situ probe hybridization has been used to confirm the presence of target species in samples, determine morphology, and locate and ennumerate organisms in environmental samples and laboratory cultures (Amann et al. 1996, Amann 1995, Burggraf et al. 1997a, Leser et al. 1995, Moller et al. 1996, Wagner et al. 1993). Due to the correlation of cellular rRNA content with growth rate, it is possible to infer the physiological activity of organisms in environmental samples based on fluorescence level relative to cultured cells (DeLong et al. 1989, Leser et al. 1995, Poulsen et al. 1993). Recent developments in the field include use of multiple probes and image analysis techniques to visualize seven individual species in a biofilm (Amann et al. 1996) and *in situ* coanalysis of phylogenetic type and metabolic activity, using simultaneous probing with oligonucleotides and cytochemical probes (Whiteley et al. 1996). In addition, organisms from enrichment cultures have been monitored and identified with fluorescent probes, then isolated using a highly focused infrared laser ("optical tweezers") and grown in pure culture (Huber et al. 1995).

Analysis of fluorescently-labelled cells in environmental samples has also been accomplished by flow cytometry (Amann et al. 1990a, Porter et al. 1996, Wallner et al. 1993). The advantages to this approach include rapid sample processing, ease of analyzing statistically significant numbers of cells, quantitative analysis of fluorescence signal, detection of multiple fluorochromes, simultaneous acquistion of phylogenetic, metabolic and morphologic information, and the ability to use cell sorting to collect organisms of interest for further analysis (Amann et al. 1990a, Porter et al. 1996). Flow cytometry can also be used for microbial population analysis using general DNA and RNA stains, light scattering properties, fluorescent antibody and lectin labeling, metabolic activity stains, and other approaches (Davey and Kell 1996, DeLeo and Baveye 1996, Edwards et al. 1996, Vesey et al. 1994).

Microscopic techniques for the examination of environmental samples

It has become increasingly important to develop microscopy techniques that enable researchers to gain a better understanding of the way in which microorganisms interact *in situ* in their environment. Some of the more established microscopic methods that traditionally have been used for this purpose include transmission electron microscopy (TEM), scanning electron microscopy (SEM), and differential interference contrast microscopy (DIC) with and without fluorescence. In recent years however, there have been significant advances in microscopic techniques for the *in situ* examination of organisms associated with environmental substrates ranging from sediments to roots to biofilms (Surman et al. 1996). Advances in microscopic techniques of most relevance have included environmental scanning electron microscopy (ESEM), atomic force microscopy (AFM) and scanning laser confocal microscopy (SCLM). The information that can be obtained from these various microscopy techniques include morphological, quantitative, identification of cell surface components (via labeling), identification of specific species within microbial consortia, distinctions between viable and non-viable cells (using vital stains), and spatial organization of microorganisms occurring within different environmental matrices.

In general, thin sections through materials, used in conjunction with TEM analyses, can provide useful information about the spatial relationships of microorganisms within a matrix. Information is also provided regarding the internal ultrastructure of the organisms. However, in some instances it is difficult, if not impossible, to obtain thin sections through a variety of hard materials. Using SEM, it is possible to examine surface topologies at high magnifications. Additionally, if SEM is used in conjunction with energy dispersive x-ray spectroscopy, it is possible to obtain information regarding the localization of various elements (including many metals). Nevertheless, the sample preparation that is required for either SEM or TEM may introduce artifacts that are not introduced with some of the more recently developed microscopic techniques.

ESEM permits the same type of high resolution, high magnification and large depth of field imaging that is obtained with standard SEM, while simultaneously allowing biological materials to be observed in a more natural state. The ESEM differs from the SEM in that it has a specimen chamber that is differentially pumped which allows it to operate with up to 10 torr of water vapor (instead of being viewed under a high vacuum). AFM is a form of scanning probe microscopy which uses a sharp probe to map the contours of a sample. Recently, there have been reports of using AFM for examination of bacterial biofilms on metal surfaces (Bremer et al. 1992, Steele et al. 1994) and from a model oil-in-water emulsion (Gunning et al. 1996). In the later study it was shown that AFM offered superior resolution to SEM, with minimal sample preparation. Additionally, high resolution DC contact mode AFM studies of bacterial surfaces were able to reveal surface features comparable in size to large proteins (Gunning et al. 1996). The advantage of both ESEM and AFM is that they allow for direct visualization of intact hydrated specimens (at high magnification). Additionally, AFM images can be easily rotated and manipulated to provide accurate measurements of individual bacterial cells.

Scanning confocal laser microscopy and image analysis has significantly increased the analytical precision of light microscopy. A major advancement presented by the development of SCLM is the ability to optically thin section and reconstruct xz-sagittal sections and 3-D images of samples. SCLM can be carried out in conjunction with a brightfield, epifluorescence or inverted microscope, thereby expanding its capabilities. SCLM can overcome problems created by autofluorescence of objects and narrow depth of focus in thick specimens. Furthermore, it allows for high-resolution analysis of the spatial distribution of bacteria within different matrices. Only recently has SCLM been used for the examination of microorganisms in environmental samples (Distel and Cavanaugh 1994, Ghiorse 1993, Ghiorse et al. 1996), in most instances the examination of biofilms (Korber et al. 1993, Lawrence et al. 1991, Lens et al. 1994). Recently, SCLM has also been used to (1) analyze Mn oxide-encrusted biofilms and particles in marine Mn-oxidizing enrichments cultures; (2) optimize fluorescence *in situ* hybridization protocols of 16S ribosomal RNA-targeted oligonucleotide signature probes for single bacterial cell identification in particles from wetlands; and (3) develop a combined immunofluorescence-microautoradiography procedure for analysis of the distribution of ^{14}C-labeled organic compounds and ^{14}C-mineralizing bacteria in groundwater seep sediments, been demonstrated (Ghiorse et al. 1996). This serves as an example of the applicability and benefits that can be derived from the use of SCLM for the examination of complex microbial assemblages.

Each of the aforementioned microscopic techniques has advantages and disadvantages. Thus, when determining which technique is most appropriate for a specific application, it is important to evaluate sample preparation time and any potential artifacts that may be caused due to sample preparation, the ability and/or necessity to view the specimen under "natural conditions," the resolution that can be obtained, and the ability to obtain quantification.

Signature lipid biomarker (SLB) profiles

A technique that provides phenotypic information about microorganisms involves signature lipid biomarker (SLB) profiles. This method has an advantage in that it does not depend on the growth or morphology of organisms for identification purposes. Instead, microbial biomass and community structure are determined based on the identification of lipid biomarkers that are characteristic of all cells.

Using SLB profiles, total lipids extracted from bacteria are separated into lipid classes by silicic acid column chromatography (White et al. 1979). The phospholipids present in bacterial membranes are converted into fatty acid methyl esters (FAME), which are separated, quantified (by gas chromatography/mass spectrometry) and identified (Guckert et al. 1985, White et al. 1979). Identification of bacterial types is made by comparing an unknown methyl ester profile with profiles from known bacteria, using a similarity index. This approach has been widely used in conjunction with sediment, bioreactor, rhizosphere, sea ice, microbial slime, and other sample types (see White 1988) from which SLBs have been shown to be readily extractable. Use of this technique is particularly valuable as it can provide insight into the viable biomass, total biomass, community structure, and nutritional/physiological status of communities, as discussed below.

Identification of cultured microorganisms and in situ *microbial community structure.* Increasingly, lipid profiles have been used for examining the phylogenetic identity of microorganisms. This is possible because specific groups or types of microorganisms often contain characteristic lipids, in particular fatty acids, which provide identifiable patterns. Fatty acid patterns are frequently used to identify bacterial isolates or to place them into groups on the basis of fatty acid relatedness. Patterns of prominent ester-linked fatty acids from cultured microorganisms have been identified for over 650 species (Microbial Identification System, MIDI, Newark, Delaware). In some instances, PLFA patterns have been useful in identifying single species (with examples including identification of individual species of methane-oxidizing and sulfate-reducing bacteria; Kohring et al. 1994, Guckert et al. 1991). In a few instances, comparisons between identifications made using PFLA profiles and 16S rRNA sequences have been shown to correlate well (Kohring et al. 1994). However, it is important to note that many species contain over-lapping PLFA patterns and therefore can not be precisely identified using SLB analyses (White 1994, White et al. 1994).

In addition to being useful for the identification of microbial isolates, SLB analyses can provide information on community structure in environmental samples (Guezennec and Fiala-Medioni 1996, Haldeman et al. 1993, Ringelberg et al. 1989). By quantification of differences between PLFA patterns obtained from individual environmental samples, it is possible to obtain information regarding shifts in overall community structure. Additionally, the analysis of other lipids, including sterols (for microeukaryotes), glycolipids (for gram-positive bacteria and phototrophs) and hydroxy fatty acids from the Lipid A portion of a lipopolysaccharide (LPS-OHFA; for gram negative bacteria), can provide more detailed information about microbial community structure. Because this type of SLB analysis is independent of culturing, there is reduced bias, with both culturable and non-culturable organisms contributing to the total community profile.

Nutritional and physiological status of microorganisms. Although of significant value for characterization of microbial community structure, the use of PFLA profiles in discerning phylogenetic relationships can be affected and or complicated by changes in lipid composition due to alterations in growth conditions. For example, exposure to toxic environments has been shown to lead to minicell formation resulting in shifts in PLFA, with an increase in *trans* versus *cis* monoenoic PLFA (Heipieper et al.

1992). In a different study, ratios of cyclopropane fatty acids to their monoenoic homologues were shown to increase in response to starvation (Guckert et al. 1986, Kieft et al. 1994). Thus, specific patterns of PLFA result from, and can be indicative of, physiological stress (Guckert et al. 1985). Additionally, it was shown that if an essential component is missing (e.g. phosphate, nitrate, trace metals, etc.), but adequate carbon and terminal electron acceptors are available, bacteria can undergo unbalanced growth (with an inability to divide) and accumulate poly B-hydroxyalkanoic acids (PHA) (Tunlid and White 1992). Thus, the examination of certain SLB ratios can provide valuable insight regarding the nutritional and physiological status of microorganisms.

Non-viable and viable biomass determinations. Viable microorganisms have intact cellular membranes that contain phospholipids and phospholipid ester-linked fatty acids (PLFA). Upon cell death, enzymes hydrolyze phospholipids within minutes to hours releasing the polar head groups (White et al. 1979). However, the lipid core remains as a diglyceride for a longer period of time. The diglyceride contains the same signature fatty acids as the phospholipids. Thus, an estimation of total non-viable and total viable (or potentially viable) biomass can be made by measuring diglyceride fatty acids and phospholipid fatty acids, and comparing the ratio between the two (White et al. 1979). More specifically, PLFA analyses can be used in this way because (1) phospholipids are found in all cellular membranes in reasonably constant amounts in a variety of bacteria, (2) they have a high natural turnover rate, and (3) they are rapidly degraded in non-viable cells. In one study examining subsurface sediments (Balkwill et al. 1988), viable biomass as determined by PFLA analyses was shown to be equivalent to estimations based on intercellular ATP, cell wall muramic acid, and acridine orange direct counts.

CASE STUDIES

Microbiology of the terrestrial subsurface—GEMHEX (Geological, Micro-biological and Hydrological Experiment at the Hanford Site, Washington)

It appears that microorganisms in deep subsurface environments must be capable of surviving and functioning in an extremely resource-limited environment. Therefore, it has been of interest to identify and learn more about the source(s) of nutrients that support the *in situ* metabolism and growth of these organisms. It has been suggested that most of the nutrients utilized by bacteria in deep subsurface environments are obtained from adjacent rocks and sediments. Initial studies resulted in only limited success in applying traditional microbiology techniques to characterize microorganisms present in deep subsurface sediments (Balkwill 1989). The population size of bacteria in subsurface (Kieft et al. 1995) or other environments, as estimated by direct microscopic counts or total microbial lipids, has typically been several orders of magnitude higher than the number of bacteria counted using plate count or most probable number estimations. These discrepancies make it difficult to establish the abundance and composition of subsurface microbial populations.

In light of these results, molecular methods that are independent of culturing have become increasingly important for investigating the size and composition of microbial communities in subsurface sediments. In the following case study, bacterial isolates from subsurface sediment samples were characterized using conventional sequencing techniques and 16S rRNA targeted phylogenetic group-specific oligonucleotide probes. Additionally, these same probes were used in conjunction with direct *in situ* hybridizations to microorganisms within subsurface sediments. The results of this study are helping us to understand the relationships between the microbiology, geochemistry and hydrology of deeply buried subsurface sediments.

In order to investigate the size and composition of microbial communities in relation to sediment geochemical and geophysical properties, thirty closely spaced cores were obtained from Miocene-aged fluvial, lacustrine and paleosol subsurface sediments ranging in depth from 173-197 m, at the U.S. Department of Energy's Hanford Site in south-central Washington State. Samples were obtained from an uncontaminated part of the aquifer that is located hydrologically up-gradient from areas affected by past site operations. All subsurface samples were cored by cable-tool percussion drilling using a split-spoon core barrel containing a sterile Lexan liner (Kieft et al. 1993). Immediately upon recovery from the borehole, cores were transferred to an on-site argon filled glove bag. The outer, potentially contaminated, sediment was pared away using sterile implements and material from the center of the cores were retained for microbiological, chemical and physical analyses.

Aerobic heterotrophic bacteria were isolated and cultured from paleosol, lacustrine, and fluvial sediments and were phylogenetically classified using a whole cell hybridization technique and group-specific probes. All major phylogenetic kingdoms were represented in the lacustrine and paleosol samples, while no members of the *Flexibacter/Cytophaga/Bacteroides* kingdom or alpha Proteobacteria were isolated from the fluvial sediments. Using this classification technique, it was noted that the high G + C gram-positive group and beta and gamma Proteobacteria were the most important phylogenetic groups for distinguishing microbial communities associated with different sediment types. In lacustrine sediments, the high G + C gram-positive group dominated while the beta and gamma Proteobacteria comprised a smaller but significant component. In the paleosols, beta Proteobacteria dominated with fewer high G + C gram-positive and gamma Proteobacteria. In the fluvial sediment samples the gamma Proteobacteria dominated. The abundance of other types of bacteria, while often numerically important, did not contribute towards distinguishing unique microbial communities between sediment types (Danielsen et al. 1997).

Microbial communities present in the same sediments were characterized directly using an *in situ* hybridization procedure using the same probes. This technique made it possible to characterize metabolically active bacteria *in situ* (without isolation). Using this approach, it was determined that the microbial communities associated with the different sediment types were most distinguished by the abundance of anaerobic delta Proteobacteria, including sulfate-reducing bacteria, and members of Archaea. Sulfate-reducing bacteria were abundant in the lacustrine sediments, but were only a minor component of the microbial community in paleosols and absent from the fluvial sediments. Culturing results from lacustrine sediments for anaerobic bacteria were in agreement with these results. The abundance of metabolically active gram-positive bacteria (high and low G + C) were approximately equivalent in all sediment types and accounted for less than 10% of the active bacteria. Additionally, 12% of the bacteria in the fluvial sediments were identified as members of Archaea (Fredrickson et al. 1995).

It is interesting that microbial communities characterized by essentially non-functional criteria, such as broad phylogenetic group-specific 16S rRNA targeted probes, should be strongly correlated with sediment lithology. However, similar results were obtained by Pedersen et al. (1996b) in the Swedish Stripa Mine subsurface environment and the same results were obtained whether isolate or *in situ* data sets were analyzed. Thus, there is increasing support that the distribution of 16S rRNA-defined phylogenetic groups is diagnostic of a geologic environment.

Total phospholipid fatty acid (PLFA) analyses showed that the greatest concentrations of microbial biomass were in low-permeability lacustrine sediments that also contained high concentrations of organic carbon. Using a conversion factor of 5×10^5 cells pmol^{-1} for biomass in subsurface sediments (Tunlid and White 1992), the concentration of cells in the lacustrine sediments were estimated to range from 2×10^5 g^{-1} to 2×10^7 g^{-1}. The concentration of total PLFA declined and was at, or below, detection throughout the remainder of the core profile. It was also possible to use lipid structures directly extracted from the sediments as indicators of microbial community composition. For example, the presence of sulfate-reducing bacteria (SRB) was possible to infer from the detection of both iso- and anteiso (terminally) branched saturated lipids, with the iso-configuration being the more abundant one. In the lacustrine sediments, increases in the molar percentages of all terminally branched saturates were seen. Thus, community structure, based on lipid analyses and *in situ* hybridization of bacterial cells with 16S rRNA-targeted oligonucleotide probes, both revealed the presence of metabolically active bacteria that respire sulfate and/or Fe (III) in the lacustrine sediments (Fredrickson et al. 1995).

Sediments were analyzed for total organic carbon (TOC), carbonate in C, pore water sulfate, and HCl-extractable Fe (Fredrickson et al. 1995). These geochemical components were selected as those likely to influence or be indicative of microbial activity. Concentrations of pore water sulfate were low (4 to 8 mg/L) and HCl-extractable Fe was predominantly Fe (II) in the same samples where total biomass and organic carbon were highest. The hydraulic conductivities of intact cores, and therefore their permeabilities, were generally low (10^{-6} to $<10^{-9}$ cm/s). Low permeability physical conditions, particularly in the lacustrine sediments, is likely to have contributed to the long term maintenance of both bacteria and organic carbon by limiting the supply of soluble electron acceptors for microbial respiration and limiting the transport of microorganisms. Thus, in many deep subsurface environments comprised of low permeability sediments, microorganisms may be remnants of populations associated with the original deposits. These results suggest that some of the organisms belonging to the microbial communities in this study are likely to be derived from organisms present during lake sedimentation approximately 6 to 8 million years ago, and which have survived by metabolizing and growing at extremely slow rates (Fredrickson et al. 1995).

Crenarchaeota in hot and not-so-hot environments

Introduction. Application of some of the SSU rRNA-based phylogenetic techniques outlined above to the study of Crenarchaeota in different environments has dramatically reshaped our ideas about the diversity and ecology of members of that kingdom. Prior to these studies, analysis of cultivated species led microbiologists to suspect that the evolutionary diversity and distribution of Crenarchaeota was restricted to a few genera living in a limited range of high-temperature habitats. These limitations meant that Crenarchaeota were unlikely to play important roles in the environment. Several recent studies using rRNA sequencing methods, however, have revealed extensive, previously unsuspected diversity within the kingdom and beyond, and has detected abundant Crenarchaeota in relatively "normal" low-temperature environments such as soils, sediments and the open ocean. Application of additional sequence-based techniques is helping to unravel some of the secrets of these novel organisms. Results already indicate that Crenarchaeota are diverse, numerous and globally distributed members of the biosphere, although the details of their physiologies remain a mystery.

Crenarchaeal diversity in Obsidian Pool hot spring. One of these studies examined the archaeal inhabitants of a boiling, black hot spring in Yellowstone National Park called Obsidian Pool (Barns et al. 1996b, Barns et al. 1994). This medium-sized (~3 × 9 m) thermal feature is located in an acidic, high sulfide area of the park. It contains several boiling (93°C) source areas and has a deep black sediment, due to the presence of obsidian sand and possibly iron sulfide. The water and sediment are approximately neutral in pH, sulfide is present, and the sediment contains an unusually high concentration of iron ($\approx$ 15,600 mg/kg).

A sediment sample was taken from a 73° to 93°C area, and microbial DNA was extracted and purified directly from it. This DNA was used to template PCR reactions, as described above, using primers specific for the rDNA genes of Archaea. PCR products were then cloned into plasmid vectors, and 239 insert containing clones were analyzed.

These clones contained almost exclusively archaeal rDNA gene copies; a total of 32 distinct sequence types were identified. Phylogenetic analysis of these sequences, together with those of cultivated Archaea, is shown in Figure 2 (clones designated "pJP" and "pSL").

As can be seen in Figure 2, a wide diversity of archaeal sequences were obtained. Of the 32 sequence types identified, about 41% were recovered in only one clone each of the 239 analyzed. This suggests that considerably greater diversity of sequence types exists in this hot spring than was detected in this analysis. In addition, none of the sequences was identical to that of any species cultivated from other sources and available in the database. Two of the sequences grouped with the euryarchaeal kingdom, specifically with members of the genus *Archaeoglobus*. This was unexpected, as all previously known members of this genus have been isolated from marine environments (Burggraf et al. 1990). The strain represented by pSL23 has recently been isolated in laboratory culture, and is physiologically very similar to *Archaeoglobus fulgidus*, except that it has a lower salt requirement for growth (Siegfried Burggraf, pers. comm.).

Most of the sequences obtained affiliated with the Crenarchaeota, and members of all previoulsy known crenarchaeal orders were identified (Burggraf et al. 1997b). However, as can be seen in Figure 2, most ot the Obsidian Pool sequences did not group specifically with any previously studied organism. Analysis of these sequences substantially increases the known phylogenetic breadth of the Crenarchaeota, adding several new orders and many new genera to the kingdom. The remaining two sequences, pJP27 and pJP78, did not associate consistently with either of the two known archaeal kingdoms, and may represent yet a third, previously unknown kingdom (Barns et al. 1996b). Although one member of this group recently has been grown in mixed culture in the laboratory (Burggraf et al. 1997a), the physiology of these novel organisms remains largely unknown.

These results indicate that high-temperature microbial communities may be just as diverse as lower temperature ones. It is not yet known what conditions foster this diversity in what seems a rather inhospitable environment. However, the abundance of sulfur, iron and volcanic glass in the sediment together with moderate pH and a variety of aerobic (due to mixing) and anaerobic (due to temperature and reducing agents) environments may create a variety of suitable microhabitats for different species to exploit. Future *in situ* analyses of these species may help to reveal their interactions with their environment.

Understanding the physiology of these organisms is most readily attained through study of laboratory cultures of isolated species. However, for reasons discussed previously, traditional cultivation techniques are unlikely to recover all, or even most, species from the environment. For study of the Obsidian Pool microbes, therefore, Karl Stetter's laboratory devised a targeted approach to obtain specific organisms identified by rRNA sequence information (Huber et al. 1995). In one such isolation, a fluorescent DNA probe was designed to target the pSL91 sequence identified in the earlier analysis. This probe was used to monitor laboratory enrichment cultures and identify the morphology of the pSL91 organism. However, none of these cultures produced the organism in very high

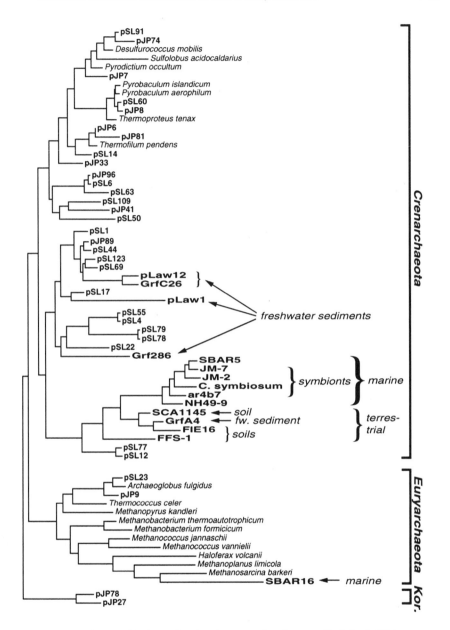

Figure 2. Phylogenetic tree of Archaea, showing placement of environmentally-derived rRNA sequences. Italicized names are those of cultivated species whose sequences were obtained from the database. Environmental rRNA sequences were obtained from the following environments: pJP and pSL sequences, hot spring (Obsidian Pool) (Barns et al. 1996); pLaw1, pLaw12, freshwater lake sediment (Schleper et al. 1997); GrfC26, Grf286, GrfA4, freshwater lake sediment (Hershberger et al. 1996); SBAR5, SBAR16, Pacific coastal waters (DeLong 1992); JM-2, JM-7, marine holothurian (McInerney et al. 1995); C. symbiosum, marine sponge (Preston et al. 1996); ar4b7, marine sponge (Stein et al. 1996); NH49-9, Pacific Ocean bacterioplankton (Fuhrman et al. 1992); SCA1145, Wisconsin soil (Bintrim et al. 1997); FIE16, agricultural soil (Ueda et al. 1995); FFS-1, forest soil (Jurgens et al. 1997). Kor. = the kingdom Korarchaeota (Barns et al. 1996).

numbers; the enrichments were always dominated by other species. Stetter's group therefore used a single-cell isolation technique employing a highly focused infrared laser beam ("laser tweezers") to micromanipulate individual cells from the enrichment culture into sterile culture medium. They then grew up a culture of the organism and sequenced its rRNA to confirm its identity as the pSL91 organism. By this technique, it is possible to isolate and characterize species identified by rRNA sequence-based surveys of natural microbial populations.

Crenarchaeota are widespread and abundant in low-temperature environments. In 1992, two laboratories investigating the microbial diversity of marine bacterioplankton reported evidence for the presence of Crenarchaeota in cool, oxygenated seawater (DeLong 1992, Fuhrman et al. 1992). In these studies, bacterioplankton were collected from Atlantic and Pacific surface and near-surface ocean waters, and DNA was extracted from the concentrated cells. Small subunit rDNA primers were then used to amplify rRNA gene sequences from the extracted DNAs, and the PCR products cloned and sequenced. Unexpectedly, phylogenetic analyses of these sequences showed that many were affiliated with the Crenarchaeota, despite the low-temperature source of the starting DNA. Two of these sequences, designated SBAR5 and NH49-9, are shown in Figure 2.

The distribution of these organisms, together with certain features of their sequences, indicated that they probably did not originate from marine hydrothermal vents, but were instead indigenous to the cold ocean waters. Archaeal abundances appear to be maximal at ~100 m depth, which corresponds to the depth of highest total planktonic abundance (Massana et al. 1997, Fuhrman et al. 1992). This suggests that these Crenarchaeota live among and compete with diverse bacterial and eukaryotic species. The analyses also recovered sequences from a novel group within the euryarchaeal kingdom, related to methanogen species (DeLong 1992, Fuhrman et al. 1993). This, too, was unexpected, as all previously known methanogens are strict anaerobes and therefore unlikely to inhabit oxygenated surface waters.

While these results suggested the presence of numerous Archaea in these environments, biases inherent in the PCR-based technique prevent inference of species abundance in the starting sample (Suzuki and Giovannoni 1996). In order to better quantitate the crenarchaeal population in marine waters, DeLong and coworkers designed DNA probes specific for the novel archaeal sequences and used them to hybridize to rRNA extracted from bacterioplankton samples. Analysis of bacterioplankton rRNAs from Pacific and Atlantic ocean waters showed that up to 3% of total rRNA was of archaeal origin (DeLong 1992). However, in a later study, analysis of frigid surface waters from Antarctica showed as much as 30% of planktonic rRNA derived from Archaea, of which up to 36% was crenarchaeal in origin (DeLong et al. 1994). These results confirmed that Archaea are indigenous to cold, oxic marine waters and indicate that they are abundant and/or metabolically active there. Although nothing is known about their physiologies, they are likely to be very different from those of the thermophilic and methanogenic species which are their closest cultivated relatives.

The finding of Crenarchaeota in cold ocean waters prompted examination of several other low-temperature environments in an effort to determine the distribution of these organisms. Each of these studies employed archaeal or crenarchaeal-specific rDNA primers in PCR amplification of DNA extracted from natural samples. In each case, sequence analysis has revealed the presence of crenarchaeal species. In the marine environment, Crenarchaeota have been identified living in specific association with temparate and deep-sea sponges (McInerney et al. 1995, Preston et al. 1996). In one case, the association was found to be specific, with only a single rDNA sequence type recovered in the analysis (Preston et al. 1996). Two studies examined freshwater lake sediments, and crenarchaeal sequences were recovered in each (Hershberger et al. 1996, Schleper et al. 1997). Several studies of both natural and agricultural soils have revealed the presence of Crenarchaeota (Bintrim et al. 1997, Jurgens et al. 1997, Schleper et al. 1997, Ueda et al. 1995). The phylogenetic positions of representative sequences from these environments are shown in Figure 2. It is clear from these studies that Crenarchaeota are indeed widespread and probably abundant in the biosphere, and play much more important ecological roles than were previously ascribed to them.

SUMMARY AND CONCLUSIONS

We have attempted to introduce the rich diversity of the microbial world and the activities of the organisms which comprise it. Subsequent chapters will discuss the geological importance of microbial processes in greater depth. It is hoped, however, that we have been able to convey some of the extraordinary variety and abundance of microorganisms in the environment and thereby intimate the profound impact they have on geochemical processes in the biosphere. In addition, we hope that the "case studies" of recent research have introduced the important concept that we still know very little about the microbial world. What we understand currently about the diversity, abundance and activities of organisms in the environment is based on the study of a minority of types. Probably greater than 90% of microbial species are yet undiscovered, and the majority of environments await exploration. The coming years should see a rapid expansion of our understanding of the variety of microbial life and its interactions with the Earth.

REFERENCES

Achenbach-Richter L, Stetter KO, Woese CR (1987) A possible biochemical missing link among archaebacteria. Nature 327:348-349

Alexander M (1977) Introduction to Soil Microbiology. John Wiley & Sons, New York

Alm EW, Oerther DB, Larsen N, Stahl DA, Raskin L (1996) The oligonucleotide probe database. Appl Environ Microbiol 62:3557-3559

Amann RI, Binder BJ, Olson RJ, Chisolm SW, Devereux R, Stahl DA (1990a) Combination of 16S rRNA-targeted oligonucleotide probes with flow cytometry for analyzing mixed microbial populations. Appl Environ Microbiol 56:1919-1925

Amann RI, Krumholz L, Stahl DA (1990b) Fluorescent-oligonucleotide probing of whole cells for determinative, phylogenetic, and environmental studies in microbiology. J Bacteriol 172:762-770

Amann RI, Snaidr J, Wagner M, Ludwig W, Schleifer K-H (1996) *In situ* visualization of high genetic diversity in a natural microbial community. J Bacteriol 178:3496-3500

Amann RI (1995) Fluorescently labelled, rRNA-targeted oligonucleotide probes in the study of microbial ecology. Mol Ecol 4:543-554

Amann RI, Ludwig W, Schleifer K-H (1995) Phylogenetic identification and *in situ* detection of individual microbial cells without cultivation. Microbiol Revs 59:143-169

Amy PS, Haldeman DL, Ringelberg D, Hall DH, Russell CE (1992) Comparison of identification systems for the classification of bacteria isolated from water and endolithic habitats within the deep subsurface. Appl Environ Microbiol 58:3367-3373

Angert ER, Clements KD, Pace NR (1993) The largest bacterium. Nature 362:239-241

Atlas RM (1984) Microbiology: Fundamentals and Applications. Macmillan, New York

Baldauf SL, Doolittle WF, Palmer JD (1996) The root of the universal tree and the origin of eukaryotes based on elongation factor phylogeny. Proc Natl Acad Sci USA 93:7749-7754

Balkwill DL (1989) Numbers, diversity and morphological characteristics of aerobic, chemoheterotrophic bacteria in deep subsurface sediments from a site in South Carolina. Geomicrobiol J 7:33-51

Balkwill DL, Leach FR, Wilson JT, McNabb JF, White DC (1988) Equivalence of microbial biomass measures based on membrane lipid cell wall components, adenosine triphosphate, and direct counts in subsurface sediments. Microbial Ecol 16:73-84

Barns SM, Delwiche CF, Palmer JD, Dawson SC, Hershberger KL, Pace NR (1996a) Phylogenetic perspectives on microbial life in hydrothermal environments. In: Evolution of Hydrothermal Ecosystems on Earth (and Mars?). Bock GR, Goode JA (eds) p 24-32. John Wiley & Sons, Chichester, UK

Barns SM, Delwiche CF, Palmer JD, Pace NR (1996b) Perspectives on archaeal diversity, thermophily and monophyly from environmental rRNA sequences. Proc Natl Acad Sci USA 93:9188-9193

Barns SM, Fundyga RE, Jeffries MW, Pace NR (1994) Remarkable archaeal diversity detected in a Yellowstone National Park hot spring environment. Proc Natl Acad Sci 91:1609-1613

Benlloch S, Acinas SG, Martinez-Murcia AJ, Rodriguez-Valera F (1996) Description of prokaryotic biodiversity along the salinity gradient of a multipond solar saltern by direct PCR amplification of 16S rDNA. Hydrobiologia 329:19-31

Bintrim SB, Donohue TJ, Handelsman J, Roberts GP, Goodman RM (1997) Molecular phylogeny of Archaea from soil. Proc Natl Acad Sci USA 94:277-282

Blochl E, Burggraf S, Fiala G, Laurerer G, Huber G, Huber R, Rachel R, Segerer A, Stetter KO, Volkl P (1995) Isolation, taxonomy and phylogeny of hyperthermophilic microorganisms. World J Microbiol Biotech 11:9-16

Boivin-Jahns V, Bianchi A, Ruimy R, Garcin J, Daumas S, Christen R (1995) Comparison of phenotypical and molecular methods for the identification of bacterial strains isolated from a deep subsurface environment. Appl Environ Microbiol 61:3400-3406

Bone TL, Balkwill DL (1988) Morphological and cultural comparison of microorganisms in surface soil and subsurface sediments at a pristine study site in Oklahoma. Microbial Ecol 16:49-64

Braun-Howland EB, Danielson SA, Nierzwicki-Bauer SA (1992) Development of a rapid method for detecting bacterial cells, in situ, using 16S rRNA-targeted probes. BioTechniques 13:928-934

Bremer PJ, Geesey CG, Drake B (1992) Atomic force microscopy examination of the topography of a hydrated bacterial biofilm on a copper surface. Curr Microbiol 24:223-230

Britschgi TB, Giovannoni SJ (1991) Phylogenetic analysis of a natural marine bacterioplankton population by rRNA gene cloning and sequencing. Appl Environ Micrbiol 57:1707-1713

Brock TD (1978) Thermophilic Microorganisms and Life at High Temperatures. Springer-Verlag, New York

Brown BJ, Leff LG (1996) Comparison of fatty acid methyl ester analysis with the use of API 20E and NFT strips for identification of aquatic bacteria. Appl Environ Microbiol 62:2183-2185

Brown JR, Doolittle WF (1995) The root of the universal tree of life based on ancient aminoacyl-tRNA synthetase gene duplications. Proc Natl Acad Sci USA 92:2441-2445

Bult CJ, et al. (1996) Complete genome sequence of the methanogenic archaeon, Methanococcus jannaschii. Science 273:1058-1073

Burggraf S, Ching A, Stetter KO, Woese CR (1991a) The sequence of Methanospirillum hungatei 23S rRNA confirms the specific relationship between the extreme halophiles and the Methanomicrobiales. System Appl Microbiol 14:358-363

Burggraf S, Heyder P, Eis N (1997a) A pivotal Archaea group. Nature 385:780

Burggraf S, Huber H, Stetter KO (1997b) Reclassification of the crenarchaeal orders and families in accordance with 16S ribosomal RNA sequence data. Intl J System Bacteriol (submitted)

Burggraf S, Jannasch HW, Nicolaus B, Stetter KO (1990) Archaeoglobus profundus sp-nov represents a new species within the sufate-reducing Archaeobacteria. Syst Appl Microbiol 13:24-28

Burggraf S, Stetter KO, Rouviere P, Woese CR (1991b) Methanopyrus kandleri: An archaeal methanogen unrelated to all other known methanogens. System Appl Microbiol 14:346-351

Cairns-Smith AG, Hall AJ, Russell MJ (1992) Mineral theories of the origin of life and an iron sulfide example. Orig Life Evol Biosphere 22:161-180

Chang S (1994) The planetary setting of prebiotic evolution. In: Early Life on Earth. Bengtson S (ed) p 10-23. Columbia University Press, New York

Clark DA, Norris PR (1996) Oxidation of mineral sulfides by thermophilic microorganisms. Minerals Eng 9:1119-1125

Cragg BA, Parkes RJ, Fry JC (1995) The impact of fluid and gas venting on bacterial populations and processes in sediments from the Cascadia Margin Accretionary System. Proc Ocean Drilling Prog, Sci Res 146:399-411

Crick FHC, Orgel LE (1973) Directed panspermia. Icarus 19:341-345

Danielsen SA, Frischer ME, Balkwill DL, Braun-Howland EB, Nierzwicki-Bauer SA (1997) Characterization of microbial communities associated with lacustrine, paleosol and fluvial deep subsurface sediments. Appl Environ Microbiol, submitted

Davey HM, Kell DB (1996) Flow cytometry and cell sorting of heterogeneous microbial populations: the importance of single-cell analyses. Microbiol Revs 60:641-696

Davies PCW (1996) The transfer of viable microorganisms between planets. In: Evolution of hydrothermal ecosystems on Earth (and Mars?). Bock GR, Goode JA (eds) p 304-314. John Wiley & Sons, Chichester, UK

DeLeo PC, Baveye P (1996) Enumeration and biomass estimation of bacteria in aquifer microcosm studies by flow cytometry. Appl Environ Microbiol 62:4580-4586

DeLong EF (1992) Archaea in coastal marine environments. Proc Natl Acad Sci USA 89

DeLong EF, Wickham GS, Pace NR (1989) Phylogenetic stains: Ribosomal RNA-based probes for the identification of single cells. Science 243:1360-1363

DeLong EF, Wu KY, Prezelin BB, Jovine RVM (1994) High abundance of Archaea in Antartic marine picoplankton. Nature 371:695-697

Devereux R, Willis SG (1995) Amplification of ribosomal RNA sequences. In: Molecular Microbial Ecology Manual. p 3.3.1-11. Kluwer Academic Publishers, Amsterdam

Distel DL, Cavanaugh CM (1994) Independent phylogenetic origins of methanotrophic and chemoautotrophic bacterial endosymbiosis in marine bivalves. J Bacteriol 176:1932-1938

Doolittle WF, Feng D-F, Tsang S, Cho G, Little E (1996) Determining divergence times of the major kingdoms of living organisms with a protein clock. Science 271:470-476

Doolittle WF, Brown JR (1994) Tempo, mode, the progenote and the universal root. Proc Natl Acad Sci USA 91:6721-6728

Edwards C (1990) Thermophiles. In: Microbiology of extreme environments. Edwards C (ed) p 1-32. McGraw-Hill, New York

Edwards C, Diaper J, Porter J (1996) Flow cytometry for the targeted analysis of the structure and function of microbial populations. In: Molecular approaches to environmental microbiology. Pickup RW, Saunders JR (eds) p 137-162. Ellis Horwood, London

Edwards U, Rogall T, Blocker H, Emde M, Bottger EC (1989) Isolation and direct complete nucleotide determination of entire genes. Nuc Acids Res 17:7843-7853

Ehrlich HL (1995) Geomicrobiology Marcel Dekker, New York

Ekendahl S, Arlinger J, Stahl F, Pederson K (1994) Characterization of attached bacterial populationss in granitic ground-water from the Stripa research mine with 16S-rRNA gene sequencing technique and scanning electron microscopy. Microbiol 140:1575-1583

Ferris MJ, Muyzer G, Ward DM (1996) Denaturing gradient gel electrophoresis profiles of 16S rRNA-defined populations inhabiting a hot spring microbial mat. Appl Environ Microbiol 62:340-346

Ferry JG (1993) Methanogenesis: Ecology, Physiology, Biochemistry and Genetics. Chapman Hall, New York

Forterre P (1997) Protein vs. rRNA: Problems rooting the universal tree of life. ASM News 63:89-95

Fox GE, Stackebrandt E, Hespell RB, Gibson J, Maniloff J, Dyer T, Wolfe RS, Balch W, Tanner R, Magrum LJ, Zablen LB, Blakemore R, Gupta R, Luehrsen KR, Bonen L, Lewis BJ, Chen KN, Woese CR (1980) The phylogeny of the prokaryotes. Science 209:457-463

Fredrickson JK, Balkwill DL, Zachara JM, Li SW, Brockman FJ, Simmons MA (1991) Physiological diversity and distributions of heterotrophic bacteria in deep Cretaceous sediments of the Atlantic Coastal Plain. Appl Environ Microbiol 57:402-411

Fredrickson JK, McKinley JP, Nierzwicki-Bauer SA, White DC, Ringelberg DB, Rawson SA, Shu-Mei L, Brockman FJ, Bjornstad BN (1995) Microbial community structure and biogeochemistry of Miocene subsurface sediments: implications for long-term microbial survival. Molec Ecol 4:619-626

Frischer ME, Floriani PJ, Nierzwicki-Bauer SA (1996) Differential sensitivity of 16S rRNA targeted oligonucleotide probes used for fluorescence *in situ* hybridization is a result of ribosomal higher order structure. Can J Microbiol 42:1061-1071

Fuchs T, Huber H, Teiner K, Burggraf S, Stetter KO (1996) *Metallosphaera prunae*, sp. nov., a novel metal-mobilizing, thermoacidophilic archaeum, isolated from a uranium mine in Germany. System Appl Microbiol 18:560-566

Fuhrman JA, McCallum K, Davis AA (1992) Novel marine archaebacterial group from marine plankton. Nature 356:148-149

Fuhrman JA, McCallum K, Davis AA (1993) Phylogenetic diversity of subsurface marine microbial communities from the Atlantic and Pacific Oceans. Appl Environ Microbiol 59:1294-1302

Garland JL, Mills AL (1991) Classification and characterization of heterotrophic microbial communities on the basis of patterns of community-level sole-carbon-source utilization. Appl Environ Microbiol 57:2351-2359

Ghiorse WC (1993) Use of the confocal laser scanning microscope for detecting viable, but unculturable subsurface bacteria. Intl Symp on Subsurface Microbiology, p.B33, Bath, UK

Ghiorse WC, Miller DN, Sandoli RL, Siering PL (1996) Applications of laser scanning microscopy for analysis of aquatic microhabitats. Microscopy Res Tech 33:73-86

Gilmour D (1990) Halotolerant and halophilic microorganisms. In: Microbiology of extreme environments. Edwards C (ed) p 147-177. McGraw-Hill, New York

Giovannoni SJ, Britschgi TB, Moyer CL, Field KG (1990) Genetic diversity in Sargasso Sea bacterioplankton. Nature 345:60-63

Gogarten JP, Kibak H, Dittrich P, Taiz L, Bowman EJ, Bowman BJ, Manolson MF, Poole RJ, Date T, Oshima T, Konishi J, Denda K, Yoshida M (1989) Evolution of the vacuolar H$^+$-ATPase: Implications for the origin of eukaryotes. Proc Natl Acad Sci 86:6661-6665

Gogarten JP, Olendzenski L, Hilario E, Simon C, Holsinger KE (1996) Dating the Cenancester of organisms. Science 274:1750-1751

Gold T (1992) The deep, hot biosphere. Proc Natl Acad Sci USA 89:6045-6049

Golding GB, Gupta RS (1995) Protein-based phylogenies support a chimeric origin for the eukaryotic genome. Mol Biol Evol 12:1-6

Goosen NK, Horemans AMC, Hillebrand SJW, Stumm CK, Vogels GD (1988) Cultivation of the sapropelic ciliate *Plagiopyla nasuta* Stein and isolation of the endosymbiont *Methanobacterium formicicum*. Arch Microbiol 150:165-170

Grassia GS, McLean KM, Glenat P, Bauld J, Sheehy AJ (1996) A systematic survey for thermophilic fermentative Bacteria and Archaea in high-temperature petroleum reservoirs. FEMS Microbiol Ecol 21:47-58

Guckert JB, Antworth CP, Nichols PD, White DC (1985) Phospholipid, ester-linked fatty acid profiles as reproducible assays for changes in prokaryotic community structure of estuarine sediments. FEMS Microbiol Ecol 31:147-158

Guckert JB, Hood MA, White DC (1986) Phospholipid, ester-linked fatty acid profile changes during nutrient deprivation of *Vibrio cholera*: Increases in the trans/cis ratio and proportions of cyclopropyl fatty acids. Appl Environ Microbiol 52:794-801

Guckert JB, Ringelberg DB, White DC, Henson RS, Bratina BJ (1991) Membrane fatty acids as phenotypic markers in the polyphasic taxonomy of methylotrophs within the Proteobacteria. J Gen Microbiol 137:2631-2636

Guezennec J, Fiala-Medioni A (1996) Bacterial abundance and diversity in the Barbados Trench determined by phospholipid analysis. FEMS Microbiol Ecol 19:83-93

Gunning PA, Kirby AR, Parker ML, Gunning AP, Morris VJ (1996) Comparative imaging of *Pseudomonas putida* bacterial biofilms by scanning electron microscopy and both DC contact and AC non-contact atomic force microscopy. J Appl Bacteriol 81:276-282

Haldeman DL, Amy PS (1993a) Diversity within a colony morphotype: implications for ecological research. Appl Environ Microbiol 59:933-935

Haldeman DL, Amy PS (1993b) Bacterial heterogeneity in deep subsurface tunnels at Rainier Mesa, Nevada Test Site. Microbial Ecol 25:183-194

Haldeman DL, Amy PS, Ringelberg D, White DC (1993) Characterization of the microbiology within a 21m^3 section of rock from the deep subsurface. Microbial Ecol 26:145-159

Han T-M, Runnegar B (1992) Megascopic eukaryotic algae from the 2.1-billion-year-old Negaunee Iron-formation, Michigan. Science 257:232-235

Hasegawa M, Fitch WM (1996) Dating the Cenancester of organisms. Science 274:1750

Hasegawa M, Hashimoto T, Adachi J, Iwabe N, Miyata T (1993) Early branchings in the evolution of eukaryotes: ancient divergence of *Entamoeba* that lacks mitochondria revealed by protein sequence data. J Mol Evol 36:380-388

Heipieper HJ, Diffenbach R, Keweloh H (1992) Conversion of cis unsaturated fatty acids to trans, a possible mechanism for the protection of phenol degrading *Pseudomonas putida* P* from substrate toxicity. Appl Environ Microbiol 58:1847-1852

Hershberger KL, Barns SM, Reysenbach A-L, Dawson SC, Pace NR (1996) Wide diversity of Crenarchaeota. Nature 384:420

Hiorns WD, Methe BA, Nierzwicki-Bauer SA, Zehr J (1997) Bacterial diversity in Adirondack Mountain lakes as revealed by 16S rRNA gene sequences. Appl Environ Microbiol 63:2957-2960

Huber R, Burggraf S, Mayer T, Barns SM, Rosnagel P, Stetter KO (1995) Isolation of a hyperthermophilic archaeum predicted by *in situ* rRNA analysis. Nature 376:57-58

Huber R, Stoffers P, Cheminee JL, Richnow HH, Stetter KO (1990) Hyperthermophilic archaebacteria within the crater and open-sea plume of erupting Macdonald Seamount. Nature 345:179-182

Ingledew WJ (1990) Acidophiles. In: Microbiology of extreme environments. Edwards C (ed) p 33-54. McGraw-Hill, New York

Iwabe N, Kuma K, Hasegawa M, Osawa S, Miyata T (1989) Evolutionary relationship of archaebacteria, eubacteria, and eukaryotes inferred from phylogenetic trees of duplicated genes. Proc Natl Acad Sci USA 86:9355-9359

Jannasch HW, Eimhjellen K, Wirsen CO, Farmanfarmaian A (1971) Matter in the deep sea. Science 171:672-675

Jimenez L (1990) Molecular analysis of deep-subsurface bacteria. Appl Environ Microbiol 56:2108-2113

Jurgens G, Lindstrom K, Saano A (1997) Novel group within the kingdom Crenarchaeota from boreal forest soil. Appl Environ Microbiol 63:803-805

Kandler O (1994) The early diversification of life. In: Early Life on Earth. Bengston S (ed) p 152-160. Columbia Univ Press, New York

Karavaiko GI, Lobyreva LB (1994) An overview of the Bacteria and Archaea involved in removal of inorganic and organic sulfur-compounds from coal. Fuel Process Tech 40:167-182

Kieft TL, Amy PS, Brockman FJ, Fredrickson JK, Bjornstad BN, Rosacker LL (1993) Microbial abundance and activities in relation to water potential in the vadose zones of arid and semiarid sites. Microbial Ecol 26:59-78

Kieft TL, Fredrickson JK, McKinley JP, Bjornstad BN, Rawson SA, Phelps TJ, Brockman FJ, Pfiffner SM (1995) Microbiological comparisons within and across contiguous lacustrine, paleosol, and fluvial subsurface sediments. Appl Environ Microbiol 61:749-757

Kieft TL, Ringelberg DB, White DC (1994) Changes in ester-linked phospholipid fatty acid profiles of subsurface bacteria during starvation and desiccation in a porous medium. Appl Environ Microbiol 60:3292

Kohring L, Ringelberg DB, Devereux R, Stahl DA, Mittelman MW, White DC (1994) Comparison of phylogenetic relationships based on phospholipid fatty acid profiles and ribosomal RNA sequence similarities among dissimilatory sulfate-reducing bacteria. FEMS Microbiol Letts 119:303-308

Kolbel-Boelke J, Tienken B, Nehrkorn A (1988) Microbial communities in the saturated groundwater environment. I. Methods of isolation and characterization of heterotrophic bacteria. Microbial Ecol 16:17-29

Korber DR, Lawrence JR, Hendry MJ, Caldwell DE (1993) Analysis of spatial variability within MOT+ and MOT- *Pseudomonas fluorescens* biofilms using representatiive elements. Biofouling 7:339-358

Lawrence JR, Korber DR, Hoyle BD, Costerton JW, Caldwell DE (1991) Optical sectioning of microbial biofilms. J Bacteriol 173:6558-6567

Lazcano A, Miller SL (1996) The origin and early evolution of life: Prebiotic chemistry, the pre-RNA world and time. Cell 85:793-798

Lee SY, Bollinger J, Bezdicek D, Ogram A (1996) Estimation of the abundance of an uncultured soil bacterial strain by a competitive quantitative PCR method. Appl Environ Microbiol 62:3787-3793

Lens P, Massone A, Rozzi A, Verstraete W (1994) The effect of sulfate reducing bacteria on the treatment performance of aerobic biofilm receptors. In: International Symposium on Environmental Biotechnology. p 58-60. Chameleon, London

Leser TD, Boye M, Hendriksen NB (1995) Survival and activity of *Pseudomonas sp.* Strain B13(FR1) in a marine microcosm determined by quantitative PCR and an rRNA-targeting probe and its effect on the indigenous bacterioplankton. Appl Environ Microbiol 61:1201-1207

Liesack W, Stackebrandt E (1992) Occurrence of novel groups of the domain Bacteria as revealed by analysis of genetic material isolated from an Australian terrestrial environment. J Bacteriol 174:5072-5078

Lowe DR (1994) Early environments: Constraints and opportunities for early evolution. In: Early Life on Earth. Bengtson S (ed) p 24-35. Columbia Univ Press, New York

Lutz RA, Shank TM, Fornari DJ (1994) Rapid growth at deep-sea vents. Nature 371:663-664

Madigan MT, Martinko JM, Parker J (1997) Biology of microorganisms. Prentice Hall, Upper Saddle River, NJ

Maidak BL, Larsen N, McCaughey MJ, Overbeek R, Olsen GJ, Fogel K, Blandy J, Woese CR (1994) The Ribosomal Database Project. Nuc Acids Res 22:3485-3487

Maidak BL, Olsen GJ, Larsen N, Overbeek R, McCaughey MJ, Woese CR (1997) The RDP (Ribosomal Database Project). Nuc Acids Res 25:109-110

Margulis L (1970) Origin of eukaryotic cells Yale University Press, New Haven, CT

Margulis L, Schwartz KV (1982) Five Kingdoms: An illustrated guide to the phyla of life on Earth. Freeman, New York

Martinez-Murcia AJ, Acinas SG, Rodriguez-Valera F (1995) Evaluation of prokaryotic diversity by restrictase digestion of 16S rDNA directly amplified from hypersaline environments. FEMS Microbiol Ecol 17:247-256

Massana R, Murray AE, Preston CM, DeLong EF (1997) Vertical distribution and phylogenetic characterization of marine planktonic Archaea in the Santa Barbara Channel. Appl Environ Microbiol 63:50-56

McInerney JO, Wilkinson M, Patching JW, Embley TM, Powell R (1995) Recovery and phylogenetic analysis of novel archaeal rRNA sequences from a deep-sea deposit feeder. Appl Environ Microbiol 61:1646-1648

Mojzsis SJ, Arrhenius G, McKeegan KD, Harrison TM, Nutman AP, Friend CRL (1996) Evidence for life on Earth before 3,800 million years ago. Nature 384:55-59

Moller S, Pedersen AR, Poulsen LK, Arvin E, Molin S (1996) Activity and three-dimensional distribution of toluene-degrading *Pseudomonas putida* in a multispecies biofilm assessed by quantitative *in situ* hybridization and scanning confocal laser microscopy. Appl Environ Microbiol 62:4632-4640

More MI, Herrick JB, Silva MC, Ghiorse WC, Madsen EL (1994) Quantitative cell lysis of indigenous microorganisms and rapid extraction of microbial DNA from sediment. Appl Environ Microbiol 60:1572-1580

Moyer CJ, Dobbs FC, Karl DM (1994) Estimation of diversity and community structure through restriction fragment length polymorphism distribution analysis of bacterial 16S rRNA genes from a microbial mat at an active, hydrothermal vent system, Loihi Seamount, Hawaii. Appl Environ Microbiol 60:871-879

Moyer CJ, Tiedje JM, Dobbs FC, Karl DM (1996) A computer-simulated restriction fragment length polymorphism analysis of bacterial small-subunit rRNA genes. Appl Environ Microbiol 62:2501-2507

Mullins TD, Britschgi TB, Krest RL, Giovannoni SJ (1995) Genetic comparisons reveal the same unknown bacterial lineages in Atlantic and Pacific bacterioplankton communities. Limnol Oceanog 40:148-158

Muyzer G, DeWaal EC, Uiterlinden AG (1993) Profiling of complex microbial populations by denaturing gradient gel electrophoresis analysis of polymerase chain reaction-amplified genes coding for 16S rRNA. Appl Environ Microbiol 59:695-700

Olsen GJ, Lane DJ, Giovannoni SJ, Pace NR, Stahl DA (1986) Microbial ecology and evolution: a ribosomal RNA approach. Ann Rev Microbiol 40:337-365

Olsen GJ, Woese CR (1996) Lessons from an archaeal genome. Trends Genet 12:377-379

Olsen GJ, Woese CR, Overbeek R (1994) The winds of (evolutionary) change: Breathing new life into Microbiology. J Bacteriol 176:1-6

Oparin AI (1938) The Origin of Life. Macmillan, New York

Oro J (1994) Early chemical stages in the origin of life. In: Early Life on Earth. Bengtson S (ed) p 48-59. Columbia Univ Press, New York

Pace NR (1991) Origin of life-facing up to the physical setting. Cell 65:531-533

Pace NR, Stahl DA, Lane DJ, Olsen GJ (1986) The analysis of natural microbial populations by ribosomal RNA sequences. Adv Microb Ecol 9:1-55

Parkes RJ, Cragg BA, Bale SK (1994) Deep bacterial biosphere in Pacific Ocean sediments. Nature 371:410-413

Pedersen K (1996) Investigations of subterranean bacteria in deep crystalline bedrock and their importance for the disposal of nuclear waste. Can J Microbiol:382-391

Pedersen K, Arlinger J, Ekendahl S, Hallbeck L (1996) 16S rRNA gene diversity of attached and unattached bacteria in boreholes along the access tunnel to the Aspo hard rock laboratory, Sweden. FEMS Microbiol Ecol 19:249-262

Pedersen K, Arlinger J, Hallbeck L, Petterson C (1996) Diversity and distribution of subterranean bacteria in ground water at Oklo in Gabon, Africa, as determined by 16S rRNA gene sequencing. Molec Ecol 5:427-436

Pierson BK (1993) The emergence, diversification, and role of photosynthetic eubcteria. In: Early Life on Earth. Bengtson S (ed) p 161-180. Columbia Univ Press, New York

Porter J, Deere D, Pickup R, Edwards C (1996) Fluorescent probes and flow cytometry: new insights into environmental bacteriology. Cytometry 23:91-96

Poulsen LK, Ballard G, Stahl DA (1993) Use of rRNA fluorescence *in situ* hybridization for measuring activity of single cells in young and established biofilms. Appl Environ Microbiol 59:1354-1360

Preston CM, Wu KY, Molinski TF, DeLong EF (1996) A psychrophilic crenarchaeon inhabits a marine sponge: *Cenarchaeum symbiosum* gen-nov., sp. nov. Proc Natl Acad Sci USA 93:6241-6246

Rappe MS, Kemp PF, Giovannoni SJ (1997) Phylogenetic diversity of marine coastal picoplankton 16S rRNA genes cloned from the continental shelf off Cape Hatteras, North Carolina. Limnol Oceanog (in press)

Reysenbach A-L, Wickham GS, Pace NR (1994) Phylogenetic analysis of the hyperthermophilic pink filament community in Octopus Spring, Yellowstone National Park. Appl Environ Microbiol 60:2113-2119

Ringelberg DB, Davis JB, Smith GA, Pfiffner SM, Nichols PD, Nickels JS, Hensen JM, Wilson JT, Yates M, Kampbell DH, Reed HW, Stocksdale TT, White DC (1989) Validation of signature polar lipid fatty acid biomarkers for alkane-utilizing bacteria in soils and subsurface aquifer materials. FEMS Microbiol Ecol 62:39-50

Rouviere PE, Wolfe RS (1988) Novel biochemistry of methanogenesis. J Biol Chem 263:7913-7916

Saiki RK, Gelfand DH, Stoffel S, Scharf SJ, Higuchi R, Horn GT, Mullis KB, Erlich HA (1988) Primer-directed enzymatic amplification of DNA with a thermostable DNA polymerase. Science 239:487-491

Sambrook J, Fritsch EF, Maniatis T (1989) Molecular cloning: a laboratory manual. Cold Spring Harbor Laboratory, Cold Spring Harbor, NY

Santegoeds CM, Nold SC, Ward DM (1996) Denaturing gradient gel electrophoresis used to monitor the enrichment culture of aerobic chemoorganotrophic bacteria from a hot spring microbial mat. Appl Environ Microbiol 62:3922-3928

Schleper C, Holben W, Klenk H-P (1997) Recovery of crenarchaeotal ribosomal DNA sequences from freshwater-lake sediments. Appl Environ Microbiol 63:321-323

Schmidt TE, DeLong EF, Pace NR (1991) Analysis of a marine picoplankton community by 16S rRNA gene cloning and sequencing. J Bacteriol 73:4371-4378

Schonheit P, Schafer T (1995) Metabolism of hyperthermophiles. World J Microbiol Biotech 11:26-57

Schopf JW (1994) The oldest known records of life: Early Archean stromatolites, microfossils and organic matter. In: Early Life on Earth. Bengtson S (ed) p 193-207. Columbia Univ Press, New York

Schwartzman D, McMenamin M, Volk T (1993) Did surface temperatures constrain microbial evolution? BioScience 43:390-393

Segerer AH, Burggraf S, Fiala G, Huber G, Huber R, Pley U, Stetter KO (1993) Life in hot springs and hydrothermal vents. Orig Life Evol Biosph 23:77-90

Shock EL (1996) Hydrothermal systems as environments for the emergence of life. In: Evolution of hydrothermal ecosystems on Earth (and Mars?). Bock GR, Goode JA (eds) p 40-52. John Wiley & Sons, Chichester, UK

Sleep NH, Zahnle KJ, Kasting JF, Morowitz HJ (1989) Annihilation of ecosystems by large asteroid impacts on the early Earth. Nature 342:139-142

Sogin ML (1994) The origin of eukaryotes and evolution into major kingdoms. In: Early life on Earth. Bengtson S (eds) Nobel Symposium No. 84. Columbia Univ Press, New York

Stackebrandt E, Liesack W, Goebel BM (1993) Bacterial diversity in a soil sample from a subtropical Australian environment as determined by 16S rDNA analysis. FASEB J 7:232-236

Stahl DA, Amann RI (1991) Development and application of nucleic acid probes in bacterial systematics. In: Nucleic acid techniques in bacterial systematics. Stackebrandt E, Goodfellow M (eds) p 205-248. John Wiley & Sons, Chichester, UK

Stanier RY, Adelberg EA, Ingraham JL (1976) The Microbial World, 4th edition. Prentice-Hall, Engelwood Cliffs, New Jersey.

Steele A, Goddard DT, Beech IB (1994) An atomic force microscopy study of the biodegradation of stainless steel in the presence of bacterial biofilms. Intl Biodeterioration Biodegradation 34:35-46

Stetter KO (1995) Microbial life in hyperthermal environments. ASM News 61:285-290

Stetter KO (1996a) Hyperthermophilc procaryotes. FEMS Microbiol Revs 18:149-158

Stetter KO (1996b) Hyperthermophiles in the history of life. In: Evolution of Hydrothermal Ecosystems on Earth (and Mars?). Bock GR, Goode JA (eds) p 1-10. John Wiley & Sons, Chichester, UK

Stetter KO, Huber R, Bloechl E, Kurr M, Eden RD, Fiedler M, Cash H, Vance I (1993) Hyperthermophilic Archaea are thriving in deep North Sea and Alaskan oil reservoirs. Nature 365:743-745

Stevens TO, McKinley JP (1995) Lithoautotrophic microbial ecosystems in deep basalt aquifers. Science 270:450-454

Stroes-Gascoyne S, Pederson K, Daumas S, Hamon CJ, Haveman TL, Ekendahl S, Jahromi N, Arlinger J, Hallbeck L, Dekeyser K (1995) Microbial analysis of the buffer/container experiment at AECL's underground research laboratory. Man. Rep. 11436, COG-95-446. Whiteshell Laboratories, Atomic Energy Canada Limited, Pinawa

Surman SB, Walker JT, Goddard DT, Morton LHG, Keevil CW, Weaver W, Skinner A, Hanson K, Caldwell D, Kurtz J (1996) Comparison of microscope techniques for the examination of biofilms. J Microbiol Meth 25:57-70

Suzuki MT, Rappe MS, Haimberger ZW, Winfield H, Adair N, Strobel J, Giovannoni SJ (1997) Bacterial diversity among small-subunit rRNA gene clones and cellular isolates from the same seawater sample. Appl Environ Microbiol 63:983-989

Suzuki MT, Giovannoni SJ (1996) Bias caused by template annealing in the amplification of mixtures of 16S rRNA genes by PCR. Appl Environ Microbiol 62:625-630

Thiele JH, Zeikus JG (1988) Control of interspecies electron transfer during anaerobic digestion: significance of formate transfer versus hydrogen transfer during syntrophic methanogenesis. Appl Environ Microbiol 54:20-29

Tindall BJ (1992) The family Halobacteriaceae. In: The Prokaryotes. Balows A, Truper HG, Dworkin M, Harder W, Schleifer K-H (eds) p 768-805. Springer-Verlag, New York

Tokura M, Ushida K, Miyazaki K, Kojima Y (1997) Methanogens associated with rumen ciliates. FEMS Microbiol Ecol 22:137-143

Tower KJ (1992) The genus *Acinetobacter*. In: The Prokaryotes. Balows A, Truper HG, Dworkin M, Harder W, Schleifer K-Z (eds) p 3137-3143. Springer-Verlag, New York

Tsai Y-L, Olson BH (1991) Rapid method for direct extraction of DNA from soil and sediments. Appl Environ Microbiol 57:1070-1074

Tunlid A, White DC (1992) Biochemical analysis of biomass, community structure, nutritional status, and metabolic activity of the microbial communities in soil. In: Soil Biochemistry. Bollag JM, Stotzky G (eds) p 229-262. Marcel Dekker, New York

Ueda T, Suga Y, Matsuguchi T (1995) Molecular phylogenetic analysis of a soil microbial community in a soybean field. Eur J Soil Sci 46:415-421

Van de Peer Y, Neefs J-M, de Rijk P, de Vos P, de Wachter R (1994) About the order of divergence of the major bacterial taxa during evolution. System Appl Microbiol 17:32-38

Vesey G, Narai J, Ashbolt N, Williams K, Veal D (1994) Detection of specific microorganisms in environmental samples using flow cytometry. Meths Cell Biol 42:489-522

Wachtershauser G (1990) Evolution of the first metabolic cycles. Proc Natl Acad Sci USA 87:200-204

Wachtershauser G (1994) Vitalysts and virulysts: A theory of self-expanding reproduction. In: Early Life on Earth, Bengtson S (ed) p 124-132. Columbia Univ Press, New York

Wagener S, Bardele CF, Pfennig N (1990) Functional integration of *Methanobacterium formicicum* into the anaerobic ciliate *Trimyema compressum*. Arch Microbiol 153:496-501

Wagner M, Amann R, Lemmre H, Schleifer K-H (1993) Probing activated sludge with oligonucleotides specific for *Proteobacteria*: inadequacy of culture-dependent methods for describing microbial community structure. Appl Environ Microbiol 59:1520-1525

Wagner M, Assmus B, Hartmann A, Hutzler P, Amann R (1994) *In situ* analysis of microbial consortia in activated sludge using fluorescently labeled rRNA-targeted oligonucleotide probes and scanning confocal laser microscopy. J Microsc 176:181-187

Wainright PO, Hinkle G, Sogin ML, Stickel SK (1993) The monophyletic origins of the Metazoa; an unexpected evolutionary link with Fungi. Science 260:340-343

Wallner G, Amann R, Beisker W (1993) Optimizing fluorescent *in situ* hybridization with rRNA-targeted oligonucleotide probes for flow cytometric identification of microorganisms. Cytometry 14:136-143

Ward DM, Bateson MM, Weller R, Ruff-Roberts AL (1992) Ribosomal RNA analysis of microorganisms as they occur in nature. Adv Microb Ecol 12:219-286

Ward DM, Weller R, Bateson MM (1990) 16S rRNA sequences reveal numerous uncultured microorganisms in a natural community. Nature 345:63-65

Ward N, Rainey FA, Stackebrandt E, Schlesner H (1995) Unraveling the extent of diversity within the order *Planctomycetales*. Appl Environ Microbiol 61:2270-2275

Weisburg WG, Barns SM, Pelletier DA, Lane DJ (1991) 16S ribosomal DNA amplification for phylogenetic study. J Bacteriol 173(2):697-703

White DC (1988) Validation of quantitative analysis for microbial biomass, community structure and metabolic activity. Adv Limnol 31:1

White DC (1994) Is there anything else you need to understand about the Microbiota that cannot be derived from analysis of nucleic acids? Microbial Ecol 28:163-166

White DC, Davis WM, Nickels JS, King JD, Bobbie RJ (1979) Determination of the sedimentary microbial biomass by extractable lipid phosphate. Oecologia 40:51-62

White DC, Ringelberg DB, Hedrick DB, Nivens DE (1994) Chapter 2. In: Rapid identification of microbes from clinical and environmental matrices. p 8-17. American Chemical Society

Whiteley AS, O'Donnell AG, MacNaughton SJ, Barber MR (1996) Cytochemical colocalization and quantitation of phenotypic and genotypic characteristics in individual bacterial cells. Appl Environ Microbiol 62:1873-1879

Whitman WB, Bowen TL, Boone DR (1991) The methanogenic bacteria. In: The Prokaryotes. Balows A, Truper HG, Dworkin M, Harder W, Schleifer KJ (eds) p 719-767. Springer Verlag, New York

Wilson KH, Blitchington RB, Greene RC (1990) Amplification of bacterial 16S ribosomal DNA with polymerase chain reaction. J Clin Microbiol 28:1942-1946

Wirsen CO, Jannasch HW (1074) Microbial transformation of some ^{14}C-labeled substrates in coastal water and sediment. Microbial Ecol 1:25-37

Woese CR (1987) Bacterial evolution. Microbiol Rev 51:221-271

Woese CR (1993) The Archaea: Their history and significance. In: The biochemistry of Archaea (archaebacteria). Kates M, Kushner DJ, Matheson AT (eds). Elsevier Science Publishers, Amsterdam

Woese CR, Fox GE (1977) Phylogenetic structure of the prokaryotic domain: the primary kingdoms. Proc Natl Acad Sci USA 74:5088-5090

Woese CR, Kandler O, Wheelis ML (1990) Towards a natural system of organisms: Proposal for the domains Archaea, Bacteria, and Eucarya. Proc Natl Acad Sci USA 87:4576-4579

Wunsche L, Babel W (1996) The suitability of the Biolog Automated Microbial Identification System for assessing the taxonomical composition of terrestrial bacterial communities. Microbiol Res 151:133-143

Yang D, Oyaizu H, Oyaizu H, Olsen GJ, Woese CR (1985) Mitochondrial origins. Proc Natl Acad Sci USA 82:4443-4447

Zablen LB, Kissel MS, Woese CR, Beutow DE (1975) Phylogenetic origin of the chloroplast and the procaryotic nature of its ribosomal RNA. Proc Natl Acad Sci USA 72:2418-2422

Zheng D, Alm EW, Stahl DA, Raskin L (1996) Characterization of universal small-subunit rRNA hybridization probes for quantitative molecular microbial ecology studies. Appl Environ Microbiol 62:4504-4513

Zhou J, Bruns MA, Tiedje JM (1996) DNA recovery from soils of diverse composition. Appl Environ Microbiol 62:316-322

Zillig W (1991) Comparative biochemistry of Archaea and Bacteria, Curr Opin Genet Develop 1:544-551

Zuckerkandl E, Pauling L (1965) Molecules as documents of evolutionary history. J Theoret Biol 8:357-366

Chapter 3

PROCESSES AT MINERALS AND SURFACES WITH RELEVANCE TO MICROORGANISMS AND PREBIOTIC SYNTHESIS

Jillian F. Banfield

Mineralogical Institute
Graduate School of Science
University of Tokyo, Tokyo, Japan

Permanent address:

Department of Geology and Geophysics
University of Wisconsin-Madison
Madison, Wisconsin 53706 U.S.A.

Robert J. Hamers

Department of Chemistry
University of Wisconsin-Madison
Madison, Wisconsin 53706 U.S.A.

INTRODUCTION

Minerals are the solids that comprise rocks. They are the substrates that microorganisms attach to, the origin of many dissolved constituents essential for metabolism, and in the case of lithotrophs, the ultimate source of their energy. In this chapter we review the nature and distribution of minerals and mineral surfaces, their compositions, defect microstructures, and reactivity. We also consider the possibility that reactions at mineral surfaces played a key role in prebiotic synthesis and the origin of life.

Many nutrients that sustain life are derived from minerals and redox reactions at mineral surfaces provide metabolic energy. For these and other reasons, microbial cells attach to and interact with atoms derived from or associated with mineral surfaces. The purpose of this chapter is to provide some insight into the bulk and surface structure and chemistry of minerals and to discuss the processes that occur at these interfaces. We review the basic concepts and reference more extensive treatments of crystal chemical and surface chemical behavior.

This chapter begins with a brief overview of the nature of solids accessible to microorganisms. We summarize the variables that control the structure, composition, and properties of minerals formed at or near the Earth's surface. In addition to noting the well understood physical and chemical variables, we draw attention to factors relevant to biologically produced minerals that may affect the phase formed, its morphology, and reactivity. The consequences of small particle size are also discussed. These are important because the vast majority of crystals produced by biologically-induced or biologically-directed mechanisms are extremely fine grained.

After reviewing some simple principles behind assembly of mineral structures, we examine the fundamental characteristics of surfaces. Surfaces are not static but are modified in response to their environment. Consequently, we consider the ways in which surfaces

0275-0279/97/0035-0003$05.00

and the solutions in proximity to them change in response to pH and other factors. The types of reactions that take place at mineral surfaces are discussed, with implications for the physical and chemical environments of microorganisms. We provide examples that relate the structures, microstructures, and compositions of typical minerals to their reactivity.

Reactions at surfaces are affected by the presence of organic molecules that complex with ions in solution and bind to surface sites. Similarly, surfaces affect the structure and composition of adsorbed organic molecules and may catalyze their reactions. This is central to the concept that mineral surfaces played a key role in prebiotic synthesis. Consequently, we briefly review current ideas on this topic.

MICROBIAL ACCESS TO MINERAL SURFACES

Geologists have traditionally grouped rocks into several categories: *Igneous* rocks, which consist predominantly of silicate minerals that crystallize from a melt, *sediments and sedimentary* rocks, which consist of chemically or biologically precipitated phases, minerals resistant to chemical weathering, and the products of chemical and physical weathering, and *metamorphic rocks*, which are rocks of sedimentary or igneous origin whose mineralogy and texture have been changed in response to temperatures, pressures, and solutions encountered at depth in the crust or in the upper mantle.

In general, igneous rocks crystallized at depth in the crust have low porosities so that the surfaces of minerals are only exposed to microbial processes where the rocks are fractured or where porosity has developed as a consequence of physical and chemical weathering. Volcanic igneous rocks are cooled more rapidly and so develop cracks, gas vesicles, and other features that facilitate both weathering and microbial colonization.

The physical characteristics of metamorphic and sedimentary rocks are more diverse, reflecting the wide range of conditions under which they can form and the materials they contain. As shown in Table 1, porosities of rocks and sediments near the Earth's surface vary dramatically, as will the chemistry of pore fluids in intimate contact with the minerals that comprise them. This variable porosity (and pore size, which is not directly shown in Table 1 but is discussed briefly below) will limit microbial colonization. Similarly the connectivity of pores in different rock types varies dramatically, as can be inferred from tabulated values of hydraulic conductivity. Hydraulic conductivity, a direct measure of the rate at which a fluid moves through a rocks or sediment, provides some indication of the expected variability in transport rates for cells as well as nutrients and other materials.

The two most basic and easily measurable geometrical parameters describing microgeometry of pore space are porosity (π), which is the ratio of void to bulk solid volume, and specific surface area (S_a), defined as the pore surface area per unit solid volume, with units of length^2/length^3. Pore structures in low-porosity crystalline rock are fairly distinct from those in porous sedimentary rock. In crystalline rock (π of ~0% to 3%),

Table 1. Porosities of common materials near the Earth's surface.

From Tables B1 and B3 of Spitz and Moreno (1996).

Material	Porosity in (l)	Hydraulic conductivity (m/s)
Granite	~0.02 - 0.002	~4 - 24 $\times$ 10^{-13}
Basalt	0.03 - 0.05	~10^{-7} - 10^{-11}
Sandstone	0.05 - 0.48	~10^{-9} - 10^{-5}
Granite weathered	0.34 - 0.57	~1 - 5 $\times$ 10^{-5}
Gravel	0.2 - 0.4	~10^{-2} - 10^{-4}

porosity often consists of sheet-like cracks and microcracks (fairly planar cracks with a high aspect ratio). For these rock types, although the overall porosity is very low, the specific surface area can be fairly high. For example, for Westerly granite with $\pi \sim 1\%$ to 2%, the S_a is ~9 mm^2/mm^3 and associated crack apertures range from < 0.05 μm to several microns (Fredrich and Wong 1986, Wong et al. 1989). In contrast to most granites and related igneous rocks, pore shapes in sedimentary rock such as sandstone are highly varied. Although three dimensional imaging technologies reveal a complete spectrum of pore shapes (e.g. Fredrich et al. 1995), the pore space is often idealized as spherical nodal pores (pores at grain vertices), sheet-like cracks (or microcracks along boundaries between two grains), and tube-like pores emanating from nodal pores and running along 3-grain junctures (Fredrich et al. 1995). Given the difference in pore shapes for crystalline versus sedimentary rocks, there is typically different scaling between porosity and specific surface area. For example, for Fontainebleau sandstone with highly varied porosity, S_a is constant and equal to about 20 mm^2/mm^3 over π ranging from 10 to 20% (Fredrich et al. 1993). Thus, the Fontainebleau sandstone has 10 to 20 times the porosity of Westerly granite but only twice the S_a. For the Fontainebleau, mean pore diameters vary from 25 to 40 μm as porosity increases from 10 to 20% (but range from sub-micron to hundreds of microns). Likewise, for the Berea sandstone with a porosity of ~20%, S_a is only about 20 to 25 mm^2/mm^3 (Fredrich, written comm. 1997). Interestingly, hydraulic conductivity in sandstones such as these is controlled by the sheet-like cracks and their connectivity, rather than by the mean size of the nodal pores (Fredrich, written comm. 1997).

The space available for microbial colonization and transport is only one factor controlling the distribution of microbes in the subsurface. Other factors include the mineralogy of the rocks, the chemistry of the solution (especially ionic strength), and the nature of bacterial cell surfaces (size, shape, hydophobicity, and charge). Detailed discussion of microbial transport is outside of the scope of this chapter. An excellent review is provided by Lawrence and Hendry (1994).

FACTORS THAT CONTROL MINERALOGY

Rocks exposed at the surface of the Earth consist of assemblages of phases formed under a diverse range of physical conditions in the crust (~35 km thick in continents, ~7 km thick under the oceans) and mantle. Mineralogy is determined by a number of factors, including the bulk chemical composition of the rocks. Table 2 summarizes the average composition of the crust as well as those of the atmosphere, oceans, and rivers.

In addition to bulk composition, there are a variety of factors that dictate the structure and chemistry of solid phases that make up rocks. Among the most important are temperature, pressure, redox potential (Eh, where $Eh = pE \times 0.059$ V at 25°C and 1 atm, e.g. see Brownlow 1979), and the abundance and composition of associated solutions. The assemblage of phases in rocks crystallized at high temperature can be successfully predicted based on measured and estimated thermodynamic quantities and knowledge of the physical and chemical characteristics of the environment. The details of mineral assemblages and conditions associated with formation of rocks are topics of active research in metamorphic, igneous, and sedimentary petrology.

The mineralogical assemblages established at depth in the crust or upper mantle are complicated by chemical reactions that occur when rocks are brought to the Earth's surface by tectonic (mountain building) activity. Once exposed to air, water, and microorganisms at or near the Earth's surface, primary minerals become unstable and are partially or completely replaced by secondary minerals. Consequently, rocks in environments colonized by microorganisms are often partially weathered so that exposed mineral surfaces are typically in intimate contact with clays and other reaction products.

Table 2. A subset of elemental abundances in the atmosphere (ppm from Richards 1965), oceans (ppm, from Turekian 1968), rivers (ppm from Livingston 1963), and continental crust (ppm and wt %; from Mason and Moore 1982).

Element	Atmosphere	Ocean water	River water	Continental crust
C		30		200
CO_2	330			
HCO_3^-		58.4		
CH_4	2			
N_2	780,840	16		
NO_3^{2-}			1.0	
O_2	209,460			46.60%
F		1.3	---	625
Na		10,800	6.3	2.83%
Mg		1,290	4.1	2.09%
Al		0.001	---	8.13%
Si		2.9	13.1	27.72%
P		0.088	---	0.10%
S		904	11.2	260
Cl		19,400	7.8	130
K		392	2.3	2.59%
Ca		411	15	3.63%
Ti		0.001	---	0.44%
Mn		0.002	---	0.09%
Fe		0.003	0.7	5.00%

The process by which primary minerals are dissolved and converted to secondary minerals (such as clays, see below) is known as chemical weathering. This acts in concert with physical degradative processes to increase the exposed surface area, providing dissolved constituents for microbial metabolism and space for colonization. A subset of elements are retained in weathering products. Elements not accommodated into these low temperature silicate and oxide minerals concentrate in sea water (e.g. Na, Cl; see Table 2).

The observed phase assemblage may differ from that predicted based on thermodynamic data due to a variety of kinetic factors (though the kinetic factors themselves are dependent on structural attributes). High activation barriers associated with development of the stable phase and rapid crystallization may result in complex metastable assemblages. The size to which a nucleus must grow before it becomes stable (critical nucleus) depends in part on the surface energy of the phase (discussed below). Less dense phases (such as amorphous materials) tend to have lower surface energies so they precipitate faster. For example, amorphous silica produced at low temperature is commonly opal-A. At ~40°C during sedimentary diagenesis (burial accompanied by mineralogical and physical changes driven by dewatering and increased temperature and pressure), this is transformed to another metastable phase (opal-CT). The stable phase only develops when temperatures become high enough to overcome activation barriers needed to crystallize the α-quartz polymorph (see below).

There are a number of factors that are less commonly considered but that can dramatically influence mineral formation, resulting in development of what would normally be considered metastable phases. The most widely recognized but perhaps not fully understood process is mediation of crystallization by organic molecules. Crystallization directed by organisms can result in formation of phases such as aragonite ($CaCO_3$) under pressures at least 2.5 kbar lower than expected, based on inorganic phase equilibria. In some cases, this may be attributed to solution chemical effects (see de Vrind-de Jong and deVrind, this volume). However, where aragonite is precipitated by macroorganisms such as mollusks, compelling evidence indicates that control of nucleation by proteins is responsible (Falini et al. 1996, Belcher et al. 1996).

A common characteristic of biologically-mediated mineralization is that the phases frequently contain highly metastable ratios of both major cations and trace constituents. The crystallization of carbonates on the surfaces of bacteria, for example, has been suggested to explain the development of dolomite under conditions where it is extremely difficult to produce inorganically (Vasconcelos et al. 1995). Similarly, the incorporation of abnormally large concentrations of elements such as Sr into calcite has been noted in several studies. This is discussed further by de Vrind-de Jong and de Vrind (this volume).

A second group of factors suggested to influence the structure of a mineral formed under low-temperature conditions are related to particle size. Both inorganic and organically-mediated crystallization can result in particles that are sufficiently small that these effects become important (few to few tens of nanometers wide). Such crystals can also be synthesized in the laboratory (e.g. by sol gel methods, see Fig. 1) and are widespread products of inorganic (Fig. 2) and organic (e.g. Bazylinski and Moscowitz, this volume) reactions.

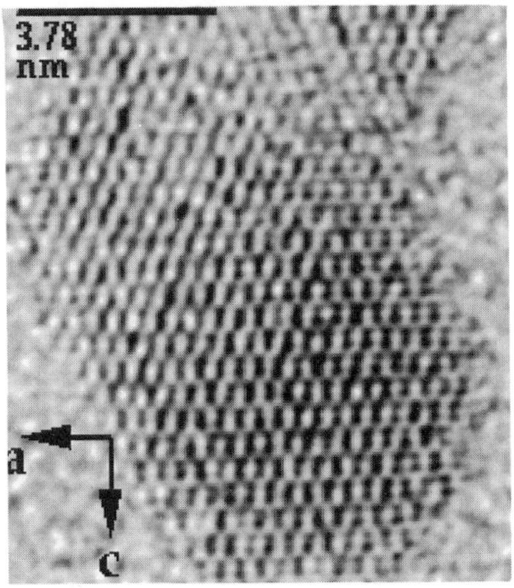

Figure 1. Atomic-resolution transmission electron micrograph down ⟨100⟩ of an anatase nanocrystal synthesized by the sol-gel method. Pairs of dark spots represent paired columns of Ti atoms. Micrograph by R. Lee Penn.

The influence of particle size on phase stability has rarely been considered, in part because it is not relevant at higher temperatures where coarsening kinetics are rapid. Langmuir (1971) postulated that the stability of hematite and goethite are reversed once particle size becomes small. This concept also has been explored for other oxides (e.g. Banfield et al. 1993, Gribb and Banfield 1997, Zhang and Banfield, in review). Gribb and Banfield (1997) present experimental evidence and calculations that support the hypothesis that in cases where the structure energies of two polymorphs are similar, surface energy differences can be significant enough to reverse stabilities. This has been analyzed quantitatively by Zhang and Banfield (in review). Similar considerations may be important in other systems where nucleation of what are normally considered metastable phases is observed.

An important (but not yet fully explored) corollary is the particle size dependence of surface tension and surface energy (see below). The 'surface tension' is defined thermodynamically as the reversible work done in creating new unit surface area from the bulk phase. It is often assumed that surface tension is approximately constant, regardless of

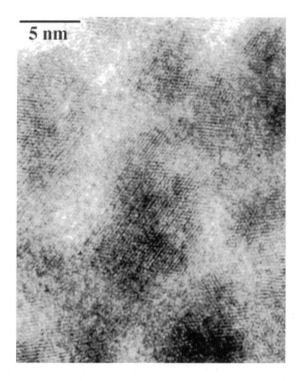

Figure 2. (a) High-resolution transmission electron micrograph of hematite (Fe_2O_3) with typical grain sizes <10 nm in diameter. This material is an important constituent of the coating on quartz sand grains in the USGS Toxic Substances-Hydrology Research site, Cape Cod, Massachusetts. Significant Al and Si, and smaller amounts of P, K, Ca, Ti, Cr, Pb, and As are associated with this material. Micrograph by Dawn Janney.

particle size. However, surface tension should be size-dependent. For a crystal of nanometer-scale dimensions or a fluid droplet, surface tension increases pressure in the interior and drives interatomic spacings away from their equilibrium values, increasing the energy of atoms in the bulk. In part because surface tension is a measure of the excess energy of the surface relative to the bulk, the surface tension decreases when particles become very small. Work by Tolman (1949) for liquids indicated that surface tension decreases by up to 60% as liquid droplets approach nanometer dimensions. Zhang and Banfield (in review) estimate a decrease of ~40% in surface tension for very small crystalline particles.

The surface tension (γ) and surface strain (ε) are related to the surface stress (g, where stress is force per unit length, strain is fractional elongation of a material in response to an applied stress). If, for simplicity, we neglect lattice strain and assume $g = \gamma$ (true only for a liquid), the variation in interatomic spacing with surface tension at a certain particle size can be estimated using the equation:

$$3(\Delta d)/d = -2\ K\gamma/r$$

where d = interatomic distance, K = compressibility, and r = radius of the particle. For some materials, this effect is sufficiently significant that change in interplanar spacings can be measured directly using X-ray diffraction.

Changes in surface tension should affect the structure of the solution at the particle/solution interface (the double layer, see below) and thus, adsorption of ions and molecules. Zhang (pers. comm. 1997) suggests that the decrease in surface tension due to reduced particle size should cause the potential of the double layer to change significantly.

This change may increase or decrease the potential (depending upon the sign of the surface charge), affecting the tendency of ions or molecules to adsorb. This affect has not yet been considered in detail, but may be of importance to theoretical analysis of processes occurring at surfaces during formation of nanometer-scale minerals, as well as to interpretation of results of experiments involving these small particles.

In addition to phase stability, particle size is well known to affect reactivity. The smaller the particles, the more prone they are to dissolution. This is explained simply because surfaces are a source of excess energy and small particles have high surface area (for example, see discussion by Dove 1995 for quartz). This effect becomes especially important for particles with diameters smaller than ~0.2 μm. Surface free energy also drives coarsening reactions and explains why many biologically produced particles are not preserved in their original form in the geologic record.

A less well understood phenomena is the connection between particle size and phase transformation rates. This has been examined for a few synthetic materials in a number of studies, all of which indicate that the rate of polymorphic phase transformation of a particle is dramatically increased when particles are small. Gribb and Banfield (1997) considered a number of possible explanations for this phenomena in titania and concluded that the increased number of surface nucleation sites is most likely to be responsible. This work is not of direct relevance to biominerals but it serves to remind us that where a reaction occurs by surface nucleation and rate is limited by nucleation (and not by the rate of growth of the nuclei), the large number of nucleation sites provided by a very small crystal may greatly accelerate reaction kinetics. This may be of significance to a subset of biomineral forming reactions and to studies concerned with the fate of biominerals (such as during diagenesis).

PRINCIPLES BEHIND ASSEMBLY OF CRYSTALS

Linus Pauling provided us with a simple set of rules which offers considerable insight into the factors that control the basic arrangement of ions in crystals (see, for example, Klein and Hurlbut 1994). These rules emphasize the importance of the relative sizes of cations (see Stone, this volume) and anions in determining the coordination number (number of cations surrounding an anion and visa versa) and the requirement for local charge balance. Approximately 94% of the volume of the Earth's crust is composed of oxygen (Mason and Moore 1982), so this is by far the dominant anion in minerals at and near the Earth's surface. Although simplistic, the general sized-based rule correctly predicts that in minerals in which oxygen is the dominant anion, Si is in 4-fold coordination (tetrahedral sites; see below), Al is in either 4- or 6-coordinated (tetrahedral or octahedral) sites, Fe, Mg, Mn, Ti favor octahedral sites, Ca and Na prefer 6- to 8-coordinated sites, while elements such as K are accommodated by larger sites. Additional details about the structure of various groups of silicate minerals are provided below. More comprehensive treatments of the detailed nature of bonds and sites in minerals and alternative views of structural arrangements can be found in the literature (e.g. Brown and Shannon 1973, Brown and Altermatt 1985, Smyth and Bish 1988, Hyde and Andersson 1989, Marfunin 1994).

Because silicon and aluminum are the dominant cations in the crust (Table 2), the most common minerals at and near the Earth's surface are aluminosilicates (however, phosphates, sulfides, sulfates, carbonates, and oxides account for large volumes, especially in chemically and biologically precipitated sediments). When coordinated by four oxygen atoms, the 4+ charge of Si is compensated and the resulting polyhedron adopts the shape of a triangular prism (a tetrahedra). Remaining charge on the oxygen atoms at the corners of the tetrahedra can be satisfied by sharing of each oxygen with an adjacent Si in tetrahedral coordination (oxygens shared between two tetrahedra are termed "bridging oxygens"). A range of fully charge balanced structures can be achieved by linking of Si tetrahedra so that all oxygen atoms are shared between a pair of adjacent tetrahedra (SiO_2). The process of linking tetrahedra through bridging oxygens is known as polymerization. Alternatively, tetrahedra may be unpolymerized. In this case, charge on the tetrahedral oxygens can be compensated by cations in other sites, or by H^+ (forming OH). A range of structures exist between these extremes. In silicates such as olivine, all of the excess charge on O associated with Si tetrahedra is balanced by Mg or Fe in octahedral sites (Figs. 3a,b). The majority of common silicates and aluminosilicates contain partially polymerized tetrahedrally coordinated Al and Si. Through partial polymerization it is possible to form minerals containing chains: pyroxenes, in which the silicate anion can be designated SiO_3^{2-} (Figs. 3c,d), double chains: amphiboles, which contain a structural unit based on $Si_4O_{11}^{6-}$; sheets: such as found in micas, chlorite, and kaolinite, based on $Si_2O_5^{2-}$ (Fig. 4); and frameworks of tetrahedra based on, or derived from, SiO_2, such as found in quartz and feldspar (Fig. 5). The surface structure and reactivity of each mineral is controlled largely by the nature of the polyhedra and the linkages between them (see below).

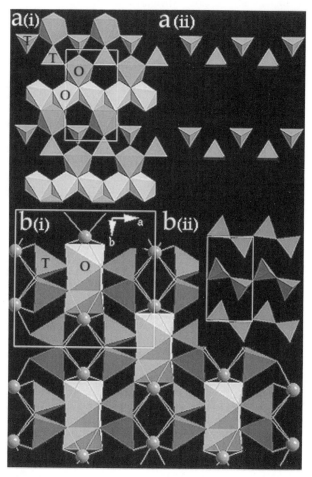

Figure 3. Diagrams illustrating details of the crystal structures of olivine: a(i) viewed down [100] showing the connectivity of tetrahedral (T) and octahedral sites (pale and darker gray octahedra (O) are symmetrically distinct); a(ii) showing only the tetrahedrally coordinated part of the structure; pyroxene: b(i) showing the arrangement of tetrahedral (T) and octahedral (O) cations into "I-beams". Larger cations can substitute into the 6- to 8-coordinated octahedral sites (shown as balls); b(ii) viewed down [100], it can be seen that tetrahedra are arranged into chains.

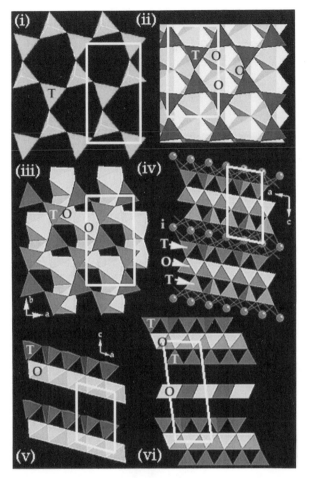

Figure 4. Diagrams illustrating aspects of the crystal structures of layer silicates. Viewed normal to the cleavage plane (i) the arrangement of tetrahedra (T) into a sheet is evident; the tetrahedral sheet can be connected to: (ii) an octahedral (O) sheet in which all sites are occupied (i.e. Mg_3); (iii) an octahedral sheet in which 2/3 of the sites are occupied (i.e. Al_2). Viewed down [010] showing the stacking of: (iv) TOT layers separated by interlayer cations 'i', as found in micas; (v) TO layers, as in kaolinite; (vi) TOT layers separated by a brucite-like interlayer (of octahedral sites), as found in chlorite.

SURFACES AND INTERFACES

Crystal form

The chemical reactivity of a mineral is virtually always determined by the properties of the solid-liquid interface. The atomic structure and composition of a surface will be affected by the crystal structure of the mineral and the orientation of the surface. Atoms at a surface or interface always have higher energy than atoms in a bulk three-dimensional solid. The higher energy of atoms at surfaces is due to the fact that these atoms have lower coordination and strongly asymmetric bonding configurations compared with atoms in a bulk solid. The total energy (G) of a three-dimensional solid can be described as: $G = EV + \gamma A$, where E is the bulk cohesive energy per unit volume, γ is the surface tension, V is the volume of the solid, and A is its surface area. Because the volume scales with the cube of the number of atoms and the area scales only with the square of the number of atoms, the first term is usually much larger than the second, and the total energy is proportional to its volume. However, the area-dependent term, surface free energy, defines the shape of the solid.

In a liquid, the surface free energy is equivalent to the surface tension. Because the surface free energy (surface tension) is always a positive number (higher energy), liquids

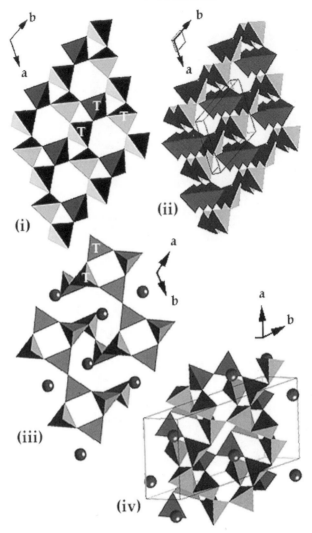

Figure 5. Diagrams illustrating aspects of the structures of framework silicates. (i) arrangement of tetrahedra (T) in α-quartz (viewed down [001]); (ii) α-quartz rotated slightly from (i) so that the three-dimensional connectivity of tetrahedra (T) is apparent; (iii) feldspar illustrating the substitution of cations (balls) into cavities in the structure to charge balance Al substitution for Si in the tetrahedral sites; (iv) feldspar rotated slightly from (iii) to emphasize the three-dimensional connectivity of tetrahedra.

and solids adopt a shape that tends to minimize the surface free energy (Herring 1951). In the case of liquid, that is achieved by forming spherical drops. In the case of solid, the situation is more complicated because of the regular arrangement of atoms. Surfaces passing through a solid in different orientations consist of different arrangements of atoms; each of these arrangements will have its own energy. Thus, the free energy of a solid is dependent on the crystallographic orientations of its surfaces. If a crystal has one particular crystal plane that has much lower energy than other crystal planes, the crystals will have very well-defined form. In contrast, minerals with more than one low energy crystal face may adopt a greater variety of forms.

Morphology modification

Adsorbates that modify the energies of different crystal planes or different reactive sites (steps, etc.) can have very strong influences on crystal form. Various molecules present in solution can modify the surface energy and thereby alter both the thermodynamics and the kinetics or surface processes. A molecule which locally bonds to a particularly reactive site, for example, can strongly decrease ("poison") the growth or dissolution rate. Adsorbed molecules can also block access of ions in solution to reactive sites. Stabilization of unusual surfaces leads to expression of forms not normally encountered in abiotically produced minerals (Berman et al. 1988, Mann et al. 1990, and review by Mann et al. 1993). Urea, for example, is well known to modify the form of various crystals grown in aqueous solution due to specific bonding of the urea molecule to certain crystal faces. Examples of diverse magnetite (see below) morphologies resulting from growth within an organic envelope are given by Bazylinkski and Moscowitz (this volume).

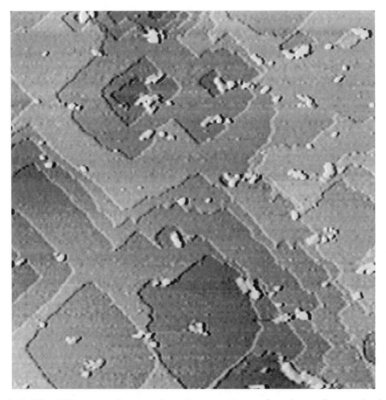

Figure 6. A 300 × 300 nm scanning tunneling microscope image of a galena surface reacting in 0.5 M Na₂SO₄ at pH = 2.7 solution (pre-purged using Ar). The tip potential was -0.15 V vs. NHE, sample potential = +0.10 V vs. NHE with a total tip-sample bias of -0.25 V. The tunneling current was 500 pA. Note the <110> surface steps. Small objects on the surface may be a composed of sulfur deposited during reaction. Image by Steven Higgins.

Defects and surface reactivity

In addition to particular crystal planes having different energies, different locations on an exposed surface will also have different energies. For example, galena surfaces are not perfectly flat, but instead have steps and other imperfections (Fig. 6). Atoms at these step

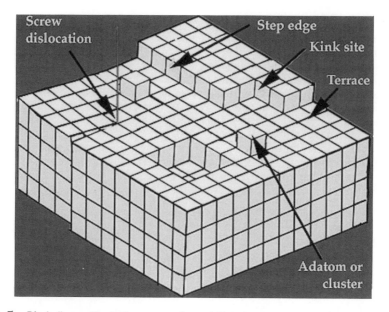

Figure 7. Block diagram illustrating terrace, edge, and kink sites and adatom (or small cluster) on a mineral surface. A screw dislocation is also shown to emphasize the importance of defects present in the bulk material for surface reactivity.

edges have even lower coordination than those on the exposed crystal planes, and therefore have even higher energy. Kink sites, where step edges turn, are even more reactive sites for reactions. Some typical surface sites are illustrated in Figure 7. Chemical kinetics tends to favor specific surface sites for many chemical reactions and dissolution processes. For example, Higgins and Hamers have shown using scanning tunneling microscopy that the chemical dissolution of galena is dominated by removal of atoms from step edges. A movie made from a sequence of images showing this is available on the world-wide web at http://www.chem.wisc.edu/~higgins/index.html.

Various defects exist in three-dimensional solids. These include point (e.g. vacancies), line (e.g. dislocations), and planar defects (e.g. stacking faults, grain boundaries). The intersection between these defects and the surface involves low-coordination atomic sites and/or local lattice strain (deviations from the ideal interatomic distances). Consequently, these areas have higher energy than "perfect" surfaces and are strongly preferred sites of reactivity. Immediately adjacent to an edge dislocation exposed at the surface we expect the atoms to have unusually high reactivity because of their unusually low coordination. However, atoms farther away from the defect might also have unusually high reactivity because of the presence of the strain field, which typically decreases rather slowly with increasing distance from the defect. The effect of lattice strain can be thought of as structural or electronic in origin. Because lattice strain drives the interatomic separations to be different from their minimum-energy values, the electronic structure (occupancy and energy of the energy levels and bands that describe the electronic structure of the solid) is modified. Because the strain field and perturbation of the electronic energy levels can propagate a long distance away from a defect, it can become difficult to determine the effective "size" of a defect in terms of its overall reactivity. In general, however, imperfections can play critical roles and, and in some cases, can dominate the overall reactivity.

CHARGE AND PROTONATION AT SURFACES

Interfaces between minerals and solution are typically characterized by an electric potential and this effects phenomena occurring there. The most important processes at surfaces are ionization of groups of surface-bound atoms and the adsorption of ions. Counter ions (with a charge opposite to that of the surface) are attracted to the surface whereas ions with the same charge will be repelled. The surface charge and associated counter ions are described as the electric double layer, which separates the mineral surface from bulk solution.

Surface charge is largely neutralized by counter ions bound to the surface but some charge neutralization occurs at greater distance, in a diffuse layer. The electrical potential with respect to bulk fluid determined at the plane separating tightly bound counter ions from those in the diffuse layer is referred to as the zeta potential. The zeta potential has a sign, which indicates the sign of charge at the surface, and a magnitude, which indicates the surface charge density. The physical and chemical nature of the double layer region differs significantly from bulk solution (see below). The surface charge is of great importance in determining how dissolved ions and molecules, as well as chemical species associated with organic polymers and microbes, interact with surfaces.

pH dependence of charge at surfaces

The pH at which the surface of a metal oxide has a net zero charge is termed the pH_{ZPC} (Parks 1965 1967). Above this pH the surface has a net negative charge, below it the surface charge is positive. The values of pH_{ZPC} are traditionally determined experimentally, but can be predicted (see below).

Table 3. The range of pH_{ZPC} calculated for a selection of minerals from Sverjensky and Sahai (1996). Column 2 lists the range of pH_{ZPC} values predicted for the constant capacitance, double layer, and triple layer models using log K expressions derived in their work. The end member composition is listed in column 3. The final column indicates the structure type, including the charge on the charge on the tetrahedral silicate anion per tetrahedral site.

Mineral	pH_{ZPC}	Composition	structure / anion
Goethite	8.8 - 9.4	FeOOH	hydroxide
Hematite	8.3 - 8.5	Fe_2O_3	oxide
Forsterite	8.3 - 9.1	Mg_2SiO_4	4.0
Grossular	7.6 - 8.1	$Ca_3Al_2Si_3O_{12}$	4.0
Enstatite	6.9 - 7.4	$Mg_2Si_2O_6$	2.0
Diopside	6.9 - 7.3	$CaMgSi_2O_6$	2.0
Hedenbergite	5.9 - 6.2	$CaFeSi_2O_6$	2.0
Tremolite	6.7 - 7.0	$Ca_2Mg_5Si_8O_{22}(OH)2$	1.50
Anthophyllite	6.4 - 6.6	$Mg_7Si_8O_{22}(OH)2$	1.50
Phlogopite	7.5 - 8.0	$KMg_3(Si_3Al)O_{10}(OH,F)_2$	1.25
Muscovite	6.3 - 6.6	$KAl_2(Si_3Al)O_{10}(OH,F)_2$	1.25
Talc	6.5 - 7.0	$Mg_3Si_4O_{10}(OH,F)_2$	1.0
Kaolinite	4.7 - 5.1	$Al_2Si_2O_5(OH)_4$	1.0
Anorthite	5.5 - 5.6	$CaAl_2Si_2O_8$	0.5
Microcline*	5.7 - 6.1	$KAlSi_3O_8$	0.25
Low albite	5.2	$NaAlSi_3O_8$	0.25
Quartz	2.9 - 3.0	SiO_2	0

* a specific structural state of K-feldspar

The pH_{ZPC} for various oxide and silicate minerals are listed in Table 3 (above). There is a very general trend of increasing pH_{ZPC} with decreasing polymerization of the silicate anion or charge on the silicate anion. The tabulation clearly demonstrates that surfaces of different minerals exposed to a single solution may have very different charges and thus, will interact differently with respect to adsorption of ionic and organic species.

Protonation of surfaces

Virtually all mineral surfaces at and near the Earth's surface have at least a monolayer of adsorbed water and many are in contact with abundant water. Surface ions, commonly oxygen, undergo protonation and deprotonation reactions. Protonation of surface species can be studied in a number of ways, including titration (e.g. James and Parks 1982) and by infrared spectroscopy of surface hydroxyl groups (e.g. Boehm and Knozinger 1983, Zeltner and Anderson 1988). Protonation reactions are generally described by equilibria between surface species containing a surface site (>S) and oxygen. Multiple acid-base reactions that actually take place on a surface. Examples include:

$$>SOH_2^+ \Leftrightarrow >SOH + H^+$$

$$>SOH_2^+ \Leftrightarrow >SO- + 2H+$$

$$>SOH_2^+ + SO^- \Leftrightarrow >SOH + >SOH$$

$$>SOH \Leftrightarrow >SO- + H+$$

There have been a number of models constructed to explain protonation reactions (e.g. the double (Dzombak and Morel 1990) and triple layer models (Davis et al. 1978, Hayes and Leckie 1987). James and Healy (1972) treated the mineral-water interface as three regions: bulk solution, an interface region, and the crystal surface and modeled protonation using the Born equation. This accounts for work done to move the charge between three regions with different dielectric constants. Note that one consequence of moving an ion into the interface region is that a fraction of the waters of solvation may be lost.

Recently, Sverjensky and Sahai (1996) presented a method for theoretical prediction of surface protonation equilibria for use in these models. Their approach predicts log K for the vth surface protonation from the Pauling bond strength for the *mineral* and the dielectric constant for the *mineral*. The dielectric constant is a measure of how much the electrons redistribute spatially in response to an electric field. For example, at atom will have its initially spherically-symmetric electron distribution distorted, with the net result being formation of an induced dipole (Δcharge/distance). The dipole moment per unit volume is the polarization, which is directly related to the dielectric constant. The Sverjensky and Sahai (1996) model and takes into consideration the Born solvation theory for the adsorbing H^+, electrostatic interactions between the proton, the oxygen ion, and the underlying cation, and the binding interactions between the proton and the surface. The conclusion that surface protonation equilibria can be predicted solely from properties of the bulk solid emphasizes the importance of mineral structural attributes (Sverjensky and Sahai 1996). To some extent, this control can be inferred from the fairly straight forward (but imperfect due to variations in the octahedral site occupancy, etc.) correlation between silicate structure type and pH_{ZPC} (Table 2).

Water molecules adjacent to surfaces may be very ordered and the effect can be highly dependent on pH. For example, Du et al., (1994) suggested that water molecules adjacent to quartz surfaces adopt ice-like arrangements at low and high pH and that the configuration of water molecules is variable and disordered at intermediate pH. This observation reinforces the conclusion that the nature of the interface between a mineral and solution is highly pH dependent and, as demonstrated by many previous workers, water close to the surface differs from bulk water significantly in its physical and chemical properties (e.g. dielectric constant (see above), viscosity, entropy, diffusional properties; see discussion in Hochella and Banfield 1995).

REACTIONS AT SURFACES

The interactions between molecules and ions in solution and species that represent the termination of the bulk crystal structure are critical in determining reactivity at solid-liquid interfaces. Two important considerations relating to processes at surfaces are the mechanisms and rates of reactions. The mechanism is the series of steps involved in the process of a reaction. One of these steps will be slower than the others and this will determine the rate (the rate determining step). Although a variety of distinct surface sites may contribute to the reaction, the overall rate is dominated by the most abundant of the most reactive sites. Rates for specific processes can be suppressed by diffusion of reactants and products to and from surfaces and are affected by the characteristics of the solid, composition (pH, saturation state, etc.) of the solution, and temperature.

The rate of a reaction is normally determined experimentally. A rate law is constructed to express the dependence of the rate of a reaction on factors that are known to influence the rate. A rate law is generally in the form:

$$\text{Rate} = k \, [\text{reactant, product, or catalyst}]^{order}$$

where k is the rate constant. Rate laws are generally in the form of differential equations and include explicit time dependence.

The order of the reaction (with respect to a relevant substance that participates in the reaction) is the power to which the concentration of that substance is raised in the equation describing the reaction rate. The order of reaction can be determined experimentally by measuring how the rate of reaction varies as the concentration of one reactant is varied. When the order is zero, the reaction rate is independent of the concentration of that substance. For a first-order reaction, there is an exponential decrease in reaction rate with time. In general, knowledge of the reaction order is not sufficient to determine a reaction mechanism, but it can be used to rule out some mechanisms. For example, if a reaction is second-order in some reactant 'A,' then it is reasonable to assume that the "transition state" involves the simultaneous attachment of two 'A's to the reactive site. Kinetics are generally measured in a "steady-state" regime in which the concentrations of most species taking place in the reaction are not changing rapidly with time.

The rate of a reaction for any particular mechanism is due to the combination of a series of elementary reactions. However, the overall rate of a reaction can be controlled by several parallel mechanisms. If these mechanisms are independent, then the total reaction rate is the sum of the rates occurring via the individual mechanisms. If the reactions involve common reactant or products, the reactions become coupled and the simple addition scheme is no longer valid. Determining the overall rate for a reaction that involves more than one step or more than one concurrent process usually involves solving a set of coupled differential equations, and these can usually not be solved analytically. In most cases, however, the coupling is sufficiently weak that the overall reaction rate can be evaluated as the simple sum of the different pathways.

The unusually high reactivity of steps and other surface defects has implications for macroscopic measurements of chemical reaction rates. For example, it is usually assumed that the rate of reaction of a bulk solid is proportional to its surface area, so that measurements of mineral reaction rates are normalized by the surface area (in units in the form of mol cm^{-2} s^{-1}). However, samples having different densities of surface defects will have correspondingly different rates. For example, under some conditions, galena dissolves by removal of atoms from step edges. Galena samples having different densities of steps will have different reaction rates for the same total surface area. Additionally, such localized reactivity can be give rise to unreasonably small pre-exponential factors in Arrhenius-type models of chemical reactionrates. In the standard form, relationship between the rate constant, k, the activation energy, ΔG^*, and temperature, T, follows the standardArrhenius relationship:

$$k = A \exp{-\Delta G^* / RT}$$

where R is the molar gas constant. The pre-exponential factor, A, represents the rate at which atoms attempt to overcome the barrier. A is normally a product of the number of moles/cm^2 (normally fixed by the crystal structure), the sample area, and a vibrational "attempt frequency" that is approximately the same as the vibrational frequency of the bond being broken (thus, A determines the absolute reaction rate). One effect of having a small number of localized reaction sites such as steps is that the pre-exponential factor in Arrhenius-type models will be too small, because only a small fraction of the number of atoms exposed at the surface is actually involved in the rate-limiting chemical reaction step. Thus, it is clear that the nature and distribution of reactive sites can directly affect measured kinetic parameters.

Dissolution

Dissolution (the process of dissolving) of solids is a competition between two competing forces: (1) the strong Coulombic interactions bonding ions to the solid, and (2) the hydration processes that occur when these ions can be surrounded by oriented water molecules in liquid solution. If the Coulombic interactions are large but the hydration energy is small, the solid is insoluble. If the Coulombic interactions are small but the hydration energy is large, the solid is very soluble. The overall solubility is determined by thermodynamics.

Thermodynamics says nothing about the rate at which a process occurs. The rate at which dissolution occurs is determined by an activation barrier created by the fact that in order for hydration to occur, the ion must be partially removed from the bulk solid. This process costs energy that is recovered once the ion is removed completely, but represents a barrier that restricts the rate of reaction. An atom at a step site has lower coordination than one embedded in a flat terrace and can form a more complete hydration sphere around it, so that its activation barrier for leaving the solid and moving into solution is smaller, and consequently its rate of reaction is higher. From a purely kinetic standpoint, we expect that atoms at kink sites are most reactive, those at straight step edges are less reactive, and those on perfect, flat extended terraces are even less reactive. Intersections between dislocations and the surface provide a source of excess energy to overcome the energy barrier and thus are sites of preferential etch pit formation (see above).

Dissolution rate analysis

Most experimental studies provide information about the macroscopic process of dissolution. Consequently, rates correlate with many macroscopic phenomena, including the structure of the solid (see below) and properties of model complexes in solution (e.g. Casey and Ludwig 1995).

If hydration or hydrolysis of a detaching surface complex at steady state controls the rate of reaction, and if a single surface complex is involved, then rates are proportional to the concentration of a single adsorbate (Furrer and Stumm 1986). Unfortunately, for silicates and other complex minerals, quantitative analysis of hydrolysis reactions is challenging because the mineral-water interface consists of a diversity of sites involving different coordination geometries and reactivities. The detailed nature of complexes at the mineral surface are rarely known.

Reactivity is strongly affected by hydrolysis of specific bonds yet surface titrations measure acid-base reactions at all surface sites (Casey et al. 1990). Thus, rate laws for dissolution can often only be expressed in terms of total analyzed concentrations of a given adsorbate. Activation energies determined for such reactions occurring via multiple pathways will only be weighted averages for the pathways, and include contributions from adsorption enthalpies, for example, proton adsorption (e.g. Casey and Sposito 1992). True rate laws should involve terms that explicitly include all surface complexes, with consideration of the ways in which protons and other species adsorb to surface groups and ligands.

At a fundamental level, dissolution rates for many minerals are determined by the reactivities of metal-oxygen bonds, which in turn are determined by bond strengths. Thus, rates can be influenced greatly by processes that modify bond strengths, such as coordination by a proton or ligand, or by redox reactions.

Rates of mineral dissolution are enhanced by adsorption of protons. Protons can either catalyze a hydrolysis reaction or induce it without catalysis, depending upon whether or not the proton is retained in the reaction product (Casey and Ludwig 1995). In general, it is assumed that rates of protonation are fast and dissolution rates are controlled by the slow rupture of the metal-oxygen bond (leading to hydration of the metal by a water molecule). Rupture is promoted by redistribution of electron charge following attachment of H^+. A protonation reaction that proceeds to equilibrium introduces a strong pH-dependence to an otherwise slow andpH-independent reaction (Casey and Bunker 1990).

Descriptions for the proton-induced rupture of surface metal-oxygen bonds in oxides at the atomic scale have been derived by analogy with ligand exchange processes, for example the rates of rupture of metal-oxygen bonds in dissolved complexes (Casey and Ludwig 1995, Casey et al. in press). However, the extrapolation of this approach to minerals containing a polymerized silicate anion is far from simple.

Other ligands (e.g. fluorine; Phillips et al. 1997) that can exist stably in the inner-coordination sphere of a surface metal can greatly increase the reactivity of metal-oxygen bonds, and thereby accelerate dissolution. Binding of ligands to these metals is an equilibrium phenomenon and the coordination number of the metal is usually constant. Rate increase due to deprotonation of water coordinating surface metals is a similar phenomenon (Wilkens 1991).

The rates of reactions (for example, where metal detachment is slow compared to diffusion and formation of surface complexes) can be modeled with transition state theory (Lasaga 1981, Aagard and Helgeson 1982, Weiland et al. 1988). The dissolution rate is related to the concentration of precursor to the activated species and has an exponential dependence on temperature (see Lasaga 1995, for details).

An important category of mineral dissolution reactions, especially from the perspective of biogeochemistry, are those that involve either reduction or oxidation of metals. Reductive and oxidative dissolution pathways are important because electron exchange changes metal-oxygen bond strengths. Iron and manganese oxides are subject to reductive dissolution whereas sulfides, silicates and some oxides are prone to oxidative dissolution. In the case of reactions proceeding via reductive or oxidative pathways, the concentration of the electron accepting or donating species as well as the activities of H^+, OH^-, and H_2O

are important in determining the rates (see Hering and Stumm 1990, for an excellent review of oxidative and reductive dissolution).

For dissolution reactions, the species that desorbs from the surface can be more complicated than the ions that constitute the original crystal structure. This is particularly true for ions that have more than one elementary charge and/or reactions involve oxidation or reduction of one or more of the species (see discussion of pyrite oxidation below).

Adsorption, precipitation, and growth

Mineral growth and surface precipitation involve attachment of ions from solution to sites on mineral surfaces (adsorption). In some adsorption reactions, one of more water molecule is lost from the hydration sphere of the ion, and a chemical bond is formed between the ion and a surface group (specific adsorption, leading to what is described as an inner sphere complex). Attachment occurs preferentially at kink sites, where atoms can form several bonds to the surface. The process by which adsorbed atoms become incorporated into the three dimensional structure of the growing crystal is known as absorption. If the adsorbed ions are the same as those in the surface, the process results in mineral growth. If the ions differ in their identity and relative abundance from those of the surface, adsorption may lead to formation of a precipitate (a three dimensional, periodic array of atoms). In other cases, the ion interacts with the surface but remains in the double layer and no chemical bond is formed to surface species (non-specific adsorption, referred to as outer sphere complexes).

Ions can be adsorbed to mineral surfaces via a number of distinctive mechanisms. Brown et al., (1995) provided the following examples:

(1) binding to a single metal site on a mineral surface:

$$>SOH + H^+ = >SOH_2^+ \quad \text{followed by} \quad >SOH_2^+ + L^{q-} = > SL^{1-q} + H_2O$$

where L^{q-} could be an oxyanion such as SeO_3^{2-} with charge q-. Note that the protonation reaction is believed to enhance the exchangeability of the surface hydroxyl group.

(2) bidentate binding to adjacent metal sites on a mineral surface:

$$2 >SOH_2^+ + L^{q-} = > S_2L^{1-q} + 2H_2O$$

(3) non-specific adsorption to a surface site:

$$>SOH_2^+ + L^{q-} = > SOH_2^+L^{q-}$$

Brown et al., (1995) noted that *in situ* microscopic measurements (using methods such as infra red, nuclear magnetic resonance, electron paramagnetic resonance, and X-ray spectroscopies) are needed in order to distinguish the details of adsorption mechanisms.

In chemical weathering reactions, secondary minerals often grow in topotactic orientations from constituents derived from the surface. For example, crystals of smectite grow at interfaces as silicate surfaces retreat (e.g. Eggleton 1984, Banfield and Barker 1994). Clay formation involves surface-mediated crystallization rather than adsorption of ions from solution followed by precipitation (ions incorporated into clays formed at interfaces have no residence time in true solution). Furthermore, complete breakdown of the primary silicate may not be necessary because structural fragments common to both minerals may be inherited by the product (e.g. Banfield et al. 1991, Banfield and Barker 1994).

A detailed treatment of dissolution and crystal growth is beyond the scope of this chapter. Additional references for more detailed treatments of geochemical reactions at mineral surface are Hochella and White (1990), White and Brantley (1995), and Vaughan and Pattrick (1995). Some additional considerations are discussed by Hochella and Banfield (1995). Sorption of organics onto surfaces is discussed by Stone (this volume).

SILICATE MINERALS AND THEIR REACTIVITY: EXAMPLES

In general, weathered surfaces of silicates have structures that can be described, to a first order approximation, as terminations of the bulk structure. The structures are probably relaxed in some or all cases, but as far as it is possible to determine from the existing literature, they are not reconstructed. Thus, a close approximation of their chemistry and atomic arrangements can be achieved through examination of the bulk crystal characteristics outlined and referenced below. The following section provides some details about reactivity of groups of silicate minerals. Greater attention is given to the minerals most abundant at the Earth's surface, those considered in detail in other chapters in this volume, and to layer silicates because of their importance as clay minerals.

Silica

Structure and composition. The most common, low temperature polymorph (structural variant) of SiO_2 is α-quartz, a mineral that consists of a completely polymerized silicate tetrahedral framework (Fig. 5a). About 20% of the crust is composed of quartz (Nesbitt and Young 1984). Other polymorphs (β-quartz, coesite, cristobalite, and tridymite) are also based on frameworks of Si tetrahedra, with different structures stabilized under different ranges of temperature and pressure (however, stishovite, a very high pressure polymorph found in meteorite impacts, contains Si in octahedral coordination). Microcrystalline silica is commonly intergrown with a metastable polymorph known as moganite (Flörke et al. 1984). Other low-temperate forms of SiO_2 include opal, which consists of disordered cristobalite/tridymite, and amorphous silica, which may be either biogenic or abiogenic in origin. Additional structural details are provided by Heaney et al. (1994).

Reactivity. Silica minerals have solubilities that increase with decreasing density (α-quartz being the least soluble). Below pH of ~8.5 (and above pH ~ 2.0) solubilities are not strongly affected by pH and quartz dissolution buffers pH to below 9.0 in most rocks (see review of Dove 1995). The rate of dissolution of quartz is slow, about 10^{-17} mol cm^{-2} sec^{-1} at 40°C at near neutral pH in pure water, because the activation energy for breakage of Si-O bonds is high (Dove 1995). A rate equation for quartz dissolution was proposed by Rimstidt and Barnes (1980), who suggest that quartz dissolution can be described by a first order rate law in silicic acid concentration. Al and Fe adsorption have been shown to decrease quartz dissolution rates (see review by Dove 1995). Organic acids have been shown to increase quartz reactivity. For example, citrate increased dissolution rates by almost an order of magnitude. This was attributed to an approximately 20% decrease in the activation energy for dissolution and increased quartz solubility (Bennett 1991). However, organic coatings can decrease dissolution rates by restricting access to surfaces (for details, see Barker et al., this volume). The effect of dislocations on dissolution rates is discussed by Brantley et al.,(1986).

Feldspars

Composition, structure, and microstructure. The most common minerals in the Earth's crust are feldspars. Feldspars are based on arrangements of Si and Al tetrahedra (in a pattern similar to that found in the SiO_2 polymorph coesite), with the lower charge of tetrahedral Al (compared to Si) compensated by substitution of cations into large sites in the structure (see Fig. 5b). The common feldspars end members have compositions ranging between albite ($NaAlSi_3O_8$) and K-feldspar ($KAlSi_3O_8$) (known as the alkali feldspars) and between albite and anorthite ($CaAl_2Si_2O_8$) (known as the plagioclase feldspars). Depending upon the temperature of crystallization, a variable but generally small quantity of Ca can substitute in alkali feldspars and a small quantity of K can substitute into the plagioclase feldspars. At low temperatures, many compositions intermediate between albite and anorthite and between albite and K-feldspar are unstable and, if they are cooled slowly enough, these intermediate feldspars unmix (exsolve) to produce fine scale intergrowths of Na-, Ca-, and K-rich phases. Cooling also leads to structural distortions and formation of twin planes and to complex ordering schemes involving Na, Ca, K, Al, and Si. Thus, although they may appear homogeneous at the hand specimen scale, many feldspars are extremely structurally and microstructurally complex. The literature on feldspar mineralogy is voluminous. For additional details see Ribbe (1981) and Brown (1984).

Reactivity. Experimentally determined rates of feldspar dissolution correlate with the concentration of adsorbed H^+ and OH^- (Blum and Lasaga 1988) and are at a minimum close to the pH_{ZPC}. Over the pH range from 1 to 4, experimental rates for the dissolution of albite range from ~10^{-14} to ~10^{-16} mol cm^{-2} s^{-1} and for K-feldspar between ~5×10^{-15} mol cm^{-2} s^{-1} to ~2×10^{-16} mol cm^{-2} s^{-1} (extrapolated from Blum and Stillings 1995). In contrast, widely reported field observations indicate that albite is far more reactive than K-feldspar. The differential weathering of albite lamellae compared to host K-feldspar and formation of a complex assemblage of weathering products is illustrated in Figure 8.

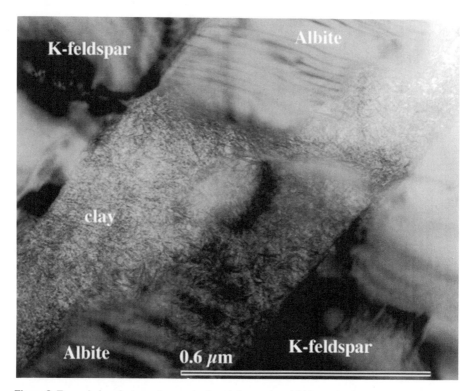

Figure 8. Transmission electron micrograph of partially weathered feldspar. the feldspar consists of a fine scale intergrowth of albite lamellae in K-feldspar. The albite lamellae has been preferentially altered to clay (smectite).

Both laboratory and field evidence points to incongruent dissolution of feldspar under some conditions. This leads to the formation of leached layers in laboratory studies conducted at low pH (e.g. Casey et al. 1990b, Schweda et al. 1997) and secondary minerals in naturally weathered samples (e.g. Banfield and Eggleton 1990). Si-rich leached layers develop at low pH because the leaching reaction involves selective removal of Al, leaving behind a partially polymerized silica residue that restructures to form amorphous silica (Casey et al. 1990b). Compositionally distinct exsolution lamellae (lamellae of phases produced by unmixing of an initially homogeneous solid solution) have been shown to contribute to unexpected elemental ratios in dissolution experiments (Inskeep et al. 1991, Stillings and Brantley 1995) and result in complex pitting and textural development at surfaces (Lee and Parsons 1995).

Layer silicates

Structures and compositions of different types of layer silicates. Layer silicates are minerals in which the tetrahedral portion of the structure is not completely polymerized (Fig. 4). In most cases, the tetrahedra are arranged in sheets linked by corner sharing (Fig. 4i), with charge associated with unshared "apical" oxygens balanced by cations in octahedral sites (Fig 4ii, iii, iv) and the interlayer (Fig. 4v). Sheet silicates generally adopt one of three basic structural types:

(1) single sheets of Si tetrahedra with apical O and OH groups forming coordination environments for Al, in the case of kaolinite: $Al_2Si_2O_5(OH)_4$ (Fig. 4iii), or Mg in the case of serpentine $Mg_3Si_2O_5(OH)_4$ (Fig. 4iv). Because these minerals are built up by stacking of a layer consisting of one tetrahedral and one octahedral sheet, these minerals are referred to as 1:1 layer silicates.

(2) two opposing sheets of Si±Al tetrahedra arranged so that planes of apical O plus OH groups form an octahedral sheet. In some cases, charge on the resulting unit (2:1 layer) is completely neutralized by the octahedral cations. Octahedral cations are Al, in the case of pyrophylite: $Al_2Si_4O_{10}(OH)_2$ and Mg in the case of talc: $Mg_3Si_4O_{10}(OH)_2$. In other cases, charge balance requires substitution of additional cations into the interlayer space. Minerals containing monovalent interlayer cations are known as micas (Fig. 4v). Examples include biotites, with compositions ranging from $KMg_3Si_3Al\ O_{10}(OH)_2$ to $KFe_3Si_3Al\ O_{10}(OH)_2$, and muscovite $KAl2Si_3AlO_{10}(OH)_2$. Because these minerals all contain a layer consisting of two tetrahedral sheets sandwiching an octahedral sheet, they are referred to as 2:1 layer silicates.

(3) 2:1 layers separated by an octahedral layer consisting of cations coordinated by OH (referred to as a brucite-like interlayer). Layer silicates based on these structural components can adopt a wide diversity of compositions and are referred to as chlorites (Fig. 4vi).

In some cases, layers are held together by H-bonding. H-bonding can be achieved when the second layer adopts one of several positions relative to the first (different options lead to different stacking sequences). Structural variants differing only in stacking sequence are referred to as polytypes (e.g. see Bailey 1988). Polytypism can also arise from stacking of essentially identical layers so that the octahedra slant in different ways (analogous to rotation of layers). The polytype provides some information about the conditions under which the mineral formed (see Bailey 1988b for details).

One common mineral whose layers are not held together by hydrogen bonding is the 1:1 layer silicate called chrysotile (one of the serpentine varieties). The layers in this mineral are curved due to misfit between the tetrahedral and octahedral portions of the structure. Curvature results in formation of tubes with submicron-scale cross sections. Halloysite (a clay mineral, see below) has a similar morphology.

Compositions and structures of clay minerals. A subset of the layer silicates are referred to as clay minerals. The term "clay" is commonly taken to indicate that the individual crystals are smaller than 2 μm in diameter. Clay minerals include kaolinite, halloysite, smectite, illite, vermiculite, and some chlorites. Smectite, illite, and vermiculite have structures that are similar to micas, except the interlayer occupancy is lower and the interlayer cations are often surrounded by water. Smectites, which are distinguished from vermiculites by lower interlayer charge, are characteristically "swelling clays" because they can accommodate variable amounts of water in the interlayer. The water molecules are apparently highly structured and arranged in an integral number of sheets (0, 1, 2, or 3), depending upon humidity. Based on the number of planes of water, the basal spacing of smectite can range from ~9.6 to ~18 Å (Moore and Reynolds 1997).

A wide variety of organic molecules can substitute into the interlayer of smectites. Positively charged organic species are localized by interlayer charge (similar to inorganic interlayer cations) whereas neutral organic molecules are adsorbed by formation of a complex with interlayer transition metal cations (Moore and Reynolds 1997). The orientations of organic molecules are dependent upon their configurations. For example, some aliphatic complexes orient so that their chains are parallel to the basal plane whereas ethylene glycol orients so that its zig-zag chain is normal to the basal plane (Moore and Reynolds 1997). Clay minerals have received special attention for their possible role as templates for early life and as substrates for synthesis of key organic molecules (see below). They are important reservoirs for water and relatively bioavailable ions and molecules in soils and sediments.

Reactivity of layer silicates. There are three distinctive groups of reactive sites important when layer silicates are exposed to solutions. Firstly, reactions can take place in the interlayer, in some cases coupled to reactions in the 2:1 layer (e.g. oxidation of octahedrally coordinated Fe). Interlayer cations, molecules, and water can be gained or lost in response to changes in the abundance and composition of solutions. In addition, the OH groups of the brucite-like interlayer in chlorite can be protonated. Weathering reactions involving biotite, muscovite, or chlorite commonly lead to the formation of clays and oxyhydroxides, including vermiculite, illite, smectite, kaolinite, and goethite. Transmission electron microscope-based studies demonstrate that weathering reactions that produce 2:1 clay minerals from micas or chlorite are focused in the interlayer and occur via a mechanism that involves inheritance of a relatively unmodified 2:1 layer (Banfield and Eggleton 1988, Kogure and Murakami 1996, Banfield and Murakami, 1998). However, only a subset of laboratory studies have been interpreted to indicate incongruent dissolution of layer silicates (for summary, see Nagy 1995). The discrepancy probably reflects differences in reaction conditions, including pH and solution saturation state.

Experimentally-measured abiotic rates for dissolution for kaolinite, muscovite, biotite, phlogopite, and chlorite range between 1×10^{-17} and 6×10^{-17} mol cm^{-2} s^{-1} at 25°C and pH 5, with faster rates observed in acidic and alkaline solutions (from tabulation by Nagy 1995). Hume and Rimsidt (1992) report the rate of dissolution of chrysotile in lung fluid as 5.9×10^{-14} mol cm^{-2} s^{-1} at 37°C. Biologically induced rates for mica alteration in soils can be much faster than abiotic rates (Hinsinger et al. 1992, Lin and Clemency 1981). To some extent, this is attributed to uptake of K from solution (the plant is a K sink). For example, Clemency and Lin (1981) report a rate for phlogopite of 2×10^{-14} mol cm^{-2} s^{-1} at 25°C (Lin and Clemency 1981) when K was removed from solution using an exchange resin, versus 3.8×10^{-17} mol cm^{-2} s^{-1} in a comparable system without a K sink (pH 5-6; Lin and Clemency 1981). This illustrates the enormous effect of solution saturation state on dissolution rate. The interlayer of mica is probably the most important source of bioavailable K in soils. The "sink effect" may be especially important in mica dissolution in the rhizosphere (see Barker et al., this volume, for further discussion).

After the interlayer, the next most reactive sites are those involving cations in sites which are charged. The terminations of layers provide a high density of such sites. The oxygen atoms can be protonated are active with respect to both adsorption and dissolution.

The third category of reactive surface sites are found on the basal planes of layer silicates. The reactivity of the basal surfaces is highly dependent on the occupancy of the tetrahedral sites or octahedral sites and the net magnitude of charge on the layer. A sheet consisting only of basal oxygens associated with Si tetrahedra will be relatively unreactive with respect to adsorption and protonation. In contrast, layers with net negative charge will adsorb cations in sites analogous to those found in the interlayer of sheet silicates.

A common measurement used in discussion of reactivity of layer silicates, especially clays, is the cation exchange capability (CEC). The CEC is a measure of the abundance of cations that can be readily exchanged (mmole per gram). The CEC consists of exchangeable cations in the interlayer plus those adsorbed to edge and other external surface sites (and thus, is pH dependent). The ease with which interlayer cations are exchanged varies with their identity. An approximate or of decreasing exchangability is Li^+, Na^+, K^+, Cs^+, Rb^+ (Eberl 1980).

Chain silicates

Structure and chemistry. In many igneous and metamorphic rocks, elements such as Mg, Fe, and Ca (as well as Al, Ti, Mn, and other elements) are found in pyroxenes (Fig. 3iii, iv) and amphiboles. Enstatite, diopside, and hedenbergite are examples of pyroxenes and tremolite and anthophyllite are examples of amphiboles (see Table 3). Pyroxenes and amphiboles are referred to as chain silicates because they contain single (pyroxene; Fig. 3iv) and double (amphiboles) chains of Si and Al tetrahedra linked through sharing of two (in the case of pyroxenes) or two alternating with three (in the case of amphiboles) oxygens by adjacent tetrahedra. In pyroxenes, chains are arranged so that the unshared zig zag strip of apical oxygens opposes a similar strip of unshared oxygens (tetrahedra point down in the first chain and up in the second), creating what is described as an "I-beam" (Fig. 3iii). I-beams are structurally similar to a 2:1 layer that is only a few octahedra and tetrahedra in length. A strip of octahedral coordination environments for cations such as Mg and Fe is created by the arrangement of oxygens with unsatisfied charge. A parallel strip of larger octahedral sites that can accommodate Ca. I-beams are stacked to form crystals (Fig. 3iii). Amphiboles have similar structures, contain OH, and large sites between the backs of adjacent I-beams that sometimes contain K and Na. Planar defects in pyroxenes and amphiboles include I-beams consisting of three or more chains (a defect of more than four chains would probably be described as a narrow strip of 2:1 layer silicate). Pyroxenes and amphiboles exhibit considerable compositional and structural variability (see Prewitt 1980, Veblen 1981). The range of stable solid solutions between compositional end members is determined by the temperature and pressure of formation, as for feldspars. Post-crystallization cooling often results in unmixing. Consequently, pyroxenes and amphiboles are often fine scale intergrowths of two or more structural and compositional variants (e.g. Smelik and Veblen 1992).

Reactivity. There have been a number of experimental studies of pyroxene and amphibole dissolution (see Brantley and Chen 1995 for summary). Chain silicate dissolution rates are dependent upon pH, generally decreasing with increasing pH. Dissolution rates increase (with some exceptions, such as for Na and Li-bearing pyroxenes) in the order: amphiboles, pyroxenes, pyroxenoids (single chain silicates with longer chain repeats). Excluding Na- and Li-bearing pyroxenes, rates for chain silicates range from about 3×10^{-13} mol cm^{-2} s^{-1} to 10^{-17} mol cm^{-2} s^{-1} at pH 2 to 3×10^{-14} mol cm^{-2} s^{-1} to $<<10^{-17}$ mol cm^{-2} s^{-1} at near neutral pH.

Studies of naturally weathered pyroxene and amphibole surfaces provide information about surface morphology, composition, and reactivity. Surfaces are typically planes that pass diagonally between the backs of the I-beams ({110}). At a microscopic scale it can be seen that these surfaces consist of I-beams steps. Because of the structural similarity between I-beams and 2:1 layers of sheet silicates, it is not surprising to find that clay minerals common grow from these stepped surfaces (Fig. 9).

Figure 9. High-resolution transmission electron micrograph of a (110) surface (in cross section) of a naturally weathered pyroxene viewed down [001]. Note the intimate association of layer silicates (lighter contrast) and the pyroxene surface, which is characterized by terraces and (010) steps (1-3 layers of I-beams; see Fig. 3b).

The topic of the surface chemistry and reactivity of amphibole surfaces with respect to organic molecules has been a subject of some interest to the medical community because some amphiboles occur as crystals with large aspect ratios (asbestiform) and are a common constituent of asbestos products (see Guthrie and Mossman 1993). To some degree, the higher danger associated with inhalation of amphibole as opposed to layer silicate (crysotile) asbestiform minerals has been attributed to its lower solubility in the human lung (Hume and Rimstidt 1992). There is some indication that pathogenic reactions may be triggered by compounds such as hydroxyl and superoxide radicals formed at the amphibole surface, or by denaturing of surface adsorbed proteins (Hansen and Mossman 1987, Giese and van Oss 1993, review by Hochella 1993, Werner et al. 1995).

Orthosilicates

Structures and compositions. A subset of important rock-forming minerals contain unpolymerized (isolated) tetrahedra (orthosilicates). Olivine and garnet (e.g. grossular; see Table 3) are common examples. In olivine, forsterite–fayalite [Mg_2SiO_4–Fe_2SiO_4], the Si tetrahedra are arranged so that the O form zig zag chains of octahedral sites occupied by Mg and Fe. Garnets have diverse compositions based on $A_3B_2(SiO_4)_3$, where A is commonly Ca, Mg, Fe^{2+}, or Mn^{2+} and B is typically Al, Fe^{3+}, Cr^{3+}. Other orthosilicates include zircon ($ZrSiO_4$) and sillimanite (Al_2SiO_5). Structural and chemical details for these minerals can be found in Ribbe (1982).

Reactivity. Measured rates of dissolution of minerals with the olivine structure scale linearly with the rates of exchange of water molecules in the hydration sphere of the appropriate metal cation, reflecting control of the metal-oxygen bond (Casey and Westrich 1992). Because olivine consists of SiO_4^{4-} groups that are unpolymerized, the dissolution of olivine can be understood simply in terms of rupture of metal-oxygen bonds, followed by release of SiO_4^{4-} (e.g. as $H_4SiO_{4(aq)}$). The dissolution of minerals with the olivine structure (but diverse occupancy of the octahedral sites) have been studied in detail, in part because the absence of polymerization simplifies the analysis (see review by Casey and Ludwig 1995).

The natural weathering of realtively Mg-rich olivine has been studied by several researchers (Eggleton 1984, Smith et al. 1987, Banfield et al. 1990, Banfield, unpublished data). In all cases it is found that Mg-rich smectite grows topotactically at the olivine-smectite interface and Fe is incorporated into nanocrystalline oxyhydroxides and oxides (e.g. goethite and hematite, see below).

Surface phenomena in the laboratory vs. natural weathering

Laboratory studies show that most silicate minerals dissolve slowly in the near-neutral or slightly acidic environments commonly encountered near the Earth's surface (Table 4).

Table 4. Summary of experimentally determined mineral dissolution rates measured under far from equilibrium conditions.

	measured rate $mol/cm^2/sec$	pH for experiment	Temp.	Ref.
Quartz	~10^{-18}	7	25°C	[1]
Albite	3×10^{-17}	5	25°C	[2]
K-feldspar	4×10^{-17}	5	25°C	[3]
Phlogopite	4×10^{-17}	5	25°C	[4]
Chlorite	3×10^{-17}	5	25°C	[5]
Montmorillonite	4×10^{-18}	5	25°C	[6]
Kaolinite	1×10^{-17}	5	25°C	[7]
Anthophyllite	3×10^{-16}	5	22°C	[8]
Enstatite	2×10^{-14}	4.8	25°C	[9]
Diopside	8×10^{-15}	4.8	22°C	[10]
Forsterite	2×10^{-14}	5	25°C	[11]

References: [1] Dove (1995), [2] Knauss and Wolery (1986), [3] extrapolation in compilation by Blum and Stillings (1995), [4] Lin and Clemency (1981), [5] May et al. (1995), [6] Furrer et al. (1993), [7] Carroll-Webb and Walther (1988), [8] Mast and Drever (1987), [9] Ferruzzi (1993); [10] Schott et al (1981), [11] Westrich et al. (1993).

To a first order approximation, rates of dissolution at near-neutral pH decrease with increasing polymerization of the tetrahedral portion of the structure (Table 4). However, experimentally determined rates may be at least three times faster than those occurring in nature (see discussion in Velbel 1993). This may be due to problems estimating the distribution and abundance of natural reactive surface area. Furthermore, the rates and mechanisms of natural reactions may be significantly affected by microorganisms (see Barker et al. this volume).

Some discrepancies between the structure and composition of surfaces formed by laboratory dissolution versus those developed during natural weathering should be noted. It is becoming increasingly apparent that leached layers form on silicates in dissolution experiments under some conditions (e.g. Casey et al. 1990, Banfield et al. 1995, Schweda et al. 1997). Similar material has been reported in proximity to some feldspar surfaces (Banfield and Eggleton 1990). However, amorphous materials are almost never detected on other naturally weathered silicate surfaces. Instead, natural surfaces are commonly in direct contact with clay minerals. In fact, there is generally bonding and structural continuity between the primary and secondary mineral and inheritance of structural polymers from the reactant is certainly possible. This raises questions regarding differences between mechanisms of reactions in nature versus the laboratory, especially in cases where the structures of primary and secondary minerals have similar structural components (e.g. for layer silicates; Banfield and Murakami, 1998). Furthermore, the physical and chemical properties of water at these interfaces may be poorly approximated by bulk solutions (see review by Hochella and Banfield 1995).

NON-SILICATE MINERAL REACTIVITY: EXAMPLES

Carbonate minerals

Structures and compositions. The two most abundant and important carbonate minerals in rocks and formed by microorganisms are calcite and aragonite. Calcite ($CaCO_3$) has a structure that can be considered as derived from that of NaCl, with Ca^{2+} in six coordinated sites in place of Na^+ and CO_3^{2-} groups in the Cl sites, with the resulting structure compressed along the cube diagonal. Related minerals include dolomite ($CaMg(CO_3)_2$ and siderite ($FeCO_3$). These and other carbonates are discussed in Reeder (1983).

Calcium in the aragonite structure is in 9-coordinated sites. The calcite-aragonite phase transformation involves changes in orientation and displacement of the CO_3^{2-} groups. At room temperatures and pressures, the aragonite structure is normally adopted only by carbonates of larger ions, such as Ba, Sr, and Pb. However, as discussed above and elsewhere in this volume, $CaCO_3$ forms with both the calcite and aragonite structure at the Earth's surface when crystallization is mediated by organisms (e.g. deVrind-de Jong and de Vrind, this volume).

Reactivity. In both solution and air, the dominant cleavage surfaces of calcite have structures closely similar to those of the bulk (Fig. 10), but possibly with rotation of alternate CO_3^{2-} groups (Hochella 1990, Stipp and Hochella 1991, Chianello and Sturcio 1995). The surface is dominated by terraces separated by monomolecular height steps that decrease in abundance over time during dissolution (Chianello and Sturcio 1995).

Carbonates are amongst the most reactive abundant minerals near the Earth's surface. They dissolve congruently at rates that are orders of magnitude faster than rates for silicate mineral dissolution. Under far from equilibrium conditions at 25°C, over the pH range of 0 to 5, experimentally measured rates decrease from approximately 10^{-2} mol cm^{-2} s^{-1} to about 10^{-7} mol cm^{-2} s^{-1} (see compilation in Morse 1983). Rates decrease with increased concentrations of Ca^{2+} and CO_3^{2-} in solution. This has been interpreted to reflect a surface

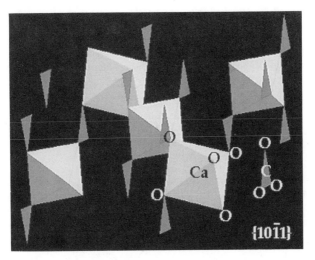

Figure 10. Diagram illustrating the Ca octahedra and CO_3^{2-} groups that may occur on a calcite surface.

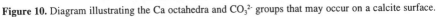

controlled reaction rate, where back reaction (precipitation of calcite) is important (Morse 1983). Other constituents in solution, adsorbed to the surface, or substituted into the carbonate structure (and the specific biological origin) can strongly effect reaction rates at calcite surfaces. The general concepts are reviewed by Morse (1983).

Oxides and oxyhydroxides

Structures and compositions. Oxides and hydroxides are common constituents of the Earth's near-surface region. Magnetite, Fe_3O_4 (Fig. 11), is a spinel that contains

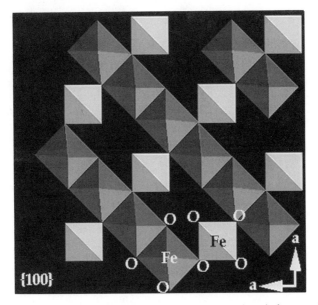

Figure 11. Diagram illustrating the arrangement of Fe tetrahedra and octahedra on a {100} magnetite surface.

Fe^{2+} and Fe^{3+} in tetrahedral and octahedral sites within a cubic closet packed oxygen framework. Magnetite is produced abundantly by some microorganisms (see Bazylinski and Moscowitz, this volume). Magnetite is ferrimagnetic due to the arrangement of Fe^{2+} and Fe^{3+}, both of which have unpaired electrons whose spins are coupled by superexchange through oxygen (see Banerjee 1991). Magnetic remanence in many rocks is carried by magnetite of both high-temperature and biogenic origin. Magnetism of magnetic minerals is described in detail by Bazylinski and Moscowitz (this volume).

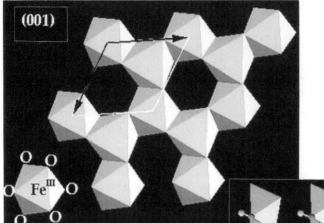

Figure 12 (above). Diagram illustrating the arrange-ment of Fe octahedra on a (001) hematite surface. A single octahedron is also shown.

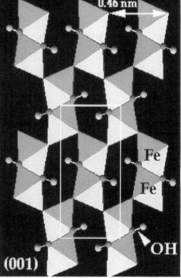

Figure 13 (right). Diagram illustrating the arrangement of Fe octahedra on a (001) goethite surface. Balls connected by short rods represent OH groups.

Fe_2O_3 commonly adopts one of two structures. Hematite (Fig. 12), a common weathering product, contains Fe^{3+} in octahedral sites in a hexagonal close packed arrangement of O atoms. Maghemite is a defect spinel that can be formed by oxidation of magnetite. Goethite (α–FeOOH, Fig. 13) and lepidocrocite (γ-FeOOH) are two of several oxyhydroxides of iron produced by chemical weathering reactions. Diaspore and boehmite are Al oxyhydroxide analogs of goethite and lepidocrocite and corundum (α–Al_2O_3) is isostructural with hematite. Ti occurs in the polymorphs anatase, rutile, brookite, and $TiO_2(B)$.

There are a number of Mn oxides (e.g. pyrolusite: MnO_2, hausmannite: Mn_3O_4) and oxyhydroxides (e.g. manganite and groutite: MnOOH). In general, Mn is in octahedral coordination and adopts oxidation states between 4+ and 2+. Many Mn oxides are tunnel structures in which tunnels are formed by linkage of octahedral strips through edge sharing. The tunnels are only one octahedra wide in pyrolusite. In structures where the tunnels are constructed by linking more than one octahedral chain, sufficient space exists to accommodate impurity cations (e.g. todorokite: $(Mn,Ca,Mg)Mn_3O_7 \cdot H_2O$). Additional details are discussed by Tebo et al . (this volume). All of the the Fe, Al, Ti, and Mn oxides and oxyhydroxide phases generally occur as fine particles with a high ability to absorb other elements (e.g. P, metals).

Reactivity.   Under oxidizing conditions near the Earth's surface, oxides of Fe, Al, Ti, and Mn are generally very insoluble. Therefore these are common residual phases in intensively weathered rocks (laterites). Oxide dissolution rates are controlled by dissociation of metal-oxygen bonds. Reactions have been modeled based on an analogy between surface sites and dissolved oxide oligomers (Casey et al. 1998). Changes in metal oxidation state affect the solubility and dissolution rates (also see Tebo et al., this volume). Dissolution can occur more rapidly via a reductive pathway due to labilization of the metal-oxygen bond and is greatly accelerated by microorganisms and microbial metabolites. A detailed treatment of metal complexation by ligands, as well as discussion of adsorption of ligands to oxide surfaces, is give by Stone (this volume). Reductive dissolution processes, including examples, are provided by Hering and Stumm (1990).

Sulfides and sulfates

Structures and compositions Sulfides and sulfates are occasionally abundant but are common minor phases in rocks. In minerals, sulfur can adopt nominal oxidation states ranging from 2- to 6+. Reduced sulfur readily forms semiconducting compounds with a variety of metals (e.g. Fe, Cu, Zn, Pb, Ni). Oxidized sulfur commonly combines with elements such as Ca, Ba, Sr, etc. There are innumerable sulfide and sulfate (hydrous and anhydrous) minerals and it is beyond the scope of this chapter to name or describe them all (see, for example, Ribbe 1974, Klein and Hurlbut 1994). The following discussion will touch only on the composition and structures of minerals mentioned elsewhere in this volume.

Minerals containing reduced sulfur include FeS troilite (FeS; only formed under very reducing conditions) and pyrrhotite (Fe(1-x)S, commonly $\sim Fe_7S_8$). Pyrrhotite is weakly magnetic. Sulfide minerals formed under low temperature conditions (e.g. in sediments and by microorganisms) include mackinawaite, Fe_9S_8, and greigite, Fe_3S_4 (which has similar structure and magnetic properties to magnetite).

The nominal oxidation state of sulfur is -1 in pyrite (FeS_2), the most common sulfide mineral. The structure of pyrite (Fig. 14i) can be considered as derived from that of NaCl, with Na replaced by Fe^{2+} and Cl replaced by S-S dumbbells (however the space group is different). Marcasite shares the composition of pyrite but has a different structure.

Elemental sulfur (S^o) forms by both inorganic and organic mechanisms, including by oxidation of sulfide minerals. Se substitutes readily for sulfur and is incorporated into biogenic sulfur crystals for this reason (Nelson et al. 1996). The minerals in which sulfur exists in its most oxidized 6+ state are sulfates. Gypsum ($CaSO_4 \cdot 2H_2O$), the most common sulfate mineral found predominantly in evaporites, dehydrates to anhydrite.

Reactivity Sulfides can dissolve by pure chemical and oxidative mechanisms. To illustrate dissolution via the generally much faster oxidative pathway, we consider the example of the most abundant sulfide phase.

The dissolution of pyrite is of great importance, both geochemically and microbiologically (see Nordstrom and Southam, this volume, for details). The pH_{ZPC} for pyrite is slightly higher than 1.0 (Fornasiero et al. 1992). The reaction in nature commonly occurs at surfaces in contact with acidic solutions because oxidative dissolution lowers the pH by the overall reaction:

$$FeS_2 + 15/4O_2 + 7/2H_2O \Rightarrow Fe(OH)_3 + 2H_2SO_4$$

The mechanism involves removal of iron and oxidized sulfur from sites such as on {100} surfaces illustrated in Figure 14(ii), 14(iii). Surface sulfur groups are slightly negatively charged due to an unshared electron pair (Luther 1987, 1990). Oxygen interacts less effectively than Fe^{3+} with surface sulfur species (Moses et al. 1987, Luther 1990). Fe^{3+} binds to S through a vacant orbital, forming a bridge through which electron transfer can occur. Luther (1987) suggested a mechanism initiated by the following steps:

(1) $Fe(H_2O)_6^{3+} \Rightarrow Fe(H_2O)_5^{3+} + H_2O$

(2) $Fe^{2+}\text{-S-S..}Fe(H_2O)_5^{3+}$

(3) $Fe^{2+}\text{-S-S..}Fe(H_2O)_5^{3+}$
$\Rightarrow Fe^{2+}\text{-S-S}^+..Fe(H_2O)_5^{2+}$

Subsequently, the hydrated ferrous iron detaches and the surface group formed in (3) reacts with another $Fe(H_2O)_6^{3+}$, further oxidizing S. This is accompanied by attachment of one O (from H_2O) to the pyrite surface and release of two protons. This process continues until Fe^{2+} and $S_2O_3^{2-}$ are formed. The $S_2O_3^{2-}$ is subsequently oxidized to sulfate. As discussed by Nordstrom and Southam (this volume), pyrite dissolution is catalyzed by microbial activity, in large part through regeneration of Fe^{3+}(aq).

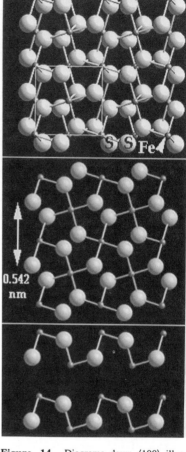

Figure 14. Diagrams down $\langle 100 \rangle$ illustrating: (i) the pyrite structure; (ii) the arrangement of Fe (small balls) and S (large balls) close to the pyrite surface; (iii) the type of Fe and S clusters very close to a hypothetical pyrite surface. Note the absence of 4-fold symmetry.

Because of their semiconducting properties, the surface structure, chemistry, and reactivity of sulfide minerals has been examined via scanning tunneling microscopy and other techniques. Surfaces can be imaged with atomic resolution (see Hochella 1995) and reveal the importance of removal of atoms from steps during dissolution (Higgins and Hamers 1996) and incipient oxidation (Fig. 15 and Eggleston 1994).

Rates of abiotic and biologically-mediated sulfide dissolution are reviewed by Nordstrom and Southam (this volume).

Phosphates

Compositions and structures. Although there are many phosphate minerals in rocks, by far the most abundant is apatite. Apatite, $Ca_5(PO_4)_3(F,Cl,OH)$, is a common

Figure 15. An atomic-resolution scanning tunneling microscope image of an approximately 10 nm × 10 nm area of galena surface immersed in 0.5 M NaClO$_4$ at pH 1.7 (with HClO$_4$) with the sample potential = +0.17 V vs. NHE, tip potential = -0.1 V vs. NHE, tunneling current = 1 nA. The tip-sample bias is -0.27 V (empty states of the surface are being imaged). Single and clusters of dark spots are interpreted to be sites of atomic vacancies or oxidation on a terrace. Image by Steven Higgins.

accessory phase (formed abiotically) in most igneous, metamorphic, and sedimentary rocks and biotically at the Earth's surface. Apatite is also formed by micro (and macro) organisms. Its structure is illustrated in Figure 16. The two most important varieties are hydroxyapatite and fluorapatite. Low temperature hydrous phosphates include brushite: CaHPO$_4$•2H$_2$O, and whitlockite: Ca$_9$(Mg,Fe)H(PO$_4$)$_7$). Vivianite is one of several iron phosphates (Fe$_3$(PO$_4$)$_2$•8H$_2$O).

Reactivity. There have been numerous studies of hydroxyapatite and fluorapatite dissolution and surface chemistry (e.g. El Gibaly et al. 1977, Inskeep et al. 1988, Thirioux et al. 1990, and references therein). Results show that apatite is sparingly soluble in water but exhibits enhanced solubility under acid conditions. Organic compounds such as oxalic, citric, tartaric, salycilic acids produced by microbial metabolism increase dissolution rates (El Gibaly et al. 1977) and fulvic, humic, and tannic acids inhibit apatite precipitation (Inskeep et al. 1988). P released by dissolution of apatite is commonly precipitated as secondary phosphates of Fe and lanthanide group elements (REE). REE and other elements such as U, Sr, and Ba can be concentrated enormously (wt % levels) in secondary phases formed at the dissolving apatite surface. These secondary phosphates (e.g. rhabdophane and florencite; Banfield and Eggleton 1989) are relatively insoluble and their formation may restrict the availability of P to organisms under some conditions.

Note: Three-dimensional animations of the crystal structures of more than thirty silicates and some other minerals can be viewed at:

http://www.geology.wisc.edu/~jill/geo360/mov.html

GEOCHEMICAL REACTIONS AND BIOAVAILABILITY

Inorganic reactions strongly affect bioavailability, and life has evolved to take advantage of what is readily available. The most accessible source of many elements needed by organisms (other than those obtained from the atmosphere) is water. The composition of water (see Table 2) reflects the balance between the rates at which elements

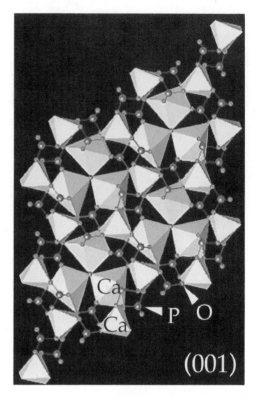

Figure 16. Diagram illustrating the structure
of apatite viewed down [001]. Note the two
distinct Ca sites and the PO_4^{3-} groups linking
octahedra.

are released from primary minerals (dissolution) and the rates at which they are removed
from solution by formation of secondary minerals. Many elements used by microorganisms
in their enzymes (e.g. K, Mg, Ca) and other organic molecules and cell structures (S, Si,
P) occur in relatively high concentrations in the oceans. It is probably no coincidence that
microorganisms generally do not require elements such as Al and Ti, because these occur in
low abundance in solutions at, and near, the Earth's surface. A key exception is Fe, which
is a critical constituent of a variety of cytochromes and enzymes. Iron is required by all life
forms, yet has it has a very low abundance in water. This apparent anomaly is explained by
the fact that Fe^{2+} was abundant in relatively reduced solutions at the Earth's surface when
life evolved (see review by Kasting 1993). The appearance of oxygenated conditions at the
Earth's surface caused precipitation of Fe^{3+} compounds, resulting in typically low
dissolved $Fe_{(aq)}$. This must have presented some difficulties for microbes adapted to
conditions where Fe was readily available. A connection must exist between the appearance
of abundant oxygen in the atmosphere and the evolution of microbes able to excrete
molecules to solubilize Fe (e.g. siderophores; see chapter by Stone, this volume).

PREBIOTIC SYNTHESIS AND THE ROLE OF MINERAL SURFACES
IN THE ORIGIN OF LIFE

This section explores current thinking about the origin of life, specifically the options
that exist for synthesis of key prebiotic molecules in solution and at mineral surfaces.
Critical molecules such as amino acids, purines, pyrimidines, and other hydrocarbons have
been found in carbonaceous chondrite meteorites (Lazcano et al. 1983). However, decades
of investigations focused on formation of prebiotic molecules in solution have been less

than successful and physically reasonable and chemically sustainable models for basic organic syntheses under early Earth conditions have not appeared. Several hypotheses exist to explain the formation of the simplest organic molecules (alcohols, nucleic acids, sugars, etc.). However, the mechanisms by which nucleic acids and sugars are assembled into nucleosides, phosphorylated nucleosides (nucleotides), and then polymerized into DNA and RNA in an environment that prevents their hydrolysis are poorly understood. Attention is now turning (and returning) to the idea that synthesis of the basic building blocks for life occurred at the mineral-water interface (e.g. Ferris et al. 1996, Huber and Wächtershäuser 1997).

This review is predicated on the idea that life originated on the Earth and was not transported to Earth by comets or other interplanetary material (the "panspermia" hypothesis of Crick and Orgel 1973, Weber and Greenberg 1985, Davies 1996). While the claim that life evolved on Earth is far from proven, the hypothesis that it did places some constraints on the range of conditions available for life-forming chemical processes. However, regardless of where life arose, minerals - specifically mineral surfaces and their associated aqueous environments - must have been important.

Prebiotic synthesis of basic building blocks

The Earth is widely believed to have formed by accretion about 4.55 billion ago. Models such as those of Chang et al. (1983) suggest that between 4.55 and 4.05 billion years ago the Earth's temperature was between 500° and 1000°C. However, Russell and Hall (1997) advocate a temperature of ~90°C at ~4.2 billion years. Current knowledge suggests the upper temperatures for bacterial life are at least 90°C (e.g. Pace 1991, Stetter et al. 1996) and may possibly be as high as 169°C (Cragg and Parkes 1994). It seems unlikely that life arose until temperatures cooled to close to this range. Even if living organisms appeared, periodic melting due to heavy bombardment of the Earth's surface during the first few hundred million years would probably volatilize any early oceans (e.g. Wardrup 1990) and periodically sterilize the surface environment, perhaps sparing organisms living in sediments or hydrothermal systems (Kasting 1993). However, current models for the bombardment history of the Earth have been questioned (Delano 1997).

The period after accretion must have been characterized by crustal cooling and rapid geological change. Today we have evidence that sediments formed in aqueous environments at ~3.85 billion years ago and these rocks have isotopic signatures that may indicate biological fractionation, suggesting life-like processes occurred at this time (Hayes et al. 1983, Mojzis et al. 1996). Complex life forms (possibly similar to cyanobacteria) existed at ~3.55 billion years ago (Schopf and Walter 1983). Thus, it appears that soon after conditions at the Earth's surface became relatively hospitable, life appeared.

The experiments of Miller (1953) and others (see Lazcano et al. 1983, for review) involving electrical discharges in a reducing NH_3, CH_4, H_2, H_2O atmosphere proved that the basic carboxylic acids and amino acid building blocks could be synthesized (Strecker synthesis). However, more recent models suggest that the primordial atmosphere was not reducing (Gillett 1985). Thus, Strecker synthesis (Oparin-Haldane hypothesis, see Barns and Nierswicki-Bauer, this volume) may not have been a practical method for production of simple organic building blocks (also see review by Kasting 1993).

Some aqueous synthesis pathways for amino acids, nucleic acid bases, and sugars are known (Miller and Orgel 1974). Adenine can be synthesized relatively easily from ammonia + hydrogen cyanide (Oró 1961, Sanchez et al. 1967). However, the aqueous synthesis of other bases, especially cytosine which is readily altered to uracil, is far more uncertain (Shapiro 1984, 1997; Joyce and Orgel 1993).

Ribose sugar, a structural component of both DNA and RNA, can be synthesized from glycoaldehyde phosphate in the presence of formaldehyde (Müller et al. 1990). The reaction between glycoaldehyde phosphate and glyceraldehyde-2-phosphate can be effectively catalyzed at pH 9.5 in the presence of layered hydroxides such as hydrotalcite (Mg-Al-hydroxide; Pitsch et al 1995). However, the relevance of this observation to a natural synthesis route at the surface of the early Earth is difficult to ascertain.

Abiotic synthesis typically yields a complex mixture of products. Minerals may serve a role in this process by selectively adsorbing and concentrating the 'useful' products from the complex mixtures that often accompany synthesis reactions.

Abiotic assembly of the first self-replicating molecules

One of the important questions facing scientists seeking to understand the origin of life has been "Which came first: DNA?, RNA? or proteins?" In modern life, proteins are needed to make both DNA and RNA and DNA and RNA are needed to make proteins. Woese (1967), Crick (1968), Orgel (1968) suggested that RNA preceded proteins and DNA. This concept was supported by the discovery that RNA has catalytic activity (can form inter-nucleotide bonds and can replicate itself without a complex ribosome factory; Cech 1987). This led to the idea that the first life was based on RNA, and the concept of an "RNA world" (Gilbert 1986).

Although it has been achieved to some extent in the laboratory, the coupling of nucleic acid bases with ribose sugar by a physically applicable method with a reasonable yield of nucleosides has not been demonstrated. Joyce and Orgel (1993) state "The only remotely plausible route to the molecular biologist's pool would involve a series of mineral catalyzed reactions, coupled with a series of subtle fractionations."

The question of how RNA was assembled has been examined by many workers. It is well known that RNA can be made from nucleotide polyphosphate (e.g. adenosine triphosphate) and that energy for polymerization (linkage of sugars via phosphodiester bonds) can come from splitting off the polyphosphate groups. Polyphosphates may be made by reaction between nucleotides and inorganic tri-metaphosphate. However, the details remain unclear.

Mineral surface-catalyzed approaches to RNA (and protein) synthesis have been considered. Joyce and Orgel (1993) write "...adsorption to a specific surface of a mineral might orient activated nucleotides rigidly and thus catalyze a highly stereo specific reaction." The surface of apatite may be a logical candidate because this is the primary mineralogical source of phosphorus. It is interesting to note that polyphosphate is a common component in many procaryotic and eucaryotic cells, occurring in quite significant concentrations in some cell walls. Polyphosphate probably served as a source and sink for activated phosphate groups (a source of energy) for early procaryotes and was possibly a precursor for ATP (Kulaev 1997).

The idea that important steps in prebiotic synthesis involved stereospecific reactions at mineral surfaces is lent some support by the recent work of Ferris and colleagues (Ferris et al. 1996). Ferris et al., (1996) showed that it is possible to synthesize oligomers up to 55 monomers in length from activated monomers by reactions at mineral surfaces. They report formation of amino acid polymers at the surfaces of illite and hydroxyapatite and of nucleotide polymers at the surface of montmorillonite. However, polymerization required activation of the nucleotide through binding of a activating group (e.g. adenine derivatives) to the phosphorus.

Problems with the "RNA world" and alternatives

As pointed out by Pace (1991) and others, the "RNA world" could not have existed in free solution because RNA is easily hydrolyzed. Pace (1991) writes: "The origin of life seems inconsistent with fully aqueous chemistry, particularly at high temperature." A suitable environment, characterized by an appropriate supply of organic and inorganic molecules, mineral surfaces, and low activity of water must be identified. Russell and Hall (1997) suggest that this environment may have been inside bubbles with colloidal FeS membranes (containing some Ni). They hypothesize that the surfaces of these bubbles catalyzed synthesis of organic anions by hydrogenation and carboxylation of hydrothermal organic species. They propose that condensation of organic molecules to polymers inside these bubbles was driven by pyrophosphate hydrolysis and involved a pH gradient across the inorganic membrane (cell wall precursor).

The hypothesis that early organic molecules for life were formed by reactions between simple compounds (water, CO, CO_2) and metal sulfide surfaces has received considerable recent attention (e.g. Wächtershäuser 1990, Huber and Wächtershäuser 1997). Chemolithotrophy requires reduction of CO_2 by H_2 to form organic molecules. H_2 can be produced abiotically by FeS oxidation (Wächtershäuser 1988) and remains accessible for subsequent chemical reactions because it is strongly adsorbed to pyrite surfaces (Rickard 1997). Both CO_3^{2-} and CO_2 bind to pyrite surfaces, the former through Fe(II)-CO_3 surface species and the later via binding of CO_2 to disulfide groups (Evangelou and Zhang 1995). Thus, pyrite surfaces bring together many of the fundamental building blocks for organic synthesis.

Huber and Wächtershäuser (1997) conducted hydrothermal experiments using aqueous mixtures of sulfides at 100°C. This temperature is consistent with most speculations about the temperature of solutions at the Earth's surface about 3.85 billion years ago, and with the molecular evidence that indicates the earliest organisms were hyperthermophiles (see Barns and Nierswicki-Bauer, this volume). Huber and Wächtershäuser (1997) reacted CO and CH_3SH (methane-thiol, proposed to be an important intermediate in the reduction of CO or CO_2 by FeS and H_2S) on a mixture of NiS FeS. They found that these converted to CH_3-CO-SCH_3 (a thioester) which is subsequently hydrolyzed to acetic acid (and CH_3SH + aniline $\Rightarrow$ acetanilide). Further-more, when CO + H_2S were reacted on Se-doped NiS + FeS, acetic acid and CH_3SH were produced. Huber and Wächtershäuser (1997) speculate that the earliest organisms metabolized CO / CO2 / H_2S at volcanic or hydrothermal vents. CH_3SH reacted to activated thioacetic acid (CH_3CO-SH) was subsequently converted to other organic compounds.

It is interesting to note that ferredoxins, considered to be ancient biological catalysts, contain Fe and S centers that can store and transfer electrons (other key enzymes contain Ni and W; see review by Russell and Hall 1997). A key enzyme in the analogous acetyl-coenzyme A pathway contains a Ni-Fe-S center. These modern molecules may preserve a record of the primordial core metabolism in their structures (Wächtershäuser 1997). Early metabolisms may have received reducing power from the oxidative formation of pyrite from FeS and H_2S (e.g. FeS + H_2S $\Rightarrow$ FeS_2 + H_2 ; Huber and Wächtershäuser 1997). A subsequent step in the life-forming pathway proposed by Huber and Wächtershäuser (1997) involves appearance of lipid vesicles that surrounded these reactions. Eventually, the cell evolved enzymes which replaced the pyrite crystal as the site of polymer synthesis.

Minerals as alternative early genetic systems

An interesting idea regarding the first replication system was presented by Bernal (1951) and advocated by Cairns-Smith (1982). These authors proposed the early genetic system was inorganic, perhaps a clay mineral (Bernal 1951, Cairns-Smith 1982, Cairns-

Smith and Hartman 1986). The initial self replicating system would have had to have been superseded by an RNA- or DNA-based system. Currently, the mechanisms of this transformation are not clear (Joyce and Orgel 1993). More recently, there has been considerable interest in precursor organic polymers as the primordial replicating molecule (Usher 1997).

Conclusion

Regardless of the details, most current hypotheses support the view that primordial life equals chemoautotrophic life and that reactions at mineral surfaces played a key role in generation of compounds necessary for the earliest metabolisms.

CONCLUDING COMMENTS REGARDING FUTURE WORK

Reactions at surfaces are of central importance to microbial and macroscopic life and control numerous steps in geochemical cycles. Over the past decades, considerable effort has been directed towards measurement of abiotic dissolution rates because these are important parameters for geochemical models. Comparatively little is known about biological dissolution of minerals, especially silicate minerals, despite the fact that microorganisms and microbial products can dramatically influence mineral weathering processes (see Nordstrom and Southam, Barker et al., and Stone, this volume).

A specific topic of great importance is the nature of reactions between organic molecules and functional groups at mineral surfaces. We know that cells attach to surfaces, dissolved organic molecules adsorb to surfaces, and surfaces catalyze organic reactions. However, the mechanisms of these processes are poorly understood. Both experimental data for organic adsorption on surfaces and modeling of the molecular interactions, especially for more complex molecules, are needed. This knowledge will be of great importance if we are to understand the role of mineral surfaces in prebiotic organic synthesis.

Future progress on the topic of mineral surface reactions relevant to geomicrobiology will require an interdisciplinary approach. Observations should be made over a range of scales of resolution and include analyses of natural systems. Information about the species of organisms, the surface structure and chemistry of the cell wall and associated organics, details of the mineral surface structure and composition (including surface chemical speciation), and chemical information for solutions in proximity to surfaces should be combined to elucidate the key interactions that occur (and have occurred over geologic time) between microorganisms and minerals.

ACKNOWLEDGMENTS

Joanne Fredrich, Sandia National Laboratories, Hengzhong Zhang, R Lee Penn, Dawn Janney, and Mary Anderson, University of Wisconsin-Madison, Steven Higgins, University of Wyoming, and William Casey, University of California-Davis are thanked for their contributions to this chapter and Michael Hochella, William Casey, and David DesMarais for review of the content. Research support from the National Science Foundation (Grants EAR-9508171, EAR-9317082 to JFB, CHE-9521731 to JFB, CHE9521731 to JFS, and CHE-9253704 to RJH) and the Department of Energy (DE-FG02-93ER14328 to JFB) is gratefully acknowledged.

REFERENCES

Aagard P, Helgeson HC (1982) Thermodynamic and kinetic reaction rates among minerals and aqueous solutions I. Theoretical considerations. Am J Sci 282:237-285

Bailey SW (1988) S-ray identification of the polytypes of mica, serpentine, and chlorite. Clays Clay Minerals 36:193-213

Bailey SW (1988b) Chlorites: structure and crystal chemistry. In: Hydrous Phyllosilicates (exclusive of micas). SW Bailey Ed. Rev Mineral 19:347-404

Banerjee SK (1991) Magnetic properties of Fe-Ti oxides. In: Oxide Minerals: Petrologic and Magnetic Significance. Rev Mineral 25:107-128

Banfield JF, Barker WW (1994) Direct observation of reactant-product interfaces formed in natural weathering reactions of exsolved, defective amphibole to smectite: Evidence for episodic, isovolumetric reactions involving structural inheritance. Geochim Cosmochim Acta 58:1419-1429

Banfield JF, Bischoff BL, Anderson MA (1993) TiO_2 accessory minerals: coarsening, and transformation kinetics in pure and doped synthetic nanocrystalline materials. Chem Geol 110:211-232

Banfield JF, Eggleton RA (1988) A transmission electron microscope study of biotite weathering. Clays Clay Minerals 36:47-60

Banfield JF, Eggleton RA (1989) Apatite replacement and rare earth mobilization, fractionation and fixation during weathering. Clays Clay Minerals 37:113-127

Banfield JF, Eggleton R A (1990) Analytical transmission electron microscope studies of plagioclase, muscovite and K-feldspar weathering. Clays Clay Minerals 38:77-89

Banfield JF, Ferruzzi GG, Casey WH, Westrich HR (1995) HRTEM study comparing naturally and experimentally weathered pyroxenoids. Geochim Cosmochim Acta 59:19-31

Banfield JF, Jones BF, Veblen DR (1991) An AEM-TEM study of weathering and diagenesis, Abert Lake, Oregon: I. Weathering reactions in the volcanics. Geochim Cosmochim Acta 55:2781-2793

Banfield JF, Murakami T (1998) Atomic-resolution transmission electron microscope evidence for the mechanism by which chlorite weathers to 1:1 semi-regular chlorite-vermiculite. Am Mineral, in press

Banfield JF, Veblen DR, Jones, BF (1990) Transmission electron microscopy of subsolidus oxidation and weathering of olivine. Contrib Mineral Petrol 106:110-123

Belcher AM, Wu XH, Christensen RJ, Hansma PK, Stucky GD, Morse DE (1996) Control of crystal phase switching and orientation by soluble mollusk-shell proteins. Nature 381:56-58

Bennett PC (1991) Quartz dissolution in organic-rich aqueous systems. Geochim Cosmochim Acta 55:1781-1797

Berman A, Addadi L, Weiner S (1988) Interactions of sea urchin skeleton macromolecules with growing calcite crystals - a study of intracrystalline proteins. Nature 331:546-548

Bernal JD (1951) The Physical Basis of Life. Routledge Kegan Paul, London

Blum AE and Lasaga AC (1988) Role of surface speciation in the low-temperature dissolution of minerals. Nature 4:431-433

Blum AE, Stillings LL (1995) Feldspar dissolution kinetics. In: Chemical Weathering Rates of Silicate Minerals, AF While, SL Brantley (eds) Rev Mineral 31:291-352

Boehm H, Knozinger H (1983) Nature and estimation of functional groups on solid surfaces. In: Catalysis Science and Technology. JR Anderson, M Boudart (eds) Springer-Verlag, Berlin, p 39-207

Brantley SA, Chen Y (1995) Chemical weathering rates of amphiboles and pyroxenes. In: Chemical Weathering Rates of Silicate Minerals, AF While, SL Brantley (eds) Rev Mineral 31:119-172

Brantley SA, Crane SR, Crerar DA, Hellmann R, Stallard R (1986) Dissolution of dislocation etch pits in quartz. Geochim Cosmochim Acta 50:2349-2361

Brown GE Jr, Parks GA, O'Day PA (1995) Sorption at mineral-water interfaces: macroscopic and microscopic perspectives. In: Mineral Surfaces. DJ Vaughan, RAD Pattrick (eds) Chapman and Hall London, 370 p

Brown ID, Altermatt D (1985) Bond-valence parameters obtained from a systematic analysis of the inorganic crystal structure database. Acta Cryst B41:244-247

Brown ID, Shannon RD (1973) Empirical bond-strength-bond-length curves for oxides. Acta Cryst A29:266-282

Brown WL (1984) Feldspars and Feldspathoids. Structures, Properties and Occurrences. NATO ASI Series C 137. D Reidel, Dordrecht, the Netherlanda, 541 p

Brownlow AH (1979) Geochemistry. Prentice-Hall, Englewood Cliffs, New Jersey, 498 p

Cairns-Smith AG (1982) Genetic Takeover and the Mineral Origins of Life. Cambridge University Press, Cambridge, UK

Cairns-Smith AG, Hartman H (1986) Clay Minerals and the Origin of Life. Cambridge University Press, Cambridge, UK 193 p

Carroll-Webb SA, Walther JV (1988) A surface complex reaction model for the pH dependence of corundum and kaolinite dissolution rates. Geochim Cosmochim Acta 52:2609-2623

Casey WH, Bunker B (1990) Leaching of mineral and glass surfaces during dissolution. In: Mineral-Water Interface Geochemistry. MF Hochella, AF White (eds) Rev Mineral 23:397-426

Casey WH, Lasaga AC, Gibbs GV (1990) Mechanisms of silica dissolution as inferred from the kinetic isotope effect. Geochim Cosmochim Acta 54:3369-3378

Casey WH, Ludwig C (1995) Silicate mineral dissolution as a ligand-exchange reaction. In: Chemical Weathering Rates of Silicate Minerals, AF While, SL Brantley (eds) Rev Mineral 31:87-114

Casey WH, Phillips BL, Nordin J (1998) Interfacial kinetics through the lens of solution chemistry: Hydrolytic processes at oxide mineral surfaces. In: Kinetics and Mechanisms of Reactions at the Mineral/Water Interface. DL Sparks, T Grundl (eds) Am Chem Soc, Washington, DC (in press)

Casey WH, Sposito G (1992) On the temperature dependence of dissolution rates. Geochim Cosmochim Acta 56:3825-3830

Casey WH, Westrich HR (1992) Control of dissolution rates of orthosilicate minerals by divalent metal-oxygen bonds. Nature 355:157-159

Casey WH, Westrich HR, Massis T, Banfield J F, Arnold GW (1990b) The surface of labradorite feldspar after acid hydrolysis. Chem Geol 78:205-218

Cech TR (1987) The chemistry of self-splicing RNA and RNA enzymes. Science 236:1532-1539

Chang S, DesMarais D, Mack R, Miller SL, Strathern GE (1983) The Earths Earliest Biosphere: Its Origin and Evolution. JW Schopf Ed. Princeton University Press, Princeton, New Jersey, 51-82

Chianello RP, Sturcio NC (1995) The calcite (10$\underline{1}$4) cleavage surface in water: Early results of a crystal truncation rod study. Geochim Cosmochim Acta 59:4557-4561

Clemency CV, Lin FC (1981) Dissolution kinetics of phlogopite. II. Open system using an ion-exchange resin. Clays Clay Minerals 29:107-112

Cragg BA, Parkes, RJ (1994) Bacterial profile sin hydrothermally active deep sediment layers from Middle Valley (NE Pacific), sites 857 and 858. In: Proc Ocean Drilling Program, Scientific Results 139: 509-516

Crick FHC (1968) The origin of the genetic code. J Mol Biol 38:367-379

Crick FHC, Orgel LE (1973) Directed panspermia. Icarus 19:341-345

Davies PCW (1996) The transfer of viable microorganisms between planets. In: Evolution of hydrothermal ecosystems on and Mars GR Bock, JA Goode Eds. John Wiley & Sons, New York, 304-314.

Davis JA, James RO, Leckie JO (1978) Surface ionization electric double layer properties in simple electrolytes. J Colloid Interface Sci 63:4480-499

Delano JW (1997) When did life become sustainable? 27th NERM American Chemical Society Meeting, Saratoga Springs, June 1997, p 63

Dove PM (1995) Kinetic and thermodynamic controls on silica reactivity in weathering environments. In: Chemical Weathering Rates of Silicate Minerals, AF While, SL Brantley (eds) Rev Mineral 31:235-282

Du Q, Freysz E, Shen YR (1994) Vibrational spectra of water molecules at quartz/water interfaces. Phys Rev Lett 72:238-241

Dzombak DA, Morel FMM (1990) Surface Complexation Modeling. John Wiley & Sons, New York

Eberl DD (1980) Alkali cation selectivity and fixation by clay minerals. Clays Clay Minerals 28:161-172

Eggleston C (1994) High resolution scanning probe microscopy: tip-surface interaction, artifacts, and applications in mineralogy and geochemistry. In: Scanning Probe Microscopy of Clay Minerals, CMS Workshop Lectures 7:1-90

Eggleton RA (1984) Formation of iddingsite rims on olivine: a transmission electron microscope study. Clays Clay Minerals 32:1-11

El Gibaly MH, El Reweiny FM, Abdel-Nasser M, El Dahtory TA (1977) Studies of phosphate-solubilizing bacteria in soil rhizosphere of different plants. Occurrence of bacterial acid producers and phosphate dissolvers. Zentrabl Bakt Parasitkde Infek Kr Hyg 132:233-239

Evangelou VP, Zhang YL (1995) A review: pyrite oxidation mechanisms and acid mine drainage prevention. Critical Reviews in Environmental Science and Technology 25:141-199

Falini G, Albeck S, Weiner S, Addidi L (1996) Control of aragonite and calcite polymorphism by mollusk shell macromolecules. Science, 271:67-69

Ferris JP, Hill AR, Rihe L, Orgel LE (1996) Synthesis of long prebiotic oligomers on mineral surfaces Nature 318:59-61

Ferruzzi GG (1993) The character and rates of dissolution of pyroxenes and pyroxenoids. MS Thesis, University of California, Davis, CA

Flörke OW, Flörke U, Giese U (1984) Moganite: a new microcrystalline silica mineral. N Jahrb Mineral Abh 149:325-336

Fornasiero D, Eijt V, Ralston J (1992) An electrokinetic study of pyrite oxidation. Colloids Surf 62:63.

Fredrich JT, Greaves KH, Martin JW (1993) Pore geometry and transport properties of Fontainebleau sandstone. Int. J. Rock Mech Mineral Sci 30:691-697

Fredrich JT, Menendez B, Wong T-F (1995) Imaging the pore structure of geomaterials. Science 268:276-279

Fredrich JT, Wong T-F (1986) Micromechanics of thermally induced cracking in three crustal rocks. J Geophys Res 91:12743-12764

Furrer G, Stumm W (1986) The coordination chemistry of weathering: I. Dissolution kinetics of δ-Al$_2$O$_3$ and BeO. Geochim Cosmochim Acta 50:1847-1860

Furrer G, Zysset M, Schlindler PW (1993) Weathering of montmorillonite: Investigations in batch and mixed flow reactors. In: Geochemistry of Clay-Pore Fluid Interactions. DAC Manning, PL Hall, CR Hughes (eds) Chapman Hall, London, p 243-262

Giese RF Jr., van Oss CJ (1993) The surface thermodynamic properties of silicates and their interactions with biological materials. In: Health Effects of Mineral Dusts. GD Guthrie, BT Mossman (eds) Rev Mineral 28:327-346

Gilbert W (1986) The RNA world. Nature 319:618

Gillett SL (1985) The rise and fall of the early reducing atmosphere. Astronomy 13:66-71

Gribb AA, Banfield JF (1997) Particle size effects on transformation kinetics and phase stability in nanocrystalline TiO$_2$. Am Mineral 82 (in press)

Guthrie GD, Mossman BT (eds) (1993) Health Effects of Mineral Dusts. Rev Mineral 28, 584 p

Hansen and Mossman BT (1987) Generation of superoxide (0^{2-}) from alveolar macrophages exposed to asbestiform and nonfibrous particles. Cancer Res 47:1681-1686

Hayes JM, Kaplan IR, Wedeking KW (1983) Precambrian organic geochemistry, preservation of the record. In: Earth's Earliest Biosphere. Its Origin and Evolution. JW Schopf (ed) Princeton University Press, Princeton, New Jersey, p 93-134

Hayes KF, Leckie JO (1987) Modeling ionic strength effects on cation absorption at hydrous oxide–solution interfaces. J Colloid Interf Sci 115:564-572

Heaney PJ, Prewitt CT, Gibbs GV (eds) (1994) Silica. Physical Behavior, Geochemistry, and Materials Applications. Rev Mineral 29, 606 p

Hering JG, Stumm W (1990) Oxidative and reductive dissolution of minerals. In: Mineral Water Interface Geochemistry. Rev Mineral 23:427-465

Herring C (1951) Some theorems on the free energies of crystal surfaces. Phys Rev 82:87

Higgins SR, Hamers RJ (1996) Chemical dissolution of the galena (001) surface observed using scanning tunneling microscopy. Geochim Cosmochim Acta 60:3067-3073

Hinsinger P, Jaillard B, Duffery JE (1992) Rapid weathering of trioctahedral mica by the roots of ryegrass. Soil Sci Soc Am J 56:977-982

Hochella MF Jr (1990) Atomic structure, microtopography, composition, and reactivity of mineral surfaces. In: Mineral Water Interface Geochemistry. Rev Mineral 23:87-132

Hochella MF Jr (1993) Surface chemistry, structure, and reactivity of hazardous mineral dust. In: Health Effects of Mineral Dusts. Rev Mineral 28:275-308

Hochella MF Jr (1995) Mineral surfaces: their characterization and their chemical, physical, and reactive nature. In: Mineral Surfaces. DJ Vaughan, RAD Pattrick (eds) Chapman and Hall, London, p 17-53.

Hochella MF Jr, Banfield JF (1995) Chemical weathering of silicates in nature: A microscopic perspective with theoretical considerations. In: Chemical Weathering Rates of Silicate Minerals. AF White, SL Brantley (eds) Rev Mineral 30:353-406

Hochella MF Jr, White AF (eds) (1990) Mineral-Water Interface Geochemistry. Rev Mineral 23, 603 p

Huber C, Wächtershäuser G (1997) Activated acetic acid by carbon fixation on (Fe,Ni)S under primordial conditions. Science 276:245-247

Hume LA and Rimstidt JD (1992) The biodurability of chrysotile asbestos. Am Mineral 77:1125-1128

Hyde BG Andersson S (1989) Inorganic Crystal Structures. John Wiley & Sons, New York, 430 p

Inskeep WP, Silver R, Tooh JC (1988) Inhibition of hydroxy apatite precipitation in the presence of fulvic, humic, and tannic acids. Soil Sci Am J 50:941-946

Inskeep WP, Nater EA, Bloom PR, Vandervoort DS, Erich MS (1991) Characterization of laboratory weathered surfaces using XPS spectroscopy and TEM. Geochim Cosmochim Acta 55:787-800

James RO, Healy TW (1972) Adsorption of hydrolyzable metal ions at the oxide-water interface III. A thermodynamic model of adsorption. J Colloid Interface Sci 40:65-81

James RO, Parks GA (1982) Characterization of aqueous colloids by their electric double-layer and intrinsic surface chemical properties. In Surface and Colloid Science. E Matijevic (ed) Plenum Press, New York, p 119-216

Joyce GE, Orgel LE (1993) Prospects for understanding the origin of the RNA world. In: The RNA World. RF Gesteland, JF Atkins (eds) Cold Springs Harbor Laboratory Press, New York, p 1-25

Kasting JF (1993) Earths early atmosphere. Science 259:920-926

Klein C, Hurlbut CS Jr. (1994) Manual of Mineralogy, 21st ed. John Wiley & Sons, New York, 596 p

Kogure T, Murakami T (1996) Direct identification of biotite/vermiculite layers in hydrobiotite using high-resolution TEM. Mineral J 18:131-137

Kulaev IS (1997) Role of polyphosphates in biochemical evolution. 27th NERM Am Chem Soc Mtg, Saratoga Springs, Florida, June 1997, p 62

Knauss KG, Wolery TJ (1986) Dependence of albite dissolution kinetics on pH and time at 25°C and 70°C. Geochim Cosmochim Acta 50:2481-2497

Langmuir D (1971) Particle size effects on the reaction goethite = hematite + water. Am J Sci 271:147-156

Lasaga AC (1981) Transition state theory. In: Kinetics of Geochemical Processes. AC Lasaga, RJ Kirkpatrick (eds) Rev Mineral 8:1-81

Lasaga AC (1995) Fundamental approaches in describing mineral dissolution and precipitation rates. In: Chemical Weathering Rates of Silicate Minerals. AF White, SL Brantley .(eds) Rev Mineral 31:23-86

Lawrence JR and Hendry MJ (1994) Transport of bacteria through geologic media. Can J Microbiol 42:410-422

Lazcano A, Oró J, Miller SL (1983) Primitive earth environments: organic synthesis and the origin and early evolution of life. In: Developments and Interactions of the Precambrian Atmosphere, Lithosphere, and Biosphere. B Nagy, R Weber, JC Guerrero, M Schidlowski (eds) Elsevier, New York, p 151-174

Lee MR and Parsons I (1995) Microtextural controls on weathering of perthitic alkali feldspars. Geochim Cosmochim Acta 59:4465-4488

Lin FC, Clemency CV (1981) Dissolution kinetics of phlogopite. I. Closed system. Clays Clay Minerals 29:101-106

Livingston DA (1963) Chemical composition of rivers and lakes. US Geol Survey Prof Paper 440-G, 64 p

Luther GW III (1987) Pyrite oxidation and reduction: molecular orbital theory consideration. Geochim Cosmochim Acta 51:3193-3215

Luther GW III (1990) The frontier-molecular-orbital theory approach in geotechnical processes. In: Aquatic Chemical Kinetics. W Stumm (ed) John Wiley & Sons, New York, 173 p

Mann S, Didymus, JM, Sanderson, NP, Heywood BR, Aso Samper EJ (1990) Morphological influence of functionalized and nonfunctionalized α, ω dicarboxylates on calcite crystallization. J Chem Soc Faraday Trans 86:1873-1880

Mann S, Archibald DD, Didymus JM, Douglas T, Heywood BR, Meldrum FC, Reeves NJ (1993) Crystallization at inorganic-organic interfaces: biominerals and biomimetic synthesis. Science 261:1286-1292

Marfunin AS (1994) Composition, structure, and properties of mineral matter: concepts, results and problems. Springer-Verlag, New York, 550 p

Mason B, Moore CB (1982) Principles of Geochemistry. John Wiley & Sons, New York

Mast A, Drever JT (1987) The effect of oxalate on the dissolution rates of oligoclase and tremolite. Geochim Cosmochim Acta 51:2559-2568

May HM, Acker JG, Smyth JR, Bricker OP, Dyar MD (1995) Aqueous dissolution of low-iron chlorite in dilute acid solutions at 25°C. Clay Mineral Soc Prog Abstr 32:88

Miller SL (1953) A production of amino acids under primitive Earth conditions. Science 117:528-529

Miller SL, Orgel LE (1974) The Origins of Life on Earth. Prentice Hall, Englewood Cliffs, New Jersey.

Mojzis SJ, Arrhenius G, McKeegan KD, Harrison TM, Nutman AP, Friend CRL (1996) Evidence for life on Earth before 3,800 million years ago. Nature 384:55-59

Moore DM, Reynolds RC Jr. (1997) X-ray Diffraction and the Identification and Analysis of Clay Minerals, 2nd ed. Oxford University Press, Oxford, UK 378 p

Morse JW (1983) Kinetics of calcium carbonat dissolution and precipitation. In: Carbonates: Mineralogy and Chemistry. RJ Reeder (ed) Rev Mineral 11:227-264

Moses CO Nordstrom DK Herman JS, Mills AL (1987) Aqueous pyrite oxidation by dissolved oxygen and by ferric iron. Geochim Cosmochim Acta 51:1561-1571

Müller D, Pitsch S, Kittaka A, Wagner E, Wintner CE, Eschenmoser A (1990) Chemie von α-Aminonitrilen. Aldomerisierung von Glykoaldehydphosphat zu racemischen hexose-2,4,6-triphosphaten und (in Gegenwart von Formaldehyd) Racemischen pentose-2,4-diphosphaten: rac.-allose-2,4,6-triphosphat und rac.-ribose-2,4-diphosphat sind die Reaktionshauptprodukte. Helv Chim Acta 73:1410-1468, cited in Joyce and Orgel (1993)

Nagy KL (1995) Dissolution and precipitation kinetics of sheet silicates. In: Chemical Weathering Rates of Silicate Minerals. AF White, SL Brantely (eds) Rev Mineral 30:173-234

Nelson DC, Casey WH, Sison JD, Mack EE, Ahmad A, Pollack J (1996) Se uptake by sulfur-accumulating bacteria. Geochim Cosmochim Acta 60:3531-3539

Nesbitt HW, Young GM (1984) Prediction of some weathering trends of plutonic and volcanic rocks based on thermodynamic and kinetic considerations. Geochim Cosmochim Acta 48:1523-1534

Orgel LE (1968) Evolution of the genetic apparatus. J Mol Biol 38:381-393

Oró J (1961) Mechanism of synthesis of adenine from hydrogen cyanide under plausible primitive earth conditions. Nature 191:1193-1194

Pace NR (1991) Origin of life—facing up to the physical setting. Cell 65:531-533

Parks GA (1965) The isoelectric points of solid oxides, solid hydroxides, and aqueous hydroxo complex systems. Chem Reviews 65:177-198

Parks GA (1967) Aqueous surface chemistry of oxides and complex oxide minerals. In: Equilibrium Concepts in Natural Water Systems. RF Gould (ed) Am Chem Soc, p 121-160

Phillips BL, Casey WH Crawford SN (1997) Solvent exchange in $AlF_x(H_2O)_{6-x})^{3+}$ (aq) complexes: ligand-directed labilization of water as an analog water as an analog. Geochim Cosmochim Acta 15:3041-3049

Pitsch S, Eschenmoser A, Gedulin B, Hui S, Arrhenius G (1995) Mineral Induced Formation of Sugar Phosphates . 15 Chemistry of Alpha-Aminonitrile from the Zürich Group. Origins of Life and Evolution of the Biosphere. 25:297-334

Prewitt CT (ed) (1980) Pyroxenes. Rev Mineral 7, 529 p

Reeder RJ (ed) (1983) Carbonates: Mineralogy and Chemistry. Rev Mineral 11, 394 p

Ribbe PH (ed) (1974) Sulfide Mineralogy. Rev Mineral 1, 284 p

Ribbe PH (ed) (1981) Feldspar Mineralogy. Rev Mineral 2, 300 p

Ribbe PH (ed) (1982) Orthosilicates. Rev Mineral 5, 450 p

Richards FA (1965) Chemical Oceanography 1. JP Riley, G Skirrow (eds) Academic Press, New York, 198 p

Rickard DT (1997) Kinetics of pyrite formation by the H_2S oxidation of iron (II) monosulfide in aqueous solutions between 25°C and 125°C: the rate equation. Geochim Cosmochim Acta 61:115-134

Rimstidt JD, Barns HL (1980) The kinetics of silica-water reactions. Geochim Cosmochim Acta 44:1683-1699

Russell MJ, Hall, AJ (1997) The emergence of life from iron monosulfide bubbles at a submarine hydrothermal redox and pH front. J Geol Soc London 154:377-402

Sanchez RA, Ferris JP, Orgel LE (1967) Studies in prebiotic synthesis. II synthesis of purine precursors and amono acids from aqueous hydrogen cyanide. J Mol Biol 30:223-253

Schott J, Berner RA, Sjoberg EL (1981) Mechanism of pyroxene and amphibole weathering. I. Experimental studies of iron-free minerals. Geochim Cosmochim Acta 45:2123-2135

Schweda P, Sjöberg L, Södervall U (1997) Near-surface composition of acid-leached labradorite investigated by SIMS. Geochim Cosmochim Acta 61:1985-1994

Schopf Walter (1983) Earths early biosphere: its origin and evolution. JW Schopf (ed) Princeton University Press, Princeton, New Jersey, p 214-239

Shapiro R (1984) The improbability of prebiotic nucleic acid synthesis. Origins Life 14:565-570

Shapiro R (1997) The prebiotic synthesis of cytosine: A critical analysis. 27th NERM Am Chem Soc Mtg, Saratoga Springs, Florida, June 1997, p 60

Smelik EA, Veblen DR (1992) Exsolution of hornblende and the solubility of Ca in orthoamphiboles. Science 257:1669-1672

Smith KL, Milnes AR, Eggleton RA (1987) Weathering of basalt: formation of iddingsite. Clays Clay Minerals 35:418-428

Smyth JR, Bish DL (1988) Crystal Structures and Cation Sites in the Rock-Forming Minerals. Allen and Unwin, Oxford, UK

Spitz K, Moreno J (1996) A Practical Guide to Groundwater and solute transport modeling. John Wiley & Sons, New York.

Stetter KO (1996) Hyperthermophilic procaryotes. FEMS Microbiol Rev 18:149-158

Stillings LL, Brantley SL (1995) Feldspar dissolution at 25°C and pH 3: Reaction stoichiometry and the effect of cations. Geochim Cosmochim Acta 59:1483-1496

Stipp SL, Hochella MF Jr. (1991) Structure and bonding environments at the calcite surface as observed with x-ray absorption spectroscopy (XPS) and low energy electron diffraction (LEED). Geochim Cosmochim Acta 55:1723-1736

Sverjensky DA, Sahai N (1996) Theoretical prediction of single-surface-protonation equilibrium constants for oxides and silicates in water. Geochim Cosmochim Acta 60:3773-3797

Thirioux L, Baillilf P, Touray JC, Ildefonse JP (1990) Surface reactions during fluorapatite dissolution–recrystallization in acid media (hydrochloric and citric acids). Geochim Cosmochim Acta 54:1969-1977

Tolman RC (1949) The effect of droplet size on surface tension. J Chem Phys 17:333-337

Turekian KK (1968) Oceans. Prentice-Hall, Englewood Cliffs, New Jersey, 92 p

Usher DA (1997) How the gene got its backbone: a perhaps so story. 27th NERM Am Chem Soc Mtg, Saratoga Springs, Florida, June 1997, p 59

Vasconcelos C, McKenzie JA, Bernasconi S, Grujic D, Tien AJ (1995) Microbial mediation as a possible mechanism for natural dolomite formation at low temperatures. Nature 377:220-222

Vaughan DJ, Pattrick RAD (eds) (1995) Mineral Surfaces. Mineralogical Society Series 5. Chapman and Hall, London, 370 p

Velbel MA (1993) Constancy of silicate mineral weathering ratios between natural and experimental weathering: implications for hydrologic control of differences in absolute rates. Chem Geol 105: 89-99

Veblen DR (ed) (1981) Amphiboles and Other Hydrous Pyriboles—Mineralogy. Rev Mineral 9A, 372 p

Wächtershäuser G (1988) Pyrite formation, the first energy source for life: A hypothesis. Systematic Applied Microbiol 10:207-210

Wächtershäuser G (1990) Evolution of the first metabolic cycles. Proc Nat Acad Sci USA 87:200-204

Wächtershäuser G (1997) The priming of the alpha cycle in an iron-sulfur world. 27th NERM Am Chem Soc Mtg, Saratoga Springs, Florida, June 1997, p 63

Wardrup MM (1990) Goodbye to the warm little pond? Science 250:1078-1080

Weber P, Greenberg JM (1985) Can spores survive in interstellar space? Nature 316:403-497

Weiland E, Wehrli B, Stumm W (1988) The coordination chemistry of weathering. III. A generalization of the dissolution rates of minerals. Geochim Cosmochim Acta 52:1969-1981

Werner AJ, Hochella MF Jr, Guthrie GD Jr., Hardy JA, Aust AE, Rimstidt JD (1995) Asbestiform riebeckite (crocidolite) dissolution in the presence of Fe chelators: Implications for mineral-induced disease. Am Mineral 80:1093-1103

Westrich HR, Cygan RT, Casey WH, Zemitis C, Arnold GW (1993) The dissolution kinetics of mixed cation orthosilicate minerals. Am J Sci 293:869-893

White AF, Brantley SL (eds) (1995) Chemical Weathering Rates of Silicate Minerals. Rev Mineral 31, 583 p

Wilkens RG (1991) Kinetics and Mechanisms of Reactions of Transition Metal Complexes. VCH Publishers, New York, 465 p

Woese, C (1967) The evolution of the genetic code. In: The Genetic Code. Harper and Row, New York, p 179-195

Wong T-F, Fredrich JT, Gwanmesia GD (1989) Crack aperture statistics and pore space fractal geometry of Westerly granite and Rutland quartzite: Implications for an elastic contact model of rock compressibility. J Geophys Res 94:10267-10278

Zeltner WA, Anderson MA (1988) Surface charge development at the goethite/aqueous solution interface. Effects of CO_2 adsorption. Langmuir 4:469-474

Zhang H, Banfield JF (in review) Thermodynamic stabilities of nanocrystalline powders. J Am Ceram Soc

Chapter 4

SPATIAL RELATIONSHIPS BETWEEN BACTERIA AND MINERAL SURFACES

Brenda J. Little and Patricia A. Wagner

Naval Research Laboratory
Stennis Space Center, Mississippi 39529 U.S.A.

Zbigniew Lewandowski

Center for Biofilm Engineering
Department of Civil Engineering
Montana State University
Bozeman, Montana 59717 U.S.A.

INTRODUCTION

In aquatic environments, microbial cells attach to solids, including minerals and metals, and initiate biomineralization reactions. "Immobilized cells grow, reproduce and produce extracellular polymers which frequently extend from the cell forming a tangled matrix of fibers which provide structure to the assemblage termed a biofilm" (Characklis and Marshall 1990). Microorganisms within biofilms are capable of maintaining environments at biofilm/surface interfaces that are radically different from the bulk in terms of pH, dissolved oxygen, and other organic and inorganic species. In some cases, these interfacial conditions could not be maintained in the bulk medium at room temperature near atmospheric pressure. As a consequence, microorganisms within biofilms produce minerals and mineral replacement reactions that are not predicted by thermodynamic arguments based on the chemistry of the bulk medium.

While it has been established that the most devastating microbiologically influenced corrosion (MIC) takes place in the presence of microbial consortia in which many physiological types of bacteria, including sulfate-reducing bacteria (SRB), acid-producing bacteria, metal-oxidizing bacteria and metal-reducing bacteria (MRB), interact in complex ways within the structure of biofilms (Fig. 1) (Pope et al. 1984, Little et al. 1991), the realization that biomineralization takes place within biofilms has received inadequate attention. Spatial relationships of microorganisms to minerals has been studied to a limited extent as related to MIC. Microorganisms influence corrosion by both forming and dissolving minerals. Biomineralization that results in mineral deposition on a metal surface can shift the corrosion potential in either a positive or negative direction, depending on the nature of the mineral. Manganese oxide biodeposition on stainless steel surfaces forces a shift in the positive, more noble direction, moving the corrosion potential above the pitting potential and making some stainless steels more vulnerable to pitting and crevice corrosion. Bioprecipitated sulfides decrease hydrogen overvoltage at cathodic sites and stimulate the cathodic reaction so that sulfide formation on all metal surfaces moves the corrosion potential in a negative, more active direction, resulting in accelerated corrosion of some metals and alloys. Iron oxide formation can initiate a sequence of events that results in underdeposit corrosion of susceptible metals. Biomineral dissolution reactions remove passive layers or force mineral replacement reactions that lead to further dissolution. In the following sections biofilm formation and spatial relationships of microorganisms and minerals as related to corrosion will be described.

0275-0279/97/0035-0004$05.00

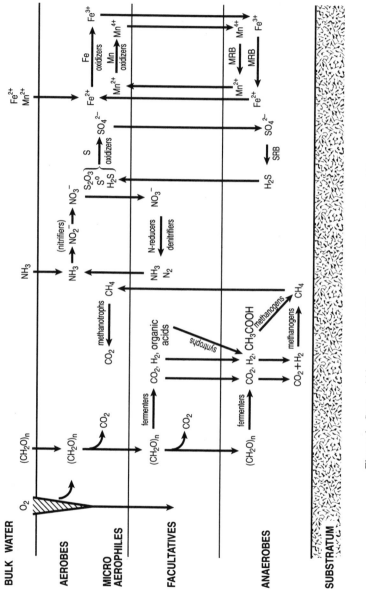

Figure 1. Strata within a typical biofilm and possible reactions within the strata.

OVERVIEW OF BIOFILM FORMATION

Initial events of biofilm formation

Biofilm formation consists of a sequence of steps and begins with adsorption of macromolecules (proteins, polysaccharides, and humic acids) and smaller molecules (fatty acids, lipids) at interfaces, including liquid/solid and gas/liquid interfaces. The following sections deal with specifics of biofilm formation on solid substrata. Adsorbed molecules form conditioning films which alter physico-chemical characteristics of the interface (Chamberlain 1992, Marshall et al. 1994). Conditioning films change surface hydrophobicity and influence surface electrical charge. Both parameters have proven effects on the extent and kinetics of microbial attachment. Surface charge may change because of the conditioning film and/or because of adsorption of ions from solution (Loeb and Neihof 1976). The amount of adsorbed organic material is a function of ionic strength, and can be enhanced on metal surfaces by polarization (Little 1985).

During initial stages of biofilm formation the major factor controlling rate of colonization is hydrodynamics (Duddridge et al. 1982). Microbial colonization begins with transport of microorganisms to the interface mediated by at least three mechanisms: (1) diffusive transport due to Brownian motion, (2) convective transport due to liquid flow, and (3) active movement of motile bacteria near the interface (van Loosdrecht et al. 1990). The influence of convection transport exceeds the other two by several orders of magnitude. Once the microbial cell is in contact with a surface it may or may not adhere. The ratio of cell number adhering to a surface to the cell number transported to this surface is termed "sticking efficiency." Sticking efficiency depends on many factors including surface properties, physiological state of organisms, and hydrodynamics near the surface (Escher and Characklis 1990).

Biofilm accumulation

Microbial cells transported with the stream of fluid above the surface interact with the conditioning films. Immediately after attachment, microorganisms initiate production of slimy adhesive substances, predominantly exopolysaccharides (EPS). Although the association of EPS with attached bacteria has been well documented (Fletcher and Floodgate 1973) there is no direct evidence suggesting that EPS participates in initial stages of adhesion. However, EPS definitely assists the formation of microcolonies and microbial films (Allison and Sutherland 1987). Recent genetic studies have clearly shown that adhesion to surfaces triggers the expression of several genes controlling polymer synthesis in *Pseudomonas* (Davies and Geesey 1995). According to Silverman et al. (1984) synthesis of adhesive substances is genetically controlled and influenced by environmental factors. For example, Little et al. (1997) documented copious EPS production by *Shewanella putrefaciens* grown on manganese oxide (birnessite), and absence of polymer when the same organism in the same medium was grown on iron oxides (hematite, goethite or ferrihydrite) (Fig. 2a,b). EPS bridges microbial cells with the substratum and permits negatively charged bacteria to adhere to both negatively and positively charged surfaces. EPS may also control interfacial chemistry at the mineral/biofilm interface.

Biofilm accumulation at surfaces is an autocatalytic process. Initial colonization increases surface irregularity and promotes further biofilm formation. Bouwer (1987) pointed out that increased surface irregularity due to biofilm formation can influence particle transport and attachment rate by (1) increasing convective mass transport near the surface, (2) providing shelter from shear forces, and (3) increasing surface area for attachment. Biofilm accumulation is the net result of the following microbial processes: attachment,

GEOMICROBIOLOGY

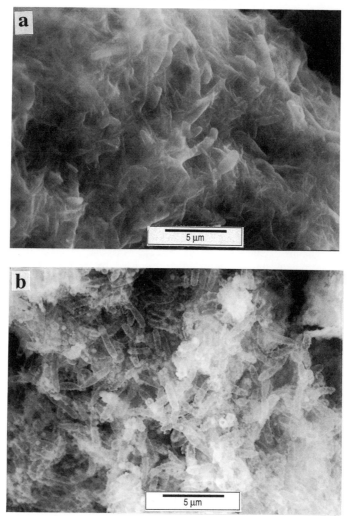

Figure 2. Mineral surfaces after exposure to *S. putrefaciens* (a) manganese oxide (birnessite) cells obscured by EPS, (b) iron oxide (goethite), no visible EPS.

growth, decay, and detachment (Fig. 3). More rigorous treatments introduce adsorption and desorption (Bryers and Characklis 1992, Escher 1986). However, because of the complexity of microbial binding to surfaces, the terms attachment and detachment are frequently used without referring to specific physical processes. Attachment is due to microbial transport and subsequent binding to surfaces. Growth is due to microbial replication and growth rate is traditionally described by Monod kinetics:

$$\mu = \frac{\mu_{max} * S}{K_s}$$

where μ_{max} = maximum specific growth rate (t^{-1}), K_s = half saturation coefficient (mole L^{-3}), S = substrate concentration (mole L^{-3}). Each species in the biofilm has its own growth parameters. Growth rate in biofilms may depend on the spatial position of

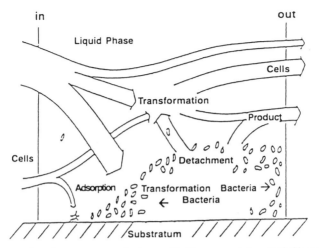

Figure 3. A composite of processes contributing to biofilm accumulation (WG Characklis, "Biofilm Processes," in *Biofilms*. ®1990 Reprinted by permission of John Wiley & Sons.)

microorganisms, unlike in suspended growth reactors. Assumption that all microorganisms of the same species have the same growth parameters in biofilms appears to be overly simplistic. Instead there is a spatial distribution of growth parameters.

Detachment includes two processes: erosion and sloughing. Sloughing is the process in which large pieces of biofilm are rapidly removed, frequently exposing the surface. Reasons for biofilm sloughing are not well understood. Biofilm erosion is defined as continuous removal of single cells or small groups of cells from the biofilm surface and is related to shear stress at the biofilm-fluid interface. Frequently detachment is identified with erosion, especially in conduits. An increase in shear stress increases erosion rate and decreases biofilm accumulation rate. Empirical observations indicate erosion rate is related to biofilm thickness and density.

Biofilm architecture

When biofilm populations in aquatic ecosystems are imaged by light and electron microscopy, they appear to be composed of bacterial cells enclosed in an EPS matrix of uniform thickness and consistency (Costerton et al. 1987). Conceptual and numerical models have traditionally treated biofilms as a layer of matrix material within which bacterial cells are randomly distributed, and these models appear to be predictive of reaction rates in bioreactors (Wanner and Gujer 1986, Rittmann and Manem 1992). With time, however, it has become apparent that multi-species biofilms form highly complex structures containing voids, channels, cavities, pores and filaments, with cells arranged in clusters. Such complex structures have been reported in a wide variety of biofilms including methanogenic films from fixed-bed reactors (Robinson et al. 1984), aerobic films from wastewater plants (Eighmy et al. 1983, Mack et al., 1975), nitrifying biofilms (Kugaprasatham et al. 1992), and pure culture biofilms of *Vibrio parahaemolyticus* (Lawrence et al. 1991) and *Pseudomonas aeruginosa* (Stewart et al. 1993). High resolution confocal laser scanning microscopy (CLSM), coupled with the sophisticated image analysis (Wilson 1990, Caldwell et al. 1992) produced detailed images of the intrabiofilm environment. Lawrence et al. (1991) presented images of a complex biofilm architecture in which cells grow in matrix-enclosed microcolonies separated by water-filled voids. This microcolony/void structure of biofilms was confirmed by Keevil and Walker (1992) who noted that the microcolonies formed stacks that extend as many as 500 micrometers away from the colonized surface. Wolfaardt et al. (1994) described this same architecture in biofilms formed by microbial populations in natural ecosystems.

Studies of the structure of biofilms using CLSM followed by studies of flow near microbially colonized surfaces using nuclear magnetic resonance imaging (NMRI), (Lewandowski et al. 1992, 1993a, 1994) and CLSM (de Beer et al. 1994a,b; Stoodley et al. 1994) delivered detailed information on the structure of biofilms and the nature of water flow in biofilm systems. The conceptual image of biofilms became much more complex (Massol-Deya et al. 1995, Gjaltema et al. 1994, Zhang and Bishop 1994 a,b) than the uniform layer with imbedded microorganisms that dominated the early studies. Lewandowski and Stoodley (1995) presented a concept of an intricate interplay between hydrodynamics and viscoelastic biopolymers leading to formation of "streamers" and characteristic biofilm matrix oscillation. Using NMRI, Lewandowski et al. (1993b) demonstrated that water flows through biofilms. This observation was corroborated and quantified by tracking fluorescent beads (0.2 μm diameter) through pores in a biofilm (Fig. 4) using CLSM (Stoodley et al. 1994, de Beer et al. 1994b). These and other experiments led the authors to conclude that biofilms form compliant surfaces which actively interact with the hydrodynamic boundary layer. It is now generally accepted that microorganisms in biofilms are aggregated in cell clusters or microcolonies separated by interstitial voids (Fig. 5) (Costerton et al. 1995). The new conceptual model assumes an inherent biofilm heterogeneity and constitutes the foundation for studying biofilm structure and the consequences of this structure to physical and chemical microenvironments. It has been speculated that biofilm structure represents an optimal arrangement for influx of nutrients but no direct evidence has been presented. There is no doubt, however, that interstitial voids participate in supplying nutrients to deeper layers of biofilms (Lewandowski et al. 1995).

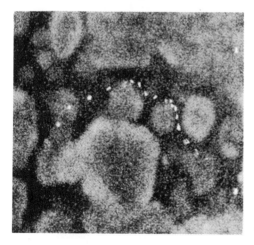

Figure 4. Superimposed time sequence image of an autofluorescing bead moving through a biofilm channel (Stoodley et al. 1994).

SPATIAL RELATIONSHIPS BETWEEN BIOFILMS AND MINERALS

Spatial relationships between bacteria and metal corrosion products have been investigated because of the economic impact to industry and military. MIC has been documented for metals exposed to seawater, fresh water, demineralized water,

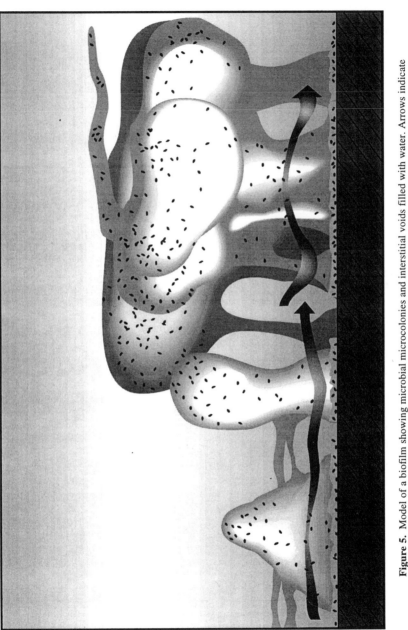

Figure 5. Model of a biofilm showing microbial microcolonies and interstitial voids filled with water. Arrows indicate convective flow (Lewandowski et al. 1995).

process chemicals, food stuffs, soils, aircraft fuels, human plasma, and sewage. The following is a brief introduction to corrosion terms (Uhlig and Revie 1985) that will be used throughout this chapter.

A metal surface is a composite of electrodes electrically short-circuited through the body of the metal. As long as the metal remains dry, local-action current and corrosion are not observed. But on exposure of the metal to water or aqueous solutions, local electrochemical cells are established and are accompanied by chemical conversion of the metal to corrosion products. The electrode at which chemical oxidation occurs is called the anode and the electrode at which reduction takes place is called the cathode. Corrosion of metals usually occurs at the anode. When a specimen is in contact with a corrosive liquid the specimen assumes a potential (relative to a reference electrode) termed the corrosion potential, E_{corr}. A specimen at E_{corr} has both anodic and cathodic currents present on its surface. However, these currents are exactly equal in magnitude so no net current can be measured. The specimen is at equilibrium with the environment even though it may be visibly corroding. E_{corr} can be defined as the potential at which the rate of oxidation is exactly equal to the rate of reduction.

An important concept that will be discussed in following sections is that of passivity. Passivity can be defined as the loss of chemical reactivity exhibited by certain metals and alloys under specific environmental conditions. That is, metals and alloys such as chromium, iron, nickel, titanium, and alloys containing these elements, become passive or essentially inert and act as if they were noble metals. Passivity is due to the formation of a surface film which acts as a barrier to further corrosion. Pitting potential is that potential required to allow penetration of the passive film. Eventually, as potential increases in the noble direction either the oxide is undermined by condensation of migrating vacancies, or cations of the oxide undergo dissolution at the electrolyte interface. In the absence of aggressive anions in the electrolyte, defects in the passive film can heal or repassivate (Szlarska-Smialowska 1986).

Experimental evidence indicates that microbial colonization changes the properties of metal surfaces and, in some cases, makes them more susceptible to corrosion. In an experiment conducted by Pendyala (1996), stainless steel coupons were exposed for 18 days to a biofilm consisting of three species: *Klebsiella pneumonia, Pseudomonas fluorescens* and *Pseudomonas aeruginosa*. Elemental and chemical composition of the surface was analyzed by X-ray photoelectron spectroscopy (XPS) and other surface sensitive spectroscopies. Figure 6a and 6b show changes in the composition of the passive film resulting from microbial colonization. The most dramatic differences are seen in the first 50 Å. Comparison of elemental composition of the passive layer of materials as-received (same as the control) and after the 18-day exposure to microorganisms indicates that microbial action depleted the relative concentrations of chromium and nickel by ~10%.

Mineral deposition

Oxides. Biomineralization of iron and manganese oxides occurs widely in natural waters, and is a dominant control in geochemical cycling of these elements (Gounot 1994). Mineralization can be carried out by a variety of organisms including bacteria, yeast, and fungi (Nealson et al. 1988), but is particularly associated with genera of the so-called iron and manganese bacteria, *Siderocapsa, Gallionella, Leptothrix, Sphaerotilus, Crenothrix,* and *Clonothrix*.

Manganese. Manganese oxidation is coupled to cell growth and metabolism of heterotrophic substrates (Arnold et al. 1988, Jung and Schweisfurth 1976a,b). While the

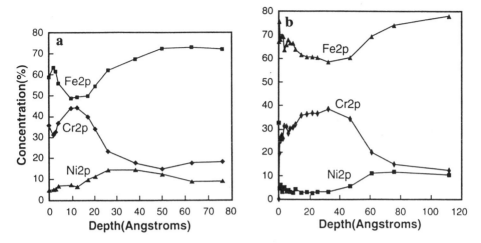

Figure 6. Depth profiles of normalized iron, chromium, and nickel concentrations in a stainless steel (a) as received and (b) after 18 days exposure to biofilm. (Courtesy of J. Pendyala and R. Avci, Dept. Physics, Montana State University.)

reduced form of manganese, generally identified as Mn^{2+} is soluble, all the various oxidized forms, Mn_2O_3, $MnOOH$, Mn_3O_4, MnO_2, are insoluble. Microbially deposited manganese oxides have an amorphous structure as MnO_2 (vernadite) and sometimes form a black precipitate of MnO_2 (birnessite) found with *Leptothrix* and spores of *Bacillus* spp. (Gounot 1994). In the *Bacillus*, however, birnessite recrystallizes to octahedral Mn_3O_4 (hausmannite) (Nealson et. al. 1988). The relationship of manganese-depositing bacteria with MnO_2 can be demonstrated with X-ray microscopy. In Figure 7 a *Pseudomonas*-like

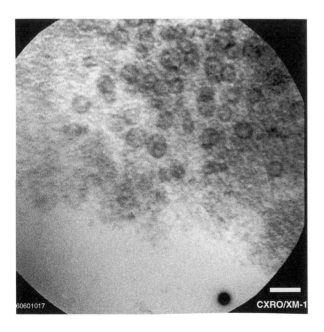

Figure 7. X-ray micrograph of manganese-depositing bacterium with Mn-oxides. Cells are dark ellipses, oxides are lighter material. (An unpublished micro-graph from the Center for X-ray Optics, Lawrence Berkeley Laboratories, Berkeley, CA, produced by B. Tonner, W. Meyer-Ilse & J. Brown from samples provided by K. Nealson, University of Wisconsin, Milwaukee, WI.)

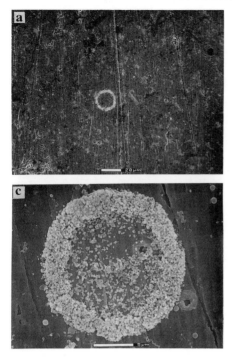
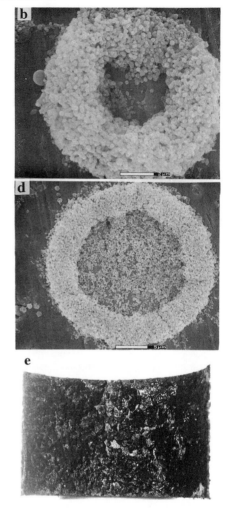
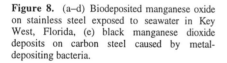

Figure 8. (a–d) Biodeposited manganese oxide on stainless steel exposed to seawater in Key West, Florida, (e) black manganese dioxide deposits on carbon steel caused by metal-depositing bacteria.

organism, originally isolated from freshwater (Jung and Schweisfurth 1976a,b), is imbedded in the manganese oxides it produced. As a result of microbial action, manganese oxide deposits are formed on submerged materials including metal, stone, glass, and plastic and can occur in natural waters with manganese levels as low as 10 to 20 ppb (Dickinson et al. 1996). Deposition rates of 1 mcoul cm^{-2} day^{-1} on stainless steel have been observed (Dickinson and Lewandowski 1996). Mature manganese deposits from both fresh water and marine sources can be identified as ring structures that become more and more numerous and dense with time (Fig. 8a–d). In other instances, surfaces become covered with manganese dioxide deposits (Fig. 8e).

It has been demonstrated that microbially deposited manganese oxide on a stainless steel (Dickinson and Lewandowski 1996) in fresh water (Fig. 9 a–c) caused an increase in E_{corr} and increased cathodic current density at potentials above –200 mV (vs saturated calomel reference electrode (SCE)). For mild steel corrosion under anodic control, the oxides can elevate corrosion current, but will cause little positive shift in E_{corr}. The increase in corrosion current may be significant, particularly for mild steel covered with

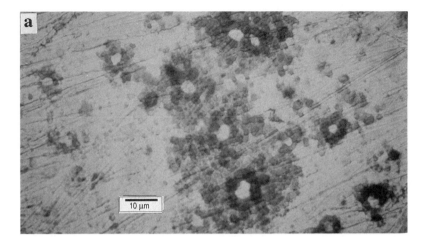

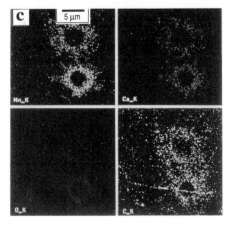

Figure 9. Annular deposits on stainless steel after 13-day exposure to fresh water, (a) reflected light micrograph (Dickinson and Lewandowski 1966), (b) SEM micrograph (Dickinson et al. 1966), and (c) EDS maps of two adjacent annular deposits confirming the presence of manganese, calcium, oxygen, and carbon in the deposits (Dickinson et al. 1966).

biomineralized oxides and tubercles that provide large mineral surface areas. Given sufficient conductivity in the tubercle, much of this material may serve as an oxide cathode to support corrosion at the oxygen-depleted anode within the tubercle. Continued biomineralization within the large tubercle may sustain a significant amount of the cathodic current. Both factors can increase the risk of stainless steel corrosion. Ennobled E_{corr} can exceed pitting potentials for low molybdenum alloys, enhancing risk of pit nucleation, while elevated cathodic current density impedes repassivation. Biomineralized manganic oxides are efficient cathodes and increase cathodic current density on stainless steel by several decades at potentials between roughly −200 and +400 mV$_{SCE}$. The extent to which the elevated current density can be maintained is controlled by the electrical capacity of the mineral reflecting both total accumulation and conductivity of the mineral-biopolymer assemblage (only material in electrical contact with the metal will be cathodically active). Oxide accumulation is controlled by the biomineralization rate and the corrosion current, in that high corrosion currents will discharge the oxide as rapidly as it is formed. This variation in accumulation causes the oxides to exert different modes of influence on the corrosion behavior of active metals compared with passive metals.

While biomineralized manganic oxides are expected to elevate corrosion current on mild steel, on passive metals they serve primarily to initiate localized attack. Passive metal corrosion currents of the order 10 nA cm^{-2} allow biomineralized material to accumulate. E_{corr} then shifts in the noble direction as increasing areal coverage anodically polarizes the metal. E_{corr} may exceed the pitting potential for low molybdenum alloys in dilute chloride media, increasing the risk of pit nucleation. Once nucleation occurs, cathodic current sustained by the MnO_2 cathode impedes repassivation by holding the corrosion potential above the protection potential. More available cathode material will support a greater number of pitting sites, increasing the probability that a metastable site will become fixed. Localized corrosion current that exceeds the biomineralization rate will discharge the oxide cathode so that eventually the corrosion rate becomes limited by the oxide biomineralization rate and by availability of other cathodic reactants (typically dissolved oxygen).

Iron. Iron-depositing bacteria produce orange-red tubercles of iron oxides and hydroxides by oxidizing ferrous ions from the bulk medium or the substratum (Fig. 10). Using environmental scanning electron microscopy (ESEM) it is possible to demonstrate iron-depositing bacteria within tubercles, twisted filaments with iron-rich deposits along their length (Fig. 11). Iron-depositing bacteria are microaerophilic and may require syner-

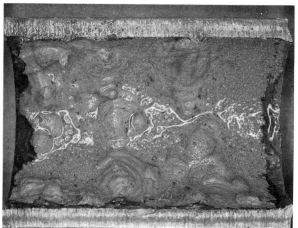

Figure 10. Iron oxide deposits on galvanized pipe (1.6×).

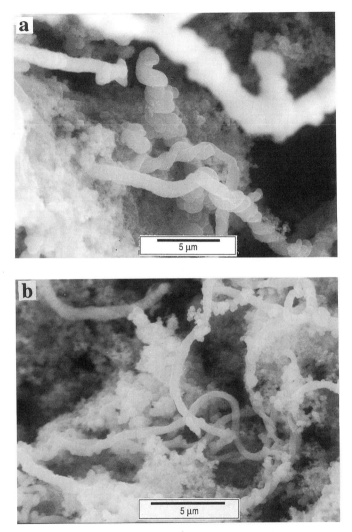

Figure 11. Iron-depositing bacteria within tubercles from Figure 10.

gistic associations with other bacteria to maintain microaerophilic conditions in their immediate environment. Bacilli and cocci were imaged on the surface of the tubercle in Figure 10. Deposits of cells and metal ions create oxygen concentration cells (Fig. 12) that effectively exclude oxygen from the area immediately under the deposit and initiate a series of events that are individually or collectively very corrosive. In an oxygenated environment, the area immediately under individual deposits becomes deprived of oxygen. That area becomes a relatively small anode compared to the large surrounding oxygenated cathode. Cathodic reduction of oxygen may result in an increase in pH of the solution in the vicinity of the metal. The metal will form metal cations at anodic sites. If the metal hydroxide is the thermodynamically stable phase in the solution, the metal ions will be hydrolyzed by water with the formation of H^+ ions. If cathodic and anodic sites are separated from one another, the pH at the anode will decrease and that at the cathode will increase. In addition, Cl^- ions from the electrolyte will migrate to the anode to neutralize any buildup of charge, forming

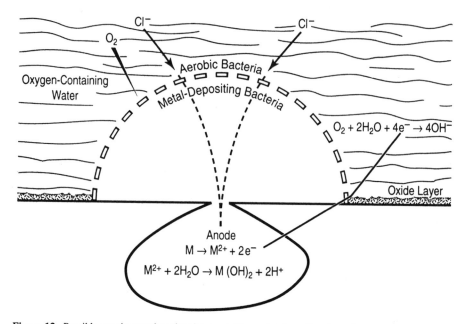

Figure 12. Possible reactions under tubercles created by metal-depositing bacteria.

heavy metal chlorides that are extremely corrosive. Under these circumstances, pitting involves the conventional features of differential aeration, a large cathode to anode surface area and the development of acidity and metallic chlorides. Pit initiation depends on mineral deposition by bacteria. Pit propagation is dependent not on activities of the organisms but on metallurgy (George 1996). This also means that attempts to kill the organisms within mineral deposits using biocides will not result in a cessation of pit propagation (Miller and Tiller 1970). The pH at the anode depends on specific hydrolysis reactions (Table 1). The largest pH decrease is found in alloys containing chromium. Stainless steels containing 6% or more molybdenum are not vulnerable to this type of attack.

Table 1. Specific hydrolysis reactions.

HYDROLYSIS REACTION	EQUILIBRIUM pH
$Fe^{2+} + 2H_2O \rightleftharpoons Fe(OH)_2 + 2H^+$	$pH = 6.64 - 1/2 \log a_{Fe}^{2+}$
$Cr^{3+} + 3H_2O \rightleftharpoons Cr(OH)_3 + 3H^+$	$pH = 1.53 - 1/3 \log a_{Cr}^{3+}$
$Ni^{2+} + 2H_2O \rightleftharpoons Ni(OH)_2 + 2H^+$	$pH = 6.5 - 1/2 \log a_{Ni}^{2+}$
$Mn^{2+} + 2H_2O \rightleftharpoons Mn(OH)_2 + 2H^+$	$pH = 7.66 - 1/2 \log a_{Mn}^{2+}$

Sulfides. Many sulfide minerals under near-surface natural environmental conditions can only be produced by microbiological action on specific precursor metals. SRB are a diverse group of anaerobic bacteria that can be isolated from a variety of environments (Pfennig et al. 1981, Postgate 1979) including seawater where the concentration of sulfate is typically 25 mM (Postgate 1979). Even though seawater is generally aerobic (typical values above the thermocline are in the range 4 to 6 ppm), anaerobic microorganisms survive in anaerobic microniches until conditions are suitable for their growth (Costerton and Geesey 1986, Staffeldt and Kohler 1973). If the aerobic respiration rate within a biofilm is greater than the oxygen diffusion rate, the metal/biofilm interface can become anaerobic and provide a niche for sulfide production by SRB (Little et al. 1990).

In the following sections, SRB sulfide production will be reviewed for iron, copper, copper alloys, silver, zinc and lead. The metal interface under the biofilm and corrosion layers will be referred to as base metal to differentiate it from layers of minerals and metal ions that have been derivatized by corrosion reactions. Mineralogical data, thermodynamic stability diagrams (Pourbaix 1966, Wagman et al. 1982) and the simplexity principle for precipitation reactions (McNeil et al. 1991) will be used to rationalize corrosion product mineralogy in fresh and saline water and to demonstrate the action of SRB.

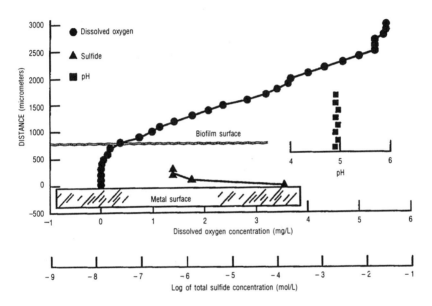

Figure 13. Concentration profiles of sulfide, oxygen, and pH in a biofilm on mild steel (Lee et al. 1993).

Iron. The corrosion rate of iron in the presence of hydrogen sulfide is accelerated by the formation of iron sulfide minerals (Wikjord et al. 1980) that stimulate the cathodic reaction through a decrease in hydrogen overvoltage at cathodic sites. Once electrical contact is established, mild steel behaves as an anode and electron transfer occurs through the iron sulfide. In the absence of oxygen, the metabolic activity of SRB causes accumulation of hydrogen sulfide near metal surfaces. This is particularly evident when metal surfaces are covered with biofilms. Figure 13 shows concentration profiles of sulfide, oxygen, and pH in a biofilm accumulated on the surface of a mild steel corrosion coupon. The concentration of sulfide is highest near the metal surface where iron sulfide forms quickly and covers the steel surface if both ferrous and sulfide ions are available. At low ferrous ion concentrations, adherent and temporarily protective films of iron sulfides are formed on the steel surface with a consequent reduction in corrosion rate. High rates of SRB-induced corrosion of mild steel are maintained only in high concentrations of ferrous ion.

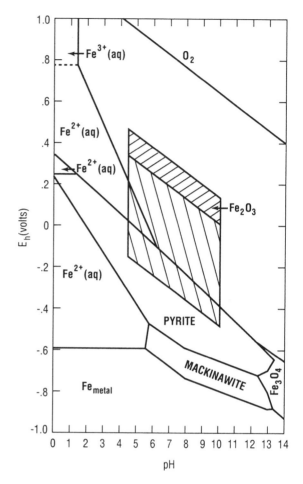

Figure 14. Iron stability diagram: water without chloride ions; total sulfide 10^{-2} M.

Figure 14 is a stability diagram for an iron-water-reduced sulfur system with lines for 10^{-6} M ferrous iron and 10^{-2} M sulfide. For clarity, pyrite (FeS_2) and mackinawite (FeS_{1-x}) are the only sulfides indicated. Parallelograms superimposed on the diagram are bounded by the highest and lowest pH values commonly found in natural fresh and saline surface waters. The upper portion of the hatched area applies to waters less than 10 m from the surface; the lower portion (oppositely hatched) represents waters at depths greater than 10 m. Conditions in the upper hatched parallelogram represent those readily achieved in stagnant waters. The lower portion indicates conditions not found in near-surface environments. Mackinawite is a tetragonal mineral that may be unstable altogether but is clearly unstable above 150°C. It cannot be produced by conventional techniques. Greigite is a thiospinel with formula Fe_3S_4. Smythite is a hexagonal compound with formula $Fe_{3-x}S_4$, $0 < x < 0.25$. Cubic FeS can be produced artificially. By applying H_2S pressures in the range of one atmosphere, Berner (1969) produced "tetragonal FeS," which has the same symmetry as mackinawite but contains somewhat less sulfur and has slight but systematic differences in lattice parameter. Presumably, further increases in H_2S pressure could produce material equivalent to natural mackinawite. Thermodynamic analyses indicate that, under redox and sulfide activity conditions in surface waters, only pyrite is stable; furthermore, pyrite forms relatively easily in nonbiological corrosion, so the preferential formation of less stable sulfides is difficult to attribute to slow pyrite formation kinetics. The region of stability of mackinawite is wholly outside the region defined by surface water conditions, excluding waters influenced by peat bogs, coal mines, volcanic activity and industrial effluents.

During corrosion of iron and steel in the presence of SRB, a thin (approximately 1 μm), adherent layer of "tarnish" is first formed. This was originally termed "kansite," but has since been identified as mackinawite. As it thickens, the layer becomes less adherent. If ferrous ion concentration in the electrolyte is low, the mackinawite alters to greigite. This alteration is not observed in nonbiological systems. If ferrous ion concentration is high, mackinawite is accompanied by green rust 2, a complex ferrosoferric oxyhydroxide. The presence of green rust 2 may be due to a solubility effect and accounts for reduced corrosion rates when the electrolyte is rich in ferrous ions and has a very low renewal rate.

In summary, mackinawite (tetragonal FeS_{1-x}) is easily produced from iron and iron oxides by consortia of microorganisms that include SRB. The presence of mackinawite in corrosion products formed in shallow water environments with the exclusions previously delineated is proof that the corrosion was SRB-induced. Recent work indicates that on continued exposure to SRB mackinawite alters to greigite (Fe_3S_4), smythite (Fe_9S_{11}) and finally to pyrrhotite (FeS_{1+x}) (McNeil and Little 1990). SRB in thin biofilms on pottery surfaces (Duncan and Ganiaris 1987, Heimann 1989) and silver (McNeil and Mohr 1993) can produce pyrite films from iron-rich waters. Pyrite is not a typical iron corrosion product, but SRB can produce pyrite from mackinawite in contact with elemental sulfur (Berner 1969). Abiotic aqueous synthesis of these minerals, with the possible exception of pyrite, requires H_2S pressures higher than those found in shallow waters.

Copper. Cuprite (Cu_2O), the first product of copper corrosion, forms epitaxially as a direct reaction product of copper with dissolved O_2 or with water molecules (North and Pryor 1970). Cuprite has a high electrical conductivity and permits transport of copper ions through the oxide layer so they can dissolve in the water and reprecipitate. If the water chemistry approximates that of seawater, copper ions reprecipitate as botallackite ($Cu_2(OH)_3Cl$) (Pollard et al. 1989) which can alter in minutes or hours to either paratacamite or atacamite (other crystal structures of $Cu_2(OH)_3Cl$) depending on local water chemistry. Paratacamite, the more common form, gives the appearance of having precipitated from solution, but may be a pseudomorph after botallackite (Kato and Pickering 1984).

The impact of sulfides on the corrosion of copper alloys has received considerable attention, including published reports documenting localized corrosion of copper alloys by SRB in estuarine environments (Little et al. 1988, 1989) and a report of the failure of copper alloys in polluted seawater containing waterborne sulfides that stimulate pitting and stress corrosion cracking (Rowlands 1965). Copper alloys suffer accelerated corrosion attack in seawater containing 0.01 ppm sulfide after 1-day exposure (Gudas and Hack 1979). A porous layer of cuprous sulfide with the general stoichiometry $Cu_{2-x}S$, $0 < x < 1$, forms in the presence of sulfide ions (Syrett 1980). Copper ions migrate through the layer, react with more sulfide, and produce a thick, black scale (Fig. 15).

Figure 16 is a stability diagram for copper and its minerals drawn for 10^{-6} M total dissolved copper and 10^{-2} M total sulfide. Parallelograms superimposed on the diagram are similar to those described for Figure 14 and are appropriate for the analysis of corrosion mineralogy under non-hydrothermal conditions. McNeil et al. (1991) used Figure 16 to interpret results from laboratory experiments. They exposed mixed cultures known to

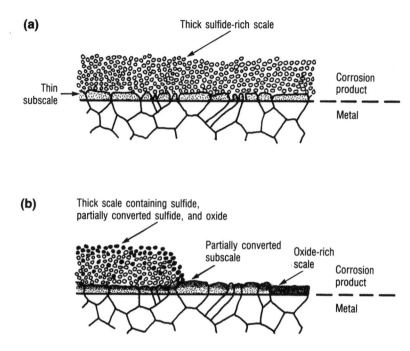

Figure 15. Schematic of (a) thick sulfide-rich scale on copper alloy and (b) disruption of sulfide film. (Reprinted from B.C. Syrett, "The Mechanism of Accelerated Corrosion of Copper-Nickel Alloys in Sulfide-Polluted Seawater," *Corrosion Science*, Vol. 21, 1981, with permission from Elsevier Science Ltd, Kidlington, UK.)

contain SRB to copper and copper/nickel alloys in a variety of natural and synthetic waters containing sulfates for 150 days. The pH values of the waters, measured after two weeks, were between 5.5 and 6.8. All copper-containing metals exposed to SRB in isolated cultures and in the natural augmented waters were covered with black sulfur-rich deposits (Fig. 17). The thickness and tenacity of the surface deposits varied among the metals and cultures. Corrosion products on commercially pure copper were consistently nonadherent. Corrosion products on copper alloys were more adherent and in some cases difficult to scrape from the surface. In all cases, bacteria were closely associated with sulfur-rich deposits (Fig. 18 a,b). Many bacteria were encrusted with deposits of copper sulfides (Fig. 19). Most scanning electron microscopy (SEM) micrographs of SRB on copper surfaces indicate a monolayer of cells overlaying a sulfide layer. Transmission electron microscopy has been used to demonstrate that bacteria are intimately associated with sulfide minerals and that on copper-containing surfaces the bacteria were found between alternate layers of corrosion products and attached to base metal (Blunn 1986).

Biomineralogy of copper sulfides has been studied for over a century (Daubree 1862, de Gouvernain 1875, Baas-Becking and Moore 1961, Mor and Beccaria 1975, Syrett 1977, 1980, 1981, McNeil and Little 1992). The complexity of the resulting observations reflects the complexity of the copper-sulfur system, especially in the presence of alloying elements or iron in the environment (Ribbe 1976, Kostov and Minceva-Stefanova 1981). Chalcocite (Cu_2S), digenite (Cu_9S_5), djurleite ($Cu_{1.93}S-Cu_{1.97}S$), anilite (Cu_7S_4), spion-kopite ($Cu_{39}S_{28}$), geerite (Cu_8S_5), and covellite (CuS) have been reported. In long-term corrosion where waters contain significant iron, chalcopyrite is a common product

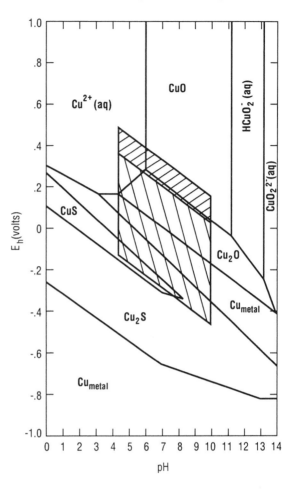

Figure 16 (above, right). Copper stability diagram: water without chloride ions; total sulfide 10^{-2} M.

Figure 17 (left). Black sulfide film on copper foil after 4-month exposure to SRB.

142   *GEOMICROBIOLOGY*

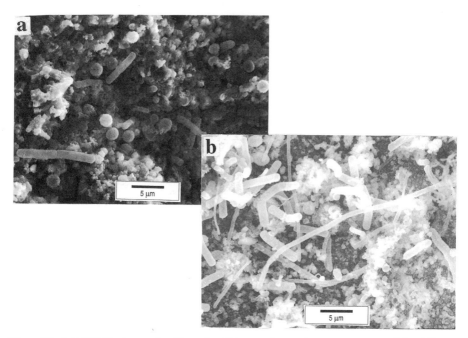

Figure 18. (a,b) ESEM micrographs of bacteria within corrosion products on copper-containing foils.

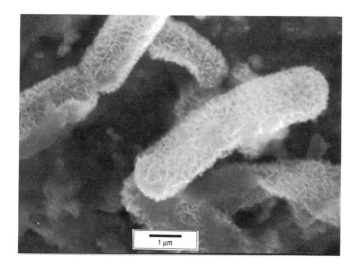

Figure 19. Encrustations of copper sulfide on bacterial cells.

(Daubree 1862, de Gouvernain 1875, McNeil and Mohr 1993). While chalcopyrite films can be formed abiotically in high sulfur concentrations (Cuthbert 1962), chalcopyrite and most other copper sulfides are not generally found as products of abiotic corrosion.

Detailed kinetics of individual reactions are not fully understood, and the consequences for corrosion depend on many factors, including mineral morphology and variations of redox and pH with time (McNeil and Mohr 1993). Discussions of alteration kinetics are contained in a number of papers (Baas-Becking and Moore 1961, Roseboom 1966, Craig and Scott 1976, Putnis 1977, Evans 1979). The general phenomenology can be understood by the following approach. Microbial consortia, that include SRB produce anoxic, sulfide-rich environments in which the conversion of copper to copper sulfides is thermo-dynamically favored at a concentration of 10^{-2} M total sulfur (Fig. 16). Reactions appear to proceed as suggested by Ostwald's rule: the first sulfur-poor compounds are converted to sulfur-rich compounds. One would expect a layering effect with covellite on the outside and chalcocite next to unreacted copper metal. This has not been studied, and indeed would be very difficult to study on pure copper because of the porous and mechanically unstable corrosion products. In short-term experiments with excess of copper over available sulfur, chalcocite with little or no covellite is formed (McNeil and Little 1991). Covellite is produced if excess sulfide is available, either deliberately provided (Baas-Becking and Moore 1961) or naturally available (Daubree 1862, Mor and Beccaria 1975).

The presence of dissolved iron leads to other complications. Not only has chalcopyrite been observed, but also digenite (Baas-Becking and Moore 1961, Mor and Beccaria 1975, North and MacLeod 1986, McNeil et al. 1991), djurleite (Macdonald et al. 1979, McNeil et al. 1991), and the hexagonal high-temperature polytype of chalcocite (McNeil et al. 1991). These observations can be interpreted in terms of the simplexity principle (Goldschmidt 1953): impurities tend to stabilize high-entropy, high-temperature polytypes. Digenite stability is promoted by iron (Craig and Scott 1976). It appears that nickel stabilizes djurleite on copper-nickel and the stabilization of djurleite has major practical consequences. McNeil et al. (1991) observed that corrosion layers showing strong digenite lines were never observed on pure copper, but frequently on copper-nickel alloys. Furthermore, corrosion products containing digenite showed substantial adherence and mechanical stability, while the corrosion products on pure copper, composed of chalcocite with only traces of other minerals, were powdery and nonadherent.

It has been argued that if the copper sulfide layer were djurelite, the sulfide layer would be protective (Nilsson et al. 1980). Even if such a sulfide film were technically passivating, the mechanical stability of the film is so poor that sulfide films are useless for corrosion protection. In the presence of turbulence, the loosely adherent sulfide film is removed, exposing a fresh copper surface to react with sulfide ions. For these reasons, turbulence-induced corrosion and sulfide attack of copper alloys cannot easily be decoupled. In the presence of oxygen, the possible corrosion reactions in a copper sulfide system are extremely complex because of the large number of stable copper sulfides (Ribbe 1976), their differing electrical conductivities, and catalytic effects. Transformations between sulfides or of sulfides to oxides result in changes in volume that weaken the attachment scale and oxide subscale, leading to spalling. Bared areas repassivate forming cuprous oxide.

The analysis of nonsulfide-induced corrosion of copper at near-room temperature in aqueous environments is extremely complex because of the numerous mineral species which can be formed. In seawater environments, chloride compounds dominate, but even in relatively low-chloride waters the strong tendency of chloride ions to migrate to anodic sites leads to formation of chloride-containing minerals. In general, corrosion of copper-nickel alloys in saline environments is less rapid than that of pure copper under the same circumstances. Earlier conjectures that this was due to changes in the electronic structure of the passive film (North and Pryor 1970) or of the base metal (Swartzendrubber et al. 1973) appear not to account adequately for the experimental data. A comprehensive review of more recent experimental work (Hack et al. 1986) indicates that, under hydrodynamic conditions that prevent accumulation of porous outer layers of hydroxychloride or hydroxycarbonate corrosion products, the critical effect of the alloying elements is to alter the cathodic oxygen reduction.

If conditions at the surface of a copper alloy permit precipitation of nantokite (CuCl) under the cuprite layer, the alloy becomes vulnerable to bronze disease (Scott 1990) or pitting corrosion (Lucey 1967) depending on mass transport conditions. A biofilm containing acid-producing bacteria could create the requisite conditions by increasing acidity in anodic areas and reducing copper ion transport from the surface, producing a higher local concentration (McNeil and Mohr 1992). While nantokite-based corrosion can also occur by nonbiological paths, the nonisometric morphology frequently observed on precipitated cuprite crystals suggests that microbiological poisoning of growth planes is a factor. The presence of alloying elements does not protect against the formation of nantokite and the consequent bronze disease corrosion (Mond and Cuboni 1893).

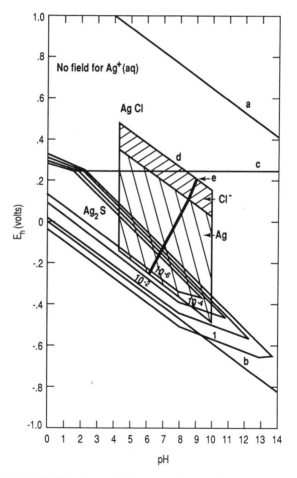

Figure 20. Silver-water-chloride-sulfur stability diagram for silver in seawater at varying reduced sulfur concentrations.

Silver. Figure 20 is a silver-water-chloride-sulfur system stability diagram. The upper diagonal line (a) is the oxygen line, above which water is thermodynamically unstable with respect to oxygen generation. The lower diagonal line (b) is the hydrogen line, below which water is thermodynamically unstable with regard to hydrogen evolution. Waters do exist outside these boundaries under special conditions. Horizontal line (c) separates the regions in which chloride corrosion of silver can and cannot take place. The upper hatched area below line (d) approximates the region of effective redox/acidity conditions for near surface waters. Strongly acidic and basic regions are characteristic of groundwaters, but not seawaters. The effective redox potential of oxygenated water is less than would be calculated thermodynamically from oxygen concentrations because of kinetic effects. The effective redox potential also depends on water pollutants, temperature and temperature/pressure history. The lower, oppositely hatched parallelogram indicates redox/acidity conditions existing in natural waters but not characteristic of near surface waters. Bulk water conditions outside the parallelograms are found in peat bogs, coal mines and at depths not considered in this chapter. The heavy diagonal line (e) through the hatched area indicates redox conditions for a series of waters sampled from a brackish, stagnant pond in an industrialized area in Scandinavia where surface waters had a high oxygen content and deep waters contained 40 mg l^{-1} H$_2$S and significant amounts of decaying organic material (Garrels and Christ 1965). Pond conditions were used to define the upper and lower Eh limits near neutrality. The upper portion of the hatched area applies for waters less than 10 m from the surface; the lower portion represents waters at depths greater than 10 m. To achieve conditions outside those defined by the two parallelograms at a metal surface requires the presence of a biofilm of maintaining conditions radically different from those in the bulk environment.

Silver and its alloys are subject to corrosion by reduced sulfur species including H$_2$S, generally of microbiological origin. In air (e.g. in a museum) H$_2$S can be the consequence of biodegradation of sulfur-containing polymeric materials, producing monoclinic acanthite (Ag$_2$S) (Banister 1952, Bauer 1988). Figure 20 indicates the possible thermodynamically stable phases for silver equilibrated with varying total sulfur compositions in 0.46 M NaCl (typical for seawater). It is assumed that reduced sulfur species (S^{2-}, HS$^-$, H$_2$S) are in equilibrium. A straightforward type of silver corrosion is conversion of silver to cerargyrite (AgCl), as indicated above line (c). Below line (c) metallic silver is stable, except for the wedge-shaped areas pointing down and to the right indicating regions of stability for monoclinic acanthite (Ag$_2$S). The region between the diagonal lines bounding the upper hatched region approximates the effective oxidizing behavior of near-surface, fully aerated seawater (Garrels and Christ 1965). Most shallow sea chemistries fall into this region. Conditions in shallow land burials where the major source of groundwater is rain or surface water percolating through soils are near this region. Cerargyrite is stable in seawater and chloride-rich shallow-land burial conditions.

There are three polymorphs of Ag$_2$S. Monoclinic acanthite is stable up to 176°C (Kracek 1946). Body-centered cubic argentite (Kracek 1946) is stable from 176°C to a temperature between 586°C and 622°C, above which the stable form is a face-centered cubic polymorph (Djurle 1958, Barton 1980). The high temperature polymorph has never been observed in corrosion. Reactions between Ag$_2$S polymorphs are very fast. Pure body-centered cubic argentite (Ag$_2$S) is not found in nature at standard temperature and pressure, and artificial argentite made with pure silver cannot be quenched to room temperature (Roy et al. 1959).

Laboratory data on sulfide and derivation of silver can be summarized as follows: (1) corrosion of silver by reduced sulfides, whether H$_2$S (Sinclair 1982, Volpe and Peterson 1989) or organic sulfides (Sinclair 1982) produces acanthite, (2) CS$_2$ does not produce corrosion, (3) the corrosivity of organic sulfides appears to be controlled by transport mechanisms and thus by vapor pressures, and (4) the rate of sulfidation is strongly affected by NH_3 and iron dissolved in the silver (Biestek and Drys 1987). Abiotic aqueous corrosion of silver in the presence of reduced sulfur species produces acanthite in bulk (Birss and Wright 1981, Campbell et al. 1982). Argentite is observed when objects made of impure silver (e.g. coins) are corroded in sediments over archaeological periods (Gettens 1963, North and MacLeod 1986). If Cl^- is present acanthite or argentite combined with cerargyrite is formed.

These observations support the hypothesis that formation of argentite is limited to precipitation of a silver-copper sulfide by reduced sulfide species. This theory is consistent with the observation that argentite corrosion products are sometimes accompanied by jalpaite ($Ag_{1.5}Cu_{.5}S$) (North and MacLeod 1986). Argentite made of pure silver is unstable at room temperature, yet there are two reasons why argentite should precipitate during MIC of archaeological objects. Argentite is usually associated with jewelry and coinage containing several percent copper. Argentite, unlike acanthite, can accommodate nearly 30% copper in its lattice (Shcherbina 1978). The phenomenon is parallel to the production of akageneite rather than goethite in the corrosion of meteorites which is attributed to the ability of the akageneite lattice to accommodate significant Cl^-, whereas the goethite lattice can accommodate little or none (Buchwald 1977). Precipitation of a mineral from an impure environment favors a loose crystal structure capable of accommodating impurity atoms (Goldschmidt 1953).

Argentite formation occurs when an object made of silver-copper alloy is in a water-saturated deposit containing SRB in a biofilm capable of maintaining reducing conditions, and bacteria (perhaps ammonia producers) capable of solubilizing silver and copper atoms. A layer of sand or soil restricts the ability of the metal ions to escape, so that concentrations of copper and silver ions within the biofilm rise to levels which cause precipitation. The precipitation of argentite is favored for the reasons given above. Jalpaite forms in regions where the copper concentration is high. Argentite and jalpaite could, in principal, be stabilized by corrosion of pure silver in a copper-rich environment (e.g. silver coins with copper coins), but for practical purposes, the presence of either argentite or jalpaite implies that the silver artifact originally contained significant copper.

Other metals. SRB-induced corrosion of zinc produces a zinc sulfide reported to be sphalerite (ZnS) (Baas-Becking and Moore 1961). SRB on lead carbonates produces galena (PbS) (McNeil and Little 1990). Galena has been found more recently as a lead corrosion product in SRB-induced corrosion of lead-tin alloys (McNeil and Mohr 1993).

Carbonates. Apparent calcium carbonate (calcite) precipitation by bacteria, algae, and yeasts has been reported by several researchers (Pentecost and Bauld 1988, Bouquet et al. 1973, Novitsky 1981, Shinano 1973). It is generally observed that calcite crystal deposition is favored by addition of Ca^{2+}, CO_3^{2-} and/or an increase in pH to 8.0 and higher. Calcification has been observed in marine, brackish, and freshwaters in the presence of $CaCo_3$ supersaturation. Bouquet et al. (1973), described calcite production by 210 soil bacterial isolates on a solid medium with added calcium.

Proposed mechanisms of calcite precipitation by microorganisms include calcium concentration from the medium by microbial binding, metabolic alteration of the medium that results in changes in bicarbonate concentration and pH, and microbial bodies acting as crystal nucleation sites. Pentecost and Bauld (1988) proposed that calcite deposition by cyanobacteria is initiated at sheath polymeric sites, on heteronuclei bound to sheath surfaces, or upon associated bacterial surfaces. However, most researchers have concluded that calcite precipitation is basically induced by chemical alteration of the medium whether by microbial activities or by abiotic reactions such as evaporation or outgassing of aqueous carbon dioxide.

Figure 21 is a stability diagram for copper that represents the condition when no sulfur is present and the electrolyte has the same chloride concentration as seawater and total carbonate content (CO_3^{2-}, HCO_3^-, H_2CO_3) is assumed to be in equilibrium with air. Stability regions for red cuprite (Cu_2O), malachite ($Cu_2(OH)_2CO_3$), and paratacamite ($Cu_2(OH)_3Cl$) are shown. Thermodynamic data (excluding malachite) are from Wagman et al. (1982); malachite data are from Woods and Garrels (1966). Azurite ($Cu_3(OH)_2(CO_3)_2$) and georgite ($Cu_5(CO_3)_3(OH)_4·6H_2O$) are not included. There are no thermodynamic data for georgite, and azurite has no stability field in water having carbonate content in equilibrium with air. If the water contains little Cl^- but is in equilibrium with the atmosphere with regard to carbonate species, malachite will precipitate on copper surfaces. The hydroxychlorides, atacamite and paratacamite, form in inhomogeneous layered structures. Malachite can form in layered structures, but sometimes forms botryoidal (Gettens 1969) or hairlike structures. The reasons for these differing morphologies are not known and all are found in association with bacteria.

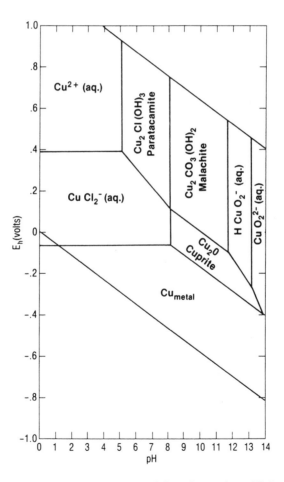

Figure 21. Copper stability diagram: seawater ($a_{Cl^-} = 0.319$), carbonates in equilibrium with air, no sulfides.

Mineral dissolution

Biofilm formation of copper alloys often results in selective dealloying. Zinc tends to be selectively removed from copper-zinc alloys, producing a spongy copper material (Walker 1977) though the incorporation of zinc in corrosion products to produce rosasite $((Cu,Zn)_2CO_3(OH)_2)$ (Gettens 1963). Dealloying of the copper from tin bronzes has been reviewed by Geilmann (1956). Partitioning of alloying elements between remaining metal ions in the corrosion product and the electrolyte has received little attention (Zolotarev et al. 1987).

Dissimilatory iron and/or manganese reduction occurs in several microorganisms, including anaerobic and facultative aerobic bacteria. Inhibitor and competition experiments suggest that Mn(IV) and Fe(III) are efficient electron acceptors similar to nitrate in redox ability and are capable of out-competing electron acceptors of lower potential, such as sulfate or carbon dioxide (Myers and Nealson 1988). Many of the recently described MRB are capable of using a variety of electron acceptors, including nitrate and oxygen (Myers and Nealson 1988) Myers and Nealson (1988) suggested that iron and manganese-reducing microorganisms must be in direct contact with oxides to reduce them. This conclusion is based on the observation that Fe(III) and Mn(IV) are not reduced if microorganisms capable of the reduction are separated from the oxides by a semi-permeable membrane that allows exchange of soluble molecules but prevents contact between the organism and the oxide. These experiments demonstrated that metal reduction by *S. putrefaciens* required cell/surface contact and the rate of reduction was directly related to surface area. ESEM and CLSM images of bacteria during both the iron and manganese reduction processes showed close contact of the cells with the oxides during the early stages of reduction. In later stages manganese oxides were coated with a layer of extracellular material that obscured the cells (Little et al. 1997).

Little et al. (1997) used synthetic iron oxides (goethite, α-FeOOH; hematite, Fe_2O_3; and ferrihydrite, $Fe(OH)_3$) as model compounds to simulate the mineralogy of passivating films on carbon steel. There is general agreement that oxide films formed on iron in air at temperatures below 200°C are composed of magnetite and hematite. Szklarska-Smialowska (1986) described the formation of hematite over a magnetite film. Ferric oxyhydroxides, including goethite and lepidocrocite (γ-FeO·OH), have also been identified in protective layers on carbon steel. Under anaerobic conditions goethite, hematite, and ferrihydrite were reduced by *S. putrefaciens* (Table 2). Rates of reduction, measured by atomic absorption spectroscopy of Fe(II) in solution as a function of time, for the three minerals indicate that after a 24-h exposure to *S. putrefaciens*, initial reduction rates for goethite and ferrihydrite were approximately the same and were 5 times faster than the reduction rate for hematite. After 22 days the integrated reduction rates for goethite and ferrihydrite were much faster than those measured at 24 h. The hematite reduction rate actually slowed over the exposure period so that after 22 days the overall integrated rate was 50 times slower than reduction rates for goethite and ferrihydrite (Arnold et al. 1988, Roden and Zachara 1996).

Table 2. Relative rates of iron reduction based on Fe(II) in solution

	24 Hours mg/L/d	% Max	22 Days mg/L/d	% Max
Goethite	3.50	100	18.20	100
Ferrihydrite	2.60	74	14.10	77
Hematite	0.58	22	0.39	2

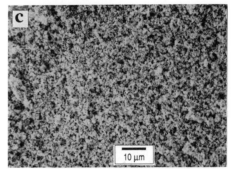

Figure 22. Iron oxides before exposure to *S. putrefaciens*: (a) goethite, (b) ferrihydrite, and (c) hematite.

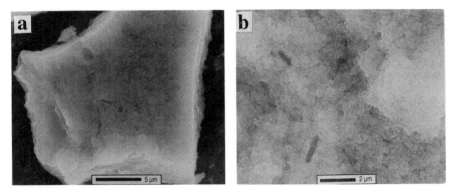

Figure 23. Bacteria on mineral surfaces after 48-h exposure to *S. putrefaciens*: (a) goethite and (b) ferrihydrite.

Differences in reduction rates may be due to the surface areas of individual oxides. Mineralogy and crystal structure must also play a role in microbial attachment and rate of biomineralization. Before exposure to *S. putrefaciens*, goethite and ferrihydrite appeared to form smooth platelet-like particles, while the hematite consisted of fine crystals (Fig. 22a–c). All iron oxides produced EDS spectra that were exclusively iron. After 48 h, isolated bacteria could be located on the surfaces of both the goethite and ferrihydrite (Fig. 23a,b), but not on the hematite. Minerals viewed with ESEM were wet and not coated with a metal coating as required for standard SEM. During initial stages of mineral dissolution, bacteria, composed of water and low atomic number elements, were not electron dense and were difficult to image. With reduction of the ferric iron, the bacterial cells became electron dense and easier to recognize. When treated with a fluorescent stain that indicated viability, approximately 50% of all observed cells appeared to be actively metabolizingwhen viewed by CLSM. Occasionally dividing cells were observed. Surface microbial populations increased markedly after 72 h.

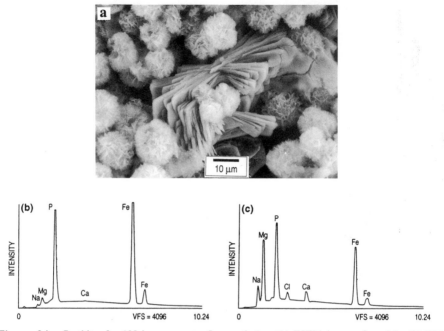

Figure 24. Goethite after 190 hr exposure to *S. putrefaciens*: (a) ESEM image of particle, (b) EDS spectrum of crystalline plate forms, and (c) EDS spectrum of globular forms.

After 190 h, ESEM images demonstrated that goethite and ferrihydrite surfaces had been altered during microbial reduction. Mineral particle size decreased and crystalline structure was transformed as the number of bacterial cells increased. Residual goethite particles consisted of both highly crystalline plates and more globular forms (Fig. 24a). EDS spectra of the two newly formed structures showed that the crystalline plate consisted of nearly stoichiometric amounts of iron and phosphorous, with a small enrichment of magnesium, suggesting that this mineral might be iron phosphate (vivianite) (Fig. 24b). Globular forms contained magnesium in addition to iron, phosphorous and magnesium (Fig. 24c). In previous experiments iron oxides were converted to siderite ($FeCO_3$) and other carbonates when experiments were conducted in closed containers from which carbon dioxide could not escape. ESEM images of residual goethite particles documented large accumulations of cells (Fig. 25a,b). The CLSM preparation of goethite particles contains a variety of sizes at different elevations within the preparation. The optical section in Figure 25c shows some particles that are completely surrounded with bright fluorescent cells. Other goethite particles have been optically cross-sectioned so that it is obvious that bacterial cells were associated with the exterior surfaces and had also created channels into the particles. Figure 26 shows the relationship between bacterial cell concentrations and Fe(II) in solution over time for goethite.

After 190 h CLSM images of optical sections indicated that individual particles of ferrihydrite could no longer be differentiated (Fig. 27). Instead, bacterial cells were

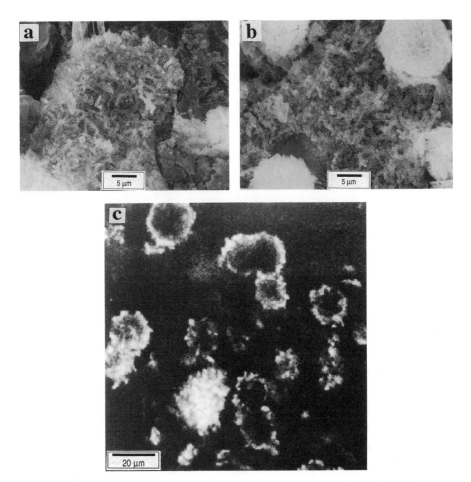

Figure 25. (a–b) Bacterial cells on goethite surface after 190-h exposure to *S. putrefaciens.* (c) CLSM image of optical cross-section of goethite particle.

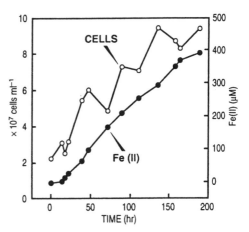

Figure 26. Bacterial reduction of goethite.

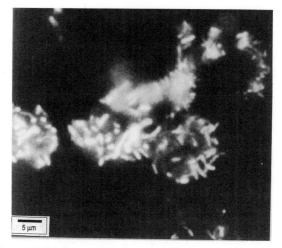

Figure 27. CLSM image of optical cross-section of ferrihydrite after 190-h exposure to *S. putrefaciens*.

distributed throughout all optical depths. Although iron reduction of hematite occurred (Table 2), no major changes in the elemental composition or crystal structure of the surface (Fig. 28a) were noted. EDS spectra of hematite indicated pure iron even after 190-h exposure. Scattered bacterial cells could be demonstrated in association with hematite surfaces (Fig. 28b).

Figure 28. Hematite surface after 190-h exposure to *S. putrefaciens*: (a) ESEM image and (b) CLSM optical cross-section.

Current noise measurements of carbon steel exposed to pure cultures of *S. putrefaciens* in media with and without the combination of sodium chloride and thiosulfate were used to monitor corrosion (Fig. 29a,b). In both cases current noise showed active surface changes. Fluctuations in noise records of a corroding metal are usually interpreted as being due to the sudden rupture of a protective oxide film followed by immediate repassivation. *S. putrefaciens* can reduce thiosulfate to produce sulfide. Obuekwe et al. (1981) evaluated corrosion of mild steel under conditions of simultaneous formation of ferrous and sulfide

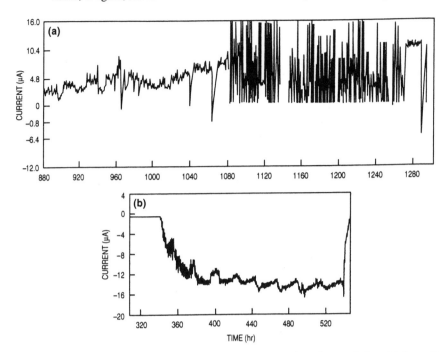

Figure 29. Electrochemical noise plots for carbon steel exposed to *S. putrefaciens* (a) in medium including thiosulfate and sodium chloride and (b) in medium without thiosulfate and sodium chloride.

ions. They reported extensive pitting when both processes were active. When only sulfide was produced, initial corrosion rates increased but later declined due to formation of a protective FeS film. High amounts of soluble iron prevent formation of protective sulfide layers on ferrous metals. Little et al. (1997) attempted to isolate the impacts of iron and thiosulfate reduction on corrosion of carbon steel by controlling the electron acceptors available in the electrolyte. In the initial experiment, thiosulfate was added to the medium, while in the second it was removed. Substantial electrochemical noise was measured in both cases and both electrodes were pitted.

ESEM examination of carbon steel electrodes revealed extensive bacterial colonization after 1300 h (Fig. 30a). The electrode was macroscopically pitted, and the location of pits coincided with colonies of bacteria (Fig. 30b). EDS analysis of the electrode surface showed that, like goethite and ferrihydrite, the modified surface, while still iron rich, had a complex mixture of elements, including phosphorous, sulfur, and chlorine (Fig. 30c). Mineral replacement reactions were documented by X-ray crystallography. After exposure to *S. putrefaciens* in anaerobic media, surface oxides were converted to ferrous phosphate.

CONCLUSIONS

Microorganisms influence corrosion of metals by both forming and dissolving minerals. Iron and manganese oxide deposition by bacteria make some metals more vulnerable to pitting and crevice corrosion by forcing the corrosion potential above the pitting potential or by initiating a sequence of events that results in underdeposit corrosion. Biomineral dissolution reactions remove passive layers or force mineral replacement reactions that lead to further dissolution. Because of the economic consequences of corrosion to both industry and military, causal relationships between bacteria and metal corrosion products have been investigated. Recently developed techniques including environmental scanning electron and X-ray microscopies make it possible to determine spatial relationships between microorganisms within biofilms and minerals as they are formed, dissolved, or replaced.

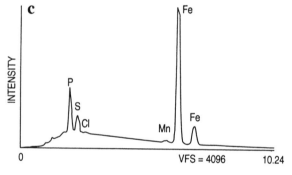

Figure 30. Carbon steel electrode after 1300-h exposure to *S. putrefaciens*: (a) bacterial colonization, (b) surface pitting (10×), and (c) EDS of surface.

Microorganisms within biofilms are capable of maintaining environments at biofilm/surface interfaces that are radically different from the bulk in terms of pH, dissolved oxygen, and other organic and inorganic species. As a consequence, microorganisms within biofilms produce minerals and mineral replacement reactions that are not predicted by thermodynamic arguments based on the chemistry of the bulk medium. For that reason, minerals within corrosion products can often be used as fingerprints for microbiologically influenced corrosion. For example, even though the region of stability predicted for mackinawite (tetragonal FeS_{1-x}) using stability diagrams is wholly outside the region defined by surface water Eh/pH conditions, excluding waters influenced by peat bogs, coal mines, volcanic activity and industrial effluents, mackinawite is easily produced from iron and iron oxides in surface waters by consortia of microorganisms that include SRB. The presence of mackinawite in corrosion products formed in shallow water environments, with the exclusions previously delineated, is proof that corrosion was SRB-induced.

ACKNOWLEDGMENTS

NRL Contribution Number NRL/BA/7333Ñ97-0001. The work for this chapter was performed under NRL Program Element 0601153N, Office of Naval Research contract N0001497WX30031 and Cooperative Agreement EEC-8907039 between the National Science Foundation and Montana State University.

REFERENCES

Allison D, Sutherland IW (1987) The role of exopolysaccharides in adhesion of freshwater bacteria. J Gen Microbiol 133:1319–1327

Arnold RG, DiChristina T, Hoffmann M (1988) Reductive dissolution of Fe(III) oxides by *Pseudomonas* sp. 200. Biotechnol Bioeng 32:1081–1096

Baas-Becking GM, Moore D (1961) Biogenic sulfides. Econ Geol 56:259–272

Banister FA (1952) An unusual synthesis of acanthite crystals. Presented at the Mineralogical Society of Great Britain Meeting. Documented in the (Reston, Virginia) Ford-Fleischer files of the U.S. Geological Survey

Barton MD (1980) The Ag-Au-S system. Econ Geol 75:303–316

Bauer R (1988) Sulfide Corrosion of Silver Contacts During Satellite Storage. U.S. Air Force Report SD-TR-88-53/AD-1196 217

Berner RA (1969) The synthesis of framboidal pyrite. Econ Geol 64:383–384

Biestek T, Drys M (1987) Corrosion products forming on silver in various corrosive environments. Powlocki Ochronne (Warsaw) 9:2–5

Birss VI, Wright GA (1981) The potentiodynamic formation and reduction of a silver sulfide monolayer on a silver electrode in aqueous sulfide solutions. Electrochim Acta 27:1–7

Blunn G (1986) Biological fouling of copper and copper alloys. In: Biodeterioration VI:567–575, CAB International, Slough, UK

Bouquet E, Boronat A, Ramos-Cormenzana A (1973) Production of calcite (calcium carbonate) crystals by soil bacteria is a general phenomenon. Nature 246:527–529

Bouwer EJ (1987) Theoretical investigation of particle deposition in biofilm systems. Water Res 21: 1489–1498

Bryers JD, Characklis W (1992) Biofilm accumulation and activity: a process analysis approach. In: LF Melo, TR Bott, M Fletcher, B Capdeville (eds) Biofilms–Science and Technology, p 221–237 Kluwer Academic Publishers, Dordrecht, the Netherlands

Buchwald VF (1977) The mineralogy of iron meteorites. Phil Trans Roy Soc (London) A286:453–491

Caldwell DE, Korber DR, Lawrence JR (1992) Confocal laser microscopy and computer image analysis in microbial ecology. In: KC Marshall (ed) Advances in Microbial Ecology, Vol. 12. Plenum Press, New York

Campbell GD, Lincoln FJ, Power GP, Ritchie IM (1982) The anodic oxidation of silver in sulfide solutions. Aust J Chem 35:1079–1085

Chamberlain AHL (1992) The role of adsorbed layers in bacterial adhesion. In: LF Melo, TR Bott, M Fletcher, B Capdeville (eds) BiofilmsÑScience and Technology, p 59–67 Kluwer Academic Publishers, Dordrecht, the Netherlands

Characklis WG (1990) Biofilm processes. In: WG Characklis, KC Marshall (eds) Biofilms p 227 John Wiley & Sons, New York

Characklis WG, Marshall KC (1990) Biofilms: a basis for an interdisciplinary approach. In: WG Characklis, KC Marshall (eds) Biofilms p 4, John Wiley & Sons, New York

Costerton JW, Cheng KJ, Geesey GG, Ladd TI, Nickel JC, Dasgupta M, Marrie TJ (1987) Bacterial biofilms in nature and disease. Ann Rev Microbiol 41:435–464

Costerton JW, Geesey GG (1986) The microbial ecology of surface colonization and of consequent corrosion. In: SC Dexter (ed) Biologically Induced Corrosion, p 223–232 NACE International, Houston, Texas

Costerton JW, Lewandowski Z, Caldwell DE, Korber DR, Lappin-Scott HM (1995). Microbial biofilms. Ann Rev Microbiol 49:711–745

Craig JR, Scott SD (1976) Sulfide phase equilibria. Rev Mineral 1:CS1-CS110

Cuthbert M (1962) Formation of bornite at atmospheric temperature and pressure. Econ Geol 57:38–41

Daubree GA (1862) Contemporary formation of copper pyrite by the action of hot springs at Bagnes-de-Bigorre. Bull Soc Geol France 19:529–532

Davies DG, Geesey GG (1995) Regulation of the alginate biosynthesis gene algC in *Pseudomonas aeruginosa* during biofilm development in continuous culture. Appl Environ Microbiol 61:860–867 de Beer D, Stoodley P, Lewandowski Z (1994b) Liquid flow in heterogeneous biofilms. Biotech Bioeng 44:636–641

de Beer D, Stoodley P, Roe F, Lewandowski Z (1994a) Effects of biofilm structures on oxygen distribution and mass transport. Biotech Bioeng 43:1131–1138

de Gouvernain M (1875) Sulfiding of copper and iron by a prolonged stay in the thermal spring at Bourbon l'Archambault. Compt Rend 80:1297–1300

Dickinson WH, Caccavo F, Lewandowski Z (1996) The ennoblement of stainless steel by manganic oxide biofouling. Corros Sci 38:1407–1422

Dickinson WH, Lewandowski Z (1996) Manganese biofouling and the corrosion behavior of stainless steel. Biofouling 10:79–93

Djurle S (1958) An X-ray study of the system Ag-Cu-S. Acta Chem Scandinavica 12:1427–1436

Duddridge JE, Kent CA., Laws JF (1982) Effect of surface shear stress on the attachment of *Pseudomonas fluorescens* to stainless steel under defined flow conditions. Biotech Bioeng 26 153–164

Duncan SJ, Ganiaris H (1987) Some sulphide corrosion products on copper alloys and lead alloys from London waterfront sites. In: EJ Black (ed) Recent Advances in the Conservation and Analysis of Artifacts , p 109–118 Summer Schools Press, University of London

Eighmy T., Maratea D, Bishop PL (1983) Electron microscopic examination of wastewater biofilm formation and structural components. Appl Environ Microbiol 45:1921–1931

Escher A, Characklis WG (1990) Modeling the initial events in biofilm accumulation. In: WG Characklis, KC Marshall (eds) Biofilms, p 445–486 John Wiley & Sons, New York

Escher AR (1986) Colonization of a smooth surface by *Pseudomonas aeruginosa*: image analysis methods. PhD Dissertation, Montana State University, Bozeman, MT

Evans HT (1979) The crystal structures of low chalcocite and djurleite. Zeits Kristallogr 150:299–320

Fletcher M, Floodgate GD (1973) An electron-microscopic demonstration of an acidic polysaccharide involved in the adhesion of a marine bacterium to solid surfaces. J Gen Microbiol 74:325–334

Garrels RM, Christ JC (1965) Solutions, Minerals, and Equilibria. Freeman Cooper, San Francisco

Geilmann W (1956) Leaching of bronzes in sand deposits. Angewandte Chemie 68:201–211

George RP (1996) Studies on the corrosion and tuberculation of carbon steel in fresh water. National Symposium of Research Scholars on Metals and Materials Research, p 118–127 Indian Institute of Metals, Madras, India

Gettens RJ (1963) The Corrosion Products of Metal Antiquities. Annual Report to the Trustees of the Smithsonian Institution for 1963, p 547–568 (1969) The Freer Chinese Bronzes. Smithsonian Institution, Washington, DC

Gjaltema PA, Arts PAM, van Loosdrecht MCM, Kuenen JG, Heijnen JJ (1994) Heterogeneity of biofilms in rotating annular reactors: occurrence, structure, and consequences. Biotechnol Bioeng 44:194–204

Goldschmidt J (1953) A simplexity principle. J Geol 61:539–551

Gounot, A. M. 1994 Microbial oxidation and reduction of manganese: consequences in groundwater and applications. FEMS Microbiology Rev 14:339–350

Gudas JP, Hack HP (1979) Sulfide-induced corrosion of copper-nickel alloys. Corros 35:67–73

Hack H, Shih H, Pickering HW (1986) Role of the corrosion product film in the corrosion protection of Cu-Ni alloys in seawater. In: E McCafferty, RJ Broadd (eds) Surfaces, Inhibition, and Passivation, p 355–367 The Electrochemical Society, Pennington, New Jersy

Heimann RB (1989) Assessing the technology of ancient pottery: the use of ceramic phase diagrams. Archeomaterials 31:123–148

Jung WK, Schweisfurth R (1976a) Manganoxydierende bakterien. II. Z fur Allg Mikrobiologie 16:133–147

Jung WK, Schweisfurth R (1976b) Manganoxydierende bakterien. III. Z fur Allg Mikrobiologie 16:587–597

Kato C, Pickering HW (1984) A rotating disk study of the corrosion behavior of Cu–9.4Ni–1.7Fe alloy in air-saturated aqueous NaCl solution. J Electrochem Soc 131:1219–1224

Keevil CW, Walker JT (1992) Normarski DIC microscopy and image analysis of biofilms. Binary 4:93–95

Kostov I, Minceva-Stefanova J (1981) Sulphide Minerals. Publishing House, Bulgarian Academy of Sciences, Sofia, Bulgaria

Kracek FC (1946) Phase relations in the system silver-sulfur and the transitions in silver sulfide. Trans Am Geophysical Union 27:367–374

Kugaprasatham S, Nagaoka H, Ohgaki S (1992) Effect of turbulence on nitrifying biofilms at non-limiting substrate conditions. Water Res 26 (12):1629–1638

Lawrence JR, Korber DR, Hoyle BD, Costerton JW, Caldwell DE (1991) Optical sectioning of microbial biofilms. J Bacteriol 173:6558–6567

Lewandowski Z, Altobelli SA, Fukushima E (1993a) NMR and microelectrode studies of hydrodynamics and kinetics in biofilms. Biotechnol Prog 9:40–45

Lewandowski Z, Altobelli SA, Majors PD, Fukushima E (1992) NMR imaging of hydrodynamics near microbially colonized surfaces. Water Sci Tech 26:577–584

Lewandowski Z, Stoodley P (1995) Flow induced vibrations, drag force, and pressure drop in conduits covered with biofilm. Biofilm structure, growth, and dynamics. Biofilm Structure, Growth and Dynamics. Meeting of the International Association on Water Quality. Leeuwenhorst, the Netherlands

Lewandowski Z, Stoodley P, Altobelli S (1995) Experimental and conceptual studies on mass transport in biofilms. Water Sci Technol 31:153–162

Lewandowski Z, Stoodley P, Altobelli S, Fukushima E (1993b) Hydrodynamics and kinetics in biofilms recent advances and new problems. Proceedings of the Second IAWQ International Specialized Conference on Biofilm Reactors, p 313–319, Paris, France

Lewandowski Z, Stoodley P, Altobelli S, Fukushima E (1994) Hydrodynamics and kinetics in biofilm systems - recent advances and new problems. Water Sci Technol 29:223–229

Little B (1985) Factors influencing the adsorption of dissolved organic material from natural waters. J Colloid Interface Sci 108:331–339

Little B, Wagner P, Hart K, Ray R, Lavoie D, Nealson K, Aguilar C (1997) The role of metal-reducing bacteria in microbiologically influenced corrosion, Paper No. 215. CORROSION/97, NACE International, Houston, Texas

Little B, Wagner P, Jacobus J (1988) The impact of sulfate-reducing bacteria on welded copper-nickel seawater piping systems. Mat Perf 27:57–61

Little B, Wagner P, Jacobus J, Janus L (1989) Evaluation of microbiologically induced corrosion in an estuary. Estuaries 12:138–141

Little B, Wagner P, Mansfeld F (1991) Microbiologically influenced corrosion of metals and alloys. Intl Mat Rev 36:253–272

Little BJ, Wagner PA, Characklis WG, Lee W (1990) Microbial corrosion. In: Biofilms, WG Characklis, KC Marshall (eds) p 635–670 John Wiley & Sons, New York

Loeb GI, Neihof RA (1976) Adsorption of an organic film at the platinum-seawater interface. J Mar Res 35:283–291

Lucey VF (1967) Mechanism of pitting corrosion of copper in supply waters. Br Corros J 2:175–185

Macdonald DD, Syrett BC, Wing SS (1979) Corrosion of Cu-Ni alloys 706 and 715 in flowing sea water—2. Effect of dissolved sulfide. Corros 35:367–378

Mack WN, Mack JP, Ackerson AO (1975) Microbial film development in trickling filters. Microb Ecol 2:215–316

Marshall KC, Power KN, Angles ML, Schneider RP, Goodman AE (1994) Analysis of bacterial behavior during biofouling of surfaces. In: G Geesey, Z Lewandowski, H-C Flemming (eds) Biofouling and Biocorrosion in Industrial Water Systems, p 15–26 CRC Press Inc, Lewis Publishers, Boca Raton, Florida

Massol-Deya AA, Whallon J, Hickey RF, Tiedje JM, (1995) Channel structures in Aerobic biofilms of fixed-film reactors treating contaminated groundwater. Appl Environ Microbiol 61:769–777

McNeil MB, Jones J, Little BJ (1991) Mineralogical fingerprints for corrosion processes induced by sulfate-reducing bacteria, paper no. 580. CORROSION/91, NACE International, Houston, Texas

McNeil MB, Little BJ (1990) Mackinawite formation during microbial corrosion. Corros 46:599–600

McNeil MB, Little BJ (1992) Corrosion mechanisms for copper and silver objects in near-surface environments. J Amer Inst for Conserv 31(3):355–366

McNeil MB, Mohr DW (1992) Interpretation of bronze disease and related copper corrosion mechanisms in terms of log activity diagrams. In: Materials Problems in Art and Archaeology II, p 1055–1063 Materials Research Society, Pittsburgh, PA

McNeil MB, Mohr DW (1993) Formation of copper-iron sulfide minerals during corrosion of artifacts and implications for pseudogilding. Geoarchaeol 8(1):23–33

Miller JDA, Tiller AK (1970) Microbial corrosion of buried and immersed metal. In: JDA Miller (ed) Microbial Aspects of Metallurgy, p 61–106 American Elsevier, New York

Mond L, Cuboni G (1893) On the nature of antique bronze patina. Atti reale Accad dei Lincei serie 52:498–499

Mor ED, Beccaria AM (1975) Behaviour of copper in artificial seawater containing sulphides. Br Corros J 10:33–38

Myers C, Nealson KH (1988) Bacterial manganese reduction and growth with manganese oxide as the sole electron acceptor. Science 240:1319–1321

Nealson K, Tebo B, Rosson R (1988) Occurrence and mechanisms of microbial oxidation of manganese. In: A Laskin (ed) Advances in Applied Microbiology 33:279–318 Academic Press, New York

Nilsson I, Ohlson S, Haggstrom L, Molin N, Mosbach K (1980) Denitrification of water using immobilized *Pseudomonas denitrificans* cells. Europ J Appl Microbiol Biotechnol 10:261–274

North NA, MacLeod ID (1986) Corrosion of metals. In: C Pearson (ed) Conservation of Archaeological Objects, p 69–98 Butterworths, London

North RF, Pryor MJ (1970) The influence of corrosion product structure on the corrosion rate of Cu-Ni alloys. Corros Sci 10:297–311

Novitsky JA (1981) Calcium carbonate precipitation by marine bacteria. Geomicrobiology 2:375–388

Obuekwe CO, Westlake DWS, Plambeck JA, Cook FD (1981) Corrosion of mild steel in cultures of ferric iron reducing bacterium isolated from crude oil I. Polarization characteristics. Corros 37:461–467

Pendyala J (1996) Chemical Effects of Biofilm Colonization on Stainless Steel. PhD Dissertation, Montana State University, Bozeman , MT

Pentecost A, Bauld J (1988) Nucleation of calcite on the sheaths of cyanobacteria using a simple diffusion cell. Geomicrobiology 6:129–135

Pfennig N, Widdel F, Truper HG (1981) The dissimulatory sulfate-reducing bacteria. In: MP Starr, et al., (eds) The Prokaryotes: A Handbook on Habitats p 926–940 Springer-Verlag, New York

Pollard AM, Thomas RG, Williams PA (1989) Synthesis and stabilities of basic copper(II) chlorides atacamite, paratacamite, and botallackite. Mineral Mag 53:557–563

Pope DH, Duquette DJ, Wayner PC, Johannes AH (1984) Microbiologically Influenced Corrosion: A State of the Art Review. Materials Technology Institute of the Chemical Process Industries, Columbus, OH

Postgate JR (1979) The Sulphate-Reducing Bacteria. Cambridge University Press, Cambridge, UK

Pourbaix M (1966) Atlas of Electrochemical Equilibria in Aqueous Solutions. National Association of Corrosion Engineers, Houston, Texas

Putnis A (1977) Electron diffraction study of phase transformations in copper sulfides. Am Mineral 62:107–114

Ribbe PH (ed) (1976) Sulfide Mineralogy. Rev Mineral Vol. 1, 284 p

Rittmann BE, Manem JA (1992) Development and experimental evaluation of a steady-state, multispecies biofilm model. Biotechnol Bioeng, 39:914–922

Robinson RW, Akin DE, Nordstedt RA, Thomas M, Aldrich HC (1984) Light and electron microscopic examinations of methane-producing biofilms from anaerobic fixed-bed reactors. Appl Environ Microbiol 48:127–136

Roden EE, Zachara JM (1996) Microbial reduction of crystalline Fe(III) oxides: influence of oxide surface area and potential for cell growth. Environ Sci Tech 30:1618–1628

Roseboom EH (1966) An investigation of the system Cu–S and some natural copper sulfides between 25 degrees and 700 degrees. Econ Geol 61:641–672

Rowlands JC (1965) Corrosion of tube and pipe alloys due to polluted seawater. J Appl Chem 15:57–63

Roy R, Majumdar AJ, Hulbe CW (1959) The Ag_2S and Ag_2Se transitions as geologic thermometers. Econ Geol 54:1278–1280

Scott DA (1990) Bronze disease; a review of some chemical problems and the role of relative humidity. J Amer Inst Conserv 29:192–306

Shcherbina VV (1978) The geochemistry of monovalent sulfides. Geokhimiya 10:1444–1451

Shinano H (1973) Studies of marine bacteria taking part in the precipitation of calcium carbonate—VII. Bull. Japanese Soc. Scientific Fisheries 39:91–95

Silverman M, Belas R, Simon M (1984) Genetic control of bacterial adhesion. Life Sci Res 31:95–107

Sinclair J D (1982) The tarnishing of silver by organic sulfur vapors. J Electrochem Soc 129:33–40

Staffeldt EE, Kohler DA (1973) Assessment of corrosion products removed from "La Fortuna," Punta del Mar, Venezia. Petrolia e Ambiente, p 163–170

Stewart PS, Peyton BM, Drury WJ, Murga R (1993) Quantitative observations of heterogeneities in *Pseudomonas aeruginosa* biofilms. Appl Environ Microbiol 59(1):327–329

Stoodley P, deBeer D, Lewandowski Z (1994) Liquid flow in biofilm systems. Appl Environ Microbiol 60:2711–2716

Swartzendruber LJ, Bennett LH, McNeil MB (1973) On the electron configuration theory of marine corrosion. Third International Conference on Marine Corrosion and Fouling, p 410–426, National Bureau of Standards, Washington DC

Syrett BC (1977) Accelerated corrosion of copper in flowing pure water contaminated with oxygen and sulfide. Corros 33:257–262

Syrett BC (1980) The mechanism of accelerated corrosion of copper-nickel alloys in sulfide polluted seawater. CORROSION/80, paper no. 33, NACE International, Houston, Texas

Syrett BC (1981) The mechanism of accelerated corrosion of copper-nickel alloys in sulphide-polluted seawater. Corros Sci 21:187–209

Szklarska-Smialowska Z (1986) Pitting Corrosion of Metals. NACE International, Houston, Texas

Uhlig HU, Revie RW (1985) Corrosion and Control, An Introduction to Corrosion Science and Engineering, John Wiley & Sons, New York

van Loosdrecht MCM, Lyklema J, Norde W, Zehnder AJB (1990) Influence of interfaces on microbial activity. Microbiol Rev 54(1):75–87

Volpe L, Peterson PJ (1989) The atmospheric sulfidation of copper in a tubular corrosion reactor. Corros Sci 29:1179–1186

Wagman DD, Evans WH, Parker VB, Schumm RH, I. Halow I, Bailey SM, Churney KL, Nuttall RL (1982) The NBS tables of chemical thermodynamic properties: selected values for inorganic and C_1 and C_2 organic substances in SI units. J Phys Chem Ref Data 11, supplement 2

Walker GD (1977) An SEM and microanalytical study of in-service dezincification of brass. Corros 33:252–256

Wanner O, Gujer W (1986) A multi-species biofilm model. Biotech Bioeng 28:314–328

Wikjord AG, Rummery TE, Doern FE, Owen DG (1980) Corrosion and deposition during the exposure of carbon steel to hydrogen sulfide water solutions. Corros Sci 20:651–671

Wilson T (1990) Confocal Microscopy. Academic Press, London

Wolfaardt GM, Lawrence JR, Robarts RD, Caldwell DE (1994) Multicellular organization in a degradative biofilm community. Appl Environ Microbiol 60:434–446

Woods TL, Garrels RM (1966) Phase relations of some copper hydroxy minerals. Econ Geol 81:1989–2007

Zhang TC, Bishop PL (1994a) Density, porosity, and pore structure of biofilms. Water Res 28:2267–2277

Zhang TC, Bishop PL (1994b) Evaluation of tortuosity factors and effective diffusivities in biofilms. Water Res 28:2279–2287

Zolotarev EI, Pchel'mikov AP, Skuratnick YaB, Dembrovskii MA, Khokhlov NI, Losev VV (1987) Kinetics of dissolution of copper-nickel alloys, anodic dissolution of Cu–30%Ni under steady state conditions. Zashchita Metallov 23:922–929

Chapter 5

SURFACE-MEDIATED MINERAL DEVELOPMENT BY BACTERIA

D. Fortin

Department of Geology
University of Ottawa
Ottawa, Ontario, K1N 6N5 Canada

F. G. Ferris

Department of Geology
Earth Science Centre
University of Toronto
Toronto, Ontario, M5S 3B1 Canada

T. J. Beveridge

Department of Microbiology
College of Biological Science
University of Guelph
Guelph, Ontario, N1G 2W1 Canada

INTRODUCTION

It is almost impossible to find natural environments where bacteria do not exist. Environmental studies over the last two decades have shown them to thrive in even the most extreme environments. They are found in the frigid depths of the oceans, underneath mineral varnishes in the driest deserts, close to thermal vents (such as "black smokers"), in the salty brines of salterns, and in the acid leachates of mine tailing dumps. They inhabit all natural waters, soils and sediments, even to great depth. We attribute this ubiquitousness of bacteria to their simplicity of cellular design and their relative ease of adaptation, mutation and selection (Beveridge 1988, 1989a). After all, they have dwelt on Earth for at least 3.6 Ga (Schopf and Walter 1983, Walter 1983) and have learned to occupy almost all imaginable environmental niches.

Bacteria are prokaryotes. They are so small they cannot afford space for a nucleus and their chromosome is free within the cytoplasm (Beveridge 1989a). They have a small volume (generally 1.5 to 2.5 μm^3; Beveridge 1989a,b) and are usually bounded by a tough, resilient cell wall (Beveridge 1981, 1989a). They have the highest surface area-to-volume ratio of any lifeform since they must depend on diffusion for their nourishment and departure of waste products (Beveridge 1988). These bacterial surfaces, then, are highly interactive with the environment, especially with the solutes which they frequently encounter. In this way, these "broad" surfaces make exceptional interfaces for the precipitation of metal ions and the development of fine-grained minerals (Beveridge 1989a, Ferris et al. 1988, Fortin and Beveridge 1997a). In addition, certain metabolic waste products, such as H_2S, also interact with metal ions to form precipitates which surround the cell (Fortin and Beveridge 1997a, Fortin et al. 1996). Or, bacteria can also pump-out ions which influence the pH of their immediate surroundings (or "micro-environment") so that precipitation of specific mineral types can be induced (Schultze-Lam et al. 1992, Thompson et al. 1997, Urrutia et al. 1992).

0275-0279/97/0035-0005$05.00

GENERAL OVERVIEW

Bacterial surfaces are highly variable and often depend on the genetic coding capacity of the genus, species or strain to which the bacterium belongs. Some possess only a cell wall, whereas others can have additional layers such as capsules, S-layers, sheaths or "slimes" (Beveridge 1981). Environmental factors may be required to activate the various genes for the production of some layers; for example, certain metal ions can induce the formation of capsular material (Appanna and Preston 1987, Corpe 1964). There is only space in this chapter for a general description of bacterial surfaces and the reader is encouraged to use the references which are supplied.

Bacterial cell walls

Bacteria have a variety of different surface formats for their cell walls. For practical purposes, we need only concern ourselves with those walls possessed by the eubacterial group and almost all of these surfaces have an overall electronegative charge at pH ~ 7 (Beveridge 1981, 1988, 1989b; Beveridge and Doyle 1989). The cell walls of eubacteria are divided into two structural formats, gram-positive and gram-negative (Beveridge 1981). [Each format reacts differently with a stain used for light microscopy and can behave positively or negatively with respect to the stain (Beveridge 1988, Schultze-Lam et al. 1993, 1995a).]

Gram-positive cell walls are exemplified by those of *Bacillus subtilis*, and are constructed of linear polymers of peptidoglycan which are rich in carboxylate groups and which are covalently linked together as they assemble around the cell (Fig. 1; Beveridge 1981). Consequently, the peptidoglycan, which encompasses the cell, is bonded together to form a single giant macromolecule (~25 nm thick) which has great strength and endurance. Usually, so-called secondary polymers (such as teichoic or teichuronic acids) are also bonded into the peptidoglycan framework, and these polymers increase the electro-

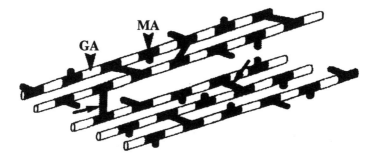

Figure 1. Peptydoglycan is an important constituent of all bacterial walls and can provide binding sites for environmentally-derived cations. The backbone of the molecule consists of repeated dimers of N-acetylglucosamine (GA) and N-acetylmuramic acid (MA). Each MA residue has a peptide stem attached to it which usually consists of four or fewer amino acids. A proportion of the four-membered peptide stems are cross-linked to those of other glycan strands (arrows). The stereochemistry of the glycan backbone is such that the peptide stems adjacent dimers are rotated 90° relative to each other. Thus, when cross-linked, a three-dimentional meshwork is formed, which encloses the entire cell, lending it shape and protection from rupture due to internal osmotic pressure. Drawing courtesy of Braden Beveridge and taken from Schultze-Lam et al. (1995a) with permission of the authors.

negative charge density of the wall with the addition of further carboxyl or phosphoryl groups. The cell wall is separated from the cell (or protoplast) by a lipid/protein bilayer called the plasma (or cytoplasmic) membrane. The end result is a gram-positive cell envelope that is highly interactive with dissolved metal ions, extremely robust, and resistant to harsh environmental conditions (Beveridge 1981, Beveridge and Doyle 1989).

Gram-negative walls have a more complex structural format and this is exemplified by that of *Escherichia coli*. These walls also contain peptidoglycan, but it is much thinner (~7.5 nm) and this polymer is devoid of secondary polymers (Beveridge 1981). The peptidoglycan is sandwiched between two lipid/protein bilayers; beneath the peptidoglycan resides the plasma membrane and above it, the outer membrane (Fig. 2). It is the outer membrane which is the outermost layer of the gram-negative cell wall and it is studded with an unusual lipid, lipopolysaccharide (LPS), which is highly anionic. Because of this, like gram-positive walls, gram-negative surfaces are also strongly electronegative and are interactive with dissolved environmental metals (Beveridge and Koval 1981).

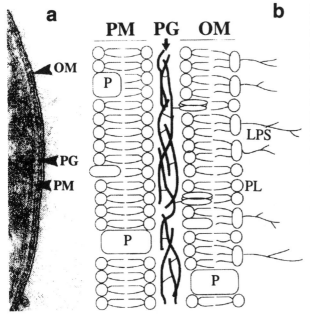

Figure 2. Gram-negative bacterial cell wall structure. TEM images (a) of thin-sectioned cells show a peptidoglycan layer (PG) flanked by bylared membranes on either side. The outer membrane (OM) has a different chemical construction than the inner, plasma memebrane (PM). Both contain protein molecules (P) which can have enzymatic and/or structural functions and phospholipids, which have polar head groups (depicted as circles in b) and two fatty acid chains. In the OM these are mostly limited to the inner leaflet. The outer leaflet is mainly composed of lipopolysaccharide (LPS) which has a higher carbohydrate chains. These can have a high negative charge density and bind metal ions, an ability demonstrated by the more intense staining of the outer leaflet on the cell in (a). Bar = 100 nm. Taken from Schultze-Lam et al. (1995a) with permission of the authors.

Additional layers above eubacterial cell walls

In natural environments there are frequently additional layers on top of gram-positive and gram-negative bacterial walls. The most common is a capsule. This is a relatively thick (25-1000 nm), amorphous, hydrated matrix consisting of acidic mucopolysaccharides (or occasionally acidic polypeptides) (Beveridge 1989a, Beveridge and Doyle 1989). Because capsules are loose, fibrillar matrices, they are readily penetrated by small solutes such as metal ions and, like cells walls, are interactive with them (Beveridge 1981, Corpe 1964).

S-layers are another frequently encountered additional layer found on top of the bacterial wall. These are formed out of protein (or glycoprotein) subunits which self-assemble into paracrystalline layers to encompass each cell (Beveridge 1981). As the S-proteins fold during self-assembly, the polar amino acid residues are usually internalized so that the completed S-layer is more hydrophobic, or non-wettable, than the underlying cell wall surface. Yet, there are enough reactive groups still exposed and strategically placed that the S-layer can be highly interactive with dissolved metals (Schultze-Lam et al. 1992). It is best to think of an S-layer as being an ordered network of protein on top of a bacterial surface with enough open channels between the subunits to resemble a molecular sieve; the channels are porous enough to allow a free-flow of simple molecules and solutes.

Biofilms

In natural environments, bacteria are frequently subjected to nutrient-limiting conditions. Because of this, bacteria will often seek out and attach to solid interfaces, thereby taking advantage of the natural tendency of such interfaces to collect and concentrate soluble nutrients. Once attached to these solid surfaces, bacteria grow and divide as they feed, often increasing their biomass to such an extent that visible films, or "biofilms," are produced containing layer after layer of bacteria.

For gram-negative bacteria the initial attachment is a surface phenomenon. The bacterium must provide a suitably reactive surface to the solid interface for attachment to occur. Hydrophilic interfaces require charged cell walls, whereas uncharged interfaces require more hydrophobic cell walls. Our studies with *Pseudomonas aeruginosa* have demonstrated that the outer membrane surface component, LPS, is most important for initial attachment. This bacterium decorates its surface with two entirely different LPSs and their O-side chains (those chemical moieties which extend from the outer face of the membrane into the external milieu; Beveridge 1988) have entirely different surface physicochemistries. The side chain of A-band LPS consists of a linear polymer of α-1, 2-, α-1,3- and α-1,3-linked rhamnose which is apolar and confers hydrophobicity to the cell surface. The side chain of B-band LPS is a longer linear polymer which consists of repeats of two manuronic acid derivatives linked to a N-acetyl fucosamine. Each of these moieties in the B-band trimer is polar. The surface of *P. aeruginosa* is, then, studded with two separate macromolecules which can separately interact with either hydrophobic (i.e. A-band LPS) or hydrophilic (i.e. B-band LPS) interfaces (Makin and Beveridge 1996). Once attached, the bacteria can exude exopolysaccharide (EPS; for *P. aeruginosa* the EPS in an alginate) to irreversibly cement the cells to the interface, and growth and division ensues. Because biofilms consist of a heterologous spectrum of bacteria and their porous (usually) acidic EPSs, biofilms are highly interactive with metallic ions in the environment (Beveridge and Doyle 1989). Overtime, they can accumulate so many metallic precipitates that they can become completely mineralized.

SORPTION OF METAL AND SILICATE IONS
TO BACTERIAL SURFACES

It is apparent that both gram-negative and gram-positive walls consist of a spectrum of different macromolecules which are situated in such a way that they can be highly interactive with the external milieu. Exposed reactive chemical groups, such as carboxylates and phosphates (thiols are relatively rare), are exposed at the cell surface and are, therefore, readily available for ionic interaction with solutes. Once a metal has complexed to such a chemical group, it serves as a nucleation site for further metal complexation; metal precipitates rapidly grow as they use available environmental counter-ions (OH^-, HCO_3^-, SO_4^{2-}, etc.) during the process (Beveridge and Koval 1981, Beveridge and Murray 1976 1980, Beveridge et al. 1982, Doyle et al. 1980, Mullen et al. 1989, Nakajima and Sakaguchi 1986). It is not unusual for metal precipitates to grow to such an extent that

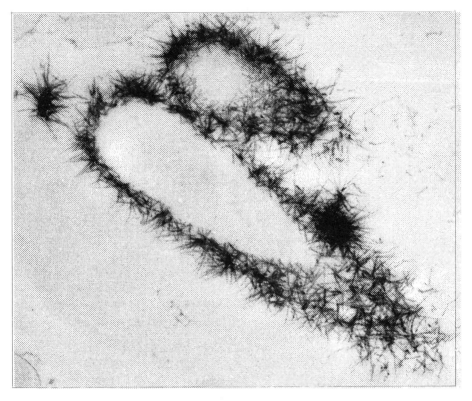

Figure 3. Thin-section of *Pseudomonas aeruginosa* after the cell has been immersed in LaCl$_3$ (1 mM, 10 min) and washed. The cell is covered with La crystals. The diameter of the cell is ~0.5 μm. With permission of the authors; taken from Mullen et al. (1989).

they approach the mass of the bacterial cell (Fig. 3). These are usually poorly ordered precipitates and are (initially) hydrous but, over time, they lose water and can become crystalline phases (Fig. 3; Beveridge 1989b, Schultze-Lam et al. 1992, 1993, 1995a; Southam and Beveridge 1992, 1994).

Most studies which have examined metal-bacteria interactions have been qualitative, determining the relative binding capacities of different metals with different types of walls (i.e. gram-positive or gram-negative) or different wall components (e.g. peptidoglycan, teichoic acids, teichuronic acids, LPS, etc.) (Beveridge and Koval 1981, Beveridge and Murray 1976, 1980; Doyle et al. 1980, Ferris and Beveridge 1984, Marquis et al. 1976, Nakajima and Sakaguchi 1986). Although some workers have applied semi-quantitative approaches (Gonçalves et al. 1987, Mullen et al. 1989), they have not differentiated between the actual surface sites chemically involved with metal binding. Because the various functional groups within the cell wall become active under different pH conditions, pH should have a profound effect on metal ion sorption. Recently, Fein et al. have taken a thermodynamic approach to help explain metal-cell wall complexation using existing speciation and transport models (Fein et al. 1997). Acid-base titrations on *B. subtilis* walls fit a three pK-model suggesting that existing functional groups (carboxyl, pK = 4.8; phosphate, pK = 6.9; and hydroxyl, pK = 9.4) account for the buffering (or complexation) power of these walls (Fig. 4). Metal adsorption (e.g. Fig. 5) increases with both increasing pH and metal concentration which is consistent with metal ion adsorption to deprotonated functional groups within the wall (Fein et al. 1997). If simple deprotonation describes all metal-wall complexation, the sorption phenomenon should be capable of being explained through the use of equilibrium thermodynamics.

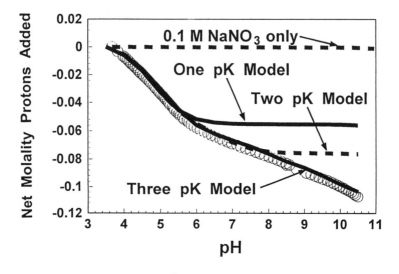

Figure 4. Acid-base titration curve of *B. subtilis* cell walls taken from Fein et al (1997), showing that a three-pK model (representing carboxyl, phosphate and hydroxyl groups) best fits the data (with permission of the authors).

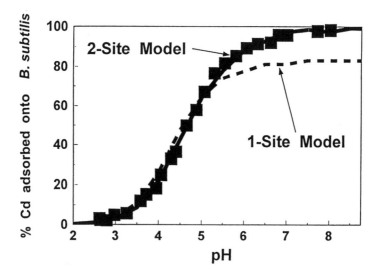

Figure 5. Cadmium sorption results taken from Fein et al. (1997) for *B. subtilis* cell walls plotted as a function of pH. In this case, the sorption data best fits a two-site model (i.e. complexation to carboxyl and phosphate groups) instead of a one-site carboxylate model (with permission of the authors).

Recently it has also been shown that bacterial surfaces also are good sorption interfaces for silicate ions (Urrutia and Beveridge 1994). Although the surface has a net electronegative charge on it, there are still some available amine groups situated throughout the cell wall matrix. These are electropositive and can interact with silicate anions at circumneutral pH (Urrutia and Beveridge 1993a). Yet, there are not enough amine groups per unit area of the wall matrix to account for the large amount of silicate which is eventually bound. This "extra" silicate is deposited through metal ion-bridging where a multivalent metal ion cross-linkings Si anions to COO^- or PO_4^{3-} by electrostatic interaction (Urrutia and Beveridge 1993a). As with the metal precipitates described previously, these silicates begin as poorly ordered phases but, over time, become crystalline. Bona fide interstratified clay phases can eventually be formed (Urrutia and Beveridge 1994). Interestingly, as silicate mineral phases develop on bacterial surfaces, they too are efficient metal ion sorption interfaces and increase the overall heavy metal sorption capacity of the cell wall (Urrutia and Beveridge 1993b, 1994). Yet, these silicates do not tend to complex the metals as tenaciously as the bacterial wall (Urrutia and Beveridge 1993b).

BACTERIAL MINERAL PRECIPITATION

An important prerequisite for the precipitation of minerals from solution, even where bacteria are involved, is that a certain degree of oversaturation must be achieved. This requirement is symptomatic of an activation energy barrier that blocks or inhibits the spontaneous formation of insoluble precipitates from solution (Stumm 1992). In thermodynamic terms, the energy needed to form a new solid-liquid interface is more easily overcome in highly saturated solutions by the energy released as a result of bond formation in the solid phase. The point at which the activation energy barrier is breached is referred to as nucleation, and involves the growth of molecular clusters (i.e. critical nuclei) that are stable relative to dissolution. Once a stable nucleus has formed, further increases in the number of ions in the solid phase are accompanied by a decrease in free energy. This process is known as crystal growth, and proceeds spontaneously providing that the concentration of ions or molecules in solution continues to exceed the solubility product of the solid mineral phase.

Nucleation is the most critical stage for mineral precipitation. It proceeds typically in one of two ways, either homogeneously or heterogeneously (Stumm 1992). In the case of homogenous nucleation reactions, stable nuclei develop simply through random collisions between ions in an oversaturated solution. On the other hand, heterogenous nucleation involves development of nuclei on surfaces of foreign solids that enhances nucleation by reducing interfacial contributions to the activation energy barrier. In effect, a heteronucleus can be envisioned as a scaffold or template of similar atomic spacing that promotes mineral precipitation. Particularly favorable heterogenous nucleation templates usually possess reactive sites where strong surface chemical interactions occur (e.g. sorption or bonding).

Bacteria intervene in mineral precipitation reactions directly as catalysts of aqueous chemical reactions (Thompson and Ferris 1990, Thompson et al. 1997, Fortin and Beveridge 1997a), and indirectly as geochemically reactive solids (Mullen et al. 1989, Fein et al. 1997). In the first instance, their metabolic activity is often significant, and can trigger changes in solution chemistry that lead to oversaturation (e.g. through the produc-tion of chemically reactive ligands like sulfide). This alone can induce mineral precipitation by lowering the activation energy barrier for both homogeneous and heterogeneous nucleation reactions. At the same time, however, reactive sites on bacterial cell surfaces that facilitate sorption of dissolved metal ions foster heterogeneous nucleation and precipitation. Thus, minerals precipitated directly from solution as a result of bacterial metabolic activity can form on the inside, outside, or even some distance away from cells. Indirect chemical precipitation as a consequence of changing geochemical conditions (e.g. groundwater degassing) is also possible and is accompanied by passive epicellular nucleation and crystal growth on the outside of living or dead bacterial cells. In natural environments, direct and indirect bacterial mineral precipitation reactions often occur at the same time and are difficult to recognize as separate processes. Yet, there are many examples of bacterially supported mineralization in present day environments, and there is considerable evidence that similar kinds of processes contributed to the development of some mineral deposits in the past.

Carbonates

Cyanobacteria have been widely implicated in the formation and deposition of carbonate minerals. The involvment of these photoautotrophic microorganisms in carbonate mineral precipitation is closely related to photosynthetic activity and biomass production. Specifically, metabolic uptake and fixation of dissolved CO_2 by cyanobacteria tends to increase pH when bicarbonate is the predominant inorganic carbon species in solution (Thompson and Ferris 1990, Schultze-Lam and Beveridge 1994, Ferris et al. 1994, Ferris et al. 1995, Stumm and Morgan 1996). This generally leads to an increase in saturation state that readily brings about carbonate precipitation.

The active metabolic role that cyanobacteria play in the formation of carbonate minerals is supported by stable carbon isotope data from natural precipitates (Burne and Moore 1987, Pentecost and Spiro 1990, Thompson et al. 1997, Ferris et al. 1997). Under conditions where dissolved inorganic carbon is not growth limiting, carbonates that precipitate in association with cyanobacteria are commonly enriched in ^{13}C relative to the bulk dissolved inorganic carbon species (i.e. positive $\Delta\partial^{13}C$ carbonate-dissolved inorganic carbon values). This difference in isotopic composition is caused by the preferential use of the lighter ^{12}C isotope during photosynthesis, which leaves the dissolved inorganic carbon that is precipitated as carbonate around cells enriched in ^{13}C by as much as 4 to 5 $^o/oo$ (Hollander and McKenzie 1991).

Carbonate precipitates commonly develop on the external surfaces of individual cyanobacterial cells (Pentecost and Bauld 1988, Thompson and Ferris 1990, Schultze-Lam and Beveridge 1994). This pattern of mineralization is consistent with the inherent ability of bacterial cells to serve as templates for heterogenous nucleation and crystal growth. Cell surface precipitation contributes further to the development of extremely fine-grain precipitates (i.e. often <1.0 μm in diameter), as is expected from rapid onset of precipitation (Stumm 1992).

In Fayetteville Green Lake, New York, *Synechoccocus* strain GL24 is a dominant cyanobacterium that typically blooms over the summer months (Thompson et al. 1997). The bacterium possesses an S-layer that exists as a hexagonal (p6) array of a 104 kDa protein on which strategic carboxyl groups are located (Schultze-Lam et al. 1992). Non-metabolizing cells acquire Ca^{2+} and SO_4^{2-} on the S-layer where gypsum aggregates develop and are subsequently shed from the cells (Fig. 6); however, as the cells become photosynthetically active, their metabolism alkalinizes the local environment that surrounds each bacterium to a pH above 8, shifting the mineral solid field from gypsum to calcite. Therefore, during the summer months when the cells are photosynthetically active, and are growing and dividing, a light rain of calcite falls to the lake bottom contributing to marl sediment deposition. In addition, calcification of cells growing on submerged solid surfaces in biofilms have contributed to the formation of large natural bioherms or freshwater reefs. Ultimately, *Synechoccocus* strain GL24 controls the calcite/gypsum mineralogy of the lake (Thompson et al. 1997). Laboratory simulations using Sr^{2+}, Mg^{2+}, or Ca^{2+} plus Sr^{2+} have shown that strontianite, celestite, hydromagnesite, and mixed calcium-strontium carbonates can also be formed (Thompson and Ferris 1990, Schultze-Lam and Beveridge 1994).

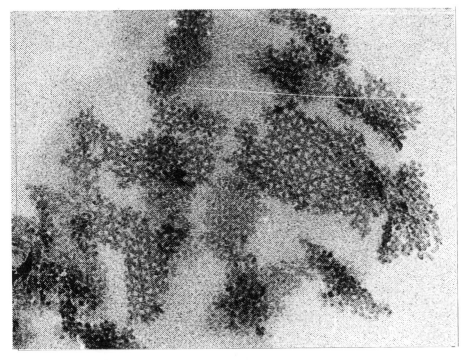

Figure 6. Whole mount of a gypsum-encrusted fragment of the S-layer from *Synechococcus* strian GL24 which has been suspended in water from Fayetteville, Green Lake, New York. No stain has been used on this specimen so that the electron density is derived only from the gypsum coating the crystallyne S-layer. The c-c spacing of the p6 array is ~20 nm. Used with permission of the authors; taken from Schultze-Lam et al. (1993).

While strontium concentrations may approach 1.0 wt % in aragonite, normal values for strontium substitution in calcites tend to be much lower (Blatt 1982, Deer et al. 1992). This has been attributed to the availability of more space in the crystal structure of aragonite in comparison to that of calcite; however, high mineral precipitation rates sustained by cyanobacterial activity apparently contribute to the partitioning of up to 1.0 wt % strontium into calcite (Ferris et al. 1995). Since the Sr^{2+} ion is expected to destabilize the crystal structure of calcite, the effect of small grain size on eliminating long-range lattice effects in fresh precipitates is considered to be an important factor contributing to the stability of strontium-calcite solid solutions (Morse and Bender 1990). This consideration implies further that solid solution formation involving trace metal species like strontium may actually be enhanced in natural waters , through rapid bacterial precipitation of fine-grained carbonate minerals (Ferris et al. 1995).

Much of the foregoing discussion has focused on cyanobacteria, however it is important to note that other types of bacteria can promote carbonate mineral precipitation. Some of the most commonly described processes involving bacteria other than cyanobacteria include ammonia production from the degradation of organic matter, nitrate-reduction, Fe(III)- and Mn(IV)-reduction, and sulfate-reduction (Kobluk and Crawford 1988, Coleman et al. 1993, Vasconcelos et al. 1995). All of these predominantly anaerobic microbial activities tend to increase pH and alkalinity (Stumm and Morgan 1996). When a state of oversaturation finally develops, carbonate mineral precipitation follows. Additional information on some of these processes is provided in the following section.

Sulfides and paragenetic minerals that form in anaerobic environments

Bacterial sulfate reduction is the main mechanism by which sulfide is produced at low temperatures (i.e. <100°C) in anoxic sediments (Berner 1980). The activity of these bacteria contributes directly to the formation of pyrite and other sulfide minerals, particularly in marine systems where sulfate is not a limiting nutrient for bacterial growth. Moreover, patterns of sulfur isotope fractionation in many stratiform sulfide mineral deposits are consistent with this form of mineralization, and support the bacteriogenic origin of reduced sulfur (Trudinger et al. 1985).

Although bacterial sulfate reduction is crucial for the precipitation of metal sulfides, laboratory and field investigations also suggest that the formation of organometallic complexes is instrumental in the partitioning of metal ions into sulfide phases (Forstner 1982). Specifically, experimental work with cultures of sulfate-reducing bacteria has shown that metal ions sorbed to bacterial cells tend to be more chemically reactive toward sulfide than when they are in solution (Mohagheghi et al. 1985). Geochemical studies further indicate that sulfide minerals deposited in anaerobic sediments tend to be closely associated with particulate organic materials. This observation is consistent with results from transmission electron microscopy (TEM) of marine, and to a lesser extent, fresh-water sediments where metal sulfide precipitates are commonly found on the external surfaces of bacterial cells (Fig. 7; Degens and Ittekkot 1982, Ferris et al. 1987).

During bacterial sulfate reduction, sedimentary organic matter is consumed as sulfide, phosphate, and metabolic CO_2 are released to the porewater system (Berner 1971). Because sulfide is a fairly strong base, this causes CO_2 to form bicarbonate. Since the reaction both removes CO_2 and increases the concentration of bicarbonate ions, equilibrium constraints on carbonate mineral solubility may be exceeded and precipitation may occur. At the same time, should dissolved phosphate concentrations in the sediment porewater system become high enough, phosphate minerals can precipitate. This often leads to the development of phosphatidic mineralization on pre-existing hard substrates, including organic matter derived from microbial cells (Soudry and Champtier 1983, Dahanayake and Krumbein 1985, Southgate 1986). In this way, sedimentary organic matter serves not only as a source for inorganic phosphate, but also as a substrate upon which phosphate minerals nucleate. Considering the interrelationships between these various processes, it is not surprising to find that pyrite formation in organic-rich sediments is often nearly coincident with the precipitation of carbonate and phosphate minerals (Glenn and Arthur 1988).

Recent studies have shown that iron-reducing bacteria are common and widely distributed in anaerobic sedimentary environments (Lovley 1991, Coleman et al. 1993, Lovley and Chapelle 1995). In fact, iron-reducing bacteria appear to be more than able to out compete sulfate-reducing and methane-producing microorganisms owing to their ability to keep the concentration of common energy yielding substrates (e.g. hydrogen and acetate) at levels that are too low for sulfate-reducers and methane-producers (Lovley 1991). On

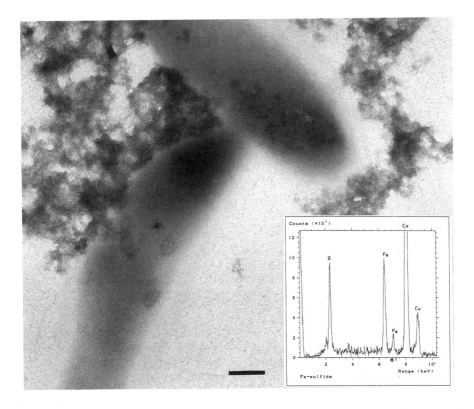

Figure 7. Whole mount of sulfate-reducing bacteria recovered from mine tailings showing the presence of amorphous Fe-sulfide precipitates (confirmed by EDS; Cu peaks are from the supporting grid) on the cell wall. Scale: 200 nm.

the other hand, iron-reducing bacteria, like sulfate-reducers, are quite capable of initiating carbonate mineral precipitation (Lovley 1991, Roden and Lovley 1993, Coleman et al. 1993). The following reactions apply when the bacteria oxidize acetate (Eqn. 1) or hydrogen (Eqn. 2) with goethite as a terminal electron acceptor (Coleman et al. 1993):

$$8\ FeOOH + CH_3COO^- + 3\ H_2O\ =\ 8\ Fe^{2+} + 2\ HCO_3^- + 15\ OH^- \tag{1}$$

$$8\ FeOOH + 4\ H_2\ =\ 8\ Fe^{2+} + 16\ OH^- \tag{2}$$

The first equation shows that bicarbonate and hydroxyl ions are produced when bacterial Fe(III)-oxide reduction proceeds using acetate as the electron donor. Conversely, Fe(III)-reduction coupled to hydrogen oxidation yields only hydroxyl ions (Reaction 2). Yet, both reaction pathways can increase alkalinity to make carbonate precipitation more favorable (Thompson and Ferris 1990, Ferris et al. 1995). An increase in pH and precipitation of siderite ($FeCO_3$) commonly occurs during growth of Fe(III)-reducing bacteria in bicarbonate-buffered culture media (Roden and Lovley 1993). Siderite accumulation in some marine sediments also appears to be a product of these processes (Aller et al. 1986, Coleman et al. 1993).

Anaerobic oxidation of organic material by iron-reducing bacteria in fractured granitic bedrock appears to be a key source of alkalinity that contributes carbonate precipitation

(Pedersen and Ekendahl 1990, Pedersen and Karlsson 1995, Lanstrom and Tullborg 1995). Studies of groundwater trace element interactions with the carbonate mineralization in these subsurface hydraulic systems show that substantial amounts of strontium (100 to 9000 ppm), barium (100 to 1000 ppm), uranium (0.3 to 3.5 ppm), and other metals become incorporated into the precipitates (Lanstrom and Tullborg 1995). Laboratory-determined values for the adsorption of the metal ions on calcite are small in comparison to the solid phase concentrations measured in the natural precipitates, again suggesting that rapid bacterial precipitation of fine-grained carbonate minerals and solid-solution formation are important for the capture and retention of trace metals in bacteriogenic calcites (Morse and Bender 1990).

Iron and manganese oxides

The formation of Fe- and Mn-oxides in natural environments mainly depends on the redox conditions of the aqueous milieu. Under oxic conditions, Mn(II) and Fe(II) are oxidized through chemical and biological reactions and precipitate as oxides, however, the oxidation rate of both iron and manganese depends strongly on pH. Near neutral pH conditions will promote a rapid chemical oxidation of Fe(II) to Fe(III), while slightly alkaline conditions (pH ~ 8-9) will enhance the Mn(II) oxidation rate (Stumm and Morgan 1996). Under a wide range of pH conditions, microorganisms affect the solubility of Fe and Mn through enzymatic and non-enzymatic reactions.

Enzymatic reactions. Enzymatic reactions leading to Fe- and Mn-oxide mineral formation have an overall rate of reaction which is faster than the non-microbial reactions (Ehrlich 1996, 1990). The oxidation of Mn(II) and Fe(II) in natural environments is often catalyzed by specific bacteria, since the pH of most natural waters (pH 5-8) slows down the rate of reaction. In the case of Fe(II) oxidation, *Thiobacillus ferrooxidans*, an iron-oxidizing bacteria, can increase the rate by as much as six orders of magnitude (Singer and Stumm 1970). *T. ferrooxidans* is an acidophilic bacterium which fixes atmospheric CO_2 and obtains its energy from the oxidation of reduced sulfur and iron species (Pronk and Johnson 1992, Harrison 1984). This bacterium is wide spread in acidic environments, such as mine tailings, where it oxidizes Fe-sulfide minerals (Fortin et al. 1995, 1996; Southam and Beveridge 1992). The main reactions in pyrite oxidation are (Singer and Stumm 1970):

$$FeS_2 + 7/2\ O_2 + H_2O\ =\ Fe^{2+} + 2\ H^+ + 2\ SO_4^{2-} \tag{3}$$

$$2\ Fe^{2+} + 1/2\ O_2 + 2\ H^+\ =\ 2\ Fe^{3+} + H_2O \tag{4}$$

$$FeS_2 + 14\ Fe^{3+} + 8\ H_2O\ =\ 15\ Fe^{2+} + 2\ SO_4^{2-} + 16\ H^+ \tag{5}$$

Reactions (3) and (4) are bacterially mediated, while Reaction (5) is strictly chemical. The widely accepted model for the enzymatic oxidation of Fe(II) by *Thiobacillus* assumes that ferrous iron is oxidized in the vicinity of the cell envelope (Ingledew et al. 1977). Electrons are transported through the periplasm to the cytochrome oxidase located on the inside surface of the plasma membrane, where they react with O_2. Other bacteria, such as *Leptospirillum ferrooxidans* and strains of *Metallogenium,* have been reported to catalyze the oxidation of Fe(II) at low pH conditions (Erhlich 1990). The role of neutrophilic bacteria in the enzymatic oxidation of Fe(II) is more difficult to ascertain since the chemical oxidation of Fe(II) is easily achieved near neutral pH conditions. The strongest evidence of enzymatic iron oxidation under such conditions is that associated with *Galionella ferruginea* (Ehrlich 1990). This bacterium does not grow in the absence of Fe(II), does not require organic carbon and appears to fix CO_2. Fe-oxides found in association with this bacterium are observed on their stalks (twisted bundle of fibrils) that are produced by the cells.

TEM observations of acidic mine tailings and of *Thiobacillus* cultures have shown that the cells serve as a substrate for poorly ordered Fe-oxide precipitation. These oxides often contain various amounts of impurities (Si, S, P, Cu, Zn, etc.; Fortin et al. 1996). The presence of impurities on the surface or within the structure of the oxides possibly stabilize them, making them less soluble than pure Fe-oxides. Various crystalline Fe-oxides have been observed in acidic environments surrounding oxidized mine tailings. Ferris et al. (1989) identified ferrihydrite (a poorly ordered Fe-oxide), goethite and hematite in association with bacteria. Impurities, such as C_{org}, S, P or Si, were detected in these oxides, which occured as finely granular material, fibrous and lath-like minerals on bacterial cell walls.

Manganese oxidation is catalized by a wide range of bacteria in natural environments (Ehrlich 1990). The formation of Mn-oxides through enzymatic oxidation of Mn(II) is seen as a two-step reaction (Ehrlich 1990, 1996):

$$2 Mn^{2+} + 1/2 O_2 + 2 H^+ = 2\{Mn^{3+}\} + H_2O \tag{6}$$

$$2\{Mn^{3+}\} + 1/2 O_2 + 3 H_2O = 2 MnO_2 + 6 H^+ \tag{7}$$

where Mn(II) is first enzymatically oxidized to Mn(III), which remains bound to the manganese-oxidizing enzyme ($\{Mn^{3+}\}$). Reaction (7) represents the catalized oxidation of bound Mn(III) to MnO_2. Different bacteria have been reported to be able to catalyze the oxidation of Mn(II): they include some *Pseudomonas* strains, *Metallogenium*, etc. (Ehrlich 1990). Some bacteria, such as *Leptothrix discophora* SS-1, can also excrete polymeric components with catalytic capacity to oxidize Mn(II) extracellularly (Adams and Ghiorse 1987, Boogerd and De Vrind 1987).

Mn(IV)-oxides (birnessite, vernadite or buserite) are the main minerals formed in the presence of bacteria, but some intermediate Mn(III) species can also be produced (Adams and Ghiorse 1988). Mn-oxides usually occur as sheets on the cell wall or on extracellular polymers.

Non-enzymatic reactions. Non-enzymatic reactions provide an indirect role for bacteria in the oxidation of Mn and Fe. As a result of microbial activity, chemical conditions (such as Eh and pH) are changed in the vicinity of the cells, favoring the chemical oxidation of Mn and Fe. In some cases, metabolic endproducts can also act as oxidizing agents (Ehrlich 1990). In addition, bacterial surfaces and exoplymers can bind oxidized Mn and Fe species and serve as substrates for mineral development. Non-enzymatic reactions in natural environments can occur simultaneously with abiotic oxidation reactions.

Any organism that produces ammonia from proteins or consumes organic salts can raise the pH of the aqueous milieu, allowing the chemical oxidation of Fe(II) (Ehrlich 1990). Photosynthetic microorganisms, such as cyanobacteria, can also promote the chemical oxidation of Fe(II) and the precipitation of Fe-oxides through pH increase. In addition, bacterial breakdown of organic ferric complexes in natural waters can free the Fe(III) ion and allow its precipitation as an oxide (Ehrlich 1990). Natural Fe-oxides, formed in marine and freshwater environments, are often found in close association with bacteria and their extracellular polymers (Cowen and Bruland 1985, St-Cyr et al. 1993, Fortin et al. 1993). These polymers become coated with Fe-oxides which often cement other particles together (Fig. 8). Bacterial cell walls and their exopolymers (e.g. EPS) serve as substrates for the sorption of oxidized Fe species. The sorption of Fe(III) is the result of an electrostatic interaction between positively charged Fe(III) ionic species and the anionic

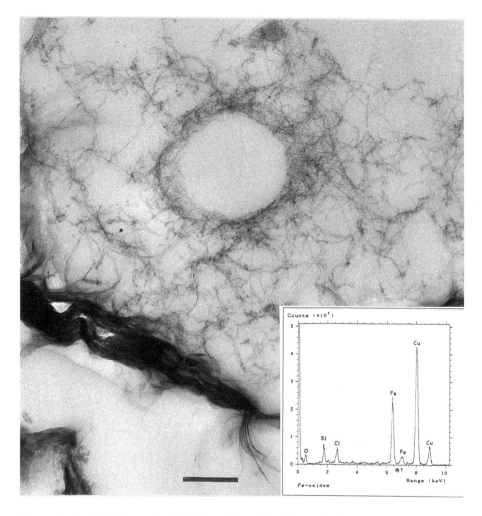

Figure 8: Unstained thin-section of Fe-rich coatings collected near an hydrothermal vent in the Pacific ocean. Fine extracellular polymers extending away from the cell are covered with amorphous Fe-oxides and attaching to detrital silicate fragments. These oxides contain small amounts of Si and Cl (Cu peaks are from the supporting grid). Scale: 200 nm.

components of the cell wall and polymers. EPS would supply oxygen atoms (from their carboxyl, hydroxyl or carbonyl groups) for the formation of basic Fe-oxide octahedral units, which in turn serve to initiate the growth of the Fe-oxide minerals (Degens and Ittekkot 1982, Mann 1988).

Non-enzymatic manganese oxidation can be caused by chemical agents produced by bacteria. The metabolic utilization of carboxylic acids (citrate, lactate, malate, gluconate and tartrate) usually leads to a pH increase in the aqueous milieu. Residual acids can then catalize the oxidation of Mn(II) (Ehrlich 1990). Some bacteria also produce proteins that can oxidize Mn(II). In addition, bacterial surfaces and extracellular polymers possibly serve as sorption surfaces for oxidized Mn species and nucleation sites for Mn-oxides.

Silicates

Various silicate minerals (silica, mixed Fe-Al-silicates) have been observed in close association with bacteria over a range of chemical conditions in natural environments. Mineralized bacterial cells were reported in hot-spring environments (Ferris et al. 1986, Schultze-Lam et al. 1995b, Hinman and Lindstrom 1996, Jones and Renaut 1996), river sediments and biofilms (Konhauser et al. 1993, 1994), mine tailings (Ferris et al. 1987, Fortin and Beveridge 1997b) and samples collected near hydrothermal sea-vents (Fig. 9). These studies showed that various types of amorphous and crystalline silicates can precipitate in the presence of bacteria. For example, fine grains and spheroidal crystallites of silica, along with poorly ordered silicates with a chemical composition similar to chamositic clays, $(Fe_5Al)(Si_3Al)O_{10}(OH)_8$, have been observed on bacterial walls (Ferris et al. 1986, Konhauser et al. 1993, Schultze-Lam et al. 1995b, Fortin and Beveridge 1997b). In some cases, more crystalline phases (such as kaolinite, $Al_4(Si_4O_{10})(OH)_4$, and glauconite-like minerals, $K(Al_{0.38}Fe_{1.28}Mg_{0.34})(Si_{3.7}Al_{0.3})O_{10}(OH)_2$) were identified (Konhauser et al. 1993, 1994). In many samples, the extent of silicification was great enough to preserve the shape of cellular structures, a first step in fossilization. According to Ferris et al (1988), who investigated the silicification of *B. subtilis* cells under laboratory conditions, the binding of metals (such as Fe) prior to silicification help preserve the original morphology of the cells. Lytic enzymes, capable of degrading the cell wall, are possibly inhibited by metals, therefore preventing a complete autolysis of the cells.

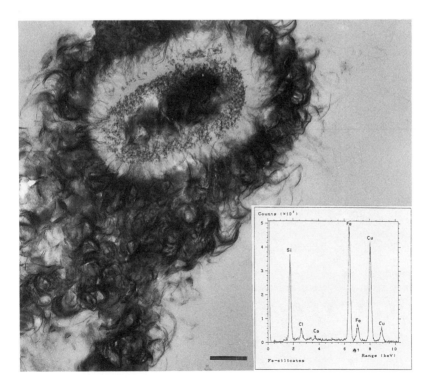

Figure 9. Stained thin-section of a sample collected near a hydrothermal vent in the Pacific where the remains of a bacterium are encrusted with Fe-silicates (Cu peaks from the supporting grid). Scale: 200 nm.

Metal sorption to bacterial walls is also seen as an important process in silicate binding and precipitation (Urrutia and Beveridge 1993b, 1994). Metals serve as a bridge between the anionic constituents of the cell wall and the silicate anions in solution. Fortin and Beveridge (1997b) showed that such a mechanism was responsible for the binding of silicate anions to the cell wall of *Thiobacillus* under acidic conditions (pH ~ 2). Energy dispersive spectroscopy (EDS) analyses of the cells indicated the presence of Fe on their surfaces prior to silica binding and precipitation. In additon, the presence of *Thiobacillus* accelerated the precipitation rate of silica, when compared to an abiotic system. The binding of small amounts of silicate anions to the bacterial walls most likely forced the dissociation of silicic acid, $Si(OH)_4$, (the dominant soluble Si species in the system) and the subsequent nucleation of silica. Even though bacteria do not play an active role (unlike the enzymatic oxidation of Fe and Mn) in silicate formation, they can accelerate the rate of precipitation, by providing sorption surfaces, which affect the equilibrium reactions of the aqueous environment.

CONCLUSIONS

The evidence is clear that prokaryotic life evolved as the first life-form on Earth and that it was present during the late Archean and Proterozoic periods. This evidence stems from abundant observations of so-called "microfossils" found in ancient, organic-rich cherts and the distinct biochemical fingerprints found in a range of sedimentary rocks (Barghoorn and Tyler 1965, Dyer et al. 1988, Eglinton et al. 1964, Saxby 1969, Schopf and Walker 1983, Walter 1983). Ancient stromatolites, because of their laminated structure, first evoked the idea of a biological origin (Barghoorn and Tyler 1965) and are believed to be the mineralized remnants of biofilms composed of filamentous bacteria similar to cyanobacteria. Certainly there are many morphological similarities between these ancient microfossils and present-day natural biofilms composed of bacteria (cf. Figs. 10a and b). The most convincing argument for mid-archean life is found in stromatolitic cherts of the Western Australia Warrawoona Group (Walter 1983). These indicate that prokaryotes inhabited this surficial environment at least 3.5 Ga ago and the complexity of their cellular remains is persuasive evidence that even simpler life must predate this period.

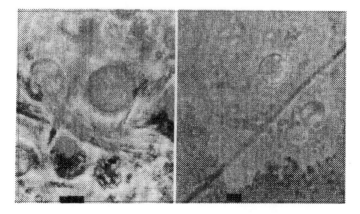

Figure 10. (a) Bright field light micrograph of a geological thin section of the Gunflint chert, Upper Superior region, Canada. This rock is filled with mineralized remains of filamentous and spherical bacteria. Although these microfossils are from ancient stromatolite, they bear a striking resemblance to the bacteria of a present-day biofilm seen in (b). The preparations were collected and prepared by F.G. Ferris. The diameters of the bacteria in both micrographs range from 1 to 20 μm. With permission of the author; taken from Beveridge and Doyle (1989).

For many years, it was difficult to understand exactly how these ancient remnants of bacteria were so well preserved. Clearly, they are mineralized and their shape is preserved in the rock matrix; but, very little organic matter is left. The mystery revolves around the unequivocal fact that bacteria are entirely composed of "soft" matter (Beveridge 1988, 1989a,b). They do not possess a hard boney endo- or exoskeleton which would help preserve their cellular structure over vast periods of time. Now that we recognize their high interaction and collection of inorganic ions from the environment, a rational answer is possible. Over their lifetime, bacteria must collect an increasing burden of minerals on their surfaces. When they die, this burden continues to grow until they become completely mineralized (Beveridge et al. 1983, Ferris et al. 1988). In fact, the very enzymes that should decompose bacteria after death are inhibited by multivalent metal ions (such as Fe^{3+}; Ferris et al. 1988) which are abundant in natural environments. Therefore, the strong cell walls (which give shape to bacteria) survive as mineralization becomes complete. In this way, although the organic matter may eventually disappear, the shape of the bacterium survives as a mineralized form possessing all of the contours and dimensions of the initial cell. These microfossils remain as such mineralized forms unless the geological horizon in which they are contained is subjected to harsh mineral-altering forces (Beveridge et al. 1983).

ACKNOWLEDGMENTS

The research reported in this article is the work of energetic former and current graduate students and postdoctoral fellows in the authors' laboratories. These are S. Makin, R. McLean, M. Urrutia, J. Thompson, S. Schultze, G. Southam, and B. Davis. All research was supported by an NSERC operating grants to TJB and FGF. The electron microscopy was done in the NSERC Guelph Regional STEM Facility which is partially maintained by an NSERC major facilities access grant.

REFERENCES

Adams LF, Ghiorse WC (1987) Characterization of extracellular Mn^{2+}-oxidizing activity and isolation of an Mn^{2+}-oxidizing protein from *Leptothrix discophora* SS-1. Bacteriol 169:1279-1285
Adams LF, Ghiorse WC (1988) Oxidation state of Mn in the Mn-oxide produced by *Leptothrix disciphora* SS-1. Geochim Cosmochim Acta 52:2073-2076
Aller RC, Macklin JE, Cox RTJ (1986) Diagenesis of Fe and S in Amazon inner shelf muds: apparent dominance of Fe reduction and implications for the genesis of ironstones. Cont Shelf Res 6:263-289
Appanna VD, Preston CM (1987) Manganese elicits the synthesis of a novel exopolysaccharide in an artic *Rhizobium*. FEBS Lett 215:79-82
Barghoorn ES, Tyler SA (1965) Microorganisms from the Gunflint chert. Science 147:563-577
Berner RA (1971) Bacterial processes effecting the precipitation of calcium carbonate in sediments. The Johns Hopkins Univ Studies in Geology 19:247-251
Berner RA (1980) Early Diagenesis, A Theoretical Approach. Princeton Univ Press, Princeton, NJ
Beveridge TJ (1981) Ultrastructure, chemistry and function of the bacterial wall. Int Rev Cytol 72:229-317
Beveridge TJ (1988) The bacteria surface: general considerations towards design and function. Can J Microbiol 34:363-372
Beveridge TJ (1989a) The structure of bacteria. In: JS Poindexter, ER Leadbetter (eds) Bacteria in Nature, p 1-65, Plenum Press, New York
Beveridge TJ (1989b) Role of cellular design in bacterial metal accumulation and mineralization. Ann Rev Microbiol 43:147-171
Beveridge TJ, Murray RGE (1976) Uptake and retention of metals by cell walls of *Bacillus subtilis*. J Bacteriol 127:1502-1518
Beveridge TJ, Murray RGE (1980) Sites of metal deposition in the cell wall of *Bacillus subtilis*. J Bacteriol 141:876-887
Beveridge TJ, Koval SF (1981) Binding of metals to cell envelopes of *Escherichia coli* K-12. Appl Environ Microbiol 42:325-335

Beveridge TJ, Doyle RJ (eds) (1989) Metal Ions and Bacteria. John Wiley & Sons, New York

Beveridge TJ, Forsberg CW, Doyle RJ (1982) Major sites of metal binding in *Bacillus licheniformis* walls. J Bacteriol 150:1438-1448

Beveridge TJ, Meloche JD, Fyfe WS, Murray RGE (1983) Diagenesis of metals chemically complexed to bacteria: laboratory formation of metal phosphates, sulfides and organic condensates in artificial sediments. Appl Environ Microbiol 45:1094-1108

Blatt H (1982) Sedimentary Petrology. W.H. Freeman, San Francisco

Boogerd FC, De Vrind JPM (1987) Manganese oxidation by *Leptothrix disciphora*. J Bacteriol 169:489-494

Burne RV, Moore LS (1987) Microbialites: Organosedimentary deposits of benthic microbial communities. Palaios 2:241-254

Coleman ML, Hedrick DB, Lovley DR, White DC, Pye K (1993) Reduction of Fe(III) in sediments by sulfate-reducing bacteria. Nature 361:36-438

Corpe W (1964) Factors influencing growth and polysaccharide formation by strains of *Chromobacterium violaceum*. J Bacteriol 88:1433-1437

Cowen JP, Bruland KW (1985) Metal deposits associated with bacteria: Implications for Fe and Mn marine biogeochemistry. Deep-Sea Res 32:253-272

Dahanayake K, Krumbein WE (1985) Ultrastructure of a microbial mat generated phosphorite. Mineral Deposita 20:260-265

Degens ET, Ittekkot IV (1982) In situ metal staining of biological membranes in sediments. Nature 298:262-264

Deer WS, Howie RA, Zussman J (1992) An Introduction to the Rock Forming Minerals. 2nd ed. Longmans, Hong Kong

Doyle RJ, Matthews TH, Streips UN (1980) Chemical basis for selectivity of metal ions by the *Bacillus subtilis* cell wall. J Bacteriol 143:471-480

Dyer B, Krumbein WE, Mossman DJ (1988) Nature and origin of stratiform kerogen seams in lower proterozoic Witwatersrand-type paleoplacers—the case for biogenicity. Geomicrobiol J 6:33-47

Eglinton G, Scott PM, Belsky T, Burkingame AL, Calvin M (1964) Hydrocarbons of biological origin from a one-billion-year-old sediment. Science 145:263-264

Ehrlich HL (1990) Geomicrobiology. 2nd ed. Marcel Dekker, New-York

Ehrlich HL (1996) How microbes influence mineral growth and dissolution. Chem Geol 132:5-9

Fein JB, Daughney CJ, Yee N, Davis T. (1997) The thermodynamics of metal adsorption onto bacterial surfaces. Geochim Cosmochim Acta (in press)

Ferris FG, Beveridge TJ (1984) Binding of a paramagnetic metal cation to *Escherichia coli* K-12 outer membrane vesicles. FEMS Microbiol Lett 24:43-46

Ferris FG, Beveridge TJ, Fyfe WS (1986) Iron-silica crystallite nucleation by bacteria in a geothermal sediment. Nature 320:609-611

Ferris FG, Fyfe WS, Beveridge TJ (1987) Bacteria as nucleation sites for authigenic minerals in a metal contaminated lake sediment. Chem Geol 63:225-232

Ferris FG, Fyfe WS, Beveridge TJ (1988) Metallic ion activity by *Bacillus subtilis*: Implications for the fossilization of microorganisms. Geology 16:149-152

Ferris G, Tazaki K, Fyfe WS (1989) Iron oxides in acid mine drainage environments and their association with bacteria. Chem Geol 74:321-330

Ferris FG, Wiese RG, Fyfe WS (1994) Precipitation of carbonate minerals by microorganisms: Implications for silicate weathering and the global carbon dioxide budget. Geomicrobiol J 12:1-13

Ferris FG, Fratton CM, Gerits JP, Schultze-Lam S, Sherwood-Lollar B (1995) Microbial precipitation of a strontium calcite phase at a groundwater discharge zone near Rock Creek, British Columbia, Canada. Geomicrobiol J 13:57-67

Ferris FG, Thompson JB, Beveridge TJ (1997) Modern freshwater microbialites from Kelly Lake, British Columbia, Canada. Palaios (in press)

Forstner U (1982) Accumulative phases for heavy metals in limnic sediments. Hydrobiologia 91:269-284

Fortin D, Tessier A, Leppard GC (1993) Characteristics of lacustrine iron oxyhydroxides. Geochim Cosmochim Acta 57:4391-4404.

Fortin D, Davis B, Beveridge TJ (1995) Biogeochemical phenomena induced by bacteria within sulfidic mine tailings. J Indust Microbiol 14:178-185

Fortin D, Davis B,Beveridge TJ (1996) Role of *Thiobacillus* and sulfate-reducing bacteria in iron biocycling in oxic and acidic mine tailings. FEMS Microbiol Ecol 21:11-24

Fortin D, Beveridge TJ (1997a) Microbial sulfate reduction within sulfidic mine tailings: formation of diagenetic Fe-sulfides. Geomicrobiol J 14:1-21

Fortin D, Beveridge TJ (1997b) Role of the bacterium, *Thiobacillus*, in the formation of silicates in acidic mine tailings. Chem Geol (in press)

Glenn CR, Arthur MA (1988) Petrology and major element geochemistry of Peru margin phosphorites and associated diagenetic minerals: Authigenesis in modern organic-rich sediments. Marine Geol 80:231-267

Gonçalves MLS, Sigg L, Reutlinger M, Stumm W (1987) Metal ion binding by biological surfaces: voltametric assessment in the presence of bacteria. Sci Tot Environ 60:105-119

Harrison AP Jr (1984) The acidophilic thiobacilli and other acidophilic bacteria that share their habitat. Annu Rev Microbiol 38:265-292

Hinman NW, Lindstrom RF (1996) Seasonal changes in silica deposition in hot spring systems. Chem Geol 132:237-246

Hollander DJ, McKenzie JA (1991) CO_2 control on carbon-isotope fractionation during aqueous photosynthesis: A paleo-pCO_2 barometer. Geology 19:929-932

Ingledew WJ, Cox JC, Halling PJ (1977) A proposed mechanism for energy conservation during Fe^{2+} oxidation by *Thiobacillus ferrooxidans*: Chemiosmotic coupling to net H^+ influx. FEMS Microbiol Lett 2:193-197

Jones B, Renaut RW (1996) Influence of thermophilic bacteria on calcite and silica precipitation in hot springs with water temperatures above 90°C: evidence from Kenya and New Zealand. Can J Earth Sci 33:72-83

Kobluk DR, Crawford DR (1990) A modern hypersaline organic mud- and gypsum-dominated basin and associated microbialites. Palaios 5:134-148

Konhauser KO, Schultze-Lam S, Ferris FG, Fyfe WS, Longstaffe FJ, Beveridge TJ (1994). Mineral precipitation by epilithic biofilms in the Speed River, Ontario, Canada. Appl Environ Microbiol 60:549-553

Konhauser KO, Fyfe WS, Ferris FG, Beveridge TJ (1993) Metal sorption and mineral precipitation by bacteria in two Amazonian river systems: Rio Solimoes and Rio Negro, Brazil. Geology 21:1103-1106

Lanstrom O, Tullborg E-L (1995) Interactions of trace elements with fracture filling minerals from the Aspo Hard Rock Laboratory. SKB Technical Report 95-13, Swedish Nuclear Fuel and Waste Management Company, Stockholm

Lovley DR (1991) Dissimilatory Fe(III) and Mn(IV) reduction. Microbiol Rev 55:259-287

Lovley DR, FH Chapelle. (1995) Deep subsurface microbial processes. Rev Geophys 33:65-381

Makin SA., Beveridge TJ (1996) The influence of A-band and B-band lipopolysaccharide on the surface characteristics and adhesion of *Pseudomonas aeruginosa* surfaces. Microbiol 142:299-307

Mann S (1988) Molecular recognition in biomineralization. Nature 332:119-124

Marquis RE, Mayzel K, Carstensen EL (1976) Cation exchange in cell walls of gram-positive bacteria. Can J Microbiol 22:975-982

Mohagheghi A, Updegraff DM, Goldhaber MB (1985) The role of sulfate-reducing bacteria in the deposition of sedimentary uranium ores. Geomicrobiol J 4:153-173

Morse JW, Bender ML (1990) Partition coefficients in calcite: examination of factors influencing the validity of experimental results and their application to natural systems. Geochim Cosmochim Acta 53:745-750

Mullen MD, Wolf DC, Ferris F., Beveridge TJ, Flemming CA, Bailey GW (1989) Bacterial sorption of heavy metals. Appl Environ Microbiol 55:3143-3149

Nakajima A, Sakaguchi T (1986) Selective accumulation of heavy metals by microorganisms. Appl Microbiol Biotechnol 24:59-64

Pedersen K, Ekendahl S (1990) Distribution and activity of bacteria in deep granitic groundwaters of southeastern Sweden. Microb Ecol 20:37-52

Pedersen K, Karlsson F (1995) Investigations of subterranean microorganisms: their importance for performance assessment of radioactive waste disposal. SKB Technical Report 95-10, Swedish Nuclear Fuel and Waste Management Company, Stockholm

Pentecost A, Bauld J (1988) Nucleation of calcite on the sheaths of cyanobacteia using a simple diffusion cell. Geomicrobiol J 6:129-135

Pentecost A, Spiro B (1990) Stable carbon and oxygen isotope composition of calcites associated with modern freshwater cyanobacteria and algae. Geomicrobiol J 8:17-26

Pronk JT, Johnson DB (1992) Oxidation and reduction of iron by acidophilic bacteria. Geomicrobiol J 10:153-171

Roden EE, Lovley DR (1993). Dissimilatory Fe(III)-reduction by the marine microorganism *Desulfuromonas acetoxidans*. Appl Environ Microbiol 59:734-742

Saxby JD (1969) Metal-organic chemistry of the geochemical cycle. Rev. Pure Appl Chem 19:131-150

Schopf JW, Walter MR (1983) Archean microfossils: new evidence of ancient microbes. In: JW Schopf (ed) Earth's Earliest Biosphere: Its Origin and Evolution, p 214-239, Princeton Univ Press, Princeton, NJ

Schultze-Lam S, Harauz G, Beveridge TJ (1992) Participation of a cyanobacterial S-layer in fine-grain mineral formation. J Bacteriol 174:7971-7981

Schultze-Lam S, Thompson JB, Beveridge TJ (1993) Metal ion immobilization by bacterial surfaces in freshwater environments. Water Poll Res J Canada 28:51-81

Schultze-Lam S, Beveridge TJ (1994) Nucleation of celestite and strontianite on a cyanobacterial S-layer. Appl Environ Microbiol 60:447-453

Schultze-Lam S, Urrutia MM, Beveridge YJ (1995a) Metal and silicate sorption and subsequent mineral formation on bacterial surfaces: subsurface implications. In: HE Allen (ed) Metal Contaminated Aquatic Sediments, p 111-147, Ann Arbor Press, Chelsea, Michigan

Schultze-Lam S, Ferris FG, Konhauser KO, Wiese RG (1995b) *In situ* silicification of an Icelandic hot spring microbial mat: implications for microfossil formation. Can J Earth Sci 32:2021-2026

Singer PC, Stumm W (1970) Acidic mine drainage. Science 167:1121-1123

Soudry D, Champtier Y (1983) Microbial processes in Negev phosphorites (Southern Isreal). Sedimentology 30:411-423

Southam G, Beveridge TJ (1992) Enumeration of *Thiobacilli* within pH-neutral and acidic mine tailings and their role in the development of secondary mineral soil. Appl Environ Microbiol 58:1904-1912

Southam G, Beveridge TJ (1994) The *in vitro* formation of placer gold by bacteria. Geochim Cosmochim Acta 58:4527-4530

Southgate PN (1986) Cambrian phoscrete profiles, coated grains, and microbial processes in phosphogenesis: Georgina basin, Australia. J Sediment Petrol 56:429-441

St-Cyr L, Fortin D, Campbell PGC (1993) Microscopic observations of the iron plaque of a submerged aquatic plant (*Vallisneria americana Michx*). Aquatic Botany 46:155-167

Stumm W (1992) Chemistry of the Solid-Water Interface. John Wiley, New York

Stumm W, JJ Morgan (1996) Aquatic Chemistry. 3rd ed. John Wiley, New York

Thompson JB, Ferris FG (1990) Cyanobacterial precipitation of gypsum, calcite, and magnesite from natural alkaline lake water. Geology 18:995-998

Thompson JB, Schultze-Lam S, Beveridge TJ, Des Marais DJ (1997). Whiting events: biogenic origin due to the photosynthetic activity of cyanobacterial picoplankton. Limnol Oceanogr 42:133-141

Trudinger PA, Chambers LA, Smith JW (1985) Low temperature sulfate reduction: Biological versus abiological. Can J Earth Sci 22:1910-1918

Urrutia M, Kemper M, Doyle R, Beveridge TJ (1992) The membrane-induced proton motive force influences the metal binding ability of *Bacillus subtilis* cell walls. Appl Environ Microbiol 58:3837-3844

Urrutia MM, Beveridge TJ (1993a) Mechanism of silicate binding to the bacterial cell wall in *Bacillus subtilis*. J Bacteriol 175:1936-1945

Urrutia MM, Beveridge TJ (1993b) Remobilization of heavy metals retained as oxyhydroxides or silicates by *Bacillus subtilis* cells. Appl Environ Microbiol 59:4323-4329

Urrutia MM, Beveridge TJ (1994) Formation of fine-grained silicate minerals and metal precipitates by a bacterial surface (*Bacillus subtilis*) and the implications in the global cycling of silicon. Chem Geol 116:261-280

Vasconcelos C, McKenzie JA, Bernasconi S, Grujic D, Tien AJ (1995) Microbial mediation as a possible mechanism for natural dolomite formation at low temperatures. Nature 377:220-222

Walter MR (1983) Archean stromatolites: evidence of Earthís earliest benthos. In: JW Schopf (ed) Earth's Earliest Biosphere: Its Origin and Evolution, p 187-213, Princeton Univ Press, Princeton, NJ

Chapter 6

MICROBIAL BIOMINERALIZATION OF MAGNETIC IRON MINERALS: MICROBIOLOGY, MAGNETISM AND ENVIRONMENTAL SIGNIFICANCE

Dennis A. Bazylinski

Department of Microbiology, Immunology and Preventive Medicine
Iowa State University
Ames, Iowa 50011 U.S.A.

Bruce M. Moskowitz

The Institute for Rock Magnetism
Department of Geology and Geophysics
University of Minnesota
Minneapolis, Minnesota 55455 U.S.A.

INTRODUCTION

Bacteria mediate the biomineralization of several magnetic iron minerals that include the iron oxide, magnetite (Fe_3O_4; $Fe^{2+}Fe_2^{3+}O_4$) and the iron sulfides, greigite (Fe_3S_4) and pyrrhotite (Fe_7S_8). These minerals can be formed in one of two fundamentally different ways or modes of biomineralization. The first is an indirect means of mineral formation called "biologically-induced mineralization" (BIM; Lowenstam 1981) while the second is considered a direct or directed mechanism of mineralization termed "biologically-controlled mineralization" (BCM) also sometimes referred to as "organic matrix-mediated mineralization" (Lowenstam 1981) or "boundary-organized biomineralization" (Mann 1986). Discussion of the structures and surface chemistries of some iron oxides and sulfides are in Banfield and Hamers (this volume).

In BIM, the biomineralization processes are not controlled by the organism. The mineral particles are usually formed extracellularly, have a broad size distribution and no defined morphology. In BCM, the organisms control the biomineralization processes to a high degree. The mineral particles formed in BCM are synthesized or deposited in a specific location with regard to the cell that is usually intracellular. In contrast to the particles formed by BIM, these crystals have a narrow size distribution and very defined, consistent morphologies.

Different physiological types of bacteria mediate BIM and BCM of magnetic minerals. Dissimilatory iron-reducing and dissimilatory sulfate-reducing (dissimilatory meaning the reduction or use of iron and sulfate as terminal electron acceptors for energy generation versus assimilatory reductions in which the iron and sulfate would be incorporated into cell material) bacteria use BIM processes to form magnetite and greigite, respectively, as well as many other iron minerals. The diverse magnetotactic bacteria form the same minerals using BCM.

There is apparently no function to the magnetite and greigite particles produced by microorganisms by BIM. However, there is great ecological, evolutionary and perhaps physiological significance to and interest in the BCM particles produced by the magnetotactic bacteria. Moreover, because the biomineralization of magnetic minerals by bacteria can be an important source of fine grained (<1 μm) magnetic material in sediment

0275-0279/97/0035-0006$05.00

and soil that can contribute to the paleomagnetic record of ancient geomagnetic field behavior or the mineral magnetic record of paleoclimate changes (Chang and Kirschvink 1989, Oldfield 1992), one must be able to tell the difference between the two types of particles to be able to recognize what microorganisms are responsible for the deposition of the magnetic minerals. The microbial production of fine-grained iron oxides with high surface areas in soils and sediments has other important geochemical consequences particularly in terms of the bioavailability of iron. Moreover, these high surface area phases may play a significant role in adsorbing metals and other species from solution. It is clear that microrganisms control the biogeochemical cycling of iron on the surface of this planet to a great degree.

The purpose of this chapter is to examine and review what mineralogical and physical features constitute BIM and BCM and what we know about these biomineralization processes and the organisms that use them in the formation of magnetic minerals. In addition, we will review the physics of magnetism of both types of particles, the magnetic criteria that are used to distinguish them from one another and the sedimentary palaeomagnetic record of biogenic magnetic minerals.

BIOLOGICALLY-INDUCED MINERALIZATION

General features

In BIM, biomineralization occurs indirectly as a result of metabolic activities of the organism and subsequent chemical reactions due to those metabolic activities. In many cases, the organisms secrete and/or produce (a) metabolic product(s) that react(s) with a specific ion or compound in the environment resulting in the production of extracellular mineral particles. Because BIM processes are not controlled by the organisms, mineral formation in this situation is equivalent to nonbiogenic mineralization under the same environmental conditions. The mineral particles are therefore an unintended byproduct of metabolic activities.

The mineral particles produced by bacteria in BIM are characterized by being poorly crystallized, having a broad particle size distribution and lacking specific crystalline morphologies. In addition, the lack of control over biomineralization can result in decreased mineral specificity and/or the inclusion of impurities in the particles. Most mineral particles produced by BIM can also be produced chemically without bacterial catalysts and have the same crystallochemical features as those produced in the abscence of bacteria (Mann et al. 1989). Therefore, particles formed by BIM are likely indistinguishable from those particles produced non-biogenically in iorganic chemical reactions under similar conditions, such as during pedogenesis (e.g. Maher 1990). The implication in BIM is that the minerals nucleate in solution or form from poorly crystallized mineral species already present. However, bacterial surfaces can also act as important sites for the adsorption of ions and mineral nucleation (see Fortin et al., this volume). The importance of bacterial surfaces in most cases of BIM has not been elucidated.

Magnetite

Magnetite can be formed by the dissimilatory-iron reducing bacteria and perhaps by other microorganisms that reduce iron (in non-dissimilatory reactions) as well. The former microorganisms respire with oxidized iron, $Fe(III)$, in the form of amorphous $Fe(III)$ oxyhydroxide (Lovley 1991) under anaerobic conditions and secrete reduced iron, $Fe(II)$, into the environment where it subsequently reacts with excess $Fe(III)$ oxyhydroxide to form magnetite. Magnetite particles are formed extracellularly by these microorganisms are (1) irregular in shape with a relatively broad size distribution, and (2) poorly crystallized (Sparks et al. 1990).

Many different species and physiological types of bacteria reduce Fe(III) (see Lovley 1987 for a complete list). However, it seems that only a handful of these organisms conserve energy and grow from the reduction of this environmentally-abundant terminal electron acceptor (Myers and Nealson 1990) and form magnetite. The most well studied of these organisms are *Geobacter metallireducens* and *Shewanella* (formerly *Alteromonas*) *putrefaciens*. *G. metallireducens* is a non-motile, freshwater, Gram-negative, obligately anaerobic rod that is phylogenetically associated with δ subdivision of the Proteobacteria (Lovley et al. 1993; see Fig. 2 in the chapter by Tebo et al., this volume), a diverse, vast assemblage of Gram-negative procaryotes in the Domain Bacteria (Zavarzin et al. 1991). *S. putrefaciens* is a motile, Gram-negative, rod-shaped, facultative anaerobe that is extremely versatile with regard to its use of terminal electron acceptors and is a member of the χ subdivision of the Proteobacteria (Myers and Nealson 1990, Lonergan et al. 1996). The most studied strain of *S. putrefaciens* is probably MR-1 isolated from anaerobic sediments of Oneida Lake (NY, USA) (Myers and Nealson 1988). Both organisms are obligate respirers, cannot ferment and have a chemoorganoheterotrophic mode of nutrition meaning that they utilize organic compounds for electron, energy and carbon sources. *S. putrefaciens* can also grow chemolithoheterotrophically using organic compounds for a carbon source but inorganic hydrogen gas as a source of energy and electrons and can couple this oxidation to the reduction of Fe(III) (Lovley et al. 1989). *S. putrefaciens* has recently been reported to be able to reduce and grow on the Fe(III) in magnetite which contains both Fe(II) and Fe(III) (Kostka and Nealson 1995) while it appears that *G. metallireducens* is unable to do so (Lovley 1991). Other organisms that have been found to reduce Fe(III) and grow on this metal as the sole electron acceptor are the anaerobic sulfur-reducing bacteria, *Desulfuromonas acetoxidans* (Roden and Lovley 1993) and *D. palmitatis* (Coates et al. 1995), and *Pelobacter carbinolicus* (Lovley et al. 1995). All of these organisms (except *Shewanella* spp.) as well as members of the genus *Desulfuromusa* appear to form a monophyletic group within the δ subdivision of the Proteobacteria (Lonergan et al. 1996). *Shewanella* and *Geobacter* appear to be very common in aquatic and sedimentary environments (DiChristina and DeLong 1993) and several new Fe(III) species of these genera have been isolated in recent years (Rossello-Mora et al. 1994, Caccavo et al. 1994) suggesting that these microorganisms may be the most environmentally-significant microbes involved in Fe(III) reduction. In addition, species of both genera can reduce and respire with metals other than Fe(III) including manganese (Mn(IV)) and uranium (U(VI)) (Myers and Nealson 1990, Lovley et al. 1993). Although magnetite formation by BIM has only been shown to occur in cultures of *Shewanella* and *Geobacter*, it is likely that magnetite could be produced by any Fe(III)-reducing bacterium, regardless of whether the organisms can grow with Fe(III), under suitable environmental conditions. The halotolerant, facultatively anaerobic, iron-reducing bacterium described by Rossello-Mora et al. (1994) may actually produce non-stoichiometric magnetite particles with a composition intermediate between magnetite and maghemite (χ-Fe_2O_3) (Hanzlik et al. 1996a).

Greigite and pyrrhotite

Certain anaerobic sulfate-reducing bacteria are known to produce particles of the magnetic iron sulfide, greigite, using BIM processes. In this case, the sulfate-reducing bacteria respire with sulfate anaerobically, releasing hydrogen sulfide. The sulfide ions react with excess iron present in the external environment (e.g. the growth medium) forming magnetic particles of greigite and pyrrhotite as well as a number of other non-magnetic iron sulfides including mackinawite (approx. FeS), pyrite (cubic FeS_2) and marcasite (orthorhombic FeS_2) (Freke and Tate 1961, Rickard 1969a,b; Hallberg 1972). The mineral species formed in these bacterially-catalyzed reactions were dependent upon the pH of the growth medium, the incubation temperature, the Eh (redox potential), the presence of specific oxidizing and reducing agents, and the type of iron source in the growth medium. For example, cells of *Desulfovibrio desulfuricans* produced greigite when grown in the presence of ferrous salts but not when the iron source was goethite, FeO(OH) (Rickard 1969a,b; Hallberg 1972). Microorganisms can modify or contribute to many of these factors such as pH, Eh, etc. Berner (1962, 1964, 1969) reported the chemical synthesis of a number of iron sulfide minerals including tetragonal FeS, marcasite, a magnetic cubic iron sulfide of the spinel type (probably greigite), pyrrhotite, amorphous FeS, and even framboidal pyrite (a globular form of pyrite that was once thought to represent fossilized bacteria, Love and Zimmerman 1961, Fabricus 1961). Not surprisingly, Rickard (1969a,b) concluded that extracellular biogenic and abiogenic particles of iron sulfides could not be distinguished from one another. Unfortunately, none of the magnetic iron sulfides produced by the sulfate-reducing bacteria have been examined in any detail using modern electron microscopy techniques.

The ubiquitous, anaerobic sulfate-reducing bacteria are a physiological group of microorganisms that are phylogenetically and morphologically very diverse and include species in the Domains Bacteria (δ subdivision of Proteobacteria and Gram-positive group) and Archaea. Most of the studies on the extracellular production of iron sulfides by sulfate-reducing bacteria including the studies described above involved the motile Gram-negative bacterium, *Desulfovibrio desulfuricans*, a member of δ subgroup of the Proteobacteria, Domain Bacteria. However, since all sulfate-reducing bacteria respire with sulfate and release sulfide ions, it is likely that all species regardless of phylogeny or classification produce iron sulfide minerals through BIM under appropriate environmental conditions when excess iron is available. It is noteworthy that the sulfate-reducing magnetotactic bacterium, strain RS-1 (see section on "The magnetotactic bacteria," below), is reported to produce extracellular magnetic iron sulfides presumably through BIM as well as intracellular particles of magnetite through BCM (Sakaguchi et al. 1993). The iron sulfides produced, however, were not identified and evidently not examined in any detail.

Other non-magnetic iron minerals produced by BIM

Other non-magnetic iron minerals formed by BIM have been observed in bacterial cultures. In these cases, Fe(II), excreted by iron-reducing bacteria during reduction of Fe(III), reacts with anions present in the growth medium. Two common examples are carbonate which reacts with Fe(II) to form siderite ($FeCO_3$) and phosphate which reacts with Fe(II) to form vivianite ($Fe_3(PO_4)_2 \cdot 8H_2O$). Siderite was observed in cultures of *Geobacter metallireducens* along with magnetite when the organism was grown in a bicarbonate buffering system (Lovley and Phillips 1988, Sparks et al. 1989) while only vivianite was produced with no production of magnetite or siderite by the same organism when Fe(III) citrate was provided as the terminal electron acceptor (Lovley and Phillips 1988). Growing cells of the magnetotactic species, *Magnetospirillum magnetotacticum* (see section on "The magnetotactic bacteria," below), produced significant amounts of extracellular, needle-like crystals of vivianite while actively reducing Fe(III) in the form of Fe(III) oxyhydroxides (Blakemore and Blakemore 1990). Although it is likely that many other iron-reducing bacteria form siderite and vivianite, most microbiologists are not looking for such compounds during routine culturing.

BIOLOGICALLY-CONTROLLED MINERALIZATION

General features

In BCM, the organisms exert a great degree of crystallochemical control over the nucleation and growth of the mineral particles. Generally, the minerals are directly synthesized at or from a specific location in or on the cell and only under specific conditions. The mineral particles produced by bacteria in BCM are characterized by being well-ordered crystals (not amorphous), having narrow size distributions and very specific particle morphologies. Because of these features, BCM processes are thought to be under specific metabolic and genetic control. In the microbial world, the most recognized example of BCM is magnetosome production by the magnetotactic bacteria.

The magnetotactic bacteria

Classification, phylogeny and general features of the magnetotactic bacteria. The magnetotactic bacteria represent a heterogeous group of procaryotes with a myriad of cellular morphologies including coccoid (spherical or more or less so), rod-shaped, vibrioid (curved), spirilloid (helical) and even multicellular (Blakemore 1982, Bazylinski 1995). All that have been examined to date are Gram-negative members of the Domain Bacteria although this does not exclude the possibility of a magnetotactic member of the Archaea. Thus, the term "magnetotactic bacteria" has no taxonomic significance and should be interpreted as a term for a collection of diverse bacteria that possess the apparently widely distributed trait of magnetotaxis (Bazylinski 1995).

Despite the great diversity of the magnetotactic bacteria, they all share several features: (1) they are all motile; (2) all exhibit a negative tactile and/or growth response to atmospheric concentrations of oxygen; and (3) they all possess a number of intracellular structures called magnetosomes. Magnetosomes are the hallmark feature of the magnetotactic bacteria and are responsible for their behavior in magnetic fields. The bacterial magnetosome consists of a single-magnetic-domain crystal of a magnetic iron mineral enveloped by a membrane (Balkwill et al. 1980). Magnetotactic bacteria generally synthesize particles of either magnetite or greigite. There are two known exceptions. In one unusual many-celled magnetotactic bacterium, nonmagnetic iron pyrite is formed in addition to greigite (Mann et al. 1990b) and a large, rod-shaped magnetotactic bacterium collected from the Pettaquamscutt Estuary was found to contain both magnetite and greigite (Bazylinski et al. 1993b, 1995).

Phylogenetic analysis, based on the sequence of the 16S ribosomal RNA gene, has been carried out on many cultured and uncultured magnetotactic bacteria. Initial studies showed that all magnetite-producing strains were associated with the α subdivision of the Proteobacteria (Burgess et al. 1993, DeLong et al. 1993, Eden et al. 1993, Spring et al. 1992), while an uncultured greigite- and pyrite-producing bacterium was found to be associated with the sulfate-reducing bacteria in δ subdivision of the Proteobacteria (DeLong et al. 1993). Because the different subdivisions of the Proteobacteria are considered to be coherent, distinct evolutionary lines of descent (Woese 1987, Zavarzin et al. 1991), DeLong et al. (1993) proposed that the evolutionary origin of magnetotaxis was polyphyletic and that magnetotaxis based on iron oxide (magnetite) production evolved separately from that based on iron sulfide (greigite) production. However, recent developments in this area have shown that not all magnetite-producing magnetotactic bacteria are associated with the α subgroup of the Proteobacteria. Kawaguchi et al. (1995) show that a cultured, magnetite-producing, sulfate-reducing magnetotactic bacterium, strain RS-1 (Sakaguchi et al. 1993), is a member of the δ subgroup of the Proteobacteria while Spring et al. (1993) describe an uncultured magnetite-producing magnetotactic bacterium called *Magnetobacterium bavaricum* which is in the Domain Bacteria but apparently not phylogenetically associated with the Proteobacteria. These results suggest that magnetotaxis as a trait may have evolved several times and, moreover, may be an indication that there is more than one biochemical-chemical pathway for the production of magnetic minerals by the magnetotactic bacteria.

Ecology of magnetotactic bacteria. Magnetotactic bacteria are ubiquitous in aquatic habitats and are cosmopolitan in distribution (Blakemore 1982). They are generally found in the highest numbers at the oxic-anoxic transition zone (OATZ) also referred to as the microaerobic zone or the redoxocline. In many freshwater habitats, the OATZ is located at the sediment water interface or just below it. However, in some brackish-to-marine systems, the OATZ is found or is seasonally located in the water column as shown in Figure 1.

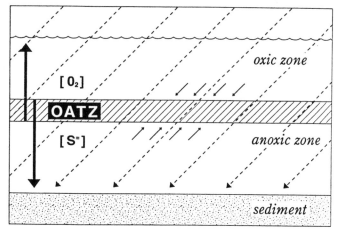

Figure 1. Depiction of the oxic-anoxic transition zone (OATZ) in the water column as typified by Salt Pond (Woods Hole, MA USA). Note the inverse double concentration gradients of oxygen ($[O_2]$) diffusing from the surface and sulfide ($[S^=]$) generated by sulfate-reducing bacteria in the anaerobic zone (vertical arrows). Magnetite-producing magnetotactic bacteria exist in their greatest numbers at the OATZ where microaerobic conditions predominate and greigite-producers are found just below the OATZ where $S^=$ becomes detectable.

When "polarly"-magnetotactic, magnetite-producing coccoid cells are above the OATZ in vertical concentration gradients of O_2 and $S^=$ (suboptimal conditions), they swim downward (small arrows above OATZ) along the inclined geomagnetic field lines (dashed lines). When they are below the OATZ (suboptimal conditions), they reverse direction (by reversing the direction of their flagellar motor) and swim upward (small arrows below the OATZ) along the inclined geomagnetic field lines. The direction of the flagellar rotation is coupled to a aerotactic sensory system that acts as a switch when cells are at a suboptimal position in the gradient as defined in the text. The magnetotactic spirilla (and other "axial" magnetotactic organisms) align along the geomagnetic field lines and swim up and down relying on a temporal sensory mechanism of aerotaxis to find and maintain position at their optimal oxygen concentration at the OATZ.

The Pettaquamscutt Estuary (Narragansett Bay, RI, USA) (Donaghay et al. 1992) and Salt Pond (Woods Hole, MA, USA) (Wakeham et al. 1984) are good examples of the latter situation. Hydrogen sulfide, produced by sulfate-reducing bacteria in the anaerobic zone and sediment, diffuses upward while oxygen diffuses downward from the surface resulting in a double, vertical chemical concentration gradient (Fig. 1) with a co-existing redox gradient. In addition, strong pycnoclines and other physical factors, probably including the microorganisms themselves, stabilize the vertical chemical gradients and the resulting OATZ.

Many types of magnetotactic bacteria are found at both the Pettaquamscutt Estuary and Salt Pond. Generally, the magnetite-producing magnetotactic bacteria prefer the OATZ proper (Fig. 1) and behave as microaerophiles. Cultural information supports this observation. Two strains of magnetotactic bacteria have been isolated from the Pettaquascutt. One is a vibrio designated strain MV-2 (DeLong et al. 1993, Meldrum et al. 1993b) and the other is a coccus designated strain MC-1 (DeLong et al. 1993, Meldrum et al. 1993a, Frankel et al. 1997). Both strains grow as microaerophiles although strain MV-2 can also grow anaerobically with nitrous oxide (N_2O) as a terminal electron acceptor. Other cultured magnetotactic bacterial strains, including spirilla (Maratea and Blakemore 1981, Schleifer et al. 1991) and rods (Sakaguchi et al. 1993), are microaerophiles or anaerobes or both. The greigite-producers appear to prefer the more sulfidic waters just below the OATZ (Fig. 1) where the oxygen concentration is zero and are likely anaerobes (Bazylinski 1995, Bazylinski et al. 1995). Unfortunately, no greigite-producing magnetotactic bacterium has been isolated and grown in pure culture.

Function of magnetotaxis. In this section, the function of magnetotaxis as it applies to the organism is discussed. There is great physical significance to the size of the magnetosome mineral phase which is reflected in its physical and magnetic properties. This is discussed in detail later in the sections on "Magnetism of magnetosomes and mechanism of magnetotaxis."

In most magnetotactic bacteria, the magnetosomes are arranged in one or more chains in which the magnetic interactions of the single particles cause their magnetic dipole moments to orient parallel to each other along the chain length. In this chain motif, the total magnetic dipole moment of the cell is simply the sum of the dipole moments of the individual particles. Therefore, the cell has maximized its magnetic dipole moment by arranging the magnetosomes in chains. The magnetic dipole moment of the cell is generally large enough so that its interaction with the Earth's geomagnetic field overcomes the thermal forces tending to randomize the cell's orientation in its aqueous surroundings (see section on "Mechanism of magnetotaxis"). Magnetotaxis results from the passive alignment of the cell along geomagnetic field lines while it swims. It is important to realize that cells are neither attracted nor pulled toward either geomagnetic pole. Dead cells, like living cells, align along geomagnetic field lines but do not move. In essence, cells behave like very small compass needles. Originally, all magnetotactic bacteria were thought to have one of two magnetic polarities, north- or south-seeking, depending on the magnetic orientation of the cell's magnetic dipole. The vertical component of the inclined geomagnetic field appeared to select for a predominant polarity in each hemisphere by favoring those cells whose polarity leads them down towards sediments and away from potentially toxic concentrations of oxygen in surface waters. This scenario appears at least partially true: north-seeking magnetotactic bacteria predominate in the Northern Hemisphere while south-seeking cells predominate in the Southern Hemisphere (Blakemore et al. 1980, Kirschvink 1980). At the Equator, where the vertical component of the geomagnetic field is zero, equal numbers of both polarities exist (Frankel et al. 1981). However, the discovery of stable populations of magnetotactic bacteria existing as "plates" in the water columns of chemically-stratified aquatic systems (Bazylinski et al. 1995) and the observation that vitually all magnetotactic bacteria in pure culture form microaerophilic bands of cells some distance from the meniscus of the growth medium (Frankel et al. 1997) is not consistent with the original simple model of magnetotaxis. For example, according to this model, persistent north-seeking magnetotactic bacteria in the Northern Hemisphere should always be found in the sediments or at the bottom of culture tubes.

Like most other free-swimming bacteria, magnetotactic bacteria propel themselves forward in their aqueous surroundings by rotating their flagella. Unlike cells of *Escherichia coli* and other chemotactic bacteria that exhibit a characteristic "run and tumble" motility, most magnetotactic bacteria move only bidirectionally, backwards and forwards, and do not change direction by tumbling (Bazylinski 1995). These bacteria were referred to as îtwo-way swimmersî and a good example would be the magnetotactic spirilla which have a polar flagellum at each end of the cell and, once aligned in the inclined geomagnetic field, can swim up or down. It is easy to understand how these bacteria form bands or plates of cells at the OATZ. Once aligned, cells swim up and down the inclined geomagnetic field lines and use aerotaxis to find their optimal oxygen concentration as shown in Figure 1. The ubiquitous microaerophilic magnetotactic cocci, which are also found at the OATZ, possess two bundles of flagella on one side of the cell and were called "one-way swimmers" because of their persistent forward motility in a preferrred direction (polarity) in wet mounts. This persistent forward swimming led to the original discovery of magnetotaxis in bacteria (Blakemore 1975). In contrast, when spirilla are observered in wet mounts, roughly equal numbers of cells swim in either direction.

One of us has recently found that cells of a cultured strain of a marine, magnetotactic coccus, MC-1, form microaerophilic bands of cells in oxygen gradients and can swim in both directions without turning around like the magnetotactic spirilla (Frankel et al. 1997). Cells of strain MC-1, like most other cultured, magnetite-producing magnetotactic bacteria, are strongly aerotactic. They move forward (north-seeking) in the Northern Hemisphere under oxic conditions which explains their persistent north-seeking behavior in wet mounts in which the aqueous medium is surely oxic. When the oxygen concentration becomes low enough (suboptimal conditions), cells reverse direction (by reversing the direction of their flagellar rotation), without turning around, using a novel aerotactic sensory mechanism that functions as a two-way switch rather than the well-recognized aerotactic sensory mechanism used by other bacteria (e.g. *Escherichia coli*) including the magnetotactic spirillum, *Magnetospirillum magnetotacticum*. When cells go above the OATZ and the oxygen concentration becomes too high (again suboptimal conditions), they reverse direction again (Fig. 1). Two types of magnetotaxis, called axial and polar, have now been distinguished for the different aerotactic mechanisms used by the magnetotactic spirilla and the magnetotactic cocci, respectively (Frankel et al. 1997). For the magnetotactic spirilla, the geomagnetic field provides an axis, not a direction, for motility along the oxygen gradient. The geomagnetic field provides both an axis and a direction of motility for the magnetotactic cocci (Frankel et al. 1997). Both mechanisms involve the passive orientation of the cellular magnetic dipole in the geomagnetic field. Thus, these bacteria use magnetotaxis in conjuction with aerotaxis (magneto-aerotaxis) to find and maintain optimal position in a vertical oxygen gradient, for many, the OATZ. Magnetotaxis is particularly advantageous to microorganisms in vertical concentration gradients because it presumably increases the efficiency of finding and maintaining an optimal position relative to the gradient by reducing a three-dimensional search problem to a one-dimensional search problem. It is likely that there are other forms of magnetically-assisted chemotaxis to molecules or ions other than oxygen, such as sulfide, or magnetically-assisted redox- or phototaxis in bacteria that inhabit the anaerobic zone (e.g. greigite-producers) in chemically-stratified waters and sediments.

It is important to realize that the function of cellular magnetotaxis described above appears to be a consequence of the cell possessing magnetosomes. Bacteria cannot think and can only react to a stimulus and therefore do not make the magnetosomes for magnetotaxis (which would be a teleological argument). Thus, it appears likely that the enormous amount of iron uptake and magnetosome production is somehow linked to the physiology of the cell and to other cellular functions as yet unknown.

Composition and morphology of the magnetosome mineral phase. As previously stated, there are two compositional types of magnetosomes in magnetotactic bacteria: iron oxides and iron sulfides. The iron oxide-type magnetosomes consist solely of magnetite. The particle morphology of magnetite varies but is consistent within cells of a single bacterial species or strain (Bazylinski et al. 1994). Three general morphologies of magnetite particles have been observed in magnetotactic bacteria using transmission electron microscopy (TEM). They include: (1) roughly cuboidal (Balkwill et al. 1980); (2) parallelepipedal (rectangular in the horizontal plane of projection) (Towe and Moench 1981, Bazylinski et al. 1988); and (3) tooth-, bullet- or arrowhead-shaped (anisotropic) (Mann et al. 1987a,b; Thornhill et al. 1994). Examples are shown in Figure 2.

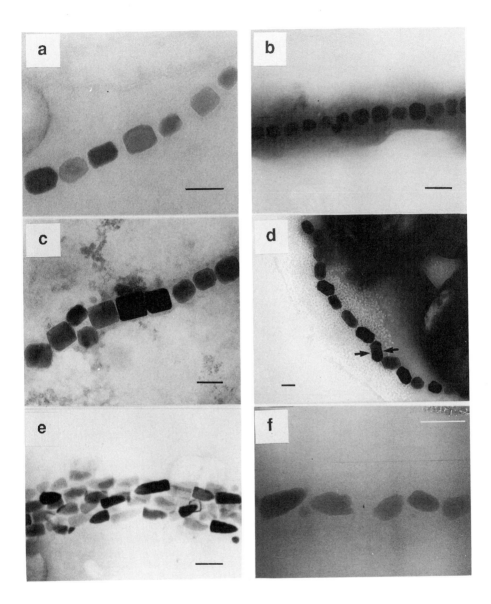

Figure 2. Morphologies of intracellular magnetite (Fe_3O_4) particles produced by magnetotactic bacteria. (a) Transmission electron micrograph (TEM) of parallelepipeds in cells of the marine vibrioid strain, MV-1; (b) Scanning-TEM (STEM) image of cubo-octahedra within a cell of *Magnetospirillum magnetotacticum*; (c) TEM of parallelepipeds within cells of the marine coccus, strain MC-1; (d) TEM of "waisted" parallelepipeds (those with a distict narrowing at the center of the long axis (arrows)) from an unidentified marine coccus; (e) TEM of tooth-shaped (anisotropic) magnetosomes from an unidentified marine rod-shaped bacterium; (f) STEM of unusual, roughly tooth-shaped magnetosomes in an unidentified marine spirillum. Rectangular prismatic and cubo-octahedral forms of greigite (Fe_3S_4) are similar to the morphologies in (a) and (b), respectively.

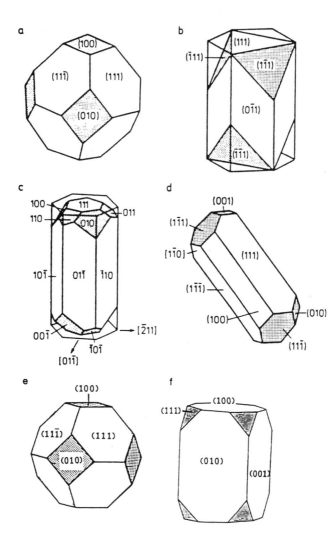

Figure 3. Idealized magnetite (a-d) and greigite (e-f) crystal morphologies of derived from high resolution transmission electron microscopy studies of magnetosomes from magnetotactic bacteria. (a) and (e) cubo-octahedrons; (b), (c), and (f) variations of hexagonal prisms; (d) elongated cubo-octahedron. Figure adapted from Mann and Frankel (1989) and Heywood et al. (1991).

High resolution analytical TEM studies have revealed that the magnetite particles within magnetotactic bacteria are of high structural perfection and have been used to determine their idealized morphologies (Matsuda et al. 1983, Mann et al. 1984a,b; 1987a,b; 1990a; Meldrum et al. 1993a,b). These studies have shown the roughly cuboidal particles to be truncated cubo-octahedra and the parallelepipedal particles to be either truncated hexahedral or octahedral prisms. Examples are shown in Figures 3a-d. The cubo-octahedral crystal morphology preserves the symmetry of the face-centered cubic spinel structure and is considered an equilibrium growth form of magnetite which simply means that this morphology can be expected to be found in chemically-produced magnetite particles. The hexa- and octahedral prismatic particles represent departures from this equilibrium form, presumably due to the acceleration or deceleration of mineral growth of certain crystal faces (Frankel and Bazylinski 1994, Mann et al. 1990a).

The synthesis of the tooth-, bullet- and arrowhead-shaped magnetite particles (Figs. 2e-f) appears to be more complex than that of the other forms. They have been examined by high resolution TEM in one uncultured organism (Mann et al. 1987a,b). The idealized morphology of these particles suggests that their growth occurs in two stages. The first, during which the length and width develop concurrently, is the formation of a well-ordered, isotropic, single-magnetic domain crystal that is cubo-octahedral. The second stage involves anisotropic growth of the particle along a preferred direction (the pointed end).

Whereas the cubo-octahedral form of magnetite is common in inorganically-formed magnetites, the prevalence of elongated hexahedral or octahedral forms in magnetosomes crystals appears to be a unique feature of the BCM process (Frankel and Bazylinski 1994). This aspect of magnetosome particle morphology forms the basis for distinguishing magnetosome magnetite from detrital or BIM-type magnetite using electron microscopy of magnetic extracts from sediments.

The iron sulfide-type magnetosomes contain either particles of greigite (Heywood et al. 1990, 1991) or a mixture of greigite and non-magnetic pyrite (Mann et al. 1990b). Only two morphologies of greigite have been observed in magnetotactic bacteria: (1) cubo-octahedral (the equilibrium form of face-centered cubic greigite); and (2) rectangular prismatic as shown in Figures 3e-f. Like that of their magnetite counterparts, the morphology of the greigite particles also appears to be species- and/or strain-specific although confirmation of this observation will require controlled studies of pure cultures of greigite-producing magnetotactic bacteria, none of which is currently available. The greigite-pyrite particles have only been observed in one microorganism to date; a many-celled "microcolony" of 20 or so similar procaryotic cells, arranged in a conglobate (spherical) manner, that is motile only as an entire unit (Bazylinski et al. 1990, 1993; Farina et al. 1993, Mann et al. 1990b, Rogers et al. 1990a,b). These particles are particularly interesting in that they are pleomorphic and lack the consistent crystalline morphologies observed in all other magnetotactic bacteria (Fig. 4). This particle pleomorphism is not understood but may be diagnostic for organisms that produce greigite-pyrite particles only. In addition, it is not known whether each mineral is biomineralized separately in this organism resulting in two particle types or whether the greigite particles are transforming into pyrite as might be expected under the strongly reducing, sulfidic conditions at neutral pH (Berner 1967) in which this organism is found.

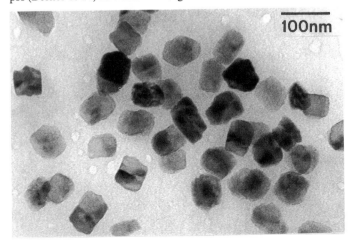

Figure 4. TEM of pleomorphic greigite-pyrite particles within cells of the many-celled magnetotactic procaryote.

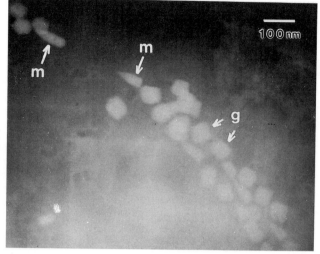

Figure 5. Darkfield STEM of rectangular prismatic greigite (short arrows; g) and arrowhead-shaped magnetite particles (long arrows; m) co-organized within the same chains of magnetosomes in an unusual rod-shaped magnetotactic bacterium collected from the Pettaquamscutt Estuary (see text).

A unique, slow-swimming, rod-shaped bacterium, collected from the OATZ from the Pettaquamscutt Estuary, was found to contain arrowhead-shaped crystals of magnetite and rectangular prismatic crystals of greigite co-organized within the same chains of magnetosomes (this organism usually contains two parallel chains of magnetosomes) as shown in Figure 5 (Bazylinski 1993b, 1995). In cells of this uncultured organism, the magnetite and greigite crystals are biomineralized with different, specific morphologies and sizes and are positioned with their long axes oriented along the chain direction. Both particle morphologies have been found in organisms with single component chains (Mann et al. 1987a,b; Heywood et al. 1990). This finding suggests that the magnetosome membranes surrounding the magnetite and greigite particles contain different nucleation templates and that there are differences in magnetosome vesicle biosynthesis. Thus, it is likely that two separate sets of genes control the biomineralization of magnetite and greigite in this organism.

All the magnetite and greigite particles found in the magnetotactic bacteria fall into a narrow size range, from about 35 to 120 nm (Bazylinski et al. 1994, Moskowitz 1995). This fact has great physical significance because the particle size falls within the single-magnetic-domain size range for both minerals (see the section "Magnetism of Magnetosomes") and is another indication that the magnetotactic bacteria exert fantastic control over the biomineralization processes involved in magnetosome synthesis (see section on "Control over BCM of magnetite").

Effect of environmental conditions on the biomineralization in magnetotactic bacteria. Can the local environmental conditions affect the biomineralization of the mineral phase of the magnetosome? This could occur as variations in the stoichiometry of the metal and/or the non-metal components of the mineral phase or the replacement of the either the metal or non-metal component of the magnetosome mineral phase. Here we discuss stoichiometry changes and specifically whether iron can be replaced with other transition metal ions and whether sulfur and oxygen can replace each other as the non-metal component in the magnetosome mineral phase.

The replacement of iron in magnetosomes has not been studied in great detail in pure cultures. Gorby (1989) showed that iron could not be replaced by other transition metal ions, including titanium, chromium, cobalt, copper, nickel, mercury, and lead, in the magnetite crystals of *Magnetospirillum magnetotacticum* when cells were grown in the presence of these ions. Results from uncultured magnetotatctic bacteria collected from natural environments have proven to be more interesting. Towe and Moench (1981) discovered very small amounts of titanium in the magnetite particles of an uncultured freshwater magnetotactic coccus collected from a wastewater treatment pond. Bazylinski et al. (1993a) reported the presence of significant amounts of copper in the greigite-pyrite particles of the many-celled, magnetotactic microorganism described earlier. The amount of copper was extremely variable and ranged from about 0.1 to 10 atomic % relative to iron. The copper appeared to be mostly concentrated on the surface of the particles and was only present in those organisms collected from a salt marsh in Morro Bay (CA, USA) and not in those collected from other sites. Interestingly, magnetosomes in rod-shaped magnetotactic bacteria that produce only greigite collected from the same site in Morro Bay did not contain copper. The presence of the copper did not appear to affect the function of the magnetosomes since the organisms were still magnetotactic but may be an indication that the biomineralization processes involved in magnetosome formation in this organism are not as controlled as they are in other species. The pleomorphism and the mixed composition (greigite-pyrite) of the magnetosome crystals may also be an indication of less control. Alternatively, these observations may indicate that the mineral phase of the magnetosomes in this organism is more susceptible to chemical and redox conditions in the external environment.

The presence of non-magnetic pyrite with greigite in the many-celled magnetotactic bacterium, discussed in the previous paragraph, may represent an example of changes in the stoichiometry of the metal and non-metal components of the magnetosome particle. Based on thermodynamic considerations, greigite would transform to pyrite under strongly reducing sulfidic conditions at neutral pH (Berner 1967), conditions that exist where this organism is found (Bazylinski et al. 1990). Pyrite formation from the iron monosulfides is considered a slow process (Berner 1967) although this may not be true for very small crystals like those present in the magnetosomes. Thus, it is not known whether this reaction could occur to any great degree during the lifetime of the cell. There is also the possibility that the reaction might actually be catalyzed by the cell. Alternatively, each mineral could be biomineralized separately in the cell which is an issue that remains to be resolved. Therefore, the function of the pyrite, if one exists, is unknown.

Cells of the slow-moving, magnetotactic rod from the Pettaquamscutt Estuary (described in the "Composition and morphology of the magnetosome mineral phase" section) that biomineralize both magnetite and greigite were found to extend well below the OATZ proper at this chemically-stratified site (Bazylinski et al. 1995). Cells collected from the more oxidized regions of the OATZ contained more or exclusively arrowhead-shaped magnetite particles while cells collected from below the OATZ in the anaerobic, sulfidic zone contained more or exclusively greigite particles. This finding suggests that environmental parameters such as local molecular oxygen and/or hydrogen sulfide concentrations and/or redox conditions somehow regulate the biomineralization of iron oxides and iron sulfides in this bacterium.

Chemistry and biochemistry of magnetosome formation. Because nothing is known of the chemistry of greigite formation in magnetosomes which contain it and there are no pure cultures of greigite-producing magnetotactic bacteria, this section is devoted to what is known of the chemistry of magnetite in magnetosomes.

The first step in magnetite synthesis in magnetotactic bacteria is the uptake of iron. Free reduced Fe(II) is very soluble (up to 100 mM at neutral pH (Neilands 1991)) and is easily taken up by bacteria usually by non-specific means. However, because free oxidized Fe(III) is so insoluble, most microbes have to rely on iron chelators which bind and solubilize Fe(III) for uptake. Microbially produced Fe(III) chelators are called siderophores which are defined as low molecular weight (<1 kDa), virtually specific ligands that facilitate the solubilization and transport of Fe(III) (Guerinot 1994; see Stone, this volume, for further discussion). Siderophores are generally produced under iron-limited conditions and their synthesis repressed under high iron conditions. Several studies, all involving magnetotactic spirilla, are focussed on iron uptake in the magnetotactic bacteria.

Paoletti and Blakemore (1986) reported the production of a hydroxamate siderophore by cells of *Magnetospirillum magnetotacticum* grown under high but not under low iron conditions. Thus, the siderophore production pattern here is the reverse of what is normally observed. Unfortunately, these results were not repeatable by others looking for siderophore production by this organism. Earlier, Frankel et al. (1983) assumed that iron uptake by this organism probably occurred via a non-specific transport system. Although iron is supplied as Fe(III) chelated to quinic acid, the growth medium also contains chemical reducing agents (e.g. thioglycollate or ascorbic acid) potent enough to reduce Fe(III) to Fe(II). Thus, both forms of iron are present in the growth medium and it is not known which form is taken up by cells.

Nakamura et al. (1993) could not detect siderophore production by *Magnetospirillum* AMB-1 and concluded that iron was taken up as Fe(III) by cells and that Fe(III) uptake was mediated by a periplasmic binding protein-dependent iron transport system. Schüler and Baeuerlein (1996) found that spent medium (medium in which the cells have already grown in) stimulated iron uptake to a high degree in cells of *M. gryphiswaldense* but found no evidence for the presence of a siderophore. They also showed that the major portion of iron for magnetite synthesis was taken up as Fe(III) and that Fe(III) uptake appears to be an energy-dependent process.

Only one study has addressed what occurs after iron is taken up by magnetotactic bacteria. Frankel et al. (1983) examined the nature and distribution of major iron compounds in *Magnetospirillum magnetotacticum* by using [57]Fe Mössbauer spectroscopy. They proposed a model in which Fe(III) is taken up by the cell (by non-specific means as described above) and reduced to Fe(II) as it enters the cell. It is then thought to be reoxidized to form a low-density hydrous Fe(III) oxide which is then dehydrated to form a high-density Fe(III) oxide (ferrihydrite) which was directly observed in cells. In the last step, one-third of the Fe(III) ions in ferrihydrite are reduced and with further dehydration, magnetite is produced. How specific shapes of magnetite are formed is unknown at present but it is thought that the final two steps occur in the magnetosome membrane vesicle which acts as a further constraint on crystal growth. It is clear that many more studies involving several different organisms are required before we will fully understand the precipitation of magnetite and greigite in magnetotactic bacteria.

Physiology of magnetotactic bacteria. Some of the qualities common to the magnetotactic bacteria were mentioned earlier. As described below, there are also some physiological features that are common to all that have been studied. However, it is important to remember that only a handful of species are in pure culture. Most of these are magnetotactic freshwater spirilla and all of these produce trucated cubo-octahedral particles of magnetite. Therefore the amount of data is limited. Although magnetite synthesis has not yet been linked to the physiology of a magnetotactic bacterium, it is important to understand the physiology of these bacteria and the conditions under which they synthesize magnetosomes in order to find this link. In the studies described in this section, one point becomes very clear; magnetite is formed by physiologically diverse magnetotactic bacteria under aerobic and anaerobic conditions.

The first magnetotactic bacterium to be isolated and grown in pure culture was *Magnetospirillum* (formerly *Aquaspirillum*; Schleifer et al. 1991) *magnetotacticum* strain

MS-1 (Blakemore et al. 1979). This species represents the most studied magnetotactic bacterium. It was isolated from a freshwater swamp and synthesizes truncated cubo-octahedral crystals of magnetite. Cells are helical and possess an unsheathed polar flagellum at each end of the cell. This organism is obligately respiratory (it cannot ferment) and is nutritionally a chemoorganoheterotroph that uses organic acids as a source of energy and carbon. Although this organism can use nitrate as a terminal electron acceptor and is a denitrifier (produces nitrous oxide and/or dinitrogen gases from "fixed" nitrate and/or nitrite), cells still require a small amount of molecular dioxygen for growth and cannot grow anaerobically and are therefore obligate microaerophiles (Bazylinski et al. 1983a). Cells of this bacterium fix atmospheric dinitrogen as evidenced by their ability to reduce acetylene to ethylene under nitrogen-limited conditions (Bazylinski and Blakemore 1983b). Interestingly, cells produce more magnetite when grown with nitrate than with oxygen as a terminal electron acceptor. Yet molecular oxygen must still be present for magnetite synthesis, with the optimal concentration for maximum magnetite yields being 1% oxygen in the headspace of cultures and concentrations greater than 5% being inhibitory (Blakemore et al. 1985). In an effort to understand the relationship between nitrate and oxygen utilization and magnetite synthesis, Fukumori and co-workers examined electron transport and cytochromes (electron carriers in the cell that contain redox active iron liganded to an organic moiety called heme) in *M. magnetotacticum*. Tamegai et al., (1993) reported a novel "cytochrome a_1-like" hemoprotein that was in greater amounts in magnetic cells than nonmagnetic cells. They did not find any true cytochrome a_1 that was once considered to be one of the terminal oxidases, as well as an *o*-type cytochrome, in *M. magnetotacticum* (O'Brien et al. 1987). A new *ccb*-type cytochrome *c* oxidase (Tamegai and Fukumori 1994) and a cytochrome cd_1-type nitrite reductase (Yamazaki et al. 1995) were isolated and purified from *M. magnetotacticum*. The latter protein was of particular interest since it showed Fe(II):nitrite oxidoreductase activity which may be linked to the oxidation of Fe(II) in the cell and thus to magnetite synthesis (Yamazaki et al. 1995).

Cells of *Magnetospirillum magnetotacticum* actively reduce Fe(III) (Blakemore and Blakemore 1990) and translocate protons when Fe(III) is introduced to them anaerobically (Short and Blakemore 1986) suggesting that cells conserve energy during the reduction of Fe(III). However, this reduction has never been linked to growth (as in the dissimilatory iron-reducing bacteria) nor to an energy-consuming process (e.g. amino acid uptake).

Other magnetotactic spirilla have been isolated in recent years. Matsunaga et al. (1991) isolated a magnetotactic spirillum physically identical to *M. magnetotacticum*, designated *Magnetospirillum* strain AMB-1, that is apparently much more oxygen tolerant than other magnetotactic species and that can form colonies on the surface of agar plates. This species appears very similar to *M. magnetotacticum* in that it synthesizes truncated cubo-octahedral crystals of magnetite, is obligately respiratory, has a chemoorganoheterotrophic mode of nutrition and uses organic acids as sources of energy and carbon. Cells of this species, like those of *M. magnetotacticum*, form more magnetosomes when grown with nitrate but unlike cells of *M. magnetotacticum*, they can grow under anaerobic conditions with nitrate and synthesize magnetite without molecular oxygen (Matsunaga and Tsujimura 1993). Growth and inhibitor studies (Matsunaga et al. 1991, Matsunaga and Tsujimura 1993) clearly show that *Magnetospirillum* AMB-1 can use nitrate as a terminal electron acceptor although the products of nitrate reduction were not reported. Another freshwater magnetotactic spirillum was isolated and described by Schleifer et al. (1991) and named *M. gryphiswaldense*. Cells of this species are very similar in morphology, ultrastructure and physiology to the *Magnetospirillum* species described above and, in addition, like all the freshwater magnetotactic spirilla, cells of *M. gryphiswaldense* produce truncated cubo-octahedral crystals of magnetite. Whether *M. gryphiswaldense* uses nitrate as a terminal electron acceptor and grows anaerobically or whether nitrate affects magnetite synthesis in this bacterium, as it does in *M. magnetotacticum*, was not reported.

A marine magnetotactic vibrioid bacterium, strain MV-1, was isolated by Bazylinski et al. (1988). Cells of strain MV-1 possess a single polar unsheathed flagellum and grow and synthesize hexahedral prismatic crystals of magnetite in their magnetosomes microaerobically and anaerobically with nitrous oxide (N_2O) as the terminal electron acceptor. Cells appear to produce more magnetite under anaerobic conditions than under microaerobic conditions (Bazylinski et al. 1988). Cells of this species are very versatile nutritionally being able to grow chemoorganoheterotrophically with organic acids as carbon and energy sources and chemolithoautotrophically with thiosulfate and sulfide as energy sources and carbon dioxide as the sole carbon source (Kimble and Bazylinski 1996). Cells use the Calvin-Benson cycle for autotrophic carbon dioxide fixation (as plants do) as cell-free extracts from thiosulfate-grown cells show ribulose bisphosphate carboxylase/oxygenase activity (Kimble and Bazylinski 1996). A virtually identical strain, designated MV-2, was isolated from the Pettaquamscutt Estuary (Meldrum et al. 1993b).

Strain RS-1 is a Gram-negative, sulfate-reducing bacterium that grows and produces bullet-shaped particles of magnetite only under anaerobic conditions (Sakaguchi et al. 1993). Cells are helicoid-to-rod-shaped in morphology and possess a single polar flagellum. Very little is known about the physiology of this strain. Cells grow chemoorganoheterotrophically using certain organic acids and alcohols as carbon and energy sources. Cells cannot use nitrate as a terminal electron acceptor.

Several other pure cultures of magnetotactic bacteria exist (Bazylinski, unpublished data) but they grow poorly and very little is known about them. Strain MC-1, discussed earlier, produces hexahedral prisms of magnetite and grows chemolithoautotrophically with thiosulfate or sulfide as an electron and energy source (Meldrum et al. 1993a). Strain MV-4, a new marine spirillum, produces elongated octahedrons of magnetite and can grow chemolithoautotrophically with thiosulfate or chemoorganoheterotrophically (Meldrum et al. 1993b).

Control over BCM of magnetite. As previously stated, the magnetotactic bacteria appear to exercise exquisite biochemical, chemical and genetic control over the biomineralization processes involved in magnetosome synthesis which is certainly reflected in the size and shape of the mineral phase of the magnetosomes. How do they do this? In all the magnetite-producing magnetotactic bacteria examined to date, the magnetosome mineral phase appears to be enveloped by a membrane. In cells of *Magnetospirillum magnetotacticum*, this "magnetosome membrane" was found to consist of a lipid bilayer containing phospholipids and numerous proteins, some of which appear to be unique to this membrane and not the outer or cell membrane (Gorby et al. 1988). The magnetosome membrane does not appear to be contiguous with the cell membrane. It is the magnetosome membrane that is presumably the locus of control over the size and morphology of the inorganic particle as well as the structural entity that anchors the magnetosome at a particular location within the cell. However, it is not known if the magnetosome membrane is premade as an "empty" membrane vesicle prior to the biomineralization of the mineral phase. Empty and partially filled vesicles have been observed in iron-starved cells of *M. magnetotacticum* (Gorby et al. 1988) but have not been commonly observed in other magnetotactic strains. The unlikely alternative would be that nucleation of the mineral phase occurs prior to its being surrounded by membrane. In any case, most biochemical and molecular biological studies directed towards the understanding of the biomineralization processes involved in magnetosome formation are focussed on aspects of the magnetosome membrane particularly on the functions of specific proteins present on this membrane.

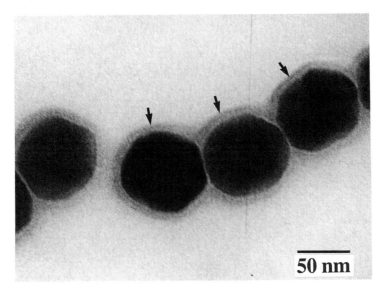

Figure 6. High magnification STEM of truncated hexagonal prismatic magnetite particles from the marine magnetotactic coccus, strain MC-1, surrounded by the 'magnetosome membrane' (denoted by arrows).

The magnetosome membrane appears to be a universal feature of at least the magnetite-producing species of magnetotactic bacteria since it has been found in virtually all cultured and some non-cultured strains (Fig. 6). Although it is not known whether greigite-producing magnetotactic bacteria actually have a magnetosome membrane like their magnetite counterparts, it seems likely that they do based on the consistent particle morphologies observed in these bacteria.

Molecular biology of magnetosome formation and genetics of magnetotactic bacteria. Very little is known about the molecular biology of magnetosome formation, for example, how many genes are required for magnetosome synthesis, how are the genes regulated, etc. Although several laboratories are trying to establish a genetic system with the magnetotactic bacteria (including ours), they have been hampered by many problems including, for example, the lack of a significant number of magnetotactic bacterial strains, the fastidiousness of the organisms in culture and the elaborate techniques required for the growth of these organisms and the inability for almost all these strains to grow on the surface of agar plates to screen for mutants etc. While the researchers involved have taken different approaches, most of the genetic and molecular studies have involved one or more of the magnetotactic spirilla.

Waleh and co-workers (Waleh 1988, Berson et al. 1989) initiated the first studies in the establishment of a genetic system with magnetotactic bacteria in order to understand the molecular biology of magnetosome synthesis in these microorganisms. Working with *Magnetospirillum magnetotacticum* strain MS-1, they showed that at least some of the genes of this organism can be functionally expressed in *Escherichia coli* and that the transcriptional and translational elements of the two microorganisms are compatible (necessary features for a good genetic system). They were able to clone, characterize and sequence the *recA* gene (a ubiquitous gene in bacteria that presumably has nothing to do with magnetite synthesis but functions in vital DNA repair mechanisms in the cell) from *M. magnetotacticum* (Berson et al. 1989, 1990). Later, now having established a genetic system that resulted in the cloning and sequencing of a specific gene from *M. magnetotacticum*, they examined iron uptake by *M. magnetotacticum*. They cloned and characterized a 2 kb DNA fragment from *M. magnetotacticum* that complemented the *aroD* (biosynthetic dehydoquinase) gene function in *E. coli* and *Salmonella typhimurium*. *AroD* mutants of these strains cannot take up iron from the growth medium. In other words, when the 2 kb DNA fragment from *M. magnetotacticum* was introduced into these mutants, the ability of the mutants to remove iron from the growth medium was restored (Berson et al. 1991) suggesting that the 2 kb DNA fragment may be important in iron uptake (and therefore possibly in magnetite synthesis) in *M. magnetotacticum*. However, although the cloned fragment restored iron-uptake deficiencies in siderophore-less, iron-uptake deficient mutants of *E. coli*, it did not mediate siderophore biosynthesis (Berson et al. 1991).

Matsunaga et al. (1991) isolated a magnetotactic spirillum, designated *Magnetospirillum* strain AMB-1, that is apparently much more oxygen tolerant than other magnetotactic species and that can form colonies on the surface of agar plates. Cells grown this way were non-magnetic, however, but by decreasing the oxygen concentration of the incubation atmosphere to 2%, cells formed black-brown colonies that were made up of magnetic cells. This feature facilitated the selection of non-magnetic mutants of *Magnetospirillum* strain AMB-1 (non-magnetic cells form white colonies rather than black) obtained by the introduction of transposon Tn5 (a piece of DNA that can incorporate itself into the genome of a host used to disrupt specific genes and render them non-functional) into the genome of *Magnetospirillum* AMB-1 by the conjugal transfer of plasmid pSUP1021 which contained the transposon. They were also able to introduce this plasmid and transposon into *M. magnetotacticum* strain MS-1 as well but could not get colony formation in this strain. They concluded that at least three regions of the *Magnetospirillum* strain AMB-1 chromosome are required for the successful synthesis of magnetosomes (Nakamura et al. 1995). One of these regions was found to contain a gene, designated *magA*, that encodes for a protein that is homologous to the cation efflux proteins, the *Escherichia coli* potassium ion-translocating protein, KefC, and the putative sodium ion/proton antiporter, NapA, from *Enterococcus hirae*. *MagA* was expressed in *E. coli* and membrane vesicles prepared from these cells that contained the *magA* gene product took up iron when ATP was supplied indicating that energy was required for the uptake of iron. Strangely, Nakamura et al. (1995) also showed that the *magA* gene was expressed to a much greater degree when wild-type *Magnetospirillum* AMB-1 cells were grown uder iron-limited conditions rather than iron-sufficient conditions in which they would produce more magnetosomes. Interestingly, the nonmagnetotactic Tn5 mutant over-expressed the *magA* gene under iron-limited conditions although it did not make magnetosomes. Thus, the role of the *magA* gene in magnetosome synthesis, if there is one, is unclear.

Okuda et al. (1996) took a different "reverse genetics" (reverse meaning to go from protein information to DNA information (i.e. a gene), rather than from gene to protein) approach to the magnetosome problem. They found three proteins with apparent molecular weights of 12, 22 and 28 kDa that appeared to be unique to the magnetosome membranes of *Magnetospirillum magnetotacticum* and are not present in the cellular membrane fraction. They were able to get enough N-terminal amino acid sequence of the 22 kDa protein and use the genetic code to make a 17 bp oligonucleotide probe for the genomic cloning of the gene encoding for that protein. They also found that the protein exhibited significant homology with a number of proteins that belong to the tetratricopeptide repeat protein family which include mitochondrial protein import receptors and peroxisomal protein import receptors. Thus, although the role of the 22 kDa magnetosome membrane protein in magnetosome synthesis has not been elucidated, it may function as a receptor interacting with associated cytoplasmic proteins (Okuda et al. 1996).

MAGNETIC PROPERTIES OF BIOGENIC MAGNETIC MINERALS

Fundamental principles

Magnetic property measurements are useful probes for magnetic mineral identification (e.g. magnetite, pyrrhotite, or greigite), magnetic granulometry (e.g. BIM-type or BCM-type magnetic particle size distributions), and relative or absolute concentrations of magnetic phases (e.g. magnetotactic bacteria cell densities). Because the magnetic fraction in natural samples is typically a minor component (<1% by volume) with particle sizes extending well below 10 μm, it is usually necessary to extract the magnetic minerals by crushing and then magnetically separating the grains before using non-magnetic methods such as X-ray diffraction, optical or electron microscopy. In contrast, magnetic methods are usually rapid allowing whole samples to be measured obviating the need for magnetic separation, which can lead to size-bias and aliasing of results.

Magnetism is a quantum mechanical phenomenon related to the orbital and spin motions of electrons. In materials like iron, magnetite, or greigite, there is a strong cooperative behavior between individual electron moments resulting in long-range magnetic ordering even in the absence of any externally applied magnetic fields. This cooperative behavior is caused by quantum exchange interactions between neighboring electrons.

For magnetic iron oxides and sulfides, the iron ions carry the net spin magnetic moments associated with the number of unpaired electrons in the $3d$ orbitals of $Fe(III)(3d^5)$ and $Fe(II)(3d^6)$. The theory of magnetic exchange provides the basis for understanding the coupling between ionic spin magnetic moments and the three main types of magnetic ordering in solids: (1) ferromagnetism, with parallel alignment of spins, resulting in a large net magnetization (e.g. Fe, Ni, and Co); (2) antiferromagnetism, with antiparallel alignment of spins between two similar sublattices, resulting in a zero net magnetization (e.g. $FeTiO_3$); and, (3) ferrimagnetism, which is a special case of antiferromagnetism, with the net magnetization of the two sublattices antiparallel but unequal, resulting in a net magnetization (e.g. Fe_3O_4). Almost all naturally-occurring magnetic minerals are ferrimagnetic, including magnetite and greigite. The ferrimagnetic properties of magnetite and other Fe-Ti oxides are reviewed in Banerjee and Moskowitz (1985) and Banerjee (1991). The magnetic properties of greigite are discussed by Snowball (1991) and Roberts (1994), but many of the basic magnetic properties are poorly known for this mineral. For further information on topics in fundamental magnetism, rock magnetism and paleo-magnetism, the reader can consult the monographs by Thompson and Oldfield (1986) and Dunlop and Özdemir (1997).

Saturation magnetization and Curie temperature

Two intrinsic magnetic properties resulting from exchange coupling are (1) a critical temperature above which the magnetic ordering breaks down, and (2) the appearance of spontaneous magnetization below the critical temperature in ferromagnetic and ferrimagnetic materials. The critical ordering temperature is called the Curie temperature for ferromagnets and ferrimagnets and the Néel temperature for antiferromagnets. The spontaneous, or saturation, magnetization (M_s) is the net magnetization that exists inside a uniformly magnetized microscopic volume, and has a magnitude that dependent on the spin magnetic moments. Above the Curie or Néel temperature, the exchange coupling is no longer effective in keeping spins aligned and each ionic spin moment is free to respond individually to an external magnetic field. In this state, the material is paramagnetic, exhibiting a weak magnetization in an applied field which goes to zero when the field is removed. Most antiferromagnetic minerals have a low Néel temperature and are paramagnetic at room temperature (e.g. siderite). Other Fe-bearing minerals (e.g. biotite, vivianite) are paramagnetic at room temperature and may not magnetically order at any temperature. Pyrite is an exception exhibiting the properties of diamagnetism, which produces a very weak negative magnetization in a positive field that goes to zero when the field is removed.

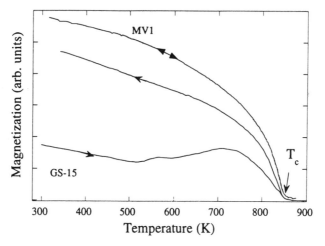

Figure 7. Thermomagnetic curves for magnetite produced by magnetotactic bacteria (strain MV-1) and dissimilatory iron-reducing bacteria (*Geobacter metallireducens*; GS-15). The magnetization curves were measured in a field of 500 mT with flowing He gas during heating and cooling to reduce the effects of alteration. The arrows on the curves indicate the heating or cooling direction. The MV-1 curve is reversible and GS-15 curve is irreversible. Both curves show a Curie temperature (T_c) of approximately 580°C.

Thermomagnetic analysis

The saturation magnetization and Curie temperature are intrinsic properties depending only on chemical composition, stoichiometry, and crystal structure (see Table 1). Thermomagnetic analysis, which consists of measuring the magnetization in a large field (typically 1 Tesla) as a function of temperature (M-T) is used to determine Curie temperatures, and hence, the particular magnetic mineral(s) present. Examples of thermomagnetic curves for BIM- and BCM-type magnetite are shown in Figure 7 and indicate Curie temperatures near 580°C, the expected value for pure magnetite.

Thermomagnetic curves can display reversible or irreversible behavior on heating and cooling, which can provide additional diagnostic information on magnetic mineralogy. Irreversible thermomagnetic behavior is usually caused by laboratory induced oxidation, dehydration, or other magnetochemical changes that lead to the formation of new magnetic phases or the transformation of the original magnetic mineralogy into non-magnetic minerals. For example, the sample containing the BIM-type magnetite in Figure 7 actually consists of a mixture of magnetite and siderite. Upon heating, the paramagnetic siderite transforms into magnetite, producing a peak in magnetization near 450°C (Fig. 7). Moreover, since small particles of magnetite can be readily converted to maghemite by oxidation, some of the irreversible behavior can be attributed to this reaction. Thermomagnetic analysis on samples containing greigite also display irreversible behavior, which can aid in its detection (e.g. Roberts 1995, Reynolds et al. 1994)

Magnetic hysteresis analysis

In contrast to paramagnetic or antiferromagnetic minerals, an important characteristic of ferro- or ferrimagnetic materials is that, when exposed to a magnetic field, they retain a memory of that field after it is removed. This property, called magnetic remanence, forms

Table 1. Magnetic properties of common magnetic minerals

Mineral name	Chemical formula	M_S (kA/m) at 300 K	Curie or Néel temperature (K)	Type of magnetic ordering at 300 K
Magnetite	Fe_3O_4	480	853	FM
Maghemite	$\gamma\text{-}Fe_2O_3$	380	863-948	FM
Hematite	$\alpha\text{-}Fe_2O_3$	≈2	948	AFM (canted)
Goethite	$\alpha\text{-}FeOOH$	≈2	393	AFM (imperfect)
Greigite	Fe_3S_4	≈125	≈603	FM
Pyrrhotite (monoclinic form)	Fe_7S_8	80	593	FM
Pyrrhotite (hexagonal form)	$Fe_{11}S_{12}, Fe_9S_{10}$	0	543,503	AFM
Pyrite	FeS_2	0		DM
Siderite	$FeCO_3$	0	20-38	PM
Vivianite	$Fe_3(PO_4)_2 \cdot 8H_2O$	0	??	PM

FM = ferrimagnetism; AFM = antiferromagnetism; PM = paramagnetism; DM = diamagnetism. Canted and imperfect AFM refer to antiferromagnetic ordering which is not exactly antiparallel resulting in a weak magnetic moment. Data from Dunlop and Özdemir (1997) and Hunt et al. (1995).

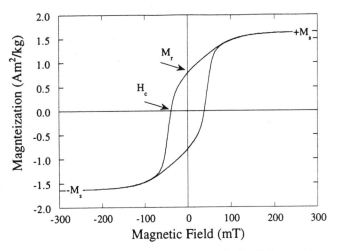

Figure 8. Magnetic hysteresis loop showing saturation magnetization (M_s), saturation remanence (M_r), and coercivity (H_c). The sample consists of freeze-dried cells of magnetotactic bacteria (strain MV-1).

the basis for magnetic recording whether we are dealing with permanent magnets, recording media, paleomagnetism, or magnetotaxis. A hysteresis loop is a plot of the variation of magnetization (M) with magnetic field (H) as the field is cycled from a large positive value to a large negative value (usually ±1-2 Tesla). Hysteresis loops are usually measured with a vibrating sample magnetometer (VSM) or an alternating gradient force magnetometer (AGFM). An example of a hysteresis loop for BCM-type magnetite plotted in Figure 8 exhibits several important magnetic parameters that can be determined from the loop. The saturation magnetization (M_s) is the maximum induced magnetic moment that can be obtained in some large field (1 to 2 Tesla); beyond this field value no further increase in magnetization occurs. Upon reducing the field to zero, the magnetization does not go to zero but persists as a saturation remanence (M_{rs}). As the field is increased in the negative direction, a point is reached where the induced magnetization becomes zero. The field at this point is called the coercivity, or coercive force (H_c). Increasing the field further in the negative direction results in saturation again but in the negative direction. Hysteresis properties are strongly dependent on grain size because they are influenced by the magnetic domain state of the samples which in turn is a function of grain size.

In bulk geological or biological samples consisting of a mixture of magnetic and non-magnetic phases, dividing the bulk sample magnetization by the M_s of the pure phase, gives the weight fraction of magnetic material. Standard laboratory measurements using a VSM or AGFM can routinely detect concentrations of magnetic phases in the ppm range. Table 1 summarizes some basic magnetic properties of common (bio)magnetic minerals (see also Hunt et al. 1993).

MAGNETISM OF MAGNETOSOMES

The particle size specificity, morphology, crystallographic orientation and chain assembly of magnetosomes in magnetotactic bacteria are optimally designed for magnetotaxis in the geomagnetic field. The bioarchitectural framework of assemblies of aligned magnetic particles makes the cell a swimming, permanent magnetic dipole, or as R.B. Frankel noted "magnetotactic bacteria are masters of permanent-magnet engineering." To see why this is so, the magnetic structure of the magnetosomes is discussed in the following section.

Particle sizes of magnetosomes

Although variations exist between species, almost all magnetosomes, regardless of whether they contain magnetite or greigite, fall within a narrow size range of about 35-120 nm when measured along their long axes (Chang and Kirschvink 1989, Vali and Kirschvink 1990, Heywood et al. 1990, 1991; Bazylinski et al. 1994). This size range is significant because it places these particles within the stable magnetic single domain (SD) size range for magnetite and greigite (e.g. Dunlop and Özdemir, p. 129, 1997). Particles within the SD size range are uniformly magnetized, which means their magnetic moment is a maximum and equal to M_s. Particles larger than about 100 to 120 nm are non-uniformly magnetized because of the formation of multiple magnetic domains, domain walls, or vortex configurations; this has the effect of making their magnetic moments significantly smaller than in SD particles. At the other extreme, SD particles smaller than about 30 nm are superparamagnetic (SPM). Although SPM particles are still uniformly magnetized, their moments are not constant in direction because of thermally-induced spontaneous reversals which produce a time-averaged moment of zero. Therefore, magnetotactic bacteria have evolved to produce the optimum particle size for maximum moment per magnetosome. Single domain, superparamagnetic, and non-single domain particles will be discussed further in the section entitled Magnetic Microstates of Biogenic Magnetic Minerals.

Crystallographic orientation of magnetosomes

The crystallographic orientation of each individual magnetosome along a chain is also significant. Magnetite magnetosomes are almost exclusively oriented with their [111] crystallographic axes aligned along the chain axis, as determined by high resolution electron microscopy (e.g. Mann and Frankel 1989). Even for elongated crystal morphologies, one of the four [111] directions is usually the axis of elongation. In contrast, greigite magnetosomes are oriented such that the [100] direction is parallel to the chain axis (Heywood et al. 1991). The significance of the [111] direction in magnetite is that it corresponds to the magnetic easy axis (defined below). Similarly, the [100] direction is also probably the magnetic easy axis in greigite. No direct determination of easy axis orientation in greigite has been made, primarily because of the difficulty of synthesizing high quality, large single crystals of this mineral usually required for magnetic anisotropy measurements.

The magnetic easy axes arise from anisotropy in the magnetocrystalline energy resulting from the interaction of spin magnetic moments with the crystalline structure. This spin-orbit coupling determines the crystallographic directions along which M_s prefers or avoids being directed. In magnetite above 120 K, the $\langle 111 \rangle$ directions are the magnetic easy axes and the $\langle 100 \rangle$ directions are the hard axes (e.g. Dunlop and Özdemir 1997, p 50). To reverse the magnetization by an applied field from one easy axis to another requires rotation through a hard axis. The magnetocrystalline anisotropy thus creates an energy barrier that pins M_s along one easy axis until a large enough field is applied to cause an irreversible jump of M_s over the anisotropy barrier. This field is related to the coercivity and is a measure of the stability of remanence against remagnetization.

Chain assembly of magnetosomes

The final design feature of the cellular dipole moment is the arrangement of the uniformly-magnetized, crystallographically-oriented magnetosomes into one or more linear chains within a cell such that the chain axis is approximately parallel to the cell's axis of motility. This configuration maximizes the dipole moment of the cell because magnetic interactions between the particles cause each magnetosome moment to align parallel to the others along the chain axis. The total moment of the cell, **M**, is thus the sum of the moments, **m** of the n individual magnetosomes.

PHYSICS OF MAGNETOTAXIS

The mechanism of magnetotaxis is described in detail by Frankel (1984) and is summarized here. Magnetotaxis results from the passive orientation of magnetotactic bacteria along the local vertical direction of the geomagnetic field by the torque exerted by the field (**B**) on the cellular dipole moment (**M**). Each cellular moment has a certain potential energy in the field given by:

$$E_m = - \mathbf{M} \cdot \mathbf{B} = - MB \cos\theta \qquad (1)$$

where θ is the angle between **M** and **B**. Thermal energy at ambient conditions will tend to misalign the swimming bacterium. In a state of thermal equilibrium at temperature T, the probability of the moment having energy E_m is proportional to the Boltzmann factor, $\exp(-E_m/kT)$, where k is Boltzmann's constant. The thermally averaged orientation of the dipole moment can be determined from the Langevin theory of paramagnetism and is given by the Langevin function

$$\langle \cos \theta \rangle = L(a) = \coth(a) - 1/a \qquad (2)$$

where $a = MB/kT$. The Langevin function is plotted in Figure 9 and asymptotically approaches 1 for large a.

For example, in *M. magnetotacticum,* a typical cell contains about 24 particles with an average particle volume of 9.6×10^{-23} m^3 (Penninga et al. 1995). Using the value of 480 kA/m for the saturation magnetization of magnetite at 295 K, the magnetic moment per cell is 1.1×10^{-15} Am^2. In a geomagnetic field of 50 μT at room temperature, $a = 13$ and $L(a) = 0.92$, meaning that the cell swims approximately parallel to the local direction of the geomagnetic field. If the number of magnetosomes is too low, then the alignment of the cell is inefficient. On the other hand, increasing the number of magnetosomes beyond a certain value will not significantly improve the alignment of the cell in the field because of the asymptotic nature of the Langevin function. Magnetotactic bacteria have optimized the

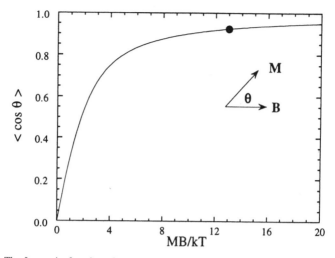

Figure 9. The Langevin function plotted as a function of MB/kT. This function gives the average alignment of a magnetotactic cell with magnetic moment (**M**) in a magnetic field (**B**). The solid symbol represent the average orientation of magnetotactic bacteria (*Magnetospirillum magnetotacticum*) in a 50 μT field at 300 K.

biomineralization process to produce just the right number of particles for efficient magnetic navigation in the geomagnetic field.

The migration velocity v_m of the bacterium in the direction of **B** is given by the component of the forward swimming velocity v_0 along the direction of the field, $v_m = v_0 L(a)$. For magnetotactic bacteria, migration velocities can be >90% of their forward velocities and are significantly faster than other motile bacteria that may use only chemotaxis or aerotaxis (Frankel 1984). Magnetotactic bacteria have effectively turned a three-dimensional biased random walk problem along a chemical concentration gradient into a one-dimensional problem of swimming up or down along the geomagnetic field as described earlier in the section "Function of Magnetotaxis."

In addition to the requirement that the magnetic energy be greater than the thermal background, another condition for magnetotaxis is that the characteristic time scale for re-orientation into the field direction after the cell is perturbed must be less than the diffusional time scale for nutrient transport (e.g,. Frankel 1984). Otherwise, magnetotaxis would lose its adaptive advantage. The re-orientation time, t_o, is given approximately by (Frankel and Blakemore 1980, Lin de Barros and Esquivel 1985),

$$t_o = 8 \left[\pi\eta R_B^3 / MB \right] \qquad (3)$$

where η is the dynamic viscosity of water ($\eta = 10^{-6} \text{ m}^2\text{s}^{-1}$) and R_B is the hydrodynamic radius of the bacterium. Typical re-orientation times are < 1 sec, and meet the condition for fast alignment time (Lin de Barros and Esquivel 1985). This equation also forms the basis for measuring the magnetic moment of individual cells in the laboratory by either measuring the time required for a cell to rotate into the new field direction after an instantaneous field reversal or in a rotating field, or the width of a "U-turn" executed by the cell after a field reversal (Kalmijn 1981, Lin de Barros and Esquivel 1985, Steinberger et al. 1994, Penninga et al. 1995).

Equation (3) also shows that the critical dipole moment of the cell scales as R_B^3 and predicts that as the hydrodynamic radius of the ' bacterium increases, additional magnetosomes are required to overcome viscous forces than would be necessary for passive alignment according to the Langevin function. There are just such examples of magnetotactic bacteria that contain hundreds of magnetosomes (Vali and Kirschvink 1990, Thornhill et al. 1994), many more than required to overcome the thermal background. One large, rod-shaped organism, *Magnetobacterium bavaricum*, contains up to 1000 bullet-shaped magnetosomes arranged in several chains traversing the cell (Spring et al. 1993). This arrangement of magnetosomes into multiple chains provides an additional means to enhance the dipole moment of the cell. Furthermore, due to mutual magnetostatic repulsion of such chains, the multiple chain design may provide additional mechanical stability and more effective coupling of the magnetic torque to the cell by forcing the chains outward toward the cell envelope (Hanzlik et al. 1996b).

MAGNETIC MICROSTATES OF BIOGENIC MAGNETIC MINERALS

To understand further the significance of the size specificity of magnetosomes in magnetotactic bacteria as well as the magnetic properties of ferrimagnetic particles in general it is instructive to review the particle size dependence of micromagnetic states in some detail. Readers interested in more in-depth discussion of magnetic domain theory, particularly as it applies to magnetite, should consult the recent textbook by Dunlop and Özdemir (1997).

As mentioned before, ferrimagnetic materials are composed of small regions of uniform magnetization called magnetic domains separated by narrow transition regions of rapidly varying spin orientation called domain walls. Domains are small in size (one to hundreds of μm), but much larger than atomic distances. The width of domain walls is typically 10 to 500 nm. Domains and domain walls can be directly observed using various magnetic imaging techniques and examples for magnetite can be found in Özdemir et al. (1995) and Pokhil and Moskowitz (1997).

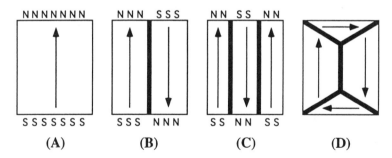

Figure 10. Idealized domain states showing (A) uniform single domain state, (B) two domain state with a domain wall separating the two domains, (C) three domain state with two domain walls, and (D) four domain state with closure domains. The arrows show the direction of magnetization and N and S indicate the formation of north or south magnetic poles where the magnetization intersects a surface. Note how the domain state in (D) produces no surface poles but at the expense of forming additional domain walls.

The main driving force for domain formation is the minimization of magnetostatic energy. In a uniformly magnetized grain (Fig. 10), uncompensated magnetic poles will form on the surfaces due to the magnetization (M_s) and are themselves a second source of magnetic field (the demagnetizing field). The magnetostatic energy is also anisotropic due to the shape of the grain. For instance, a long thin needle magnetized along its long axis will have lower magnetostatic energy because this configuration induces fewer poles on the two end surfaces which are farther apart from each other compared to the case when the same grain is magnetized perpendicular to its long axis. If a grain is elongated, it is shape anisotropy rather than magnetocrystalline anisotropy that inhibits remagnetization during a hysteresis cycle and is another source of coercivity in materials with high M_s such as magnetite.

Single domain and multidomain states

A SD grain is uniformly magnetized with M_s along an easy axis, thus minimizing exchange and anisotropy energies (Fig. 10A). However, depending on the grain volume, the magnetostatic energy of this configuration may not minimized. The magnetostatic energy can be reduced if the grain divides into two domains each magnetized with M_s in opposite directions (Fig. 10B), thus reducing the remanent moment of the particle. This subdivision into more and more domains (Fig. 10C-D) can not continue indefinitely because the domain walls separating domains require additional exchange and magnetocrystalline energy to be produced. In the domain walls, the magnetization changes smoothly from that of one domain to the other. Eventually, an equilibrium number of domains will be reached which reflects the balance between the decrease in magnetostatic energy and the increase in wall energy. for a given grain size. Therefore, unlike SD grains, a multidomain (MD) grain is not uniformly magnetized. Inside each domain, the magnetization $M_i = M_s$ but the directions of M_is are different in different domains which results in the grain as a whole having a small net remanence S $M_i \ll M_s$. On average, as the particle size increases, the number of domains also increases.

Superparamagnetic state

The SD state is not the only important microstate for biogenic magnetic minerals. As particle size continues to decrease within the SD range, another critical threshold is reached, at which M_r and H_c go to zero. When this happens, the grain becomes superparamagnetic (SPM). For magnetite, the SPM transition at room temperature is about 30 nm (e.g. Dunlop and Özdemir 1997, p 131). In SPM particles, the magnetization, although uniform, is not constant in direction. Thermal energy causes spontaneous transitions of the magnetization over anisotropy barriers between energetically equivalent easy axis directions in the grain resulting in a time-average remanent moment of zero; thus, SPM particles do not exhibit remanence.

An SPM particle with volume V will approach an equilibrium value of magnetization in zero field with a characteristic relaxation time given by

$$1/\tau = f_o \exp(KV/kT) \qquad (4)$$

where f_o is the frequency factor ($\sim 10^9$ sec^{-1}; Moskowitz et al. 1997), KV is the anisotropy energy, and T is absolute temperature. Because of the exponential dependence of the relaxation time on V and T, particles display sharp blocking transitions between stable SD with long relaxation times ($\tau \gg$ measurement time), and magnetically unstable SPM with very short relaxation times ($\tau \ll$ measurement time). At a particular temperature (the blocking temperature), τ equals the characteristic time (t) of the measurement and the particles become superparamagnetic and no longer carry remanence. At sufficiently low temperature, SPM particles block and exhibit stable SD behavior. The relaxation behavior of SPM particles as a function of temperature can be used to detect the presence of these particularly small particles (<30 nm) in biological and geological samples by measuring the temperature dependence of magnetic remanence or using variable temperature Mössbauer spectroscopy (e.g. Moskowitz et al. 1989, 1997; Sparks et al. 1990, Hanzlik et al. 1996a).

Theoretical domain calculations: Butler-Banerjee model

As the grain size decreases, a critical size (d_0) will be reached where the grain can no longer accommodate a domain wall. Below this critical size, the single domain structure is

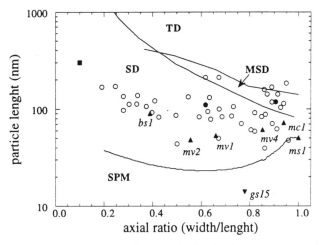

axial ratio (width/lenght)

Figure 11. Theoretical domain state diagram for magnetite showing the superparamagnetic (SPM), single domain (SD), two domain (TD), and metastable single domain (MSD) stability ranges for parallelepiped-shaped particles. Cubic-shaped grains correspond to an axial ratio of 1.0 and elongated particles have axial ratios (width/length) < 1.0. The SPM-SD and SD-TD curves are from Butler and Banerjee (1975) and the MSD curve is from Fabian et al. (1996). The symbols represent the size and shape distributions of magnetite particles from magnetotactic bacteria and dissimilatory iron-reducing bacteria. Open circles: uncultured magnetotactic bacteria from lake sediments (Petersen et al. 1989); solid triangles: cultured strains of magnetotactic bacteria (MS-1, MV-1, MV-2, MV-4, MC-1, BS-1); solid square: highly elongated magnetosome in magnetotactic bacteria from the Moorsee in Bavaria (Vali and Kirschvink 1989); solid circles: magnetosomes in magnetotactic bacteria from Brazil (Farina et al. 1993); solid inverted triangle: Geobacter (GS-15; Sparks et al. 1990).

the lowest energy microstate. The critical size for the SD state depends on several parameters (M_s, magnetocrystalline anisotropy, exchange energy, grain shape). Theoretical estimates of size and shape ranges of single-domain, two-domain, and superparamagnetic states for magnetite at 290 K were calculated by Butler and Banerjee (1975) and are shown in Figure 11. These calculations have been used almost exclusively to show that magnetosomes are SD particles (e.g. Kirschvink and Lowenstam 1979).

The Butler-Banerjee model is one-dimensional with the wall width being the only adjustable parameter and compares the energies between the SD and two-domain (TD) states as a function of grain size for idealized parallelepiped-shaped particles. It implicitly assumes that the lowest energy non-SD state is the simple TD state. In other words, the SD state will transform only into a TD state with a lamellar domain wall separating the domains. The model makes two predictions: (1) for an equidimensional cube, the SD-TD threshold size is 76 nm; and (2) with increasing particle elongation, d_0 increases and becomes greater than 1000 nm for an axial ratio (width/length) < 0.2. Because of shape anisotropy, the SD threshold increases with decreasing axial ratio for non-equidimensional particles.

Experimental confirmation for these numerical predictions are difficult because the predicted SD threshold size in magnetite is below 1 μm and direct domain observations are limited at this length scale. Instead of domain observations, remanence and coercivity data obtained from sized dispersions of natural and synthetic magnetites that span the predicted SD threshold are used to infer the actual SD size range. Results from these types of experiments are generally consistent with the theoretical predictions that the critical threshold size for SD behavior is <1 μm (e.g. Merrill and Halgedahl 1995, Dunlop and Özdemir 1997).

Originally, the Butler-Banerjee calculations were used to confirm that magnetosomes were indeed uniformly magnetized, SD particles (e.g. Kirschvink and Lowenstam 1979, Frankel and Blakemore 1980), as would be required for optimizing magnetotaxis in the geomagnetic field. Almost all magnetite magnetosomes (both cultured and uncultured varieties) that have been characterized by TEM fall within the theoretical SD size range (Fig. 11). Because there are biological arguments for magnetosomes to be SD as well as independent experimental evidence for the SD nature of magnetosomes from magnetic measurements on bulk samples of freeze-dried cells (Moskowitz et al. 1988) and on individual cells (Proksch et al. 1995, Penninga et al. 1995), the best evidence for the general validity of the Butler-Banerjee model in fact comes from the size and shape distributions of magnetosomes found in magnetotactic bacteria.

Local energy minima and metastable SD states: micromagnetic models

As mentioned above the dimensions of almost all magnetosomes places them within the theoretical SD size range, but some appear to be much larger than SD and plot within the theoretical TD size range. In some cases, the particle dimensions may be just an experimental artifact resulting from the determination of three-dimensional shape from a two-dimensional projection on a TEM photomicrograph. Other examples are more unambiguous, such as the large magnetosome found in coccoid cells from Lagoa de Itaipu, near Rio De Janerio (Lins et al. 1994, Farina et al. 1994) and plotted in Figure 11. Can magnetosomes be non-uniformly magnetized in a TD state and still exhibit magnetotaxis? Is there possibly another function besides magnetotaxis for TD magnetosomes? Is the theoretical groundstate SD-TD transition size underestimated and the magnetosomes are in fact really SD? Or is there another explanation related to the micromagnetic structures in non-SD particles?

One problem with the Butler-Banerjee model is that it is a one-dimensional model. Constrained by the assumptions that the magnetization is uniform within domain walls and that the SD state transforms into a TD state. In fact the TD state may not be lowest energy ground state configuration for particles sizes larger than SD. More recent calculations solving the nonlinear equations of micromagnetism in three dimensions using supercomputers remove many of the constraints necessitated by earlier modeling techniques (Dunlop and Özdemir 1997, p 171). In these later models, the grain is divided into cells in which the magnetization direction of each cell is varied independently until a minimum energy structure is obtained. Unlike the Butler-Banerjee model, domains and walls are not imposed upon the model initially but instead become model predictions. These micromagnetic calculations predict that a variety of microstates with lower energy configurations than the simple TD state can develop initially as the non-SD state at d_0 (e.g. Williams and Dunlop 1989, Merrill and Halgedahl 1995, Fabian et al. 1996). Furthermore, these alternative microstates can be local energy minima (LEM) rather than global energy minima or ground states. This means that particles can be trapped in higher energy LEM states until perturbed by magnetic fields which subsequently initiate the transformation to a lower energy state. An important LEM state is the metastable SD state, in which the nearly uniformly magnetized SD state metastably persists in larger sized particles whose ground state configurations are predicted to be non-SD. Metastable SD states and alternative LEM states for the same grain are not mere numerical curiosities but have actually been experimentally observed in magnetite, titanomagnetite, and pyrrhotite for grain sizes >1 μm, well above their equilibrium values for d_0 (see Dunlop and Özdemir (1997) for a recent review).

While three-dimensional micromagnetic calculations have predicted a variety of non-SD structures forming just above the SD transition size the equilibrium, ground state transition size d_0 does not change significantly from the original Butler-Banerjee calculations. For an equidimensional cube of magnetite, $d_0 \sim 80$ nm. However, the fundamental new insight obtained from the micromagnetic models is the existence of a metastable SD region that effectively extends the particle size range for SD states. At the metastable SD boundary, the particle will spontaneously transform into a non-SD state, either by nucleating a domain wall or transforming into a vortex-like state. The metastable SD transition boundary as a function of axial ratio, obtained from a recent unconstrained three-dimensional micromagnetic model (Fabian et al. 1996) is shown in Figure 11. For an equidimensional cube of magnetite, a metastable SD state can exist in particle sizes up to 140 nm, significantly higher than the ground state value of 80 nm.

Magnetosomes and micromagnetism

The existence of the metastable SD state provides a possible explanation for the particle dimensions of the "anomalously-sized" magnetosomes (Moskowitz 1995, Fabian et al. 1996). These magnetosomes are considered "large" because their crystal dimensions place them outside the theoretical ground state SD size range. As one explanation, Lins et al. (1994) and Farina et al. (1994) suggested that large magnetosomes found in coccoid cells were indeed consistent with the theoretical SD size range but only if their particles are considered prolate spheroids based on the model calculations by Diaz Ricci and Kirschvink (1992), instead of the parallelepiped-shaped particles modeled by Butler and Banerjee (1975). However, while prolate spheroids may be a first order approximation for other bullet-shaped particles, the large magnetosomes described by these authors are pseudo-hexagonal prismatic crystals and are clearly not shaped like prolate spheroids. More simply, using the results from three-dimensional micromagnetic models, the magnetosomes in the coccoid cells from Brazil as well as other large magnetosomes fall within the predicted metastable SD range and therefore can quite naturally possess an SD structure. Because the metastable SD magnetosome has a larger volume than one in the SD ground state, it will have a larger particle moment ($m = vM_s$). Therefore, instead of producing more particles along a chain or additional chains to increase the dipole moment of the cell, an organism may instead produce metastable SD magnetosomes to accomplish the same goal. It is noteworthy that the coccoid cells described from Brazil (Lins et al. 1994, Farina et al. 1994) containing "large" magnetosomes are found near the magnetic equator where the geomagnetic field is low. The authors suggest that the higher moment particles help to compensate for this low geomagnetic field.

As initially uniformly-magnetized magnetosomes nucleate and grow in size from the SPM state to the stable SD state and beyond, it may be energetically favorable for the particles to retain a uniform SD state into the metastable SD range instead of reverting to a non-SD state because the additional activation energy needed for the transformation is not available. Magnetic interactions between magnetosomes along a chain may also help to stabilize the SD structure (Fabian et al. 1996). If this speculation is true, bacteria that make metastable SD magnetosomes can provide critical confirmation of micro-magnetic models as well as provide a source of metastable SD magnetite particles for study. This hypothesis can be tested by exposing cells with large magnetosomes to high magnetic fields (>20 µT) sufficient to transform the metastable SD states back to their lower-energy non-SD states. If this happens, the cells would be demagnetized and lose their magnetotactic response in the geomagnetic field.

MAGNETIC CHARACTERISTICS OF BCM- AND BIM-TYPE MAGNETITE

Room temperature magnetic measurements

Room temperature hysteresis loops for BCM-type (from magnetotactic bacteria) and BIM-type (from dissimilatory iron-reducing bacteria) magnetite are shown in Figures 8 and

12. BCM magnetite exhibits classical SD behavior. The saturation remanence to saturation magnetization ratio, M_r/M_s, is approximately 0.5, which is the theoretical value for a randomly oriented assemblage of SD particles with uniaxial anisotropy (Dunlop and Özdemir 1997, p 205). The chain structure effectively removes the equivalence among the different $\langle 111 \rangle$ easy directions and produces a unique easy axis coinciding with the one particular $\langle 111 \rangle$ direction aligned along the chain axis. For this particular sample, the coercivity (H_c) is 40 μT, which is much larger than the geomagnetic field (0.05 μT). This demonstrates that changes in the geomagnetic field are not sufficient to remagnetized the polarity of the chain.

The BIM-type magnetite shows an entirely different hysteresis response that is more typical of superparamagnetic materials. The loop is narrow, with very low values of M_r and H_c, and its approach to saturation in high fields is more gradual than in the SD loop for BCM-type magnetite. These features are hallmarks of superparamagnetism. For a pure SPM sample, both M_r and H_c would be zero and the loop closed; however, the low but finite values of M_r and H_c indicate that there is a broad particle size distribution with a tail that extends up into the SD size range at room temperature (Moskowitz et al. 1989). For a mixture of SD and SPM particles, the SPM fraction may be determined from the M_r/M_s ratio of the bulk sample using the relationship $f_{spm} = 1$ to 2 M_r/M_s. The measured values of M_r/M_s from Figure 12 are less than 0.02 and indicate that over 96% of the particle diameters are less than the SPM threshold size at room temperature. This is consistent with electron microscopic observations of particles produced by *Geobacter metallireducens*.

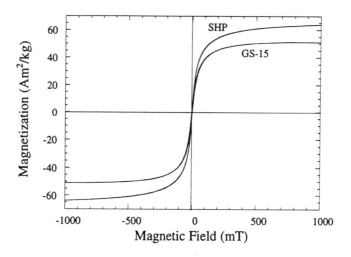

Figure 12. Room temperature hysteresis loops for BIM-type magnetite produced by Geobacter (GS-15) and a marine strain of *Shewanella putrefaciens* (SHP). The hysteresis behavior ($H_c \sim 0$, $M_r \sim 0$) is typical for superparamagnetic particles (cf. the SD loop in Fig. 8).

Low-temperature magnetic properties

Among the numerous magnetic methods available, low-temperature ($T < 300$ K) remanence measurements are particularly useful. Unlike traditional high temperature ($T > 300$ K) methods such as Curie-temperature analysis (Fig. 7), a low temperature approach avoids permanent chemical alteration to the minerals under study. For example, low-temperature (LT) magnetometry has been used as a diagnostic tool for identifying minute amounts of magnetite and pyrrhotite in bulk rocks and sediments (Rochette et al. 1990);

ultra-fine grained superparamagnetic phases (Moskowitz et al. 1989, Banerjee et al. 1993), mixtures of magnetite and maghemite (Özdemir et al. 1993), and, magnetochemical alteration caused by biogeochemical changes associated with the oxidation of organic carbon in marine sediments (Tarduno 1995). In addition, Moskowitz et al. (1993), have shown that low-temperature remanence measurements may provide a method to identify and quantify BCM magnetite produced by magnetotactic bacteria in bulk sediment and soil samples.

Two features related to mineral composition and grain size make low-temperature remanence measurements useful for investigating the biogenesis of iron oxide and iron sulfide particles in terrestrial and possibly extraterrestrial materials. First, magnetite and pyrrhotite have magnetic isotropic points at 120 K and 34 K, respectively (e.g. Banerjee 1991, Rochette et al. 1990), where magnetocrystalline anisotropy becomes zero and the easy directions change orientation. In magnetite, this magnetic transition is associated with a crystallographic phase transition, known as the Verwey transition (T_v) at 100-120 K. For both magnetite and pyrrhotite, a remanence given above or below their respective transition temperatures will be abruptly reduced when thermally cycled through these transitions. When observed this remanence behavior provides a diagnostic magnetic signature for these two phases. Because it is a remanence signature, these transitions are only observed in SD and larger particles (i.e. those with stable remanence). Unfortunately, greigite and maghemite, which are two other important magnetic phases, lack low-temperature remanence transitions (e.g. Roberts 1995, Özdemir et al. 1993). As an example of this method, low-temperature remanence measurements for inorganic phases of magnetite, pyrrhotite, greigite, and maghemite are shown in Figure 13. The apparent lack of a low-temperature transition in greigite can actually be used to discriminate between the two magnetic iron sulfide phases, pyrrohite and greigite (Roberts 1995, Tori et al. 1996); however, the lack of a transition by itself can not be considered conclusive evidence for the presence of greigite in any given sample.

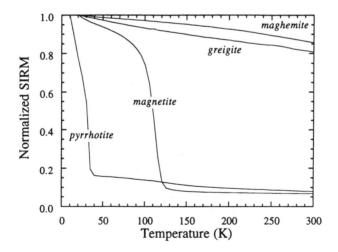

Figure 13. Normalized thermal demagnetization of saturation remanent magnetization (SIRM) for inorganic pyrrhotite, greigite, magnetite, and maghemite. Samples were given an SIRM in 2.5 T at 20 K and then remanence was measured while warming to 300 K in approximately zero field. Curves are normalized to the initial SIRM value at 20 K. The sharp drop in remanence at 34 K for pyrrhotite and 110 K for magnetite are associated with magnetic isotropic points in these phases. Greigite and maghemite do not show remanence transitions at low temperatures.

GEOMICROBIOLOGY

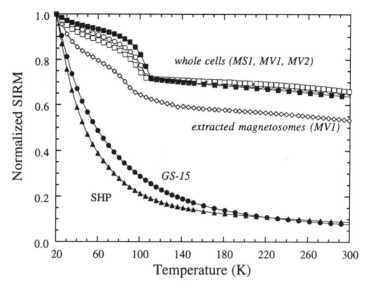

Figure 14. Normalized thermal demagnetization of saturation remanent magnetization (SIRM) for BCM-type and BIM-type magnetite. BCM-type magnetite from magnetotactic bacteria (strains MV-1, MV-2, MS-1) show the characteristic magnetite transition near 110 K, expected for particle sizes greater than 30 nm. BIM-type magnetite produced by *Geobacter* (GS-15) and a marine strain of *Shewanella putrefaciens* (SHP) show typical superparamagnetic behavior. Whole cells are freeze-dried samples of cells containing intact magnetosome chains. Extracted magnetosomes from strain MV-1 where obtained by cell rupture and then dispersed in a non-magnetic matrix.

The second useful feature of LT measurements is that they can be used to discriminate superparamagnetic (<30 nm) particles of iron oxides and iron sulfides from stable single domain (30 to 100 nm) and larger multi-domain particles (>100 nm). In biogenic samples, the superparamagnetic behavior would be indicative of BIM particles produced by iron- and sulfate-reducing bacteria and single domain behavior would be indicative of BCM-type particles produced by magnetotactic bacteria (Moskowitz et al. 1989, 1993). These two types of magnetic behavior for biogenic magnetite produced by magnetotactic bacteria and by dissimilatory iron-reducing bacteria (*Geobacter metallireducens* and *Shewanella putrefaciens*) are shown in Figure 14. Unlike whole cells of magnetotactic bacteria, magnetite produced by a BIM process shows an initially rapid decrease in intensity up to 60 K and then a more gradual decrease to 300 K. There is no evidence of a remanence transition near 100 K. This behavior is typical of superparamagnetism and consistent with the presence of SPM particles in the magnetite produced by the dissimilatory iron-reducing bacteria. The shape of the BIM remanence curves reflect the distribution of blocking temperatures for the particular particle sizes present in the samples which results in the progressive unblocking of magnetization as SD grains become SPM with increasing temperature. Both *G. metallireducens* and *S. putrefaciens* produces such a small fraction of grains that are within the stable SD size range above 120 K that the remanence behavior associated with the Verwey transition at 100 K is swamped by the SPM response. The thermal decay of remanence for magnetosomes extracted from whole cells of the magnetotactic strain, MV-1, exhibits an initially rapid decrease in SIRM up to 50 K and a broad remanence transition at 100 K. This type of thermal behavior is intermediate between the sharp transition in the whole cells with BCM magnetite and the SPM behavior of BIM magnetite and is characteristic of fine-grained magnetites that have been partially oxidized to SPM maghemite (Özdemir et al. 1993, Moskowitz et al. 1993). SPM particles of iron sulfides produced by BIM processes would also produce a remanence curve similar to the

SPM curves shown in Figure 14. Therefore, the shape of the SPM thermal unblocking curve is diagnostic of particle size but not mineral composition.

Magnetic and geochemical study of magnetotactic bacteria in chemically-stratified environments

As described in the section on "Ecology of Magnetotactic Bacteria," magnetotactic bacteria are found in the greatest numbers at the OATZ of aquatic habitats. One of the locations we have been studying in detail is Salt Pond, a well-characterized, seasonally chemically-stratified coastal pond that is about 5.5 m in depth and has marine and freshwater input (Wakeham et al. 1984, 1987). The anaerobic hypolimnion of Salt Pond has high concentrations of hydrogen sulfide generated from anaerobic sulfate-reducing bacteria. During the summer months, the anoxic (anaerobic) zone rises to about 3 m from the surface and the OATZ becomes quite defined with steep oxygen and hydrogen sulfide gradients (Wakeham et al. 1984). A small anoxic zone (up to 0.5 m) without detectable oxygen or hydrogen sulfide is also present (Wakeham et al. 1987). There is a "plate" of microorganisms that extends from the OATZ proper to the top of the anoxic hypolimnion where hydrogen sulfide becomes detectable. At least 5 different morphological types of magnetotactic bacteria can be found in this plate. The magnetite-producing types are found in the highest numbers in the OATZ proper where oxygen is still present in low concentrations. The greigite-producers are found in the highest numbers in the anoxic hypolimnion both where hydrogen sulfide is absent and present (Bazylinski et al. 1990).

By filtering water (at 0.2 μm) collected from specific, discreet depths by pumping (Wakeham et al. 1987), it is possible to determine the depth-distribution of magnetic materials. Preliminary chemical and magnetic remanence profiles from Salt Pond are shown in Figure 15. The peak in magnetic remanence near a depth of 3.4 m occurs within the OATZ and corresponds to the location where magnetotactic bacteria were observed to occur in great number and to the location of maximum available dissolved Fe(II). Low-temperature magnetic results from our study of Salt Pond are show in Figure 16 for water depths corresponding to the oxic zone (2.9 m), just below the OATZ interface (3.3 m), and the deepest sample collected within the anoxic zone (4.5 m). The 2.9 m curve shows a strong SPM-type response with a smeared-out, but weaker magnetite transition. The room-temperature remanence is also very weak at this depth (Fig. 15), indicating a low concentration of magnetic phases. Very few magnetotactic bacteria were observed at this depth and therefore this signal may represents wind-blown pedogenic material from the catchment area consisting of SPM particles of magnetite/maghemite. At the 3.4 m depth, the magnetite transition is well developed (Fig. 16) indicating a predominance of BCM-type magnetite and only a small SPM contribution. Magnetite-producing magnetotactic bacteria are also found in great numbers at this depth consistent with the magnetic remanence results. Finally, in the anoxic zone at the bottom of the section at 4.5 m, the magnetic results suggest a mixture of BCM-type magnetite, possibly BCM-type greigite, and SPM particles. The reappearance of SPM particles at this depth may be related to the presence of iron- and sulfate-reducing bacteria producing BIM-type particles, or simply to progressive dissolution of magnetite. Although the amount of BCM-type magnetite at this depth is significantly less than at 3.3 m, the normalized remanence remaining at 300 K (Fig. 16) is greater and the data above 100 K shows less of a thermal decay, suggesting the presence of a significant fraction of non-SPM material, possibly single-domain greigite (cf. Figs. 13 and 16). Preliminary observations shows that the magnetotactic bacteria occurring at this depth are almost exclusively greigite producers. However, dead cells of magnetite producers from shallower depths may also contribute to the magnetic signal at this depth.

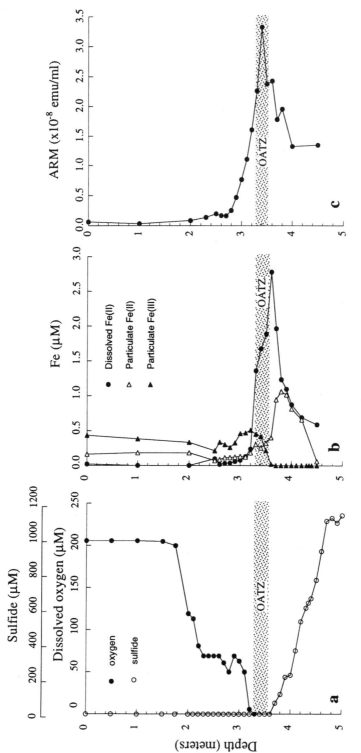

Figure 15. Geochemical and magnetic profiles as a function of water depth in Salt Pond, a chemically-stratified, semi-anaerobic basin near Woods Hole, Massachusetts (July 1995). (a) Dissolved oxygen (solid symbol) and sulfide (open symbol) concentrations. The OATZ is at about 3.5 m and shown as a shaded zone. (b) Fe species concentration: dissolved Fe(II)-solid circle; particulate Fe(II)-open triangle; particulate Fe(III)-closed triangle. Note the peak in dissolved Fe(II) at the OATZ. (c) Anhysteretic remanent magnetization (ARM) of water filtered at 0.45 μm. ARM is a laboratory-induced magnetization sensitive to SD particles. Note that the peak in ARM intensity correlates with the high density of magnetotactic bacteria at the OATZ. The geochemical data are from B.H Howes and D.R. Schlezinger (unpublished).

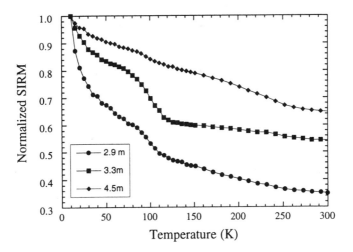

Temperature (K)

Figure 16. Normalized thermal demagnetization of saturation remanent magnetization (SIRM) for filtered water samples collected at Salt Pond in July 1995. The data are from three different water depths which sample material in the oxic zone (2.9 m), near the OATZ interface (3.3 m), and from the bottom of the section in the anoxic zone (4.5 m). The results from above the OATZ indicate an SPM contribution possibly from wind-blown magnetic soil particles. The results near the OATZ indicate a strong BCM-type magnetite contribution, whereas the 4.5 m sample indicates predominately BCM-type greigite with a small magnetite contribution. See text for details.

BIOGENIC MAGNETIC MINERALS IN SEDIMENTS

Magnetofossils and natural remanent magnetization

When magnetotactic bacteria die, their magnetosomes can be deposited as magnetofossils (Kirschvink and Chang 1984) and preserved in sediments, contributing to the sedimentary paleomagnetic and mineral magnetic records. The SD size of magnetosomes makes them excellent recorders of the paleomagnetic field and their unique hexagonal crystal forms provide a means for identifying them as magnetosomes via electron microscopy of magnetic extracts from sediments. SD sized particles (<0.1 μm) of magnetite of possible biogenic origin have been identified in sediments from several different depositional environments spanning the Phanerozoic including: (1) present day sediments from lakes, ponds, and estuaries (Vali et al. 1987, Petersen et al. 1989); (2) Holocene lake and marine surface sediments (Stolz et al. 1986, Chang et al. 1987, McNeill 1990, Peck and King 1996); (3) Quaternary to Eocene deep-ocean sediments (Kirschvink and Chang 1984, Petersen et al. 1986, Hess 1994); and, (4) Late Jurassic and Cambrian limestones (Vali et al. 1987, Chang et al. 1987). In some cases, such as in limestone deposits where there is little detrital input of magnetic material, there is a clear association between the occurrence of SD-sized biogenic magnetite and a stable, primary paleomagnetic signal (Chang et al. 1987, McNeil 1990).

BIM-type magnetic minerals produced by dissimilatory iron- and sulfate-reducing bacteria lack the SD size specificity of magnetosomes and, at least for magnetite produced by *Geobacter metallireducens*, resemble inorganic magnetite particles produced during pedogenesis (Lovley 1990, Maher and Taylor 1988). Hence, crystal morphology alone is not a useful criterion for identifying BIM-type magnetite in sediments. Furthermore, the mean particle size produced by *G. metallireducens* is much smaller than the SD size of magnetosomes and is within the superparamagnetic size range for magnetite; thus, BIM-type magnetic minerals are not expected to make a significant contribution to the natural remanent magnetization. Nevertheless, SPM grains, because of their high intrinsic magnetic susceptibility, could contribute to the susceptibility variations in sediments. Although *G. metallireducens* produces copius amounts of magnetite in laboratory cultures, there is little direct evidence that BIM-type magnetite contributes to the magnetization of sediments.

Besides contributing to stable remanence, the extent to which biogenic magnetic minerals contribute to the fine-grained (<0.1 μm) magnetic mineral fractions in sediments is important for interpreting the magnetic proxy records of paleoenvironmental change (King and Channell 1991, Oldfield 1992). Oldfield (1992) summarizes what he calls the "detrital" and "biomagnetic" interpretations of the source(s) of fine-grained magnetite in Quaternary sediments. For example, the biogenic fraction is produced by bacterial activity in the water column or at the sediment-water interface. Variations in the concentration of biogenic magnetic minerals therefore reflect temporal changes affecting bacterial paleoecology. In contrast, the nonbiogenic fine-grained fraction is primarily derived from terrigenous sources, which is then transported and eventually deposited in lakes and oceans. Variations in the concentration and grain size of this fraction reflect temporal changes in the source areas and transport processes, many of which are climate-driven. Failure to identify correctly the sources of fine grained magnetic minerals in sediments can lead to misinterpretation of the magnetic proxy record of paleoclimatic changes. Therefore it is important to develop magnetic tests that can be used to detect and quantify the biogenic vs. detrital magnetic signal in sediments and soils and establish sediment-source linkages. Several techniques have been proposed by Oldfield (1994) and Moskowitz et al. (1993).

Magnetochemical alteration and preservation of biogenic magnetic minerals

Although magnetotactic bacteria are ubiquitous in many present day aquatic environments, the eventual fate of magnetosomes and their relative contributions to remanent magnetization and the mineral magnetic record in sediments is not so obvious. Microbially-mediated chemical reactions involved in the oxidation of organic carbon in anoxic sediments can cause the dissolution of iron oxides and growth of secondary iron sulfides such as pyrite or greigite (Canfield and Berner 1987, Leslie et al. 1990, Roberts and Turner 1993, Tarduno 1995) or the formation of new magnetic phases including magnetite (Karlin et al. 1987, Karlin 1990). Direct dissolution and reduction of magnetite by bacteria has been shown to occur at near-neutral pH conditions in laboratory cultures of *Shewanella purtrefaciens* (Kostka and Nealson 1995), which may also contribute to the magnetochemical alteration in marine sediments. It has even be suggested that siderophores (iron-chelating compounds) produced by magnetotactic bacteria may help in the dissolution of their own magnetosomes after deposition (Vali and Kirschvink 1989). Because of the high surface area to volume ratio of submicron-sized BCM- and BIM-type magnetic minerals, they are more susceptible to dissolution than larger detrital grains (Canfield and Berner 1987, Karlin 1990).

Several studies show that, though biogenic magnetite is present, it either does not contribute to the sedimentary paleomagnetic record (Schwartz et al. 1997) or maybe removed by reduction diagenesis at depth (Snowball 1994, Peck and King 1996, Schwartz et al. 1997). For example, in a study of Lake Greifen (Switzerland) sediments spanning the past 300 years, Hawthorne and McKenzie (1993) conclude that dissolution and sulfidization of both detrital and biogenic magnetite in the upper 30 cm occurred in response to a change in the depositional environment of the lake from aerobic to anoxic due to eutrophication associated with agricultural and industrial development in the area since 1887. In contrast, Snowball (1994) documents a high concentration of biogenic magnetite in the upper sediment levels with progressive dissolution of biogenic magnetite at depth in sediments from a subarctic Swedish lake (Pajep Njakjaure). A biogenic magnetite contribution was confirmed in both studies based on TEM observations and comparative magnetic studies on catchment and sediment samples.

Modern magnetotactic bacterial activity is often closely linked to Fe-redox changes in sediments (Bazylinski and Frankel 1992, Petermann and Bleil 1993, Snowball 1994, Bazylinski et al. 1995), and variations in the abundance of magnetofossils in marine sediments spanning the last few glacial stages have been attributed to climatically-mediated redox changes (Hess 1994, Tarduno and Wilkison 1996). The occurrence of the modern Fe-redox boundary at depths ranging from the sediment-water interface to several tens of meters below the seafloor suggests that modern magnetotactic bacterial populations may be distributed over a wide range of sub-seafloor depths. Variations in the depth-range of living magnetotactic bacteria can have profound effects on the depth of remanence lock-in of magnetofossils possibly leading to a smearing out or low-pass filtering of the geomagnetic signal recorded by sediments (e.g. Tarduno and Wilkison 1996). These possible effects of variations in magnetotactic bacterial activity with depth on magnetic records have not yet been thoroughly evaluated.

MAGNETOSOMES, MAGNETITE AND MARS

Recently, McKay (1996) et al. noted a number of mineralogical and other features from the Mars meteorite ALH84001, collected in Antarctica, that could be explained by biological processes and therefore could represent evidence for ancient life on Mars. Included in these features was the presence of ultrafine-grained magnetite, pyrrhotite and probably greigite. The magnetite crystals ranged from about 10 to 100 nm, were in the superparamagnetic and single-magnetic domain size ranges, and were cuboid, teardrop and irregular in shape. The iron sulfide particles varied in size and shape, the pyrrhotite particles ranging up to 100 nm. Both mineral particles were embedded in in a fine-grained carbonate matrix on the rim of carbonate inclusions within the meteorite.

The magnetite particles were used as evidence of life in particular, because of the similarities in size distribution and morphology between these particles and those produced by the magnetotactic bacteria on Earth. In addition, magnetite particles in ALH84001 are very different from those found in other meteorites. Other nanocrystalline forms of magnetite, including platelets and whiskers, are also present in ALH84001. Some of these contain specific crystalline defects (e.g. axial screw dislocations) indicating that they were formed at high temperature, non-biologically, by vapor phase growth (Bradley et al. 1996). These results again show the importance of being able to distinguish between biogenic and non-biogenic froms of iron minerals. Although discussion and debate on the interpretation of these fascinating findings is far from being over, it is intriguing to think that microbes may be or have been responsible for the geochemical cycling of iron on other planets as well as our own Earth.

ACKNOWLEDGMENTS

We thank P.L. Donaghay, A.L. Hanson and J.W. King for continued collaboration on the Pettaquamscutt Estuary project; B.H. Howes and D.R. Schlezinger for continued collaboration on the Salt Pond project; and J. Kostka for magnetite samples. We are especially indebted to R.B. Frankel for collaboration and inspiration. DAB is supported by U.S. Office of Naval Research grant ONR N00014-91-J-1290 and U.S. National Science Foundation grant OPP-9714101. This is contribution 9707 of the Institute for Rock Magnetism (IRM). The IRM is supported by grants from the Keck Foundation and the National Science Foundation.

218 *GEOMICROBIOLOGY*

REFERENCES

Balkwill DL, Maratea D, Blakemore RP (1980) Ultrastructure of a magnetic spirillum. J Bacteriol 141:1399-1408
Banerjee SK (1991) Magnetic properties of Fe-Ti oxides. Rev Mineral 25:107-128
Banerjee SK, Hunt CP, Liu X-P (1993) Separation of local signals from the regional paleomonsoon record of the Chinese Loess Plateau: a rock-magnetism approach. Geophys Res Lett 20:843-846
Banerjee SK, Moskowitz BM (1985) Ferrimagnetic properties of magnetite. In Kirschvink JL, Jones DS, MacFadden BJ (eds) Magnetite Biomineralization and Magnetoreception in Organisms, p 17-41. Plenum Press, New York
Bazylinski DA (1995) Structure and function of the bacterial magnetosome. ASM News 61:337-343
Bazylinski DA, Blakemore RP (1983a) Denitrification and assimilatory nitrate reduction in *Aquaspirillum magnetotacticum*. Appl Environ Microbiol 46:1118-1124
Bazylinski DA, Blakemore RP (1983b) Nitrogen fixation (acetylene reduction) in *Aquaspirillum magnetotacticum*. Curr Microbiol 9:305-308
Bazylinski DA, Frankel RB (1992) Production of iron sulfide minerals by magnetotactic bacteria from sulfidic environments. In Skinner HCW, Fitzpatrick RW (eds) Biomineralization Processes of Iron and Manganese: Modern and Ancient Environments, p 147-159. Catena, Cremlingen
Bazylinski DA, Frankel RB, Garratt-Reed AJ, Mann S (1990) Biomineralization of iron sulfides in magnetotactic bacteria from sulfidic environments. In Frankel RB, Blakemore RP (eds) Iron Biominerals, p 239-255. Plenum Press, New York
Bazylinski DA, Frankel RB, Heywood BR, Mann S, King JW, Donaghay PL, Hanson AK (1995) Controlled biomineralization of magnetite (Fe_3O_4) and greigite (Fe_3S_4) in a magnetotactic bacterium. Appl Environ Microbiol 61:3232-3239
Bazylinski DA, Frankel RB, Jannasch HW (1988) Anaerobic production of magnetite by a marine magnetotactic bacterium. Nature 334:518-519
Bazylinski DA, Garratt-Reed AJ, Abedi, A, Frankel RB (1993a) Copper association with iron sulfide magnetosomes in a magnetotactic bacterium. Arch Microbiol 160:35-42
Bazylinski DA, Heywood BR, Mann S, Frankel RB (1993b) Fe_3O_4 and Fe_3S_4 in a bacterium. Nature 366:218
Bazylinski DA, Garratt-Reed AJ, Frankel RB (1994) Electron microscopic studies of magnetosomes in magnetotactic bacteria. Microsc Res Tech 27:389-401
Berner RA (1962) Synthesis and description of tetragonal iron sulfide. Science 137:669
Berner RA (1964) Iron sulfides formed from aqueous solution at low temperatures and atmospheric pressure. J Geol 72:293-306
Berner RA (1967) Thermodynamic stability of sedimentary iron sulfides. Am J Sci 265:773-785
Berner RA (1969) The synthesis of framboidal pyrite. Econ Geol 64:383-393
Berson AE, Hudson DV, Waleh NS (1991) Cloning of a sequence of *Aquaspirillum magnetotacticum* that complements the *aroD* gene of *Escherichia coli*. Mol Microbiol 5:2261-2264
Berson AE, Peters MR, Waleh NS (1989) Cloning and characterization of the *recA* gene of *Aquaspirillum magnetotacticum*. Arch Microbiol 152:567-571
Berson AE, Peters MR, Waleh NS (1990) Nucleotide sequence of *recA* gene of *Aquaspirillum magnetotacticum*. Nucl Acids Res 18:675
Blakemore RP (1975) Magnetotactic bacteria. Science 190:377-379
Blakemore RP (1982) Magnetotactic bacteria. Annu Rev Microbiol 36:217-238
Blakemore RP, Blakemore, NA (1990) Magnetotactic magnetogens. In Frankel RB, Blakemore RP (eds) Iron Biominerals. 51-67. Plenum Press, New York
Blakemore RP, Frankel RB, Kalmijn AJ (1980) South-seeking magnetotactic bacteria in the southern hemisphere. Nature 236:384-385
Blakemore RP, Maratea D, Wolfe RS (1979) Isolation and pure culture of a freshwater magnetic spirillum in chemically defined medium. J Bacteriol 140:720-729
Blakemore RP, Short KA, Bazylinski DA, Rosenblatt C, Frankel RB (1985) Microaerobic conditions are required for magnetite formation within *Aquaspirillum magnetotacticum*. Gemicrobiol J 4:53-71
Bradley JP, Harvey RP, McSween HYJr (1996) Magnetite whiskers and platelets in the ALH84001 martian meteorite: evidence for vapor phase growth. Geochim Cosmochim Acta 60:5149-5155
Burgess JG, Kanaguchi R, Sakaguchi T, Thornhill RH, Matsunaga T (1993) Evolutionary relationships among *Magnetospirillum* strains inferred from phylogenetic analysis of 16S rRNA sequences. J Bacteriol 175:6689-6694
Butler RF, Banerjee SK (1975) Theoretical single-domain grain size range in magnetite and titanomagnetite. J Geophys Res 80:4049-4058
Caccavo, FJr., Lonergan DJ, Lovley DR, Davis M, Stolz JF, McInerny MJ (1994) *Geobacter sulfurreducens* sp. nov., a hydrogen- and acetate-oxidizing dissimilatory metal reducing microorganism. Appl Environ Microbiol 60:3752-3759
Canfield DE, Berner RA (1987) Dissolution and pyritization of magnetite in anoxic marine sediments. Geochim Cosmochim Acta 51:645-659
Chang S-BR, Kirschvink JL (1989) Magnetofossils, the magnetization of sediments, and the evolution of magnetite biomineralization. Ann Rev Earth Planet Sci 17:169195

Chang S-BR, Kirschvink JL, Stolz JF (1987) Biogenic magnetite as a primary remanence carrier in limestone deposits. Phys Earth Planet Inter 46:289-303

Coates JD, Lonergan DJ, Lovley DR (1995) *Desulfuromonas palmitatis* sp. nov., a long-chain fatty acid oxidizing Fe(III) reducer from marine sediments. Arch Microbiol 164:406-413

DeLong EF, Frankel RB, Bazylinski DA (1993) Multiple evolutionary origins of magnetotaxis in bacteria. Science 259:803-806

Diaz Ricci JC, Kirschvink JL (1992) Magnetic domain state and coercivity predictions for biogenic greigite (Fe3S4): a comparison of theory with magnetosome observations. J Geophys Res 97:17309-17315

DiChristina TJ, DeLong EF (1993) Design and application of rRNA-targeted oligonucleotide probes for the dissimilatory iron- and manganese-reducing bacterium *Shewanella putrefaciens*. J Bacteriol 59:4152-4160

Donaghay PL, Rines HM, Sieburth JM (1992) Simultaneous sampling of fine scale biological, chemical and physical structure in stratified waters. Arch Hydrobiol Beih Ergebn Limnol 36:97-108

Dunlop DJ, Özdemir Ö (1997) Rock Magnetism: Fundamentals and Frontiers. Cambridge University Press, Cambridge, UK 573 p

Eden PA, Schmidt TM, Blakemore RP, Pace NR (1991) Phylogenetic analysis of *Aquaspirillum magnetotacticum* using polymerase chain reaction-amplified 16S rRNA-specific DNA. Int'l J Syst Bacteriol 41:324-325

Fabian K, Kirchner A, Williams W, Heider F, Leibl T, Huber A (1996) Three-dimensional micromagnetic calculations for magnetite using FFT. Geophys J Int'l 124:89-104

Fabricus F (1961) Die Strukturen des "Rogenpyrits" (Kossener Schichten, Rat) als Betrag zum Problem der "Vererzten Bakterien". Geol Rundshau 51:647-657

Farina M, Kachar B, Lins U, Broderick R, De Barros HL (1994) The observation of large magnetite crystals from magnetotactic bacteria by electron and atomic force microscopy. J Microsc 173:1-8

Frankel RB (1984) Magnetic guidance of organisms. Annu Rev Biophys Bioeng 13:85-103

Frankel RB, Bazylinski DA, Johnson M, Taylor BL (1997) Magneto-aerotaxis in marine, coccoid bacteria. Biophys J 73:994-1000

Frankel RB, Blakemore RP (1980) Navigational compass in magnetic bacteria. J Magn Magn Mater 15-18:1562-1564

Frankel RB, Blakemore RP, Torres de Araujo FF, Esquivel DMS, Danon J (1981) Magnetotactic bacteria at the geomagnetic equator. Science 212:1269-1270

Frankel RB, Papaefthymiou GC, Blakemore RP, O'Brien W (1983) Fe$_3$O$_4$ precipitation in magnetotactic bacteria. Biochim Biophys Acta 763:147-159

Freke AM, Tate D (1961) The formation of magnetic iron sulphide by bacterial reduction of iron solutions. J Biochem Microbiol Technol Eng 3:29-39

Gorby YA, Beveridge TJ, Blakemore RP (1988) Characterization of the bacterial magnetosome membrane. J Bacteriol 170:834-841

Gorby YA (1989) Regulation of magnetosome biogenesis by oxygen and nitrogen. 72-88. PhD Thesis, University of New Hampshire, Durham

Guerinot ML (1994) Microbial iron transport. Annu Rev Microbiol 48:743-772

Hanzlik MM, Petersen N, Keller R, Schmidbauer E (1996a) Electron microscopy and ^{57}Fe Mössbauer spectra of 10 nm particles, intermediate in composition between Fe$_3$O$_4$-χ-Fe$_2$O$_3$, produced by bacteria. Geophys Res Lett 23:479-482

Hanzlik MM, Winklhofer M, Petersen N (1996b) Spatial arrangement of chains of magnetosomes in magnetotactic bacteria. Earth Planet Sci Lett 145:125-134

Hawthorne TB, McKenzie JA (1993) Biogenic magnetite: authigenesis and diagenesis with changing redox conditions in Lake Greifen, Switzerland. In Assaoui DM, Hurley NF, Lidz BH (eds) Applications of Paleomagnetism to Sedimentary Geology. Soc Sed Geol Spec Publ 49:3-15

Hess PP (1994) Evidence for bacterial paleoecological origin of mineral magnetic cycles in oxic and sub-oxic Tasman Sea sediments. Mar Geol 117:1-17

Heywood BR, Bazylinski DA, Garratt-Reed AJ, Mann S, Frankel RB (1990) Controlled biosynthesis of greigite (Fe$_3$S$_4$) in magnetotactic bacteria. Naturwiss 77:536-538

Heywood BR, Mann S, Frankel RB (1991) Structure, morphology and growth of biogenic greigite (Fe$_3$S$_4$). In Alpert M, Calvert P, Frankel RB, Rieke P, Tirrell D (eds) Materials Synthesis Based on Biological Processes, p 93-108. Materials Research Society, Pittsburgh, Pennsylvania

Hunt CP, Moskowitz BM, Banerjee SK (1995) Magnetic properties of rocks and minerals. In Ahrens, TJ (ed) Rock Physics and Phase Relations: A Handbook of Physical Constants 3:189-204. American Geophysical Union, Washington, DC.

Kalmijn AJ (1981) Biophysics of geomagnetic field detection. IEEE Trans Mag MAG-17:1113-1123

Karlin R (1990) Magnetic mineral diagenesis in suboxic sediments at Bettis Site W-N, NE Pacific Ocean. J Geophys Res 95:4421-4436

Karlin R, Lyle M, Heath GR (1987) Authigenic magnetite formation in suboxic marine sediments. Nature 326:490-493

Kawaguchi R, Burgess JG, Sakaguchi T, Takeyama H, Thornhill RH, Matsunaga T (1995) Phylogenetic analysis of a novel sulfate-reducing magnetic bacterium, RS-1, demonstrates its membership of the δ-Proteobacteria. FEMS Microbiol Lett 126:277-282

Kimble LK, Bazylinski DA (1996) Chemolithoautotrophy in the marine magnetotactic bacterium, strain MV-1. Abstr 96th Annu Meet Am Soc Microbiol K-174

King J, Channell JET (1991) Sedimentary magnetism, environmental magnetism, and magnetostratigraphy. Rev Geophys Suppl:358-370

Kirschvink JL (1980) South-seeking magnetic bacteria. J Exp Biol 86:345-347

Kirschvink JL, Lowenstam HA (1979) Mineralization and magnetization of chiton teeth: paleomagnetic, sedimentologic, and biologic implications of organic magnetite. Earth Planet Sci Lett 44:193-204

Kirschvink JL, Chang S-BR (1984) Ultrafine-grained magnetite in deep-sea sediments: possible bacterial magnetofossils. Geol 12:559-562

Kostka JE, Nealson KE (1995) Dissolution and reduction of magnetite by bacteria. Environ Sci Technol 29:2535-2540

Leslie BW, Lund SP, Hammond DE (1990) Rock magnetic evidence for dissolution and authigenic growth of magnetic minerals in anoxic marine sediments. J Geophys Res 95:4437-4452

Lins U, Solórzano G, Farina M (1994) High volume magnetite crystals from magnetotactic bacteria. Bull Instit Océanograph, Monaco, (no. spécial, 14) 1:95-104.

Lins de Barros HGP, Esquivel DMS (1985) Magnetotactic microorganisms found in muds from Rio de Janeiro: a general view. In Kirschvink JL, Jones DS, MacFadden BJ (eds) Magnetite Biomineralization and Magnetoreception in Organisms, p 289-309. Plenum Press, New York

Lonergan DJ, Jenter HL, Coates JD, Phillips EJP, Schmidt TM, Lovley DR (1996) Phylogenetic analysis of dissimilatory Fe(III)-reducing bacteria. J Bacteriol 178:2402-2408

Love LG, Zimmerman DO (1961) Bedded pyrite and microrganisms from the Mount Isa Shale. Econ Geol 56:873

Lovley DR (1987) Organic matter mineralization with the reduction of ferric iron: a review. Geomicrobiol J 5:375-399

Lovley DR (1990) Magnetite formation during microbial dissimilatory iron reduction. In Frankel RB, Blakemore RP (eds) Iron Biominerals, p 151-166. Plenum Press, New York

Lovley DR (1991) Dissimilatory Fe(III) and Mn(IV) reduction. Microbiol Rev 55:259-287

Lovley DR, Giovannoni, SJ, White DC, Champine JE, Phillips EJP, Gorby YA, Goodwin S (1993) *Geobacter metallireducens* gen. nov. sp. nov., a microorganism capable of coupling the complete oxidation of organic compounds to the reduction of iron and other metals. Arch Microbiol 159:336-344

Lovley DR, Phillips EJP (1988) Novel mode of microbial energy metabolism: organic carbon oxidation coupled to dissimilatory reduction of iron or manganese. Appl Environ Microbiol 54:1472-1480

Lovley DR, Phillips EJP, Lonergan DJ (1989) Hydrogen and formate oxidation coupled to dissimilatory reduction of iron and manganese by *Alteromonas putrefaciens*. Appl Environ Microbiol 55:700-706

Lovley DR, Phillips EJP, Lonergan DJ, Widman PK (1995) Fe(III) and S^{o} reduction by *Pelobacter carbinolicus*. Appl Environ Microbiol 61:2132-2138

Lovley DR, Stolz JF, Nord GLJr., Phillips EJP (1987) Anaerobic production of magnetite by a dissimilatory iron-reducing microorganism. Nature 330:252-254

Lowenstam HA (1981) Minerals formed by organisms. Science 211:1126-1131

Maher BA (1990) Inorganic formation of ultrafine-grained magnetite. In Frankel RB, Blakemore RP (eds) Iron Biominerals, p 179-192. Plenum Press, New York

Maher BA, Taylor RM (1988) Formation of ultra-fine grained magnetite in soils. Nature 336:368-370

Mann S (1986) On the nature of boundary-organised biomineralization. J Inorg Chem 28:363-371

Mann S, Frankel RB (1989) Magnetite biomineralization in unicellular organisms. In Mann S, Webb J, Williams RJP (eds) Biomineralization: Chemical and Biochemical Perspectives, p 389-426. VCH Publishers, New York

Mann S, Frankel RB, Blakemore RP (1984a) Structure, morphology and crystal growth of bacterial magnetite. Nature 310:405-407

Mann S, Moench TT, Williams RJP (1984b) A high resolution electron microscopic investigation of bacterial magnetite. Implications for crystal growth. Proc R Soc London B 221:385-393

Mann S, Sparks NHC, Blakemore RP (1987a) Ultrastructure and characterization of anisotropic inclusions in magnetotactic bacteria. Proc R Soc London B 231:469-476

Mann S, Sparks NHC, Blakemore RP (1987b) Structure, morphology and crystal growth of anisotropic magnetite crystals in magnetotactic bacteria. Proc R Soc London B 231:477-487

Mann S, Sparks NHC, Board RG (1990a) Magnetotactic bacteria: microbiology, biomineralization, palaeomagnetism, and biotechnology. Adv Microb Phys 31:125-181

Mann S, Sparks NHC, Couling SB, Larcombe MC, Frankel RB (1989) Crystallochemical characterization of magnetic spinels prepared from aqueous solution. J Chem Soc Faraday Trans 85:3033-3044

Mann S, Sparks NHC, Frankel RB, Bazylinski DA, Jannasch HW (1990b) Biomineralization of ferrimagnetic greigite (Fe_3O_4) and iron pyrite (FeS_2) in a magnetotactic bacterium. Nature 343:258-260

Maratea D, Blakemore RP (1981) *Aquaspirillum magnetotacticum* sp. nov., a magnetic spirillum. Int'l J Syst Bacteriol 31:452-455

Matsuda T, Endo J, Osakabe N, Tonomua A, Arii T (1983) Morphology and structure of biogenic magnetite particles. Nature 302:411-412

Matsunaga T, Nakamura C, Burgess JG, Sode K (1992) Gene transfer in magnetic bacteria: transposon mutagenesis and cloning of genomic DNA fragments required for magnetite synthesis. J Bacteriol 174:2748-2753

Matsunaga T, Sakaguchi T, Tadokoro, F. (1991) Magnetite formation by a magnetic bacterium capable of growing aerobically. Appl Microbiol Biotechnol 35:651-655

Matsunaga T, Tsujimura N (1993) Respiratory inhibitors of a magnetic bacterium *Magnetospirillum* sp. AMB-1 capable of growing aerobically. Appl Microbiol Biotechnol 39:368-371

Mckay, DS, Gibson EKJr, Thomas-Keprta KL, Vali H, Romanek CS, Clemett SJ, Chillier XDF, Maechling CR, Zare RN (1996) Search for past life on Mars: possible relic biogenic activity in Martian meteorite ALH84001. Science 273:924-930

McNeill DF (1990) Biogenic magnetite from surface Holocene carbonate sediments, Great Bahama Bank. J Geophys Res 95B:4363-4372

Meldrum FC, Heywood BR, Mann S, Frankel RB, Bazylinski DA (1993a) Electron microscopy study of magnetosomes in a cultures coccoid magnetotactic bacterium. Proc R Soc London B 251:231-236

Meldrum FC, Heywood BR, Mann S, Frankel RB, Bazylinski DA (1993b) Electron microscopy study of magnetosomes in two cultured vibrioid magnetotactic bacteria. Proc R Soc London B 251:237-242

Merrill RT, Halgedahl SL (1995) Theoretical and experimental studies of magnetic domains. Rev Geophys Supp 137-144

Moskowitz BM (1995) Biomineralization of magnetic minerals. Rev Geophys Supp 123-128

Moskowitz BM, Frankel RB, Flanders PJ, Blakemore RP, Schwartz BB (1988) Magnetic properties of magnetotactic bacteria. J Magn Magn Mater 73:273-288

Moskowitz BM, Frankel RB, Bazylinski DA, Jannasch HW, Lovley DR (1989) A comparison of magnetite particles produced anaerobically by magnetotactic and dissimilatory iron-reducing bacteria. Geophys Res Lett 16:665-668

Moskowitz BM, Frankel RB, Bazylinski DA (1993) Rock magnetic criteria for the detection of biogenic magnetite. Earth Planet Sci Lett 120:283-300

Moskowitz BM, Frankel RB, Walton SA, Dickson DPE, Wong KKW, Douglas T, Mann S (1997) Determination of the pre-exponential frequency factor for superparamagnetic maghemite particles in magnetoferritin. J Geophys Res in press

Myers CR, Nealson KH (1988) Bacterial manganese reduction and growth with manganese oxide as the sole electron acceptor. Science 240:1319-1321

Myers CR, Nealson KH (1990) Iron mineralization by bacteria: metabolic coupling of iron reduction to cell metabolism in *Alteromonas putrefaciens* MR-1. In Frankel RB, Blakemore RP (eds) Iron Biominerals. 131-149. Plenum Press, New York

Nakamura C, Burgess JG, Sode K, Matsunaga T (1995) An iron-regulated gene, *magA*, encoding an iron transport protein of *Magnetospirillum* AMB-1. J Biol Chem 270:28392-28396

Nakamura C, Sakaguchi T, Kudo S, Burgess JG, Sode K, Matsunaga T (1993) Characterization of iron uptake in the magnetic bacterium *Aquaspirillum* sp. AMB-1. Appl Biochem Biotechnol 39/40:169-176

Neilands JB (1984) A brief history of iron metabolism. Biol Metals 4:1-6

Okuda Y, Denda K, Fukumori Y (1996) Cloning and sequencing of a gene encoding a new member of the tetratricopeptide protein family from magnetosomes of *Magnetospirillum magnetotacticum*. Gene 171:99-102

Oldfield F (1992) The source of fine-grained magnetite in sediments. Holocene 2:180-182

Oldfield F (1994) Toward the discrimination of fine grained ferrimagnets by magnetic measurements in lake and near-shore marine sediments. J Geophys Res 99:9045-9050

Özdemir Ö, Dunlop DJ, Moskowitz BM (1993) The effect of oxidation on the Verwey transition in magnetite. Geophys Res Lett 20:1671-1674

Özdemir, Ö, Xu S, Dunlop DJ (1995) Closure domains in magnetite. J Geophys Res 100:2193-2209

Paoletti LC, Blakemore RP (1986) Hydroxamate production by *Aquaspirillum magnetotacticum*. J Bacteriol 167:73-76

Peck JA, King JW (1996) Magnetofossils in the sediments of Lake Baikal, Siberia. Earth Plant Sci Lett 140:159-172

Penninga I, deWaard H, Moskowitz BM, Bazylinski DA, Frankel RB (1995) Remanence curves for individual magnetotactic bacteria using a pulsed magnetic field. J Magn Magn Mater 149:279-286

Petermann,H, Bleil U (1993) Detection of live magnetotactic bacteria in South Atlantic deep-sea sediments. Earth Planet Sci Lett 117:223-228

Petersen N, von Dobeneck T,Vali H (1986) Fossil bacterial magnetite in deep-sea sediments from the South Atlantic Ocean. Nature 320:611-615

Petersen N, Weiss DG, Vali H (1989) Magnetic bacteria in lake sediments. In Lowes FJ et al. (eds) Geomagnetism and Paleomagnetism. 231-241. Kluwer Academic Publishers, Dordrecht, Netherlands

Pokhil TG, Moskowitz BM (1997) Magnetic domains and domain walls in pseudo-single-domain magnetite studied with magnetic force microscopy. J Geophys Res in press

Proksch R, Moskowitz BM, Dahlberg ED, Bazylinski DA, Frankel RB (1995). Magnetic force microscopy of the submicron magnetic assembly in a magnetotactic bacterium. Appl Phys Lett 66:2582-2584

Reynolds RL, Tuttle ML, Rice CA, Fishman NS, Karachewski JA, Sherman DM (1994) Magnetization and geochemistry of greigite-bearing cretaceous strata, north slope basin, Alaska. Am J Sci 294:485-538

Rickard DT (1969a) The microbiological formation of iron sulfides. Stockholm Contrib Geol 20:50-66

Rickard DT (1969b) The chemistry of iron sulfide formation at low temperatures. Stockholm Contrib Geol 20:67-95

Roberts, AP (1995) Magnetic properties of sedimentary greigite (Fe$_3$S$_4$). Earth Planet Sci Lett 134:227-236Roberts AP, Turner GM (1993) Diagenetic formation of ferrimagnetic iron sulfide minerals in rapidly deposited marine sediments, South Island, New Zealand. Earth Planet Sci Lett 115:257-273

Rochette P, Fillon G, Mattéi J-L, Dekkers MJ (1990) Magnetic transition at 30-34 Kelvin in pyrrhotite: insight into a widespread occurrence of this mineral in rocks. Earth Planet Sci Lett 98:319-328

Roden EE, Lovley DR (1993) Dissimilatory Fe(III) reduction by the marine microorganism *Desulfuromonas acetoxidans*. Appl Environ Microbiol 59:734-742

Rogers FG, Blakemore RP, Blakemore NA, Frankel RB, Bazylinski DA, Maratea D, Rogers C (1990a) Intercellular structure in a many-celled magnetotactic procaryote. Arch Microbiol 154:18-22

Rogers FG, Blakemore RP, Blakemore NA, Frankel RB, Bazylinski DA, Maratea D, Rogers C (1990b) Intercellular junctions, motility and magnetosome structure in a multicellular magnetotactic procaryote. In Frankel RB, Blakemore RP (eds) Iron Biominerals. 239-255. Plenum Press, New York

Rossello-Mora RA, Caccavo FJr., Osterlehner K, Springer N, Spring S, Schüler D, Ludwig W, Amann R, Vannacanneyt M, Schleifer K-H (1994) Isolation and taxonomic characterization of a halotolerant, facultative anaerobic iron-reducing bacterium. Syst Appl Microbiol 17:569-573

Sakaguchi T, Burgess JG, Matsunaga T (1993) Magnetite formation by a sulphate-reducing bacterium. Nature 365:47-49

Schleifer K-H, Schüler D, Spring S, Weizenegger M, Amann R, Ludwig W, Kohler M (1991) The genus *Magnetospirillum* gen. nov., description of *Magnetospirillum gryphiswaldense* sp. nov. and transfer of *Aquaspirillum magnetotacticum* to *Magnetospirillum magnetotacticum* comb. nov. Syst Appl Microbiol 14:379-385

Schüler D, Baeuerlein E (1996) Iron-limited growth and kinetics of iron uptake in *Magnetospirillum gryphiswaldense*. Arch Microbiol 166:301-307

Schwartz M, Lund SP, Hammond DE, Schwartz R, Wong K (1997) Early sediment diagenesis on the Blake/Bahama Outer Ridge, North Atlantic Ocean, and its effects on sediment magnetism. J Geophys Res 102:7903-7914

Snowball IF (1991) Magnetic hysteresis properties of greigite (Fe$_3$S$_4$) and a new occurrence in Holocene sediments from Swedish Lappland. Phys Earth Planet Inter 68:32-40

Snowball IF (1994) Bacterial magnetite and the magnetic properties of sediments in a Swedish lake. Earth Planet Sci Lett 126:129-142

Sparks NHC, Mann S, Bazylinski DA, Lovley DR, Jannasch HW, Frankel RB (1990) Structure and morphology of magnetite anaerobically-produced by a marine magnetotactic bacterium and a dissimilatory iron-reducing bacterium. Earth Planet Sci Lett 98:14-22

Spring S, Amann R, Ludwig W, Schleifer K-H, Petersen N (1992) Phylogenetic diversity and identification of non-culturable magnetotactic bacteria. Syst. Appl Microbiol 15:116-122

Spring S, Amann R, Ludwig W, Schleifer K-H, van Gemerden H, Petersen N (1993) Dominating role of an unusual magnetotactic bacterium in the microaerobic zone of a freshwater sediment. Appl Environ Microbiol 59:2397-2403

Steinberger B, Petersen N, Petermann H, Weiss D (1994) Movement of magnetic bacteria in time-varying magnetic fields. J Fluid Mech 273:189-211

Stolz JF, Chang S-BR, Kirschvink JL (1986) Magnetotactic bacteria and singledomain magnetite in hemipelagic sediments. Nature 321:849-850

Tamegai H, Fukumori Y (1994) Purification, and some molecular and enzymatic features of a novel *ccb*-type cytochrome c oxidase from a microaerobic denitrifier, *Magnetospirillim magnetotacticum*. FEBS Lett 347:22-26

Tamegai H, Yamanaka T, Fukumori Y (1993) Purification and properties of a "cytochrome a$_1$"-like hemoprotein from a magnetotactic bacterium, *Aquaspirillum magnetotacticum*. Biochim Biophys Acta 1158:237-243

Tarduno JA (1995) Superparamagnetism and reduction diagenesis in pelagic sediments: enhancement or depletion. Geophys Res Lett 22:1337-1340

Tarduno JA, Wilkison SL (1996). Non-steady state magnetic mineral reduction, chemical lock-in, and delayed remanence acquisition in pelagic sediments. Earth Planet Sci Lett 144:315-326

Thompson R, Oldfield F (1986) Environmental Magnetism. Allen and Unwin, London, 227 p

Thornhill RH., Burgess JG, Sakaguchi T, Matsunaga T (1994) A morphological classification of bacteria containing bullet-shaped magnetic particles. FEMS Microbiol Lett 115:169-176

Torii M, Fukuma K, Houng C-S, Less T-Q (1996) Magnetic discrimination of pyrrhotite- and greigite-bearing sediment samples. Geophys Res Lett 23:1813-1817

Towe KM, Moench TT (1981) Electron-optical characterization of bacterial magnetite. Earth Planet Sci Lett 52:213-220

Vali H, Forster O, Amarantidis G, Petersen N (1987) Magnetotactic bacteria and their magnetofossils in sediments. Earth Planet Sci Lett 86:389-400

Vali H, Kirschvink JL (1989) Magnetofossil dissolution in a paleomagnetically unstable deep-sea sediment. Nature 339:203-206

Vali H, Kirschvink JL (1990) Observations of magnetosome organization, surface structure, and iron biomineralization of undescribed magnetic bacteria: evolutionary speculations. In: Frankel RB, Blakemore RP (eds) Iron Biominerals, p 97-115. Plenum Press, New York

Wakeham SG, Howes BL, Dacey JWH (1984) Dimethyl sulphide in a stratified coastal salt pond. Nature 310:770-772

Wakeham SG, Howes BL, Dacey JWH, Schwarzenbach RP, Zeyer J (1987) Biogeochemistry of dimethylsulfide in a seasonally stratified coastal salt pond. Geochim Cosmochim Acta 51:1675-1684
Waleh NS (1988) Functional expression of *Aquaspirillum magnetotacticum* genes in *Escherichia coli* K12. Mol Gen Genet 214:592-594
Williams W, Dunlop DJ (1989) Three dimensional micromagnetic modelling of ferromagnetic domain structure. Nature 337:634-637
Woese CR (1987) Bacterial evolution. Microbiol Rev 51:221-271
Yamazaki T, Oyanagi H, Fujiwara T, Fukumori Y (1995) Nitrite reductase from the magnetotactic bacterium *Magnetospirillum magnetotacticum*; a novel cytochrome cd_1 with Fe(II):nitrite oxidoreductase activity. Eur J Biochem 233:665-671
Zavarzin GA, Stackebrandt E, Murray RGE (1991) A correlation of phylogenetic diversity in the Proteobacteria with the influences of ecological forces. Can J Microbiol 37:1-6

Chapter 7

BACTERIALLY MEDIATED MINERAL FORMATION: INSIGHTS INTO MANGANESE(II) OXIDATION FROM MOLECULAR GENETIC AND BIOCHEMICAL STUDIES

Bradley M. Tebo[1], William C. Ghiorse[2], Lorraine G. van Waasbergen[1,*]
Patricia L. Siering[2] and Ron Caspi[1,†]

[1]*Marine Biology Research Division and*
Center for Marine Biotechnology and Biomedicine
Scripps Institution of Oceanography
University of California, San Diego
9500 Gilman Drive
La Jolla, California 92093 U.S.A.

[2]*Section of Microbiology*
Division of Biological Sciences
Cornell University, Wing Hall
Ithaca, New York 14853 U.S.A.

INTRODUCTION

Microorganisms that oxidize divalent manganese [Mn(II)] have been known from the beginning of the 20th century (Jackson 1901). However, even after almost a century of study, the mechanisms of Mn(II) oxidation and the biological functions it serves remain enigmatic. In the last six years with the application of modern molecular biological and genetic approaches, new insights into the mechanism(s) of microbial Mn(II) oxidation have emerged. This knowledge has, in turn, allowed us to consider the possible functions and selective advantages gained by bacteria that oxidize Mn(II). In addition, it may even allow us to consider the role of Mn(II)-oxidizing bacteria in controlling the properties of the Mn minerals they form, such as their surface area and charge, and composition.

The focus of this chapter is to review the recent progress that has been made in studies of microbial (primarily bacterial) Mn(II) oxidation with an emphasis on molecular genetics and biochemistry. The beginning of this chapter presents some background material on the geochemistry, mineralogy, and biology of Mn(II) oxidation. An overview of the state of knowledge concerning the rates of Mn(II) oxidation in natural systems and microbially-produced Mn minerals is provided. This introductory material is not intended to be an exhaustive review of the field, but instead is intended to present enough information for the reader to appreciate the recent advances that have been made and to provide sufficient references for the interested reader to consult. The rest of the chapter describes in more detail three model bacterial Mn(II) oxidation systems and the new insights that have been gained in these systems. Through comparisons of the different systems we could not resist making some general speculations concerning the

*current address: *Carnegie Institution*
of Washington
290 Panama Street
Stanford, CA 94305

†current address: *Department of Biology*
University of California, San Diego
9500 Gilman Drive
La Jolla, CA 92093

0275-0279/97/0035-0007$05.00

mechanisms and functions of Mn(II) oxidation by microbes. Despite a century of study, the field of microbial Mn(II) oxidation is only in its infancy. Continued investigations into this area are warranted because, not only is Mn(II) oxidation an important environmental process in which microbes play a major role, but Mn(II)-oxidizing microorganisms themselves represent exciting systems for study with potential applications of benefit to society.

BACKGROUND

Manganese geochemistry

Manganese, the second most abundant transition metal in the earth's crust, exists in a number of oxidation states, among which the II, III and IV oxidation states are of greatest environmental importance. Mn(II), in its soluble form that is available to organisms, is stable in the pH 6 to 9 range of natural waters. Mn(II) also exists in a variety of minerals such as rhodochrosite ($MnCO_3$), alabanite (MnS), and reddingite ($Mn_3(PO_4)_2 \cdot 3H_2O$). Mn(III) is thermodynamically unstable and does not occur in soluble form except in the presence of strong complexing agents such as humic or other organic acids (Morgan and Stumm 1965, Kostka et al. 1995). Mn(III) and Mn(IV) primarily form insoluble oxides and oxyhydroxides. The solubility of MnO_2 is so low that soluble Mn(IV) is undetectable within the pH range 3 to 10 (Morgan and Stumm 1965) (e.g. the solubility product of MnO_2 (pyrolusite) is 10^{-41}; Stumm and Morgan 1996). Mn(III,IV) minerals that have highly negative surface charges, open crystal structures, and contain a variety of charge-balancing, low valence cations (e.g. Na, Mg, Ca, Cu, Ni, etc.) are chemically referred to as manganates (Giovanoli and Arrhenius 1988). For the purposes of this review, the Mn(III,IV) oxides, oxyhydroxides, and manganates will be collectively referred to as Mn oxides or MnO_x (where $1 < x < 2$).

Mn(II) oxidation results in the formation of Mn oxides that are important mineral phases in soils, sediments, and waters (Huang 1991, Ehrlich 1996a). In these environments Mn oxides occur as coatings on soil and sediment particles and as discrete particles. They are very reactive components of natural environments and affect the fate, transport, and bioavailability of a variety of heavy metals and organic compounds (Jenne 1967, Kinniburgh and Jackson 1981, Davis and Kent 1990). In the environment, Mn oxides are the strongest oxidizing agent encountered other than oxygen (Laha and Luthy 1990). At pH 7 they are considerably more oxidizing than Fe oxides (Table 1). For example, they promote the oxidation of As(III), Co(II), Cr(III), Pu(III), and possibly other trace metals (Huang 1991). Mn oxides also catalyze the degradation of organics (Stone 1983, Stone and Morgan 1984a,b; Stone 1987, Stone and Ulrich 1989) and the formation of humic substances and organic N complexes (Huang 1991). Recently it has

Table 1. Oxidation reduction potentials of important oxidants in soils.

Reaction	E_0	E' (pH 7, $[Me^{2+}] = 10^{-6}M$)
$\frac{1}{4} O_2 + H^+ + e^- = \frac{1}{2}H_2O$	+1.23	+0.82
$\frac{1}{2}MnO_2 + 2 H^+ + e^- = \frac{1}{2}Mn^{2+} + H_2O$	+1.29	+0.64
$MnOOH + 3 H^+ + e^- = Mn^{2+} + 2 H_2O$	+1.50	+0.61
$FeOOH + 3 H^+ + e^- = Fe^{2+} + 2 H_2O$	+0.67	−0.22

Oxide data from Stone and Ulrich (1989).

been shown that Mn oxides can oxidatively degrade natural humic and fulvic acids, resulting in the formation of a variety of biologically utilizable low molecular weight organic compounds (Sunda and Kieber 1994).

Mn oxides are known to adsorb a variety of cations on their surfaces and incorporate other metals in their crystal structure. The highly charged surfaces scavenge a variety of trace elements, such as Cu, Co, Cd, Zn, and certain radionuclides such as ^{210}Pb, ^{60}Co, and Ra and Th isotopes. The distributions of several naturally-occurring radionuclides (^{210}Pb, ^{234}Th, ^{228}Th, ^{228}Ra, and ^{226}Ra) as well as Fe and Mn have been investigated in detail in several environments (Bacon et al. 1980, Todd et al. 1988, Wei and Murray 1991). In all of these cases it was concluded that the geochemical cycling of Mn- and Fe-oxides controls the distributions of the radionuclides. The scavenging by Mn oxides of several trace elements can be so efficient that it results in reduction in the concentration of soluble trace metals by several orders of magnitude.

Manganese mineralogy

Field and laboratory based studies have shown that Mn oxides are often poorly crystallized or amorphous (Mandernack et al. 1995b, Wehrli et al. 1995). Because of the geochemical importance of these minerals in freshwater and marine systems, knowledge of their identities (i.e. structures and compositions) and reactivities has been of interest to geochemists and mineralogists for a long time. However, characterization of these Mn precipitates with X-ray diffraction techniques is difficult and frequently incomplete (Manceau and Combes 1988, Manceau et al. 1992a,b; Mandernack et al. 1995b). This results mostly from the fact that many of these precipitates are poorly crystalline or consist of very small crystallite sizes with mixed Mn oxidation states. High resolution transmission electron microscopy (HRTEM) and selected-area electron diffraction (SEAD) have frequently been used to identify some of the more crystallized Mn oxide phases (Turner and Buseck 1981, Golden et al. 1986, Drits et al. 1997). X-ray absorption fine structure (XAFS) spectroscopy is a powerful tool that can provide information regarding the local structures of Mn in amorphous precipitates and the identity of nanocrystalline phases (Manceau and Combes 1988, Manceau et al. 1992a, Manceau et al. 1992b, Muller et al. 1995, Drits et al. 1997, Silvester and Manceau 1997).

Table 2. Mn(II) oxidation reactions and Mn minerals.

Reaction	Oxidation state of mineral	Examples of mineral formed
1) $Mn^{2+} + \frac{1}{2}O_2 + H_2O \rightarrow MnO_2 + 2\,H^+$	4	pyrolusite (β-MnO_2) vernadite (δ-MnO_2). todorokite ([Ca,Na,K][Mg,Mn]$Mn_5O_{12}\cdot H_2O$) buserite ([Na,Ca,K][Mg,Mn]$Mn_6O_{14}\cdot 5H_2O$) birnessite ([Na,Ca]$Mn_7O_{14}\cdot 2.8H_2O$)*
2) $Mn^{2+} + \frac{1}{4}O_2 + \frac{3}{2}H_2O \rightarrow MnOOH + 2\,H^+$	3	manganite (γ-MnOOH) groutite (α-MnOOH) feitknechtite (β–MnOOH)
3) $3\,Mn^{2+} + \frac{1}{2}O_2 + 3\,H_2O \rightarrow Mn_3O_4 + 6\,H^+$	2.67	hausmannite (Mn_3O_4)
4) $Mn_3O_4 + 2\,H^+ \rightarrow 2\,MnOOH + Mn^{2+}$	3	see Reaction 2
5) $Mn_3O_4 + 4\,H^+ \rightarrow MnO_2 + 2\,Mn^{2+} + 2\,H_2O$	4	see Reaction 1
6) $2\,MnOOH + 2\,H^+ \rightarrow MnO_2 + Mn^{2+} + 2\,H_2O$	4	see Reaction 1

*Natural phase from Birness, Scotland (Post and Veblen 1990). Composition is highly variable and other cations (e.g. H^+, Mg^{2+}, K^+) can serve as counterions.

Mn(II) oxidation proceeds via a number of reactions yielding various oxidation states and mineral forms (Table 2, above). Mn(II) can be oxidized directly to Mn(IV) or to mixed Mn(II,III) and Mn(III) oxides and oxyhydroxides which may undergo protonation or disproportionation reactions to form Mn(IV) oxides and manganates (Hem and Lind 1983, Murray et al. 1985). The Mn(IV) minerals thus formed can be categorized into three major groups based on crystal structure (Burns and Burns 1979). These are: (1) the phyllomanganates, which consist of layered sheets of manganese octahedra; (2) the chain structures, which consist of cross-linked single or double rows of octahedra running parallel to the **c** axis; and (3) the tunneled structure manganates, consisting of cross-linked chains of octahedra, of two or more unit width, running parallel to the **c** axis in minerals with tetragonal symmetry or parallel to the **b** axis in minerals having monoclinic symmetry. While Burns and Burns (1979) distinguish between chain and tunnel type Mn minerals, the primary difference between them is the tunnel diameter. Examples of phyllomanganates include the minerals birnessite and buserite, which, by powder X-ray diffraction, show characteristic d-values between their octahedral sheets of 7.3 Å and 10 Å, respectively. The structures of these two phyllomanganates are very similar, the major difference being an extra layer of water molecules lying between the octahedral sheets of buserite that gives it a wider d-spacing. Dehydration of buserite results in the loss of this water and the formation of a birnessite type structure.

An example of a tunneled structure manganate is todorokite, a mineral commonly found in marine Mn deposits and reported to be a product of microbial Mn(II) oxidation (Burns and Burns 1979, Piper et al. 1984, Jannasch and Mottl 1985, Ferris et al. 1987, Giovanoli and Arrhenius 1988, Takematsu et al. 1988). Todorokite can be described as having octahedral sheets similarly spaced to those of buserite, but containing cross rows of octahedral chains of varying length running along the **a** axis which bridge the sheets at a fixed 9.6 Å spacing and prevent it from collapsing even at elevated temperatures (Usui et al. 1989). The fixed structure of todorokite can be distinguished from the collapsible buserite structure because the latter freely expands its octahedral sheets depending on the size of the countercation. However, metal cations such as Mg, Ca, and Ni, can intercalate between the buserite layers and stabilize its 10 Å spacing, making it difficult to distinguish from todorokite unless careful electron diffraction or infrared spectroscopy is used (Potter and Rossman 1979, Arrhenius and Tasi 1981, Turner and Buseck 1981). Baking at 110°C, which collapses most 10 Å phyllomanganates (e.g. buserite), can sometimes be used to distinguish buserite from todorokite (Mandernack and Usui, personal communication) .

Although characterization of todorokite from the cation-stabilized phyllomanganates usually requires a multi-technique approach, non-collapsible manganates, as determined by powder X-ray diffraction, are commonly referred to as todorokite (Piper et al. 1984, Ferris et al. 1987, Takematsu et al. 1988) although in fact they may be cation-stabilized buserite. Mn minerals formed microbially are initially amorphous, i.e. lacking crystal structure. These amorphous minerals have frequently been referred to as vernadite (δ-MnO_2), however recent work suggests that vernadite is identical to hexagonal birnessite (H+-birnessite, Drits et al. 1997, Friedl et al. 1997, Silvester and Manceau 1997). The types of Mn minerals formed through microbial activity is a subject of a separate section of this chapter (see below).

Abiotic Mn(II) oxidation

The concentration of dissolved Mn in groundwaters and surface waters is, to a great extent, controlled by redox reactions between Mn(II) and Mn(III, IV) and governed by pH (Fig. 1). Mn(II) oxidation in such natural systems is thermodynamically favorable but often proceeds at very slow rates in the absence of microbes (Diem and Stumm 1984,

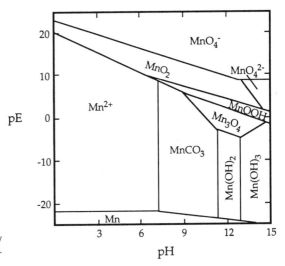

Figure 1. Manganese pE/pH stability diagram (redrawn from Nealson et al. 1988).

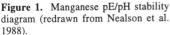

Nealson et al. 1988). Only under the more extreme conditions (pH > 8.5; $pO_2 \sim 1$ atm; or $[Mn_T] > 450$ μM) does Mn(II) homogeneously oxidize within a few weeks to months (Morgan and Stumm 1964, Hem and Lind 1983, Murray et al. 1985). Mn(II) oxidation is autocatalytic with the Mn (oxyhydr)oxide products adsorbing Mn(II) and catalyzing its oxidation (Davies and Morgan 1989, Stumm and Morgan 1996). Mn(II) oxidation is also catalyzed by a variety of other surfaces including Fe oxides and silicates (Sung and Morgan 1981, Davies and Morgan 1989, Junta and Hochella 1994). The final products at neutral pH and with O_2 present are Mn oxides containing the most thermodynamically stable Mn oxidation state, Mn(IV).

Based on laboratory experiments, the abiotic reaction sequence from Mn(II) to Mn(IV) has been reported to involve two steps (Hem 1981, Hem et al. 1982, Hem and Lind 1983, Murray et al. 1985). Mn(II) oxidation begins with the precipitation of Mn(III)-bearing oxides (e.g. Mn_3O_4) or oxyhydroxides (MnOOH), which subsequently disproportionate slowly in a second step to form Mn(IV) oxides (MnO_2) (Table 2). The disproportionation reaction is believed to be the rate controlling step in the abiotic formation of MnO_2 (Ehrlich 1996b). The products of the heterogeneous oxidation of Mn(II) at hematite, goethite, and albite surfaces (see Banfield and Hamers, this volume) after six months at room temperature in aerated solutions containing 4.0 to 26.7 ppm Mn(II) at pHs ranging from 7.8 to 8.7 consisted of Mn(III)-bearing oxyhydroxides, predominantly β-MnOOH (feitknechtite) (Junta and Hochella 1994). Tian et al. (1997) recently reported abiotic formation of mixed-valent, semiconducting, catalytically-active Mn oxides, which have mesoporous structures, high thermal stability, and catalyze the oxidation of cyclohexane and n-hexane in aqueous solutions. These oxides include Mn_2O_3 and Mn_3O_4 in which the crystal structures are based on various arrangements of MnO_6 octahedra.

Biological Mn(II) oxidation

Microorganisms can accelerate the rate of Mn(II) oxidation by up to five orders of magnitude compared to abiotic Mn(II) oxidation and thus are responsible for much of the Mn(II) oxidation observed in the environment (Nealson et al. 1988, Tebo 1991, Wehrli et al. 1995). Traditionally, much of the interest in Mn(II)-oxidizing microbes focused on the so called iron- and manganese-depositing bacteria, such as *Leptothrix discophora* and *metallogenium* (e.g. see Nealson 1983 and Ghiorse 1984). With the isolation of more

Mn(II)-oxidizing bacteria over the years, it has become clear that rather than being limited to a few specialized strains, Mn oxidation is widespread. Bacteria and fungi have been reported to oxidize Mn(II) in fresh- and marine waters, soils, and sediments. Both direct and indirect mechanisms of Mn(II) oxidation can occur (Nealson et al. 1989). Indirect oxidation occurs via modification of the redox conditions of the local aqueous environments (e.g. release of oxidants, acids, or bases), while direct oxidation occurs via binding to cell macromolecules, such as proteins, cell wall components, or extracellular polymers, with concomitant oxidation (Ghiorse 1984b, Nealson et al. 1989). Mn binding and oxidizing activities in cell extracts have been reported for a number of bacteria (Ehrlich 1968, Douka 1977, Jung and Schweisfurth 1979, Douka 1980, Ehrlich 1984). Microbial Mn(II) oxidation appears to be exclusively an extracellular process. Mn oxides precipitate around cells or accumulate on slime layers or sheaths (Ghiorse 1984b). Often the cells become completely encrusted with Mn deposits (Nealson and Ford 1980).

Presumably, the ability to mediate the extracellular oxidation of Mn(II) must impart some selective advantage to the microorganisms, although in all cases studied thus far the underlying benefits of this activity remains elusive. In some instances, it has been proposed that the oxidation of Mn(II) yields energy for mixotrophic or lithotrophic growth (Ehrlich 1976, Kepkay and Nealson 1987b). However, this mode of energy generation has not been proved unequivocally, and for many Mn(II)-oxidizing microorganisms, there is little or no evidence that the oxidation provides energy for growth (Ghiorse 1984b, Nealson et al. 1988, Ehrlich 1996a, Caspi et al., submitted). One Mn(II)-oxidizing bacterium (marine strain SI95-9A1) has been shown to possess the genes for Ribulose-1,5-bisphosphate Carboxylase/Oxygenase (Caspi et al. 1996), the primary CO_2-fixing enzyme found in aerobic chemolithoautotrophic bacteria. Although no enzyme activity could be detected in the strain, the genes are fully functional in *Escherichia coli* (Caspi et al. 1996). Nealson and coworkers (1988) proposed that several axioms must be established in order to prove that Mn(II) provides energy for autotrophic growth: (1) the true oxidation rate (rather than Mn^{2+} binding rate) needs to be determined and it should correlate with the rate of CO_2 fixation and growth; (2) the oxidation state and structure of the Mn minerals must be known; and (3) the concentration of substrates, Mn^{2+} and O_2, and the pH must be known. Furthermore, cell-free extracts should possess both Mn(II) oxidation activities and the enzymes associated with CO_2 fixation. In no case have all these criteria been fulfilled.

A number of alternative benefits that microbes may receive by catalyzing Mn(II) oxidation have been proposed. The oxidation of Mn(II) may be a mechanism whereby microorganisms convert soluble, toxic Mn(II) into extracellular particulate MnO_x, or Mn(II) oxidation may help protect the cell from toxic oxygen species (Ghiorse 1984b). The metal-encrusted surfaces of Mn(II)-oxidizing bacteria may also provide increased resistance to ultraviolet radiation, predation and viral attack (Emerson 1989). Mn oxides may scavenge trace metals required for growth due to their adsorptive capacity (Nealson et al. 1988) or immobilize toxic metals by adsorption. For the particular case of the marine *Bacillus* sp. strain SG-1, whose spores oxidize Mn(II) while the vegetative cells are known to reduce Mn oxides, it has been suggested that the oxides may function as electron acceptors during germination under anaerobic conditions (Tebo 1983, de Vrind et al. 1986b).

It was recently proposed that the oxidation of Mn(II) by microorganisms in humic-rich environments may be a mechanism whereby bacteria can indirectly utilize the large, biologically recalcitrant pools of carbon contained in humic substances (Sunda and Kieber 1994). In most oxidizing environments, the dissolved natural organic carbon pool consists primarily of biologically refractory compounds such as humic and fulvic acids. Sunda and Kieber (1994) showed that Mn oxides abiotically oxidize humic and fulvic acids and release low molecular weight organic compounds such as pyruvate, acetone, formaldehyde, and acetaldehyde as products (Sunda and Kieber 1994). Pyruvate was the major low molecular weight product identified in this study. Interestingly, pyruvate is also the major product of photolysis of humic compounds by ultraviolet radiation, and its photochemical production in seawater has been shown to be closely coupled to its use by the microbiota in seawater (Kieber et al. 1989). There have been no published investigations that show enhanced utilization of natural organic matter by Mn(II)-oxidizing bacteria as a result of the oxidation of the organics by the microbially-produced Mn oxides, although results of growth experiments with *L. discophora* growing in humic media suggest that this may be so (W. Ghiorse and P. Hirsch, unpublished observations). Recent work has also suggested that some marine Mn(II)-oxidizing bacteria obtain carbon and energy via this mechanism (Tebo et al. 1997). Clearly, this interesting hypothesis has profound implications for our understanding of the role of Mn(II)-oxidizing bacteria in the biogeochemical cycling of carbon.

Mn(II)-oxidizing bacteria are phylogenetically diverse. The phylogeny based on sequencing the 16S small subunit ribosomal genes indicates that the Mn(II) oxidizers sequenced to date all fall within the Gram-positive or Proteobacteria branches of the Domain Bacteria (Fig. 2). The Gram-negative Mn oxidizers sequenced so far fall within the α, γ, and β Proteobacteria. A variety of fungi have also been reported to oxidize Mn(II) (Glenn et al. 1986, Emerson et al. 1989, Golden et al. 1992, Perez and Jeffries 1992, Souren 1997).

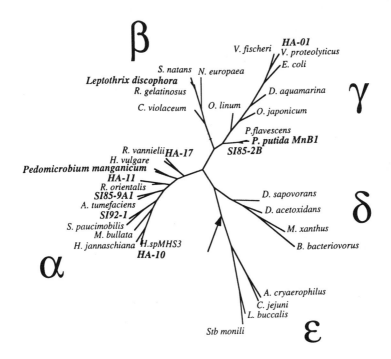

Figure 2. Phylogenetic tree of Proteobacteria based on ribosomal small subunit (16S) genes showing the diversity of Mn(II)-oxidizing bacteria (boldface). Many recently isolated Mn(II)-oxidizing bacteria from the marine environment, designated by strain number only (e.g. HA-01 or SI85-9A1), probably represent news species of bacteria. The tree was constructed by maximum parsimony using the branch and bound search algorithm in PAUP (Swofford 1990). The arrow indicates the outgroup, *Chlamydia psittaci*.

The mechanisms of Mn(II) oxidation by microorganisms have not been conclusively determined and are a matter for debate. One speculation is that microorganisms may first enzymatically oxidize Mn(II) to Mn(III), which remains bound to the Mn(II)-oxidizing enzyme ({Mn^{3+}}; Ehrlich 1996a,b):

$$2 Mn^{2+} + \frac{1}{2} O_2 + 2 H^+ \Rightarrow 2 \{Mn^{3+}\} + H_2O. \tag{1}$$

The enzyme then catalyzes an oxidation of the enzyme-bound Mn(III) to Mn(IV), which precipitates as MnO_2:

$$2 \{Mn^{3+}\} + 3 H_2O + \frac{1}{2} O_2 \Rightarrow 2 MnO_2 + 6 H^+. \tag{2}$$

So far no experimental results have been obtained to substantiate this hypothesis. In fact, if we calculate the $\Delta G^{o\prime}$ (pH 7) for Reaction (1) assuming free Mn^{3+} is formed (i.e. using the Mn^{3+}/Mn^{2+} redox couple), the reaction is highly unfavorable. For this reaction to proceed, the Mn(III)–enzyme complex must have a redox potential sufficiently lower than the Mn^{3+}/Mn^{2+} couple, and lower than the O_2/H_2O couple, or the reaction would require an input of energy (i.e. coupled to another reaction).

An alternate two step mechanism that seems more feasible from a thermodynamic perspective is the oxidation of Mn^{2+} by hydrogen peroxide to Mn^{3+} which remains bound to the enzyme, and the subsequent oxidation of Mn^{3+} to MnO_2 by O_2:

$$2 Mn^{2+} + H_2O_2 + 2 H^+ \Rightarrow 2 \{Mn^{3+}\} + 2 H_2O \tag{3}$$

$$2 \{Mn^{3+}\} + O_2 + 4 H_2O \Rightarrow 2 MnO_2 + H_2O_2 + 6 H^+. \tag{4}$$

Although the H_2O_2/H_2O redox couple ($E'_0 = 1.35$ v) is lower than the Mn^{3+}/Mn^{2+} couple ($E'_0 = 1.54$ V), when Mn^{3+} is complexed with an enzyme, the Mn^{3+}/Mn^{2+} redox potential may be sufficiently low for H_2O_2 to be the oxidant. In fact, a reaction similar to Equation (3) is catalyzed by the fungal enzyme, Mn peroxidase (see below, Aitken and Irvine 1990). One difference is that, instead of Mn(III) remaining bound to the fungal enzyme, Mn^{3+} is complexed by organic acids and thereby stabilized in solution.

Recent results (van Waasbergen et al. 1996, Corstjens et al. 1997) indicate that two completely different species of Mn(II)-oxidizing bacteria appear to possess Mn(II)-oxidizing proteins that are related to the multicopper oxidase family of proteins (see below). In the multicopper oxidases, multiple copper ions serve as redox cofactors for the oxidation of a variety of different substrates. Three types of copper ions based on their spectroscopic properties can be distinguished in multicopper oxidases. The type 1 or 'blue' copper is usually involved in a one electron oxidation of a substrate. The type 1 copper then transfers the electron to the type 2 and type 3 coppers ions which form a cluster within the protein and are involved in the binding and reduction of molecular oxygen. Multicopper oxidases and cytochrome oxidases are the only proteins known to catalyze the four-electron reduction of O_2 to H_2O. Although it is possible that Mn(II) oxidation could occur by two electrons, the well-characterized multicopper oxidases oxidize their substrates by one electron transfers. Thus, it seems likely that Mn(II) oxidation by these proteins proceeds in successive one electron steps in which Mn^{3+} is only a transient intermediate. In this case, a variety of oxygen intermediates could serve as the oxidant and oxygen source to the Mn oxides.

On the other hand, it is theoretically possible that Mn(II) is oxidized microbially to Mn(IV) in a single step without intermediate formation of Mn(III):

$$Mn^{2+} + \frac{1}{2} O_2 + H_2O \Rightarrow MnO_2 + 2 H^+. \tag{5}$$

Ehrlich (1996b) suggests, however, that this one-step Mn(II) oxidation mechanism is not very likely because the two electrons removed from Mn(II) are not equivalent energetically. Nevertheless, many enzymes catalyze two-electron transfer reactions, so this one-step Mn(II) oxidation mechanism is entirely feasible.

Several species of fungi possess a number of enzymes that are capable of oxidizing Mn(II) (Glenn et al. 1986, Perez and Jeffries 1992, Souren 1997). Because many of these enzymes are important in the degradation of wood products (depolymerization of lignin), bleaching of wood pulp, and for bioremediation of xenobiotics, the mechanism of fungal Mn(II) oxidation is in many ways better understood (Barr and Aust 1994). Fungi produce at least two types of extracellular enzymes capable of oxidizing Mn(II): the peroxidases (including Mn peroxidase and lignin peroxidase) and laccase (Souren 1997, Tebo 1997). While Mn(II) oxidation by lignin peroxidase and laccase seems to be a side reaction, Mn peroxidase is absolutely dependent on Mn(II) for activity (Aitken and Irvine 1990). Mn(II) is also known to regulate the production of these three enzymes in the white rot fungi (Perez and Jeffries 1992). These enzymes catalyze the oxidation of Mn(II) to Mn(III) which is chelated by organic acids. The Mn(III) complex then diffuses away from the active site of the enzyme and acts as a diffusible oxidant for a variety of different phenolic compounds. The general mechanism of Mn peroxidase (MnP) can be written as

$$MnP + H_2O_2 \Rightarrow compound\ I \tag{6}$$

$$compound\ I + Mn(II) \Rightarrow compound\ II + Mn(III) \tag{7}$$

$$compound\ II + Mn(II) \Rightarrow MnP + Mn(III). \tag{8}$$

The overall reaction can be written as

$$2\ Mn^{2+} + H_2O_2 + 2\ H^+ \Rightarrow 2\ Mn^{3+} + 2\ H_2O \tag{9}$$

which is essentially identical to the Reaction (4) proposed to occur in bacteria. Organic acids play an important role in this scheme because they stabilize Mn(III) and facilitate its diffusion from the active site, thereby stimulating MnP activity.

The reactions carried out by lignin peroxidase (LiP) and laccase are similar to MnP except that Mn(II) is not absolutely essential and the presence of an organic (phenolic) substrate is required (Popp et al. 1990, Archibald and Roy 1992). In these reactions Mn(III) is generated indirectly through production of phenolic radicals or superoxide. Like the MnP reaction, these reactions are favored by the presence of a suitable organic compound to complex the Mn(III) that is formed. In all the fungal Mn(II) oxidation reactions the primary product is not Mn oxide, but rather a soluble Mn(III) chelate that is eventually reduced back to Mn(II) upon reaction with organic matter (e.g. lignin). However, in the absence of suitable complexing agents the Mn(III) that is formed can disproportionate and Mn precipitates as MnO_2 (Glenn et al. 1986, Perez and Jeffries 1992, Souren 1997). By analogy, perhaps bacterial Mn(II) oxidation may, in some cases, proceed by a similar pathway and serve a similar role.

RATES OF MN OXIDATION / MINERAL FORMATION

Although abiotic Mn(II) oxidation occurs on mineral surfaces, such as Fe and Mn oxides and silicates (Davies and Morgan 1989, Junta and Hochella 1994), biological processes seem to be responsible for the majority of Mn(II) oxidation that occurs in natural waters and sediments (Nealson et al. 1988, Tebo 1991, Thamdrup et al. 1994). Rates of Mn(II) oxidation follow Michaelis-Menten kinetics (Tebo and Emerson 1986, Sunda and Huntsman 1987, Moffett 1994b) and are extremely variable from environment to environment. For example, the rates of Mn(II) oxidation in different marine waters ranges from extremely low values, of around 2 $pM \cdot h^{-1}$ in deep-sea hydrothermal vent plumes (Cowen et al. 1990, Mandernack and Tebo 1993), to greater than 65 $nM \cdot h^{-1}$ at the oxic/anoxic interface in nearshore waters of the Black Sea (Tebo 1991). Mn(II) oxidation rates in freshwaters can be even higher, with reported values ranging upwards of 350 $nM \cdot h^{-1}$ (Tipping 1984). As a consequence of these high rates of Mn(II) oxidation, the residence time of dissolved Mn(II) in most natural waters is usually on the order of hours to days. However, in certain environments such as oceanic surface waters where photodissolution of Mn oxides occurs (Sunda et al. 1983, Sunda and Huntsman 1988), or deep sea buoyant hydrothermal plumes with high pressures and cold temperatures, the residence times can be on the order of weeks to months, or even years.

A variety of different parameters including oxygen, temperature, pH, light, and metal concentration affect the rate of microbial Mn oxidation in the environment, (Tebo et al. 1984, Tebo and Emerson 1985, Sunda and Huntsman 1987, Richardson et al. 1988, Sunda and Huntsman 1988). As a result, Mn(II) oxidation rates vary diurnally and seasonally (Sunda and Huntsman 1990, Miyajima 1992, Thamdrup et al. 1994). Whereas all experimental evidence so far suggests that Mn(II) oxidation is an obligately aerobic process (Nealson et al. 1988), in certain environments, such as the suboxic zone of the Black Sea, chemical profiles suggest that Mn(II) (and NH_4^-) could be oxidized anaerobically (Codispoti et al. 1991, Lewis and Landing 1991, Tebo 1991). From thermodynamic considerations Mn(II) oxidation to Mn(IV) oxides could be coupled to denitrification (NO_3^- to N_2). There is some experimental data that suggest this may occur in sand filters where nitrate additions were shown to stimulate Mn(II) removal (Vandenabeele et al. 1995). Surprisingly, the process was not inhibited by oxygen, although there are a number of reports that suggest denitrification might occur in the presence of low concentrations of oxygen (see discussion in Vandenabeele et al. 1995).

As is typical of biological processes, environmental measurements of Mn(II)-oxidizing activity display distinct temperature and pH optima (Tebo and Emerson 1985, Sunda and Huntsman 1987). Mn(II) oxidation generally increases with small increases in Mn(II) concentration but is frequently inhibited by high Mn(II) concentrations (Chapnick et al. 1982, Tebo et al. 1984, Tebo and Emerson 1985). High light intensity has also been shown to inhibit rates of Mn oxide formation because light inhibits Mn(II)-oxidizing bacteria and causes photodissolution of Mn oxides (Sunda et al. 1983, Sunda and Huntsman 1988, Sunda and Huntsman 1990).

MINERALOGY OF MICROBIALLY-PRODUCED MANGANESE OXIDES

The question arises whether there are any distinctions in the Mn oxides formed abiotically from those formed through microbial action. Whereas the Mn oxidation states in synthetic Mn oxides formed chemically in the laboratory at neutral pH are low, 3.1 (Hem 1981, Murray et al. 1985, Junta and Hochella 1994), in many natural waters microbial activities lead to the formation of Mn oxides with Mn oxidation states significantly higher, often greater than 3.4 (Kalhorn and Emerson 1984, Tebo et al. 1984, Tipping et al. 1984, Tipping et al. 1985, Friedl et al. 1997). It is tempting to speculate that abiotic oxidation does not result in the same oxidation products as those found in nature (Tipping 1984).

Since environmental conditions of pH, pE, Mn(II) concentration, and temperature affect the Mn minerals that form (Hem and Lind 1983, Hastings and Emerson 1986, Mandernack et al. 1995b), it is not surprising that a variety of different Mn oxidation states or mineral forms have been observed in microbial precipitates (Ferris et al. 1987, Adams and Ghiorse 1988, Mann et al. 1988, Takematsu et al. 1988, Mita et al. 1994,

Mandernack et al. 1995b). Because most Mn oxides produced microbially (including those found in nature) are difficult to identify by X-ray or electron diffraction analyses, relatively little is known about the mineralogy of microbially produced Mn oxides. In several cases todorokite has been tentatively identified as a mineral of microbial origin (Jannasch and Mottl 1985, Ferris et al. 1987, Takematsu et al. 1988, Beveridge 1989, Mandernack et al. 1995b).

Using *Bacillus* sp. strain SG–1 spores as a model system (see below) for the microbial production of Mn oxides, Hastings and Emerson (1986) observed the formation of Mn(IV) minerals in seawater (pH 7.5) containing moderate (2 μM) concentrations of Mn(II). These authors observed hausmannite (Mn_3O_4) as an initial precipitate in some of their experiments and gradual increases in the oxidation state with time. These results were consistent with Hem and Lind's model (1983) that Mn(IV) minerals form by a two step process involving first the formation of Mn_3O_4 which subsequently disproportionates under low Mn(II) concentrations (Hastings and Emerson 1986).

Mandernack et al. (1995b), noting that Mn(IV) minerals were reported to form at Mn(II) concentrations too high to favor Mn_3O_4 disproportionation (Takematsu et al. 1988, Greene and Madgwick 1991), extended the studies of Hastings and Emerson (1986) and more systematically examined the Mn oxidation state and mineralogy of the Mn oxides formed in low ionic strength buffer and in natural seawater under various environmentally-relevant temperatures and Mn(II) concentrations. Experiments were conducted with SG–1 spores in HEPES buffer and HEPES buffered seawater, at pH 7.4-8.0, in 10 μM to 10 mM Mn(II), and at 3°C, 25°C, and 50° to 55°C. After two weeks at high Mn(II) concentrations or high temperature, mixed phases of lower valence state minerals (Mn_3O_4; feitknechtite, β-MnOOH; and manganite, γ-MnOOH) formed in both seawater and buffer (Fig. 3). At lower Mn(II) concentrations Mn(IV) minerals precipitated. In low ionic strength buffer the Mn(IV) minerals most often resembled sodium buserite, as evidenced by collapse of a 10-Å to 7-Å phase upon drying (Fig. 4). In seawater, both buserite and noncollapsible 10-Å manganates formed (Fig. 5). In general, the Mn oxidation states of the minerals formed in seawater were higher than those formed in the buffered medium. In many cases Mn(IV) mineral formation occurred at pH and Mn(II) concentrations where the lower valence state minerals (Mn_3O_4, MnOOH) are thermodynamically stable with respect to disproportionation. These results suggested that SG–1 spores are capable of oxidizing Mn(II) to Mn(IV) without the formation of lower valence state solid phase intermediates.

That Mn(II) could be oxidized directly to Mn(IV) oxides by SG–1, as well as another marine Mn(II)-oxidizing bacterium (strain SI85-9A1), was more elegantly demonstrated by oxygen isotopic ($\delta^{18}O$) studies of chemically and microbially produced Mn oxides (Mandernack et al. 1995a). Theoretically, the maximum amount of oxygen in Mn oxides that is derived from O_2 can be estimated from the stoichiometry of the Mn(II) oxidation reactions (see Table 2). Thus, for oxidation reactions involving the formation of lower (II,III) Mn oxide intermediates, the maximum amount of oxygen that could be derived from O_2 is 25%. Since no further O_2 is incorporated upon disproportionation of these oxides to Mn(IV), the maximum signal from O_2 in Mn(IV) oxides formed by this two step pathway is 25%. In contrast, Mn(IV) oxides formed directly (Reaction 1 in Table 2) could have up to 50% of the oxygen derived from O_2. Mandernack et al. (1995a) showed that hausmannite (Mn_3O_4), synthesized either chemically or microbially, did not incorporate oxygen atoms from molecular O_2. Since disproportionation of Mn_3O_4 also would not incorporate oxygen atoms from O_2 (see Reaction 5 in Table 2), a Mn(IV) oxide formed by this pathway would have a $\delta^{18}O$ signature solely reflecting the water in which it formed. In contrast, the Mn(IV) oxides produced by both SG–1 and strain

236   GEOMICROBIOLOGY

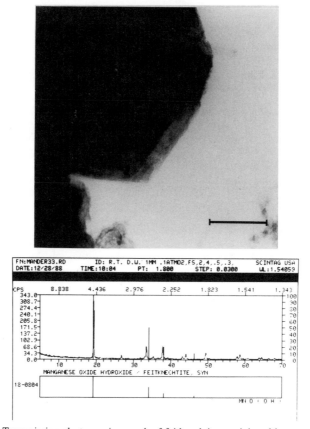

Figure 3. Top: Transmission electron micrograph of feitknechtite precipitated by spores of the marine *Bacillus* sp. strain SG–1 (bar = 0.1 μm). Bottom: Representative powder XRD pattern of the feitknechtite minerals produced by SG–1. For comparison, the feitknechtite index pattern from the Joint Committee of Powder Diffraction Studies (JCPDS) is shown beneath the bacterially-produced mineral. Units of left and right vertical axes are counts per second and % relative intensity, respectively; units of top and bottom horizontal axes are ångströms (Å) and °2θ, respectively. [Used by permission of the editor of *Geochimica Cosmochimica Acta*, from Mandernack et al. (1995b), Appendix A, p 4406.]

Figure 4. Opposite page. Top: Transmission electron micrograph of a buserite-like Mn(IV) manganate precipitated by spores of the marine *Bacillus* sp. strain SG–1 in HEPES buffer (bar = 0.1 μm). Middle: XRD pattern from a slow scan of the buserite-like mineral dried a room temperature. Bottom: XRD patterns from fast scans of SG–1 Mn(IV) manganates both wet (see 10 Å peak of pattern labeled "MANDER9") and dry (7 Å peak of pattern labeled "MANDER67"). Note collapse of 10-Å buserite to 7-Å birnessite. For the XRD patterns, units of left and right vertical axes are counts per second and % relative intensity, respectively; units of top and bottom horizontal axes are ångströms (Å) and °2θ, respectively. [Used by permission of the editor of *Geochimica Cosmochimica Acta*, from Mandernack et al. (1995b), Fig. 1, p 4402.]

Figure 5. Second next page. Top: Transmission electron micrograph of a todorokite-like Mn(IV) manganate precipitated by spores of the marine *Bacillus* sp. strain SG–1 in seawater (bar = 0.1 μm). The circular void space resulted from extraction of the bacterial organic material. Middle: XRD pattern from a slow scan of the todorokite-like mineral dried a room temperature. Bottom: XRD patterns XRD patterns of the todorokite-like mineral dried at room temperature, 100°, and 150°C. For the XRD patterns, units of left and right vertical axes are counts per second and % relative intensity, respectively; units of top and bottom horizontal axes are ångströms (Å) and °2θ, respectively. [Used by permission of the editor of *Geochimica Cosmochimica Acta*, from Mandernack et al. (1995b), Fig. 2, p 4402.]

Figure 4.
Caption on
opp. page.

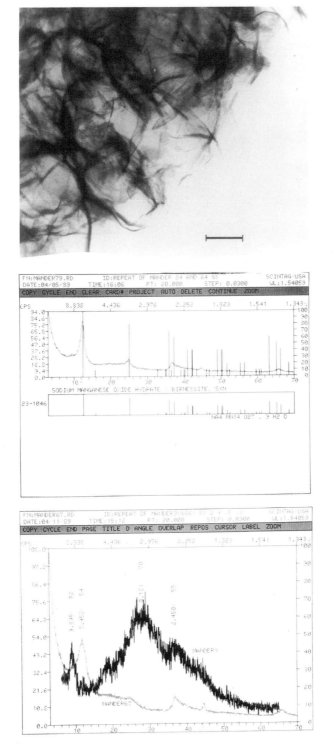

Figure 5. Caption on second previous page.

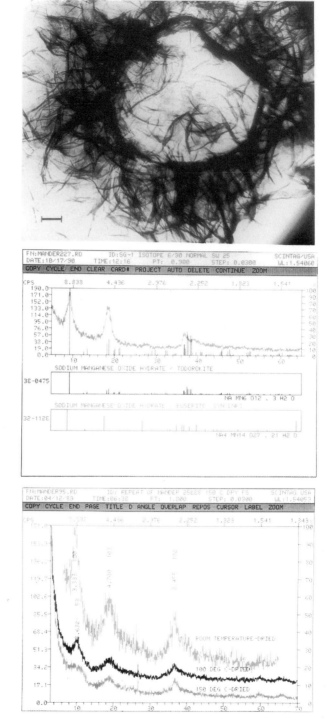

SI85-9A1 showed that 40 to 50% of the oxygen in the Mn(IV) minerals is derived from O_2, consistent with the oxidation of Mn(II) to MnO_x by Reaction (1) (Table 2). These results suggest that Mn(II) oxidation by these bacteria occurs via a metal-centered oxidation in which Mn(II) is directly attacked by dissolved O_2 (or an intermediate derived from O_2) and electron transfer proceeds via an inner-sphere mechanism, as has been proposed on theoretical grounds (Luther III 1990, Wehrli 1990).

Environmental studies lend support to the laboratory studies that suggest that microbial Mn(II) oxidation results in the formation of Mn(IV) oxides without lower valence intermediates. Through analysis of a lake sediment, Wehrli et al. (Wehrli et al. 1995) found that the sediment is dominated by vernadite ($\delta-MnO_2$), an X-ray-amorphous Mn(IV) oxide, and contained no reduced forms of Mn oxides (Mn_3O_4), hydroxides [$(Mn(OH)_2$], or oxyhydroxides (MnOOH), as revealed by X-ray absorption spectroscopy. They concluded that the direct oxidation of Mn(II) to Mn(IV) by microorganisms provides an efficient transfer of oxidizing equivalents to the sediment surface. More recently Friedl et al. (1997) examined the Mn oxide microbially formed at the oxic/anoxic boundary in the water column of a eutrophic lake. These authors identified the Mn mineral by extended X-ray absorption fine structure (EXAFS) spectroscopy as H^+–birnessite. The rapid formation of particulate Mn with high Mn oxidation state (> 3) above the oxic/anoxic interface of lakes and fjords also supports the direct Mn(II) → Mn(IV) mechanism (Kalhorn and Emerson 1984, Tebo et al. 1984, Tipping et al. 1984, Tipping et al. 1985). Measurements of the $\delta^{18}O$ composition of Mn oxides in a freshwater ferromanganese nodule from Lake Oneida, New York indicate that 50% of the oxygen atoms in the Mn oxides are derived from O_2 providing evidence that they were microbially formed by the direct pathway without lower Mn oxidation state solid phase intermediates (Mandernack et al. 1995a).

MODEL SYSTEMS FOR THE STUDY OF GENETICS AND BIOCHEMISTRY OF MN(II) OXIDATION

In the last 5 to 6 years modern molecular biological approaches have been applied to gain an understanding of the mechanism(s) of bacterial Mn(II) oxidation. To date, significant progress has been made in three phylogenetically distinct Mn(II)-oxidizing bacteria, *Pseudomonas putida* strain MnB1, marine *Bacillus* sp. strain SG–1, and *Leptothrix discophora* strain SS-1 (see Fig. 2). This section reviews the state of knowledge of the mechanisms and possible functions of Mn(II) oxidation in these bacteria.

Pseudomonas putida strain MnB1

Many Mn(II)-oxidizing bacteria of the genus *Pseudomonas* have been isolated from different environments, including fresh- and salt-water, and soils (Zavarzin 1962, Jung and Schweisfurth 1976, Douka 1977, Nealson 1978, Schutt and Ottow 1978, Douka 1980, Gregory and Staley 1982, Kepkay and Nealson 1987a, Abdrashitova et al. 1990). The best studied Mn(II)-oxidizing *Pseudomonas* sp. to date is *P. putida* strain MnB1, a fresh water bacterium that was isolated in Trier, Germany (Schweisfurth 1973) from a Mn crust that accumulated in drinking-water pipes. It was first named *P. manganoxidans*, but was recently found to be a strain of the ubiquitous soil and freshwater species *P. putida* (DePalma 1993, Caspi 1996). MnB1 is an attractive choice for a model system for Mn(II) oxidation because *P. putida* has an established set of tools for genetic manipulation and is fairly well understood biochemically.

Biochemistry. P. putida strain MnB1, upon reaching stationary phase, oxidizes Mn(II) in liquid and solid media to Mn(IV) oxides that precipitate on the cell surface.

Within two days of growth on a Mn(II) containing agar, the colonies, which are originally cream colored, become brown. Jung and Schweisfurth, who isolated the organism, reported Mn(II) oxidation in cell-free extracts of strain MnB1 that was expressed at the onset of stationary phase. The oxidation appeared to be non-catalytic, and did not require O_2 in the reaction (Jung and Schweisfurth 1979). The Mn(II)-oxidizing activity was heat labile, and could be inactivated by protease treatment. They therefore attributed the activity to a Mn(II)-oxidizing protein. The oxidation activity did not depend on presence of Mn(II) in the growth medium prior to the assay, and seemed to depend mainly on the nutritional condition of the culture. Cells had to be starved for 2 hours prior to Mn(II) oxidation assays, in order to induce Mn(II) oxidation. This phenomenon has been demonstrated before with different Mn(II)-oxidizing bacteria (Bromfield 1956, van Veen 1972). Mn(II) oxidation was detected only in the soluble fraction of cell-free extracts, indicating that the activity is not associated with the cell wall. The assays originally performed by Schweisfurth were repeated by DePalma (DePalma 1993) and the results were somewhat different. DePalma found that oxygen was indeed required for Mn(II) oxidation, and that Mn(II) oxidation did display a catalytic activity, suggesting an enzyme was responsible. DePalma confirmed that the Mn(II) oxidation activity was preferentially induced by multiple-nutrient or glucose starvation. It should be noted that these earlier works did not identify a protein that was specifically responsible for Mn(II) oxidation. Rather, it was an interpretation by the authors of their findings.

Genetics. In a recent study, Caspi et al (Caspi 1996, Caspi et al., submitted) used transposon mutagenesis to obtain mutants of strain MnB1 that have lost the ability to oxidize Mn(II). In transposon mutagenesis, a transposon (transposable genetic element) is inserted randomly into different loci in the chromosome of the organism. The insertions disrupt the nucleotide sequence of the chromosome, thus mutating and/or inactivating genes at the location of insertion or downstream from the insertion point in the case of an operon (genes that are co-transcribed). The transposon also serves as a tag for easy identification of the mutated gene. Mutagenesis was performed so that only one transposon was inserted into each mutant and a large number of mutants were generated, ideally representing the whole genome (except for lethal mutations). The resulting mutants were screened for loss of Mn(II) oxidation, and about 30 non-Mn(II)-oxidizing mutants were isolated. The mutated genes in 21 mutants were cloned and partially sequenced, and the partial sequences were used to identify the genes based on similarity to genes in the Genbank database.

The mutants could be divided into two general groups, based on their response (positive or negative) to a cytochrome c oxidase assay. (Cytochrome c oxidase is the terminal enzyme complex that transfers electrons from the electron transport chain to O_2 resulting in the reduction of O_2 to water. Electron transport chains, which are central to aerobic and anaerobic respiration, are composed of membrane-associated electron carriers that function to transfer electrons from electron donors (energy sources) to a terminal electron acceptor. Whereas the carriers can vary among different organisms and types of metabolism, the electron transport chains generally function to conserve some of the energy released during the electron transfer reactions for synthesis of ATP, the cellular energy currency.) While the wild type was clearly positive for cytochrome c oxidase activity, 17 out of the 21 mutants were negative. These mutants could be further classified into three sub-groups, namely the cytochrome c biosynthesis (*cyc*) group (5 mutants), the TCA cycle group (9 mutants) and the pyrimidine biosynthesis (*pyr*) group (3 mutants).

The *cyc* group mutants had insertions in different parts of an operon that is crucial for the assembly of mature c-type cytochromes (Fig. 6). These mutants tested negative in a cytochrome c oxidase assay, and spectrophotometric studies indicated that they are lacking *c*-type cytochromes. When a fragment of the chromosome containing an unmodified version of this operon was introduced into one of these mutants, cytochrome *c* oxidase activity, as well as Mn(II) oxidation activity, were restored. The TCA cycle group mutants had insertions in genes (*sdhA, sdhB, sdhC, aceA* and *icd*) encoding different key enzymes of the TCA cycle, a central biochemical pathway that is involved in the break down of metabolites and the generation of energy and biosynthetic precursors in the cell. Interestingly, while all these mutants were lacking the activity of the particular enzyme encoded by the disrupted genes, eight of them also tested negative in the cytochrome *c* oxidase assay, and spectrophotometric studies showed that they were lacking *c*-type cytochromes. The pyrimidine biosynthesis mutants had insertions in genes that are involved in the biosynthesis of pyrimidines (*carB* and *pyrE*). These mutants were cytochrome *c* oxidase negative as well. Only four mutants were cytochrome *c* oxidase positive. Two of those had insertions in the *trpE* gene, which encodes anthranilate synthetase, an enzyme involved in the biosynthesis of the amino acid tryptophan.

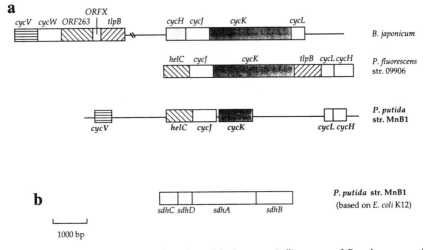

Figure 6. (a) The cytochrome *c* (*cyc*) and succinate dehydrogenase (*sdh*) operons of *Pseudomonas putida* strain MnB1 aligned to similar operons in *Bradyrhizobium japonicum* and *Pseudomonas fluorescens* strain 09906. Regions of MnB1 that have been sequenced are boxed. (b) Organization of the *sdh* operon in MnB1 based on sequence from *E. coli* K12. Figure redrawn from Caspi et al. (submitted).

These findings give genetic support to an old hypothesis: a correlation between Mn(II) oxidation and the electron transfer chain (in this case, through *c*-type cytochromes). In the case of the *cyc*- mutants, which can not synthesize any *c*-type cytochromes, the correlation is obvious. As for the other cytochrome *c* oxidase negative mutants, it was speculated that Mn(II) oxidation deficiency in these mutants was actually a secondary phenomenon: the transposon mutations in these mutants impaired a central pathway, which abolished, among other functions, the synthesis of *c*-type cytochromes, and it was the lack of cytochromes that was responsible for loss of Mn(II) oxidation ability (Caspi 1996). It is easiest to explain this idea for the TCA cycle mutants: all cytochromes contain heme, and the first committed step in heme biosynthesis is the synthesis of α-levulinate (Ferreira and Gong 1995). In *Pseudomonas* α-levulinate is synthesized from glutamate (Avissar et al. 1989), which in turn is synthesized largely from 2-oxoglutarate, an intermediate of the TCA cycle. It is possible that the shut-down of the TCA cycle by the elimination of essential enzymes such as pyruvate dehydrogenase, isocitrate dehydrogenase, and succinate dehydrogenase deplete the concentration of available 2-oxoglutarate to a degree that heme biosynthesis would stop or be greatly diminished. Under these conditions the bacteria will not be able to synthesize active c-type cytochromes.

If electrons from Mn(II) are indeed channeled through the respiratory chain, then the oxidation of Mn(II) would result in the conservation of energy. While strain MnB1 is clearly not an autotroph, it is possible that it is capable of mixotrophic growth, utilizing Mn(II) as an energy source when organic carbon resources become limiting. As noted before, Mn(II) oxidation activity in strain MnB1 is preferentially induced by multiple-nutrient or glucose starvation.

Mn(II) oxidation and cytochromes. The notion that c-type cytochromes are involved in Mn oxidation is not new. Hogan (1973) investigated electron transport in *L. discophora*, and found that the bacteria possess cytochromes of the a, b and c types. Rotenone, an inhibitor of the oxidation of NAD-linked substrates, as well as urethane and antimycin A, inhibitors of cytochrome b, did not inhibit Mn oxidation, while cytochrome c oxidase inhibitors such as cyanide and azide did. This indicates that the entry point of electrons from Mn to the system must be above the cytochrome b level. The author investigated the possibility of a direct participation of cytochrome c in the oxidation of Mn, but the reduction of c-type cytochromes by Mn(II) could not be demonstrated.

Somewhat conflicting results were obtained by Ehrlich (1976), who reported that Mn(II) oxidation by a Gram-negative marine strain (*Oceanospirillum* strain BIII 45), was inhibited by dicumarol (an inhibitor of coenzyme Q) and antimycin A, as well as azide and cyanide, suggesting that in that organism the entry point of electrons from Mn(II) must be at a much lower level (see below for a discussion of the redox potential problem). Arcuri and Ehrlich (1979) reported the reduction of c-type cytochromes by Mn(II) in the same strain and a similar one (*Oceanospirillum* strain BIII 82). However, the c-type cytochromes appeared to have been reduced by an indigenous electron donor as soon as azide was added to the samples, and the addition of Mn(II) merely enhanced this phenomenon. A complication in this work was the fact that neither of these strains was capable of oxidizing Mn(II) unless it was bound to the surface of Mn oxides which had to be present in all assays. Graham et al. (1987) isolated several components of the electron transfer chain of another marine Mn(II)-oxidizing isolate (strain SSW$_{22}$). This strain has an inducible Mn(II) oxidation system that was inhibited by rotenone, antimycin A, 2-n-nonyl-4-hydroxyquinoline-N-oxide (NOQNO), cyanide and azide (Ehrlich 1983). The authors isolated two soluble c-type cytochromes, a soluble non-heme iron protein, and a membrane-bound cytochrome bc-type complex. However, none of these components was induced in Mn-grown cultures, which does not agree with the inducible nature of Mn(II) oxidation in this strain. The authors suggested that a key enzyme of the system may not have been isolated.

Possible mechanisms for the oxidation of Mn(II) by strain MnB1. In order to speculate on possible mechanisms for Mn(II) oxidation in which electrons from Mn(II) are passed to the electron transfer chain, generating energy in the process, we need to address two key problems. The first is the question of the available energy in Mn(II) oxidation. The reaction that is traditionally proposed for the direct oxidation of Mn(II) to Mn(IV) (as suggested from $\delta^{18}O$ data for other bacteria) is

$$Mn^{2+} + {}^1/_2 O_2 + H_2O \Rightarrow MnO_2 + 2\,H^+ \tag{5}$$

The standard free energy at pH 7 ($\Delta G°´$) of this reaction is $= -70.9$ kJ/mole. This energy, while small, is not unusual for the oxidation of inorganic energy sources, and is enough to generate two moles of ATP per mole Mn^{2+} oxidized, assuming more than 80% efficiency. The hypothesis that Mn(II) oxidation can result in generation of ATP in bacteria is supported by the findings of Ehrlich and Salerno (1990), who performed experiments with everted membrane vesicles from several marine isolates, and showed that Mn(II) oxidation can be coupled to ATP production.

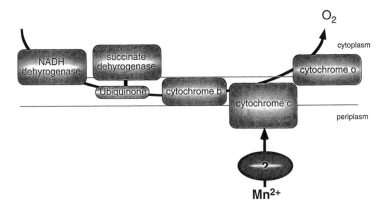

Figure 7. Hypothetical electron transport chain involved in Mn(II) oxidation in *P. putida* strain MnB1. Electrons from Mn(II) are hypothesized to enter the electron transport chain at the level of cytochrome *c*. Mn(II) may interact either directly with cytochrome *c* or via an intermediate carrier molecule (shown as '?).

The second question concerns the redox potential of the electron donors and acceptors. Typically, the respiratory chains of heterotrophic organisms obtain electrons at low redox potential from primary dehydrogenases in which $NADH_2$ or $FADH_2$ serve as coenzymes, and channel them through components that have standard redox potentials ($E^{\circ\prime}$, pH 7) in the range of -400 mV to +400 mV (Fig. 7 and Table 3). However, the $\delta MnO_2/Mn^{2+}$ redox couple has a high positive potential ($E^{\circ\prime}$ = 406 mV), and a respiratory chain that interacts with these electrons faces a redox problem.

A possible solution to this problem may be similar to the one used by *Nitrobacter*, which oxidizes nitrite to nitrate. In that case, electrons from nitrite are transferred to a *c*-type cytochrome, which passes them to a cytochrome *c* oxidase. Even though the standard redox potential of the *c*-type cytochrome is lower than that of the nitrate/nitrite pair, the high oxidase activity keeps the cytochrome *c* almost completely oxidized, and this shifts up the cytochrome's redox potential, so it becomes actually higher. This process is described by the Nernst equation:

$$\Delta E_h{}^\prime = \Delta E_m{}^\prime + \frac{RT}{nF} \frac{[ox]}{[red]}$$

where $E_h{}^\prime$ is the actual redox potential at pH = 7, $E_m{}^\prime$ is the midpoint redox potential at pH = 7, R is the gas constant, T is the absolute temperature, n is the number of electrons transferred, F is the Faraday constant, [ox] is the concentration of the oxidized species and [red] is the concentration of the reduced species.

Another solution to the redox problem would be to oxidize Mn(II) in a one electron transfer reaction, to a Mn(III) intermediate. Such an intermediate could theoretically be either the manganic ion Mn^{3+}, or a mineral form, such as manganite (MnOOH) and hausmannite (Mn_3O_4). Regarding the first option, the Mn^{3+}/Mn^{2+} couple has a very high redox potential (Table 3), and the oxidation of Mn(II) to free Mn(III) requires an oxidant stronger than O_2. The participation of a *c*-type cytochrome in such a reaction poses even more difficulties than the MnO_2/Mn^{2+} reaction.

Table 3. Oxidation/reduction potentials (E°′) for selected couples.

Couple	E°′ (mV)
Ferredoxin ox/red	-430
NAD/NADH$_2$	-320
FeOOH/Fe^{2+}	-296
FAD/FADH$_2$	-220
fumarate/succinate	30
cytochrome c ox/red	(-260) - (+390)
ubiquinone ox/red	100
◆Mn$_3$O$_4$/Mn^{2+}	160
cytochrome c_1 ox/red	230
◆MnOOH/Mn^{2+}	260
cytochrome a (heart muscle) ox/red	290
cytochrome $c552$ (*Pseudomonas*)	300
cytochrome f	360
◆δMnO$_2$(s)/Mn^{2+}	410
NO$_3^-$/NO$_2^-$	420
◆MnOOH/Mn$_3$O$_4$	460
◆MnO$_2$/Mn$_3$O$_4$	620
Fe^{3+}/Fe^{2+}	650^a
◆MnO$_2$/MnOOH	670
Fe^{3+}/Fe^{2+}	770
O$_2$/H$_2$O	820
◆Mn^{3+}/Mn^{2+}	1540

^{a} at pH 2.0 in 200 mM SO$_4^{2-}$ (Ingledew et al. 1977)
Data from the CRC Handbook of Biochemistry, 3rd ed. (1975), Physical and Chemical Data, except for Mn, which was calculated from data from the CRC Handbook of Chemistry and Physics, 77th ed. (1996) and from data in Hem and Lind (1983). Values for Mn redox pairs are indicated by an arrow.

The other alternative solution involving Mn(III) is the oxidation of Mn(II) to certain Mn(III)-containing minerals. While this oxidation does not pose a redox problem (the couples Mn$_3$O$_4$/Mn^{2+} and MnOOH/Mn^{2+} have standard redox potentials of 159 and 259 mV, respectively), further biological oxidation of these minerals to Mn(IV) minerals is unlikely since the standard redox potentials for the MnO$_2$/Mn$_3$O$_4$ and MnO$_2$/MnOOH couples, which are much higher than that of the MnO$_2$/Mn^{2+} couple, would again pose a redox problem. The standard free energies associated with these reactions are lower than the energy in the Mn^{2+} MnO$_2$ reaction ($\Delta G°′$ is -42.3 kJ/mole Mn for the Mn^{2+} $\Rightarrow$ Mn$_3$O$_4$ oxidation, and -53.7 kJ/mole Mn for the Mn^{2+} $\Rightarrow$ MnOOH oxidation). Such energy gain is certainly a less attractive option for the bacteria, but is still favorable and would be useful, particularly if the bacteria oxidize enough of the substrate.

A final consideration concerns how the electrons from Mn(II) enter the electron transport chain. Are electrons transferred directly from Mn(II) to a cytochrome (e.g. cytochrome c) or is there some intermediate carrier(s) between Mn(II) and the cytochrome chain (Fig. 7)? That there are two classes of Mn(II) oxidation mutants, those that are cytochrome c oxidase positive and those that are cytochrome c oxidase negative, suggests that there is probably an intermediate carrier that interacts with Mn(II). If Mn(II) interacted with a cytochrome component directly, then all the Mn(II) oxidation mutants should also be cytochrome c oxidase negative. Clearly, more genetic and biochemical studies are needed to identify and characterize the Mn(II) oxidation system of *P. putida* strain MnB1.

Marine *Bacillus* sp. strain SG–1

A variety of Gram-positive bacteria belonging to the genus *Bacillus* are known to oxidize Mn(II) (Vojak et al. 1984, Ehrlich 1996a). *Bacillus* spp. that are aerobic or facultatively aerobic are characterized by their ability to form endospores: heat- and chemically-resistant, thick-walled, differentiated cells that develop inside the cell and, when mature, allow the bacteria to survive unfavorable conditions. The marine *Bacillus* sp. strain SG-1 is a sporeformer that was isolated from a Mn(II)-oxidizing enrichment culture from a marine sediment sample (Nealson and Ford 1980). While some Mn(II)-oxidizing *Bacillus* spp. oxidize Mn(II) during vegetative growth, in SG-1 it is the mature, free spores (those released from the cell) that oxidize Mn (Rosson and Nealson 1982). Mn oxides accumulate on the spore surface. As the spores become encrusted with Mn oxide, the rates of Mn(II) oxidation decrease, apparently as a result of masking the active Mn(II) binding sites (de Vrind et al. 1986b, Hastings and Emerson 1986). Although Mn(II) oxidation is not unique to SG–1 spores, spores from most other *Bacillus* species do not oxidize Mn(II). Bacterial spores are dormant structures, and SG–1 spores rendered non-germinable in the laboratory continue to oxidize Mn(II) (Rosson and Nealson 1982), therefore Mn(II) oxidation is not a result of energy generation by the organism nor is it a direct result of germination processes. The presence of surfaces, such as sands or clays, during the cultivation of SG–1 enhance sporulation, and hence enhance Mn(II) oxidation (Nealson and Ford 1980, Kepkay and Nealson 1982).

Interestingly, there are many species of *Bacillus* that also reduce Mn oxides during cellular (vegetative) growth (i.e. not as spores) (Ghiorse and Ehrlich 1974, Ghiorse and Ehrlich 1976, Ehrlich 1996a). SG–1 is also capable of reducing Mn oxides, presumably using them as an electron acceptors during growth under oxygen limiting conditions (de Vrind et al. 1986a). The ability of SG–1 to both oxidize and reduce Mn in different stages of its life cycle gives rise to one possible function for Mn(II) oxidation in SG-1 spores (Tebo 1983): The spores aerobically oxidize Mn(II) in the water column, fall to the sediment, germinate, and under low oxygen or anaerobic conditions the newly formed cells are able to utilize the Mn oxides for vegetative growth.

Biochemistry. Early studies indicated that Mn(II) oxidation by SG-1 spores is catalyzed by a protein from the spore coat (the protein layers around the spore) or a protein from the exosporium (a membrane-like layer found external to the spore coat, compositionally distinct from the spore coats, but not present in all species). Studies on whole SG-1 spores (Rosson and Nealson 1982) showed that Mn(II) oxidation by the spores was inhibited by pretreatment with temperatures above 80°C or by treatment with metalloprotein inhibitors, including mercuric chloride, which interacts with sulfur groups in proteins. This work indicated that something associated with the spores that is sensitive to heat, such as a protein, was responsible for Mn(II) oxidation. The work also suggested that this putative protein is a metalloprotein with sulfur groups involved in Mn(II) oxidation. Transmission electron microscopy of the spores showed Mn oxide is precipitated on the outermost spore layer that is characterized by ridges (Fig 8; Tebo 1983). Purified spore coat preparations, processed to retain all the outer spore layers and remove the spore contents, retain full oxidizing activity (de Vrind et al. 1986b). Similar to the whole spores, Mn(II) oxidation by the spore coat preparations is inhibited by potassium cyanide, sodium azide, and mercuric chloride. Oxygen is also required for Mn(II) oxidation (de Vrind et al. 1986b, Hastings and Emerson 1986). Measurements of the consumption of O_2 and the production of protons during Mn(II) oxidation suggest that Mn(II) oxidation follows the stoichiometry of the reaction typically written for Mn(II) oxidation to MnO_2 (Eqn. 5, de Vrind et al. 1986b). Recent studies have shown that Mn(II)-oxidizing activity is localized in the exosporium, which can be removed from the spores by passage through a French pressure cell (Fig. 8, Francis et al. 1997). These studies, taken in total, suggest that a protein present in the exosporium is responsible for Mn(II) oxidation.

Attempts were made to isolate the Mn(II)-oxidizing protein by extracting proteins from the spores, separating them by polyacrylamide gel electrophoresis (PAGE), and incubating the gels with Mn(II) (Tebo et al. 1988). Occasionally a high molecular weight

Figure 8. Transmission electron micrographs of *Bacillus* sp. strain SG–1 spores incubated for 3 h with Mn(II) prior to fixation and embedding for transmission electron microscopy. Left: SG–1 spore with its exosporium intact. Right: SG–1 spores after passage through a French pressure cell which partially removes the exosporium. Note that the ridge structure, characteristic of SG–1 spores, has been removed. Remnants of the exosporium can be visualized by the Mn oxides that are present (the electron dense fibrous-looking material on the outer spore surface). Bars = 0.25 μm.

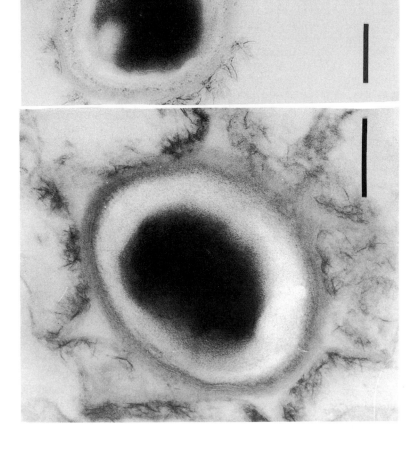

Mn(II)-oxidizing (brown) protein band (~205 kDa) could be seen in the gels. But these experiments were hard to reproduce, perhaps because the Mn(II)-oxidizing factors were damaged during extraction, or perhaps because there are several components necessary for oxidation that are separated during electrophoresis.

Genetics. Because of the inherent difficulty in working with proteins from SG–1 spores, a genetic approach to the problem was devised (van Waasbergen et al. 1993, van Waasbergen et al. 1996). Some of the most revealing information to date into mechanism of Mn(II) oxidation by the spores comes from these genetic studies. Techniques for transposon mutagenesis and genetic manipulation of SG-1 were developed, and mutants strains of SG-1 were produced that form spores that do not oxidize Mn(II). The genes that were mutated in the strains were identified, and a number of them were sequenced. In addition, the mutant spores were examined for clues into the mechanism of Mn(II) oxidation. Many of these mutants had their lesions at various locations within a series of seven genes which appear to be in an operon, termed the *mnx* operon (Fig. 9). In the region upstream of the operon, regulatory sequences were present that suggested that the genes would be transcribed by the mother cell (the cell within which the endospore is forming) during sporulation. As expected, transcription of the *mnx* genes was found to take place at mid-sporulation (during formation of the endospore) at the time when spore coat proteins begin to be produced by the mother cell. These characteristics are consistent with the possibility that one or more of the genes within the *mnx* operon is a spore coat or exosporium protein. No differences were seen in germination and resistance properties between the mutant, non-oxidizing spores and the wild-type. So, apparently neither Mn(II) oxidation nor the missing proteins function in these capacities. However, it is possible that in the mutants there is loss of resistance to something not tested or that other redundant protective factors are still active. Transmission electron microscopy of the mutant spores showed some disruption of the outermost spore layers, suggesting that the missing protein factors, in addition to being involved in Mn(II) oxidation, may be components of the outermost spore coat and/or something involved in its assembly. This latter possibility raises another possible function for the Mn(II)-oxidizing factor: perhaps in addition to being able to oxidize Mn(II), the factor is a more general oxidase that is present in the outermost spore coat and is responsible for its assembly.

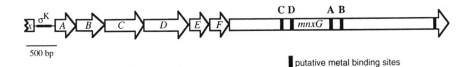

500 bp

putative metal binding sites

Figure 9. The organization of the Mnx operon of *Bacillus* sp. strain SG–1 based on DNA sequence analysis. *mnxG* is the gene that codes for the putative Mn(II)-oxidizing protein which shows significant similarity to the multicopper oxidase family of proteins, particular in the regions of copper binding (boxed areas). The amino acid sequences of the copper binding sites designated with the letter A–D are shown in Figure 10. σ^K represents the site upstream of the operon where RNA polymerase binds to initiate RNA synthesis (DNA transcription).

Possible mechanism for the oxidation of Mn(II) by SG–1. From examination of the sequences of the *mnx* genes, clues into the mechanism of Mn(II) oxidation have been found. The putative protein products from two of the *mnx* genes, MnxG (138 kDa) and MnxC (22 kDa), have sequence similarity to other proteins present in the sequence databases. The other *mnx* genes have not been found to share significant sequence similarity with any other known genes, but the MnxA, MnxB, and MnxE proteins are predicted to be extremely hydrophobic, suggesting that they may be spore coat or membrane bound proteins. The MnxG protein shares sequence similarity with members of the family of multicopper oxidases (Fig. 10). In addition, searches against the protein motif library PROSITE, identified in MnxG the signature region for copper binding members of the family (H-C-H-x(3)-H-x(3)-G-[LM], where x = any amino acid). Multi-copper oxidases are a group of very diverse proteins that occur in a wide variety of organisms from fungi to humans. Most bind multiple copper cofactors, and these cofactors are responsible for the redox activities of the proteins (Huber 1989, Ryden and Hunt 1993). Members of the multicopper oxidase family include laccase (from plants and fungi), ascorbate oxidase (from cucumber & squash), human ceruloplasmin, CopA (a copper resistance protein from *Pseudomonas syringae*), and Fet3 (a Fe(II) oxidizing protein involved in Fe transport in yeast). Multicopper oxidases, along with cytochrome *c* oxidase, are the only proteins known that catalyze the four-electron reduction of O_2 to H_2O with the oxidation of substrate. In general, most multicopper oxidases oxidize organic substrates. However, human ceruloplasmin and Fet3 are both known to oxidize Fe(II) (Solomon et al. 1996, Stearman et al. 1996, Eide and Guerinot 1997, Kaplan and O'Halloran 1997).

A

MnxG	527	M	H	I	H	F	V
MofA	304	I	H	L	H	G	G
Asox	94	I	H	W	H	G	I
Lacc	78	V	H	W	H	G	L
Hcer	119	F	H	S	H	G	I
CopA	99	I	H	W	H	G	I
Fet3	8	M	H	F	H	G	L
			2		3		

B

MnxG	572	F	F	H	D	H	L
MofA	384	W	Y	H	D	H	T
Asox	137	F	Y	H	G	H	L
Lacc	121	W	Y	H	S	H	Y
Hcer	178	I	Y	H	S	H	I
CopA	140	W	Y	H	S	H	S
Fet3	52	W	Y	H	S	H	T
				3		3	

C

MnxG	281	H	V	F	H	Y	H	V	H
MofA	1174	H	P	V	H	F	H	L	L
Asox	480	H	P	W	H	L	H	G	H
Lacc	508	H	P	I	H	K	H	G	N
Hcer	994	H	T	V	H	F	H	G	H
CopA	542	H	P	I	H	L	H	G	M
Fet3	341	H	P	F	H	L	H	G	H
		1			2		3		

D

MnxG	334	H	C	H	L	Y	P	H	F	G	I	G M
MofA	1279	H	C	H	I	L	G	H	E	E	N	D F
Asox	542	H	C	H	I	E	P	H	L	H	M	G M
Lacc	585	H	C	H	I	A	S	H	Q	M	G	G M
Hcer	1039	H	C	H	V	T	D	H	I	H	A	G M
CopA	590	H	C	H	L	L	Y	H	M	E	M	G M
Fet3	411	H	C	H	I	E	W	H	L	L	Q	G L
		3	1	3				1				1

Figure 10. Amino acid sequence alignment of copper-binding sites in MnxG, MofA, and other multicopper oxidases. The letters A–D correspond to the copper binding sites shown in Figure 9. Abbreviations: Asox, ascorbate oxidase (cucumber and squash); Lacc, laccase (fungi); Hcer, human ceruloplasmin; CopA, a copper resistance protein from *Pseudomonas syringae*; and Fet3, an iron oxidizing/transport protein in yeast. The amino acid residues conserved among the different proteins are shaded and the copper-binding residues are numbered according to the type of copper they potentially help coordinate. The one-letter codes for the amino acids are given in Table 4.

Multicopper oxidases contain coppers of three spectroscopically distinct types, 'blue' copper (or Type 1), Type 2, and Type 3. In the three-dimensional protein structure, the amino acids (histidine, cysteine, and methionine, see Table 4) that make up each type of copper center come into close contact and form a copper binding site. The Type 1 center accepts the initial electron from the substrate and passes it to the Type 2 and Type 3 centers that bind and reduce molecular oxygen:

$$\text{Substrate} \xrightarrow{e^-} \text{Type 1} \xrightarrow{e^-} \left\{ \text{Type 2} + (\text{Type 3})_2 \right\} \xrightarrow{e^-} O_2$$

The type 2 and 3 centers are now known from crystallography to be structurally associated in a roughly triangular cluster. From amino acid sequence alignments and

Table 4. Natural protein amino acids

Name	1-letter symbol	Linear structural formula[*]	Character, Class	Metal binding
Alanine	A	CH_3-$CH(NH_3^+)$-COO^-	Neutral, Hydrophobic	
Arginine	R	$HN=C(NH_2)$-NH-$(CH_2)_3$-$CH(NH_3^+)$-COO^-	Basic, hydrophilic	
Asparagine	N	H_2N-CO-CH_2-$CH(NH_3^+)$-COO^-	Neutral, hydrophilic	
Aspartic acid	D	$HOOC$-CH_2-$CH(NH_3^+)$-COO^-	Acidic, hydrophilic	M^{2+}, M^{3+}
Cysteine	C	HS-CH_2-$CH(NH_3^+)$-COO^-	Neutral, hydrophilic	Zn, Cu, Fe, Ni, Mo
Glutamine	Q	H_2N-CO-$(CH_2)_2$-$CH(NH_3^+)$-COO^-	Neutral, hydrophilic	
Glutamic acid	E	$HOOC$-$(CH_2)_2$-$CH(NH_3^+)$-COO^-	Acidic, hydrophilic	M_2^+
Glycine	G	(NH_3^+)-CH_2-COO^-	Non polar, hydrophobic	
Histidine	H	NH-$CH=N$-$CH=C$-CH_2-$CH(NH_3^+)$-COO^-	Neutral-basic hydrophilic	Zn, Cu, Mn Fe, Ni
Isoleucine	I	CH_3-CH_2-$CH(CH_3)$-$CH(NH_3^+)$-COO^-	Non polar, hydrophobic	
Leucine	L	$(CH_3)_2$-CH-CH_2-$CH(NH_3^+)$-COO^-	Non polar, hydrophobic	
Lysine	K	H_2N-$(CH_2)_4$-$CH(NH_3^+)$-COO^-	Basic, hydrophilic	
Methionine	M	CH_3-S-$(CH_2)_2$-$CH(NH_3^+)$-COO^-	Non polar, hydrophobic	Cu, Fe
Phenylalanine	F	Ph-CH_2-$CH(NH_3^+)$-COO^-	Non polar, hydrophobic	
Proline	P	NH_2^+-$(CH_2)_3$-CH-COO^-	Non polar, hydrophobic	
Serine	S	HO-CH_2-$CH(NH_3^+)$-COO^-	Neutral, hydrophilic	
Threonine	T	CH_3-$CH(OH)$-$CH(NH_3^+)$-COO^-	Neutral, hydrophilic	
Tryptophan	W	Ph-NH-$CH=C$-CH_2-$CH(NH_3^+)$-COO^-	Non polar, hydrophobic	
Tyrosine	Y	HO-p-Ph-CH_2-$CH(NH_3^+)$-COO^-	Neutral, hydrophobic	Mn^{3+}, Fe^{3+}
Valine	V	$(CH_3)_2$-CH-$CH(NH_3^+)$-COO^-	Non polar, hydrophobic	

Source: da Silva and Williams (1991).
[*] Ionic forms that predominate under physiological conditions (pH 7).

X-ray crystallography, multicopper oxidases possess a distinctive subdomain structure (Solomon et al. 1996). Laccase, ascorbate oxidase, and Fet3 have three domains while human ceruloplasmin has six. In all the multicopper oxidases there is significant internal homology among the subdomains suggesting they all arose from a common gene ancestor by gene duplication (Ryden and Hunt 1993, Solomon et al. 1996).

The similarity that MnxG shares with the multicopper oxidases includes the copper binding sites (Fig. 10). On the basis of subdomain structure, MnxG appears to be most similar to human ceruloplasmin, containing six subdomains. Azide, which inhibits Mn(II) oxidation in SG-1, is known to inhibit multicopper oxidases by bridging the type 2 and type 3 copper atoms (da Silva and Williams 1991). Additions of small amounts of copper(II) enhance Mn(II) oxidation rates, further supporting the idea that *mnxG* codes for the Mn(II)-oxidizing protein. In recent work (Francis et al. 1997), MnxG has been localized to the exosporium. Although genetic analysis strongly suggests that MnxG participates in Mn(II) oxidation, no Mn(II)-oxidizing activity could be detected in the protein expressed in *E. coli* (Francis and Tebo, unpublished), possibly due to incorrect protein folding.

At present we do not know what the role of MnxG is in vivo. One possibility is that it helps to cross-link the outermost spore coat. By analogy to human ceruloplasmin and Fet3, a second possibility is that it plays a key role in iron metabolism, perhaps by oxidizing Fe(II) to facilitate its uptake and storage in the spore. A third possibility is that MnxG plays a role in conferring resistance to toxic chemicals or oxidative damage to the spores (although no loss of resistance was seen in mutant spores). Finally, perhaps the real role of MnxG is to oxidize Mn(II). Establishing whether the oxidation of Mn(II) is the actual function of MnxG or a fortuitous side reaction, requires further investigation.

	50										60										70				
MnxC	H	D	-	-	Q	K	I	K	L	V	V	F	F	Y	T	H	C	P	D	I	C	P	M	T	L
C.rum. imdom.	K	D	F	L	G	K	H	M	L	V	L	F	G	F	S	S	C	K	T	I	C	P	M	E	L
A.marg. MSP5	G	D	F	G	G	K	H	M	L	V	I	F	G	F	S	A	C	K	Y	T	C	P	T	E	L
P.stutz. ORF193	E	D	L	K	G	R	W	H	I	L	F	F	G	F	T	A	C	P	D	I	C	P	T	T	L
S.aur. 18K mer	P	D	-	-	E	K	P	T	L	I	Y	F	M	A	T	W	C	P	S	-	C	I	Y	N	-
S.cerev. SCO2	E	D	L	K	G	K	F	S	I	L	Y	F	G	F	S	H	C	P	D	I	C	P	E	E	L
S.cerev. SCO1	K	D	L	L	G	K	F	S	I	I	Y	F	G	F	S	N	C	P	D	I	C	P	D	E	L

Figure 11. Amino acid sequence alignment of regions surrounding CxxxC containing area (asterisks, *) in MnxC-like proteins. Shaded amino acids represent identical residues or conservative replacements. Abbreviations: C.rum. imdom., the immuno dominant protein from *Cowdria ruminantium*; A.marg. MSP5, a major surface protein from *Anaplasma marginale*; P. stutz. ORF193, an open reading frame (unknown gene) from *Pseudomonas stutzeri*; S.aur. 18K mer, an 18-kDa protein from the *Staphylococcus aureus mer* operon; S.cerev.SCO1 and SCO2, proteins from *Saccharomyces cerevisiae* (see text). The one-letter codes for the amino acids are given in Table 4.

Mnx C shares similarity with a growing list of various cell surface proteins of unknown function and multicomponent redox-active proteins in the sequence databases (Fig. 11). The amino terminus (beginning sequence) of MnxC contains a signal peptide, a sequence pattern that suggests that the protein is associated with a membrane. One of the proteins that MnxC is similar to is a redox-active protein, the 18 kDa *mer* protein, that is involved in mercury resistance in the bacterium *Staphylococcus aureus* (Laddaga et al. 1987). The region of similarity includes a pair of cysteines (amino acids that contain sulfide groups, Table 4) that are, in MnxC, separated by three other amino acids. In the *mer* protein these cysteines are separated by only two amino acids, and the cysteines form part of a thioredoxin motif (a sequence pattern of C-x-x-C, Ellis et al. 1992). This raises the possibility that these two cysteines in MnxC also comprise a thioredoxin motif and that MnxC functions as a thioredoxin (a redox-active protein that transfers electrons by forming disulfide bonds between the cysteines). However, in light of a recent report (Glerum et al. 1996) concerning another pair of proteins with which MnxC shares similarity, the yeast mitochondrial membrane proteins SCO1 and SCO2, an intriguing alternative is that these cysteines play a role in copper binding. In Glerum et al. (1996), SCO1 and SCO2, which were known to play a role in cytochrome oxidase assembly, were identified as having a possible copper binding site that involves these two cysteines (separated by three amino acids). Glerum et al. (1996) hypothesize that SCO1 and SCO2 deliver copper to cytochrome oxidase or are involved in copper uptake by the mitochondrion. By analogy, perhaps MnxC delivers copper to MnxG or is involved in copper transport into the cell during sporulation. Copper uptake in *B. megaterium* has been shown to increase at certain times during sporulation, presumably so that copper may serve as a cofactor in various enzymes during spore formation (Krueger and Kolodziej 1976). Since we have recently localized MnxC to the exosporium (Francis and

Tebo, unpublished) and MnxG is also localized to the exosporium, we hypothesize that MnxC delivers copper to MnxG as the spore coats and exosporium are being assembled. MnxG may then use copper as a cofactor for some function related to sporulation or the properties of mature spores themselves.

Surface chemistry of SG–1 spores. Recently, the surface chemistry and Cu(II) adsorption properties of SG-1 spores have been characterized (He and Tebo, submitted). The specific surface area of freeze-dried SG-1 spores was 6.3–6.9 m^2 g^{-1} of spores as determined by a BET surface area analyzer or calculated from spore dimensions determined in the electron microscope. A value of 74.7 m^2 g^{-1}, however, was obtained with wet spores using a methylene blue adsorption method. Since BET analysis required freeze drying whereas the dye adsorption method uses wet spores, the difference in the surface area suggests that the spore surface is porous rather than smooth and that the porous structure collapses after freeze drying. The surface exchange capacity measured by the proton exchange method was found to be 30.6 $mmol \cdot m^{-2}$, equal to a surface site density of 18.3 $sites \cdot nm^{-2}$, based on the methylene blue adsorption surface area. The SG-1 spore surface charge characteristics were obtained from the acid-base titration data. The surface charge density varied with pH, with the point of zero charge (PZC) being 4.5. The surface was dominated by negatively charged sites, believed to be primarily carboxylate but also phosphate groups. Copper adsorption by SG-1 spores was rapid and complete within minutes and was low at pH 3, increasing with pH. The high surface area and surface site density, comparable to Fe and Al mineral colloids, give SG-1 spores a high capacity for binding Cu(II) and other metals on their surface. Whether Cu(II) adsorption is due to nonspecific binding or specific binding related to the presence of MnxC or MnxG on the spore surface is yet to be determined.

Leptothrix discophora strains SS-1 and SP-6

Gram-negative sheathed bacteria belonging to the genus *Leptothrix* oxidize both iron and manganese directly, and deposit the oxidized minerals in exopolymers associated with their sheath (Fig. 12). Closely related sheathed bacteria in the genus *Sphaerotilus* (of which the only described species is *S. natans*) are distinguished from *Leptothrix* spp. by their lack of Mn(II) oxidation. *S. natans* deposits only iron oxides in its sheath; the question of a biochemical versus abiotic Fe oxidation mechanism has not been studied. *Leptothrix discophora* is a model organism for the study of the iron- and manganese-deposition by sheathed bacteria; it is the only organism in which both the iron- and Mn(II)-oxidizing factors have been identified (Adams and Ghiorse 1987, Boogerd and de Vrind 1987, Corstjens et al. 1992, Corstjens et al. 1997). *Leptothrix* spp. (Fig. 12) are ubiquitous in wetlands and iron seeps and springs around the world. They have been isolated from or observed in other diverse environments such as deep sea and deep lake hydrothermal vents, water wells, sand filters in groundwater treatment systems, and coatings on *Lemna* spp. roots in shallow ponds (Ghiorse and Ehrlich 1992, Juniper and Tebo 1995).

Biochemistry. The first Mn(II)-oxidizing protein to be purified and partially characterized was that of the extracellular Mn(II)-oxidizing system of a sheathless strain of *Leptothrix discophora*, strain SS-1 (Adams and Ghiorse 1987, Boogerd and de Vrind 1987). The Mn(II)-oxidizing protein was also identified in a sheath-forming strain of *L. discophora* (SP-6), and determined to be similar to the protein purified from strain SS-1 with respect to its apparent molecular weight (after PAGE), and the effects of several inhibitors on its activity (Table 5; Emerson and Ghiorse 1992). In all these studies, the high activity in unpurified exopolymer and the decrease in specific activity during purification indicated that there may be other factors, such as acidic polysaccharides, required for optimum Mn(II) oxidation in situ. The Mn(II) oxidation factor of SS-1 may

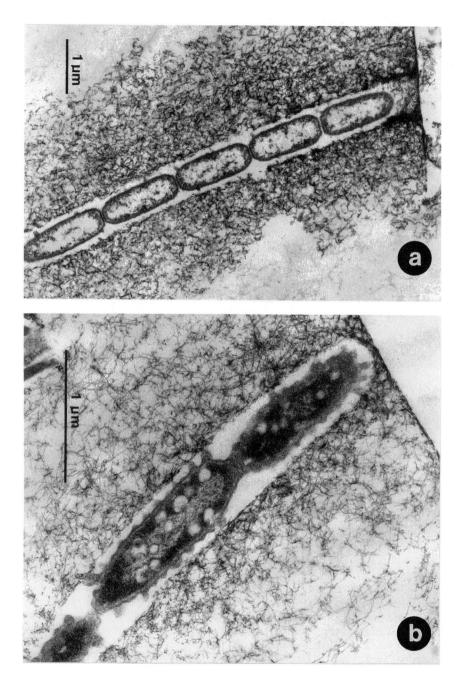

Figure 12. Transmission electron micrographs showing thin sections of unidentified *Leptothrix* spp. filaments in their natural habitats: (A) a forest pond near Kiel, Germany, (B) Sapsucker Woods wetland in Ithaca, New York. The sheaths appear as a matrix of ferromanganese-encrusted electron dense filaments surrounding the cells. The flattened top ends of the sheath mark holdfast regions at the air-water interface.

be a glycoprotein associated with anionic heteropolysaccharides excreted by the cells. In SP-6 these components apparently are associated with anionic sheath polymers (Emerson and Ghiorse 1992). Indeed, in natural environments, Mn oxides are found to be associated with the acidic surface polysaccharides on the surfaces of Mn-depositing microorganisms (Ghiorse 1984b, Nealson et al. 1989).

Table 5. Characteristics of the extracellular Mn(II)-oxidizing system of *Leptothrix discophora.*

General description:	Manganese-oxidizing protein(s) associated with acidic heteropolysaccharides and sheath components.
Temperature optimum :	28°C
pH optimum:	7.3
Inactivation by:	Pronase E Heat (80°C)
Inhibition by:	KCN, o-phenanthroline, NaN_3, elevated [Mn^{2+}], heavy metals, EDTA
Mn^{2+}-oxidizing protein(s) (SDS-PAGE):	M_r 100-110 kD (M, 50 kD not always detected)
Kinetic parameters:	$K_m \sim 6\ \mu M\ Mn^{2+}$, $V_{max} \sim 1.0$ nmoles Mn^{2+} oxidized·min^{-1} μg $protein^{-1}$
Oxidation reaction:	$Mn^{2+} + {}^1/_2\,O_2 + H_2O \Rightarrow MnO_x + 2\,H^+$
Oxidation product:	Mn(III,IV) oxide containing buserite, birnessite, or vernadite after 90 days
Putative gene:	*mofA*, encodes Mn-oxidizing protein with leader sequence and copper domains
MofA gene distribution:	Found in *Leptothrix* environment, apparently not homologous to Mn(II) oxidation genes of other Mn(II)-oxidizing bacteria
Proposed Mn(II) oxidation mechanisms:	Mn^{2+} binds to heteropolysaccharide-*MofA* protein complex in sheath matrix; *MofA* protein catalyzes the oxidation of Mn(II) forming Mn(III) oxide which is further oxidized to Mn(IV) by O_2.

Note: Data taken from Adams (1986), Emerson (1989), Corstjens (1993), and Siering (1996).

Genetics. The gene *mofA*, encoding the putative Mn(II)-oxidizing protein of *L. discophora* strain SS-1, was recently cloned and sequenced (Corstjens 1993, Corstjens et al. 1997). Although Corstjens and coworkers demonstrated in vitro transcription and translation of the cloned gene (Corstjens et al. 1997), expression of the MofA protein in an alternate host has not been successful due to the lack of a transformation system, and suitable available mutants. Clearly, the lack of such genetic tools for use in *Leptothrix* spp. has hindered the understanding of the molecular genetics of this organism and Mn(II) oxidation system. However, as reported recently (Corstjens et al. 1997), some progress has been made. For example, sequence analysis of the *mofA* protein shows that the N-terminus of the protein probably is a signal peptide which helps transport the protein out of the cell to the sheath. Also, its derived amino acid sequence indicates a molecular wt of 174 kD rather than 110 kD indicated by gel electrophoresis (Table 5). This discrepancy cannot be accounted for by loss of the signal peptide (36 amino acids). However, it could be due to polysaccharide components that may alter the electrophoretic migration of the protein in the gel or to denaturation of the larger protein under conditions of electrophoresis (i.e. presence of SDS, a strong detergent). Indeed, two Mn(II)-oxidizing protein bands (110 kD and 50 kD) are frequently detected by SDS-PAGE (Adams 1986, Adams and Ghiorse 1987, Boogerd and de Vrind 1987, Corstjens et al. 1997).

There are no proteins with significant similarity to the entire sequence of MofA in the SwissProt and GenEMBL databases. However, searches of the SwissProt and GenEMBL databases using the BLAST program (Altschul et al. 1990) revealed short regions (10-40 amino acids in length) of identity between the MofA protein and 2 multicopper oxidase proteins (phenoxazinone synthase and bilirubin oxidase; Corstjens et al. 1997, Siering and Ghiorse 1997). In a recent paper by Hsieh and Jones (1995), the authors aligned two short regions of the MofA protein (residues 305-308 and 382-388) to two of the four putative copper binding domains of phenoxazinone synthase and several other proteins in the multicopper oxidase family. At that time, the MofA protein did not appear to contain either of the signature sequences for this family of proteins (Bairoch et al. 1995) including the histidine residues associated with the type 1 copper centers (see above) which are believed to be involved in the initial oxidation of the substrate. However, it has recently been shown that the *mofA* coding sequence is longer than originally believed, and that two additional potential copper binding sites have been found (Corstjens et al. 1997). The amino acid sequence similarity between the MofA protein and the multicopper oxidase proteins may be due to the role of these residues in the binding and reduction of molecular oxygen, or in parallel mechanisms of catalysis. Although copper has been show to be inhibitory to the Mn(II)-oxidizing activity of *L. discophora* SS-1 (Adams and Ghiorse 1987) when present at high concentrations (100 µM) (Adams 1986), no studies have been conducted to determine if Mn(II) oxidation by MofA is stimulated by low concentrations of copper (<10 µM).

Although MofA and MnxG from *Bacillus* sp. strain SG-1 share no significant sequence similarity, it is interesting that both of these proteins contain regions similar to the copper-binding motifs of the multicopper oxidases (Fig. 10). Whether either of these proteins contain bound copper ions awaits spectroscopic analysis or determination of their respective crystal structures. Such studies are not likely in the near future since it has been impossible to obtain preparative amounts of either of these proteins in a pure form (unpublished results; Adams and Ghiorse 1987, Corstjens et al. 1997). As other Mn(II) oxidation proteins are identified, perhaps the use of a copper cofactor will prove to be a common thread which unites the Mn(II) oxidation systems from diverse microorganisms (see discussion below).

Tryptophan is an amino acid with a bulky aromatic side chain (Table 4). It is relatively rare in proteins and infrequently substituted with other amino acids between related proteins. Consequently it is regarded as structurally important where it does occur. Sequence analysis of the primary structure of the MofA protein revealed a striking tryptophan-rich region containing eight tryptophan residues over a stretch of 66 amino acids (Corstjens 1993, Corstjens et al. 1997). Also, just outside the tryptophan-rich region were 2 of the 6 cysteine residues. It was originally speculated that this tryptophan-rich region of the protein may be involved in binding of the Mn^{2+} substrate (Corstjens 1993, Corstjens et al. 1997), however, tryptophan is not normally thought of as binding metals (see Table 4).

A molecular ecological study of the mofA gene in its native habitat. The unusual nature of the tryptophan-rich region of MofA suggests that it might be useful as an indicator of the presence or absence of *Leptothrix* species in ecological studies. The tryptophan-rich region of the MofA protein was targeted for a polymerase chain reaction (PCR)-based method to detect the *mofA* gene in the available strains of *Leptothrix* spp. and in nucleic acids extracted directly from environmental samples taken from the Sapsucker Woods wetland in Ithaca, NY (Siering 1996, Siering and Ghiorse 1997). Using this approach, the *mofA* gene was detected in nucleic acids extracted from two distinct types of surface films where *Leptothrix* species normally occur but not in nucleic acids extracted from wetland sediment where *Leptothrix* is normally absent. These results confirmed previous observations that the actively growing *Leptothrix* is restricted to the ferromanganese surface film and the root zone of *Lemna* spp. in the wetland water column (Ghiorse and Chapnick 1983, Ghiorse 1984a). This gene-based screening method can be used to rapidly screen for the presence and absence of *Leptothrix* in different environments.

Variable stringency hybridization analysis has recently shown that the Mn(II)-oxidizing factor encoded by *mofA* is similar (but not identical) to the *mofA* genes present in known strains of *Leptothrix* (Siering 1996, Siering and Ghiorse 1997). However, it bears no detectable similarity to the Mn(II)-oxidizing factors present in several other Mn(II)-oxidizing bacteria examined (Siering 1996). These included: *Pedomicrobium* sp. 868 and 869; uncharacterized Mn(II)-oxidizing *Pseudomonas* spp. strains B1, B6, 11A, 11A1, B14, 18A, 18B, 104A, 104B (Kindly provided by K. H. Nealson, University of Wisconsin, Milwaukee), and an uncharacterized, gram negative Mn(II)-oxidizing marine isolate strain FB-1 (Isolated by W. C. Ghiorse).

Possible mechanisms for oxidation of Mn(II) by Leptothrix spp. There are great uncertainties associated with both the composition and mechanism of the Mn(II) oxidation systems of *Leptothrix* spp. Clearly, the Mn(II) oxidation takes place in a complex matrix of heteropolysaccharides containing at least one protein that catalyzes the oxidation reaction with molecular oxygen. Presumably the protein-polysaccharide complex is responsible for Mn^{2+} binding and oxidation. Molecular analysis generally supports this model, but as yet, no definitive biochemical mechanisms have been elucidated. Many additional questions remain. Is there a Mn(II) oxidation factor operon in *L. discophora* as suggested by Corstjens et al. (1997)? How many different Mn(II) oxidation genes are involved in the *Leptothrix* system? How are the other known Mn(II) oxidation systems related to one another and to the MofA- and MnxG-associated systems of *Leptothrix* and *Bacillus* (see below for further discussion)? If the systems are related, what are the most conserved regions of the genes?

COMPARATIVE ASPECTS OF MN(II)-OXIDIZING PROTEINS

Although the evidence to date suggests that Mn(II) oxidation in two unrelated organisms, *Bacillus* sp. strain SG–1 and *L. discophora*, is carried out by multicopper oxidase-like proteins, absolute proof that MnxG of SG–1 and MofA of *L. discophora* are copper-containing proteins is not available. Nevertheless, because these proteins contain the signature copper binding domains that are so highly conserved among other multicopper oxidases (Fig. 10), it is intriguing to speculate on a more universal role of copper in bacterial Mn(II) oxidation. For example, it will be extremely interesting to see if copper plays a direct role in Mn(II) oxidation in *P. putida* strain MnB1. Copper is probably involved indirectly in Mn(II) oxidation in MnB1 because cytochrome *c* oxidase uses copper as a cofactor. Although studies of the Mn(II)-oxidizing proteins (genes) in other Mn(II)-oxidizing bacteria are clearly needed, a number of interesting points arise by comparing known (or strongly suspected) Mn oxidizing proteins.

One of the first comparisons that needs to be made is between the bacterial and fungal enzymes known to oxidize Mn(II). One of the proteins MnxG and MofA resemble is laccase, a multicopper oxidase found in plants and fungi that is involved in the oxidation of phenolic compounds and in fungi is believed to be involved in lignin degradation. Laccase is known to oxidize Mn(II) to Mn(III) chelates in the presence of phenolic compounds (Archibald and Roy 1992). As mentioned in the background section, other fungal enzymes are also capable of oxidizing Mn(II), most notably manganese peroxidase, but also lignin peroxidase. These peroxidases use heme rather than copper ions as a cofactor and hydrogen peroxide rather than O_2 as the oxidant. Comparison of

Table 6. Comparison of different Mn(II)-oxidizing proteins.

Protein:	Bacteria			Fungi
	MnxG	MofA	Laccase	Mn Peroxidase (MnP)
Active site(s)	Multiple copper ions	Multiple copper ions	Multiple copper ions	heme
e^- acceptor	O_2	O_2	O_2	H_2O_2
Oxidation product	Mn(III, IV) Oxides	Mn(III, IV) Oxides	Mn(III)–chelate	Mn(III)–chelate
Affinity for Mn(II)	high	high	low	high
MW	138 kDa	174 kDa	64 kDa	50–65 kDa

some properties of the fungal and bacterial Mn(II)-oxidizing proteins is given in Table 6. Azide is a potent inhibitor of all these Mn(II)-oxidizing enzymes, which probably accounts for the fact that azide is such a good inhibitor of Mn(II) oxidation in environmental studies (Tebo 1997). In the case of the fungal systems, the main products of Mn(II) oxidation are soluble Mn(III) chelates. These Mn(III) chelates are strong oxidants capable of oxidizing a variety of phenolic compounds (Aitken and Irvine 1990). In the absence of complexation, Mn(III) can disproportionate to Mn(II) and Mn(IV), the latter precipitating as Mn oxides (Souren 1997). It is probably this reaction that accounts for the formation of Mn oxides by fungi in decaying wood (Blanchette 1984) and laboratory experiments (Glenn et al. 1986, Perez and Jeffries 1992, Souren 1997). Because disproportionation of Mn(III) would not incorporate molecular O_2, one would predict that the $\delta^{18}O$ of the Mn oxides formed by fungi would reflect solely the oxygen isotopic composition of the water in which it formed (Tebo 1997). Since two different Mn(II)-oxidizing bacteria have been shown to incorporate oxygen from O_2 into their Mn oxides, this predicts that the $\delta^{18}O$ of Mn oxide might be useful for distinguishing Mn deposits formed by fungi from those formed by bacteria. However, a number of other factors, including isotopic fractionation or exchange and other abiotic Mn(II) oxidation reactions, can obscure an O_2-derived $\delta^{18}O$ signature (Mandernack et al. 1995a) and make those distinctions difficult.

Should MnxG and MofA indeed turn out to be multicopper oxidases as the available data suggest, then they would represent a new class of multicopper oxidase, i.e. one that oxidizes Mn(II). [Although laccase is known to oxidize Mn(II), that is not believed to be its main function (Souren 1997).] Only two other metal-oxidizing multicopper oxidases have been identified so far: human ceruloplasmin and Fet3 of yeast. These proteins are important in iron uptake. In humans, Fe(II) enters the bloodstream from liver stores and the intestine, and Fe(III) is delivered to other tissues in the body. Because most of the oxygen in the blood is bound to hemoglobin, and plasma contains significant concentrations of ascorbate which inhibit nonenzymatic oxidation of Fe(II), without ceruloplasmin to oxidize Fe(II), free Fe(II) in the bloodstream could promote Fenton chemistry-induced oxidative damage of important macromolecules. Ceruloplasmin appears to function both in the mobilization of Fe(II) from tissues and oxidation of Fe(II) which facilitates Fe uptake by transferrin in the bloodstream. In yeast, Fet3 is part of a high affinity Fe transport system in which Fet3 functions to oxidize Fe(II) to Fe(III) (Eide and Guerinot 1997, Kaplan and O'Halloran 1997). Fe(III) is then handed off to a second protein Ftr1 which transports the Fe(III) across the plasma membrane into the cytoplasm. Thus ceruloplasmin controls iron egress from cells while Fet3 controls iron entry into cells. In bacteria a multicopper oxidase (CopA) is part of a copper resistance operon involved in copper egress. By analogy, could MnxG and MofA play a role in metal ion transport (e.g. during spore formation in SG–1 or through the sheath in *L. discophora*)?

Perhaps some clue into the role of MnxG and MofA can be inferred from the affinity of these proteins (or the cells) for Mn(II). In the multicopper oxidase family of proteins, the interactions between the enzymes and their substrates can be broadly classified into two classes, those with low substrate specificity and those with high specificity (Solomon et al. 1996). For example, the plant and fungal laccases fall into the first category. Laccases non-specifically oxidize a variety of aromatic substrates with K_ms generally in the range of 1 to 10 mM. (K_m is the substrate concentration at which the enzyme functions at half its maximal rate, V_{max}.) This indicates that there is probably no binding pocket for substrate and that the oxidation is strictly outer-sphere (Solomon et al. 1996). In other multicopper oxidases there is a significant degree of substrate specificity ($K_m <$ 1 mM), which suggests there is a substrate binding pocket (Solomon et al. 1996). Both ceruloplasmin and Fet3 show high specificities for Fe(II). In a similar manner, the apparent K_m for Mn(II) of the Mn(II)-oxidizing protein of *L. discophora* is in the range of 7 to 13 μM (Adams and Ghiorse 1985, Boogerd and de Vrind 1987) and SG–1 has a similarly high affinity for Mn(II) (Tebo unpublished results, Rosson and Nealson 1982, Lee and Tebo 1994). This would argue that the role of MnxG and MofA are indeed for Mn(II) oxidation. A possibility that needs to be considered, however, is that the role these proteins play are not so much for Mn(II) oxidation but for some other form of metal metabolism. For example, metals of similar size, like Fe(II), that could fit into the substrate binding pocket, might be the real substrates for these proteins which might function in iron storage, transport, or oxidation. It is interesting to note that *L. discophora* also makes an Fe(II)-oxidizing protein that appears to be distinct from the Mn(II)-oxidizing protein, although the activities of both proteins appear to be linked in some fashion (Corstjens et al. 1992). Alternatively, the proteins may really be for copper uptake or detoxification.

The role of the multicopper oxidases in SG–1 and *L. discophora* might also not have anything to do with metals (other than using Cu as a cofactor) but rather might play some sort of structural role. Interestingly, in searching the protein sequence databases for similarities to MofA, the protein that shows the greatest similarity is CotA, a *Bacillus subtilis* spore coat protein that is involved in the formation of a brown pigment on sporulating colonies (but does not appear to be Mn oxide, van Waasbergen et al. 1993). This pigment is somehow related to manganese as it becomes darker as more Mn(II) is provided in the growth medium (Iichinska 1960). The connection between this relationship and MnxG or MofA is unclear at this time, however, because CotA is a spore coat protein, a structural function is implied. One possibility is that MnxG (and CotA) and MofA play similar structural and assembly roles in the cross-linking of the spore coats or exosporium and the *Leptothrix* sheath, respectively. This is supported by the fact that another copper protein (albeit not a multicopper oxidase), lysyl oxidase, is involved in cross-linking of collagen in animals.

As mentioned above, the high affinity of SG–1 and *L. discophora* for Mn(II) suggests that Mn(II) oxidation is the real function of the MnxG and MofA proteins. If so, what purpose does Mn(II) oxidation serve the cells? There is no indication that Mn(II) oxidation provides cellular energy for these organisms. In *L. discophora* Mn oxide formation has been demonstrated to prolong the survival of stationary phase cells (Adams and Ghiorse 1985) possibly by reacting with toxic H_2O_2 or other toxic oxygen intermediates or sequestering toxic cations. Similarly, Mn oxide formation by SG–1 spores could enhance the spore's resistance to harsh chemicals, however, spores produced by mutants in the *mnx* operon did not appear to have altered resistances (van Waasbergen et al. 1996). Alternatively, perhaps activities related to Mn(II) oxidation enhance metal uptake during formation of SG-1 spores. Mn oxide formation could also benefit the bacteria by scavenging essential nutrients (Adams and Ghiorse 1985) that could be used by *Leptothrix* cells or by SG-1 cells upon germination of the spores. Although, as discussed above, it has also been suggested that Mn(II) oxidation by SG-1 spores may serve as a means of storing an electron acceptor for germination and outgrowth of SG-1 spores under oxygen-limiting conditions (Tebo 1983). It is unlikely that *L. discophora* uses the oxides in this manner, as the members of the *Sphaerotilus-Leptothrix* group appear to be obligate aerobes (van Veen et al. 1978). Clearly, further work addressing the function of Mn(II) oxidation in these bacteria is needed.

INTERACTIONS OF MN(II)-OXIDIZING BACTERIA WITH OTHER METALS

There have been a number of environmental studies that suggest that metals other than Mn are acted upon by Mn(II)-oxidizing bacteria, most notably cobalt and cerium (Tebo 1983, Tebo et al. 1984, Moffett 1994a,b; Moffett and Ho 1996). We also know from surface chemical studies of Mn oxide minerals that a variety of different metals and radionuclides are adsorbed to the mineral surface (and sometimes oxidized) or intercalated within the crystal structure. One needs only look at the enrichment of trace metals in deep sea ferromanganese nodules to gain an appreciation for the scavenging properties of these oxides. Because of these properties, Mn(II)-oxidizing bacteria have the potential to mediate the removal and transport of many metals in the environment.

There have been few studies that examined the direct interaction (i.e. in the absence of MnO_x) between Mn(II)-oxidizing bacteria or their cell components and other metals. Clearly, most bacterial surfaces have negative charges, so a variety of metals will passively adsorb or bind to the surface. For example, the sheath of *L. discophora* is composed of a matrix of anionic heteropolysaccharides and proteins which are associated with the outer membrane of the Gram-negative cell wall (Emerson and Ghiorse 1993). The negatively charged capsular polymers readily bind cationic colloidal iron and ferritin. The Mn(II)- and Fe(II)-oxidizing proteins of *L. discophora* are found in the sheath, on which the Mn and Fe oxides precipitate. Whether these proteins oxidize other metals or perhaps organics, such as phenolic compounds in association with Mn(II) and Fe(II) oxidation, has not been examined.

The *Bacillus* sp. strain SG-1 spores have the capability of oxidizing Co(II) (Tebo and Lee 1993, Lee and Tebo 1994) and Fe(II) (Tebo and Edwards, unpublished) in addition to Mn(II). These oxidations occur even in the absence of Mn(II) or preformed MnO_x. The Co(II)-oxidizing properties of SG-1 are very similar to those for Mn(II) oxidation. Co(II) oxidation occurs over a wide range of pH, temperature, and Co(II) concentrations, and is oxygen dependent. Co(II) oxidation occurs optimally around pH 8 and 55° to 60°C. Co(II) is oxidized from the trace levels found in seawater to 100 mM and the oxidation follows Michaelis-Menton kinetics. An Eadie-Hofstee transformation of the data suggests that SG-1 spores have two oxidation systems, a high-affinity–low rate system (K_m, 3.3×10^{-8} M; V_{max}, 1.7×10^{-15} M·spore^{-1}·h^{-1}) and a low-affinity–high rate system (K_m, 5.2×10^{-6} M; V_{max}, 8.9×10^{-15} M·spore^{-1}·h^{-1}). Because mutants in the *mnx* operon that do not oxidize Mn(II) also do not oxidize Co(II), it is suspected that Co(II) oxidation occurs at the same active site as for Mn(II) oxidation. Interestingly, the same phenomenon has been observed for the manganese peroxidase from fungi (Souren 1997).

SG-1 spores are also capable of binding a variety of other metals, including Cd(II) and Zn(II), in the absence of Mn oxides. Although we cannot be certain that these metals do not bind within the substrate binding site of the Mn(II)-oxidizing protein, it is likely that this metal binding occurs in a non-specific manner. As mentioned previously, Cu(II) is also rapidly and extensively bound, however, the relation between Cu binding and the MnxG multicopper oxidase-like protein in the exosporium is currently unknown.

CONCLUSIONS

The application of molecular biology to studies of bacterial Mn(II) oxidation has revolutionized how we think about Mn oxide biomineralization and opened the doors for a number of new and exciting research opportunities. Molecular biological approaches may someday be used to modify active sites of the proteins to alter metal binding affinity or specificity. Proteins can also be engineered to increase expression of Mn(II)-oxidizing activity. Genetics can be used to clone the Mn-oxidizing genes into other organisms, generating novel metabolic capabilities. New insights into the oxidation mechanism can be gleaned by spectroscopic (e.g. X-ray absorption spectroscopic) studies of the Mn(II)-oxidizing proteins. This technique can be used to examine the role of copper in Mn(II) oxidation, analyze Mn speciation and its coordination environment, and determine the solid phase products that are formed. It will also be extremely exciting to characterize the microenvironments of the microbe–water and microbe–mineral interface using modern surface chemical and imaging techniques. Knowledge of these processes are a necessary starting point for understanding the complex suites of chemical reactions and transformations that occur at microbe-water interfaces and essential to the development of realistic geochemical models of metal cycling in the environment. The potential applications of Mn(II)-oxidizing bacteria and their proteins for new material syntheses (e.g. catalysts) or metal removal and recovery (e.g. in bioremediation or water treatment) are just two examples of the many possible societal uses for these bacteria (Ghiorse 1986, Tebo 1995).

The development of an understanding of Mn(II) oxidation at the molecular level will be of great use in environmental studies. During the past 10-15 years there have been advances allowing measurement of biologically-mediated Mn(II) oxidation activity in natural samples (Tebo and Emerson 1985, 1986; Sunda and Huntsman 1987, 1988; Juniper and Tebo 1995). More recently, culture-independent methods facilitating the detection of Mn(II)-oxidizing microorganisms, their genes, and gene expression are currently being developed to address distribution, abundance and activity questions (Siering and Ghiorse 1997). Future work should include demonstrating the presence of certain genes in samples where we can measure or detect Mn(II)-oxidizing gene transcripts, and determination of the rates of biologically mediated Mn(II) oxidation activity (Tebo and Emerson 1985, 1986; Sunda and Huntsman 1987, 1988) Unfortunately, the lack of detailed knowledge of the genetics and biochemistry of specific bacterial Mn(II)-oxidizing systems prevents a comprehensive understanding of the relative contribution of particular bacteria to the overall manganese cycling in any given environment.

ACKNOWLEDGMENTS

B.M. Tebo, L.G. van Waasbergen and R. Caspi are grateful for funding from the National Science Foundation (OCE94-168944 and MCB94-07776) and the University of California Toxic Substances Research and Teaching Program. Funding has also been provided by the National Sea Grant College Program, National Oceanic and Atmospheric Administration, U.S. Department of Commerce (#NA66RG0477 Project R/CZ 123) through the California Sea Grant College System and the California State Resources Agency. BMT acknowledges the contributions of many former and current undergraduate and graduate students, post-docs and technicians, including K. Mandernack, Y. Lee,

L. He, L. Park, L. Knight, K. Casciotti, C. Francis, and D. Edwards, without whom the work would not have been accomplished.

W.C. Ghiorse and P.L Siering acknowledge support from USDA Hatch Act funds, the National Science Foundation under Grant BCS-91-00209, and the National Institute of Environmental Health Sciences, Superfund Basic Research Program.

REFERENCES

Abdrashitova SA, Abdullina GG, Mynbaeva BN, Ilyaletdinov AN (1990) Oxidation of iron and manganese by arsenic-oxidizing bacteria. Mikrobiologiya 59:85-89
Adams LF (1986). Physiology, cytology,and manganese-oxidizing activity of *Leptothrix discophora* SS-1. PhD Dissertation, Cornell University, Ithaca, New York
Adams LF, Ghiorse WC (1985) Influence of manganese on growth of a sheathless strain of *Leptothrix discophora* SS-1. Appl Environ Microbiol 49:556-562
Adams LF, Ghiorse WC (1987) Characterization of extracellular Mn^{2+}-oxidizing activity and isolation of an Mn^{2+}-oxidizing protein from *Leptothrix discophora* SS-1. J Bacteriol 169:1279-1285
Adams LF, Ghiorse WC (1988) Oxidation state of Mn in the Mn oxide produced by *Leptothrix discophora* SS-1. Geochim Cosmochim Acta 52:2073-2076
Aitken MD, Irvine RL (1990) Characterization of reactions catalyzed by manganese peroxidase from *Phanerochaete chrysosporium*. Arch Biochem Biophys 276:405-414
Altschul SF, Ghish W, Miller W, Myers EM, Lipman DJ (1990) Basic local alignment search tool. J Mol Biol 215:403-410
Archibald F, Roy B (1992) Production of manganic chelates by laccase from the lignin-degrading fungus *Trametes (Coriolus) versicolor*. Appl Environ Microbiol 58:1496-1499
Arcuri EJ, Ehrlich HL (1979) Cytochrome involvement in Mn(II)-oxidation by two marine bacteria. Appl Environ Microbiol 37:916-923
Arrhenius G, Tasi AG (1981) Structure, phase transformation and prebiotic catalysis in marine manganate minerals. In: SIO Reference Series, p 1-19 Scripps Institution of Oceanography, La Jolla, California
Avissar YJ, Ormerod JG, Beale SI (1989) Distribution of δ-aminolevulinic acid biosynthesis pathways among phototrophic bacterial groups. Arch Microbiol 151:513-519
Bacon MP, Brewer PG, Spencer DW, Murray JW, Goddard J (1980) Lead-210, polonium-210, manganese and iron in the Cariaco Trench. Deep-Sea Res 27:119-135
Bairoch A, Butcher P, Hofmann K (1995) The PROSITE database, its status in 1995. Nucleic Acids Res 24:189-196
Barr DP, Aust SD (1994) Mechanisms white rot fungi use to degrade pollutants. Environ Sci Technol 28:78A-87A
Beveridge TJ (1989) Role of cellular design in bacterial metal accumulation and mineralization. Ann Rev Microbiol 43:147-71
Blanchette RA (1984) Manganese accumulation in wood decayed by white rot fungi. Phytopathology 74:725-730
Boogerd RC, de Vrind JPM (1987) Manganese oxidation by *Leptothrix discophora*. J Bacteriol 169:489-494
Bromfield SM (1956) Oxidation of manganese by soil microorganisms. Austr J Biol Sci 9:238-252
Burns RG, Burns VM (1979) Manganese oxides. In: RG Burns (ed) Marine Minerals. Rev Mineral 6:1-46
Caspi R (1996). Molecular biological studies of manganese oxidizing bacteria. PhD Dissertation, University of California, San Diego
Caspi R, Haygood MG, Tebo BM (1996) Unusual ribulose-1,5-bisphosphate carboxylase/oxygenase genes from a marine manganese-oxidizing bacterium. Microbiology 142:2549-2559
Caspi R, Tebo BM, Haygood MG (submitted) c-type cytochromes and manganese oxidation in *Pseudomonas putida* strain MnB1.
Chapnick S, Moore WS, Nealson KH (1982) Microbially mediated manganese oxidation in a freshwater lake. Linmol Oceanogr 17:1004-1014
Codispoti LA, Friederich GE, Murray JW, Sakamoto C (1991) Chemical variability in the Black Sea: Implication of data obtained with a continuous vertical profiling system that penetrated the oxic/anoxic interface. Deep-Sea Res 38:S691-S710
Corstjens P (1993). Bacterial oxidation of iron and manganese—a molecular biological approach. PhD Dissertation, Rijksuniversiteit te Leiden, the Netherlands
Corstjens PLAM, de Vrind JPM, Goosen T, de Vrind-de Jong EW (1997) Identification and molecular analysis of the *Leptothrix discophora* SS-1 *mofA* gene, a gene putatively encoding a manganese-oxidizing protein with copper domains. Geomicrobiol J 14:91-108

Corstjens PLAM, de Vrind JPM, Westbroek P, de Vrind-de Jong EW (1992) Enzymatic iron oxidation by *Leptothrix discophora*: Identification of an iron-oxidizing protein. Appl Environ Microbiol 58:450-454

Cowen JP, Massoth GJ, Feely RA (1990) Scavenging rates of dissolved manganese in a hydrothermal vent plume. Deep-Sea Res 37:1619-1637

da Silva JJRF, Williams RJP (1991) The Biological Chemistry of the Elements. Clarendon Press, Oxford

Davies SHR, Morgan JJ (1989) Manganese(II) oxidation kinetics on metal oxide surfaces. J Colloid Interface Sci 129:63-77

Davis JA, Kent DB (1990) Surface complexation modeling in aqueous geochemistry. In: MF Hochella, AF White (eds) Mineral-Water Interface Geochemistry. Rev Mineral 23:177-260

de Vrind JPM, Boogerd FC, deVrind-deJong EW (1986a) Manganese reduction by a marine *Bacillus* species. J Bacteriol 167:30-34

de Vrind JPM, de Vrind-de Jong EW, de Voogt J-WH, Westbroek P, Boogerd FC, Rosson RA (1986b) Manganese oxidation by spores and spore coats of a marine *Bacillus* species. Appl Environ Microbiol 52:1096-1100

DePalma SR (1993). Manganese oxidation by *Pseudomonas putida*. PhD Dissertation, Harvard University, Cambridge, Massachusetts

Diem D, Stumm W (1984) Is dissolved Mn^{2+} being oxidized by O_2 in absence of Mn-bacteria or surface catalysts? Geochim Cosmochim Acta 48:1571-1573

Douka C (1977) Study of bacteria from manganese concretions. Soil Biol. Biochem. 9:89-97

Douka C (1980) Kinetics of manganese oxidation by cell-free extracts of bacteria isolated from manganese concretions from soil. Appl Environ Microbiol 39:74-80

Drits VA, Silvester E, Gorshkov AI, Manceau A (1997) The structure of synthetic monoclinic Na-rich birnessite and hexagonal birnessite. Part 1. Results from X-ray diffraction and selected area electron diffraction. Am Mineral (in press)

Ehrlich HL (1968) Bacteriology of manganese nodules II. Manganese oxidation by cell-free extract from a manganese nodule bacterium. Appl Microbiol 16:197-202

Ehrlich HL (1976) Manganese as an energy source for bacteria. In: O Nriagu (ed) Environmental Biogeochemistry 2:633-644, Ann Arbor Science Publishers, Ann Arbor, Michigan

Ehrlich HL (1983) Manganese oxidizing bacteria from a hydrothermally active area on the Galapagos Rift. Ecol Bull (Stockholm) 35:357-66

Ehrlich HL (1984) Different forms of bacterial manganese oxidation. In: WR Strohl, OH Tuovinen (eds) Microbial chemoautotrophy p 47-56, The Ohio State University Press, Columbus, OH

Ehrlich HL (1996a) Geomicrobiology. Marcel Dekker, New York

Ehrlich HL (1996b) How microbes influence mineral growth and dissolution. Chem Geol 132:5-9

Ehrlich HL, Salerno JC (1990) Energy coupling in Mn^{2+} oxidation by a marine bacterium. Arch Microbiol 154:12-17

Eide D, Guerinot ML (1997) Metal ion uptake in eurkaryotes. ASM News 63:199-205

Ellis LBM, Saurugger P, Woodward C (1992) Identification of the three-demensional thioredoxin motif: related structure in the ORF3 protein in the *Staphylococcus aureus mer* operon. Biochemistry 31:4882-4891

Emerson D (1989). Ultrastructural organization, chemical composition, and manganese-oxidizing properties of the sheath of *Leptothrix discophora* SP-6. PhD Dissertation, Cornell University, Ithaca, New York

Emerson D, Garen RE, Ghiorse WC (1989) Formation of *Metallogenium*-like structures by a manganese-oxidizing fungus. Arch Microbiol 151:223-231

Emerson D, Ghiorse WC (1992) Isolation, cultural maintenance, and taxonomy of a sheath-forming strain of *Leptothrix discophora* and characterization of manganese-oxidizing activity associated with the sheath. Appl Environ Microbiol 58:4001-4010

Emerson D, Ghiorse WC (1993) Ultrastructure and chemical composition of the sheath of *Leptothrix discophora* SP-6. J Bacteriol 175:7808-7818

Ferreira GC, Gong J (1995) 5-Aminolevulinate synthase and the first step of heme biosynthesis. J Bioenerg Biomembr 27:151-158

Ferris FG, Fyfe WS, Beveridge TJ (1987) Manganese oxide deposition in a hot spring microbial mat. Geomicrobiol J 5:33-42

Francis CA, Casciotti KL, Tebo BM (1997) Manganese oxidizing activity localized to the exosporium of spores of the marine *Bacillus* sp. strain SG–1. In: Ann Mtg American Society of Microbiology Miami, FL, May 1997, American Society of Microbiology, Washington, DC

Friedl G, Wehrli B, Manceau A (1997) Solid phases in the cycling of manganese in eutrophic lakes: New insights from EXAFS spectroscopy. Geochim Cosmochim Acta 61:275-290

Ghiorse WC (1984a) Bacterial transformations of manganese in wetland environments. In: MJ Klug, CA Reddy (eds) Current Perspectives in Microbial Ecology, p 615-622 American Society of Microbiology, Washington, DC

Ghiorse WC (1984b) Biology of iron- and manganese-depositing bacteria. Ann Rev Microbiol 38:515-550

Ghiorse WC (1986) Applicability of ferromanganese-depositing microorganisms to industrial metal recovery processes. Biotechnology and Bioengineering Symp 16:141-148

Ghiorse WC, Chapnick SD (1983) Metal-depositing bacteria and the distribution of manganese and iron in swamp waters. Environmental Biogeochemistry. Ecol Bull (Stockholm) 35:367-376

Ghiorse WC, Ehrlich HL (1974) Effects of seawater cations and temperature on manganese dioxide-reductase activity in a marine *Bacillus*. Appl Microbiol 28:785-792

Ghiorse WC, Ehrlich HL (1976) Electron transport components of the MnO_2 reductase system and the location of the terminal reductase in a marine *Bacillus*. Appl Environ Microbiol 31:977-985

Ghiorse WC, Ehrlich HL (1992) Microbial biomineralization of iron and manganese. In: HCW Skinner, RW Fitzpatrick (eds) Biomineralization Processes of Iron and Manganese: Modern and Ancient Environments 21:75-99, Catena Supplement, Germany

Giovanoli R, Arrhenius G (1988) Structural chemistry of marine manganese and iron minerals and synthetic model compounds. In: P Halbach (ed) The Manganese Nodule Belt of the Pacific Ocean: Geological Environment, Nodule Formation, and Mining Aspects, p 20-37 Ferdinand Enke Verlag

Glenn JK, Akileswaran L, Gold MH (1986) Mn(II) oxidation is the principal function of the extracellular Mn-peroxidase from *Phanerochaete chrysosporium*. Arch Biochem Biophys 251:688-696

Glerum DM, Shtanko A, Tzagoloff A (1996) SCO1 and SCO2 act as high copy suppressors of a mitochondrial copper recruitment defect in *Saccharomyces cerevisiae*. J Biol Chem 271:20531-20535

Golden DC, Chen CC, Dixon JB (1986) Synthesis of todorokite. Science 231:717-719

Golden DC, Zuberer DA, Dixon JB (1992) Manganese oxides produced by fungal oxidation of manganese from siderite and rhodochrosite. In: HCW Skinner, RW Fitzpatrick (eds) Biomineralization Processes of Iron and Manganese: Modern and Ancient Environments 21:161-168, Catena Supplement, Germany

Graham LA, Salerno JC, Ehrlich HL (1987) Electron transfer components of manganese-oxidizing bacteria. In: CH Kim (ed) Advances in membrane biochemistry and bioenergetics. p 267-272, Plenum Press, New York

Greene AC, Madgwick JC (1991) Microbial formation of manganese oxides. Appl Environ Microbiol 57:1114-1120

Gregory E, Staley JT (1982) Widespread distribution of ability to oxidize manganese among freshwater bacteria. Appl Environ Microbiol 44:509-511

Hastings D, Emerson S (1986) Oxidation of manganese by spores of a marine Bacillus: kinetic and thermodynamic considerations. Geochim Cosmochim Acta 50:1819-1824

He LM, Tebo BM (submitted) Surface characterization of and Cu(II) adsorption by spores of a marine *Bacillus*. Appl Environ Microbiol

Hem JD (1981) Rates of manganese oxidation in aqueous systems. Geochim Cosmochim Acta 45:1369-1374

Hem JD, Lind CJ (1983) Nonequilibrium models for predicting forms of precipitated manganese oxides. Geochim Cosmochim Acta 47:2037-2046

Hem JD, Roberson E, Fournier RB (1982) Stability of β-MnOOH and manganese oxide deposition from springwater. Water Resource Res 18:563-570

Hogan VC (1973). Electron transport and manganese oxidation in *Leptothrix Discophorus*. Ohio State University Press, Columbus, OH

Hsieh C-J, Jones GH (1995) Nucleotide sequence, transcriptional analysis, and glucose regulation of the phenoxazinone synthase gene (*phsA*) from *Streptomyces antibioticus*. J Bacteriol 177:5740-5747

Huang PM (1991) Kinetics of redox reactions on manganese oxides and its impact on environmental quality. In: DL Sparks, DL Suarez (eds) Rates of Soil Chemical Processes, p 191-230 Soil Science Society of America, Madison, Wisconsin

Huber R (1989) A structural basis of light energy and electron transfer in biology. Angew Chem Intl Ed Engl 28:848-869

Iichinska E (1960) Some physiological features of asporogenous mutants of bacilli. Mikrobiologiya 29:147-150

Ingledew WJ, Cox JC, Halling PJ (1977) A proposed mechanism for energy conservation during Fe^{2+} oxidation by *Thiobacillus ferrooxidans*: chemiosmotic coupling to net H^+ influx. FEMS Microbiol Let. 2:193-197

Jackson DD (1901) The precipitation of iron, manganese, and aluminium by bacterial action. J Soc Chem Ind 21:681-684

Jannasch HW, Mottl MJ (1985) Geomicrobiology of deep-sea hydrothermal vents. Science. 229:717-725

Jenne EA (1967) Controls on Mn, Fe, Co, Ni, Cu and Zn concentrations in soils and water: the signficant role of hydrous Mn and Fe oxides. In: Trace Inorganics in Water, p 337-387 American Chemical Society, Washington, DC,

Jung WK, Schweisfurth R (1976) Manganoxydierende bakterien III. Wachstum und manganoxydation bei *Pseudomonas manganoxidans* Schw. Z Allg Mikrobiol 16:587-597

Jung WK, Schweisfurth R (1979) Manganese oxidation by an intracellular protein of a *Pseudomonas* species. Z Allg Mikrobiol 19:107-115

Juniper SK, Tebo BM (1995) Microbe-metal interactions and mineral deposition at hydrothermal vents. In: DM Karl (ed) The Microbiology of Deep-Sea Hydrothermal Vents, p 219-253 CRC Press, Boca Raton, Florida

Junta JL, Hochella Jr MF (1994) Manganese(II) oxidation at mineral surfaces: A microscopic and spectroscopic study. Geochim Cosmochim Acta 58:4985-4999

Kalhorn S, Emerson S (1984) The oxidation state of manganese in surface sediments of the Pacific Ocean. Geochim Cosmochim Acta 48:897-902

Kaplan J, O'Halloran TV (1997) Iron metabolism in eukaryotes: Mars and Venus at it again. Science 271:1510-1512

Kepkay P, Nealson KH (1987a) Growth of a manganese oxidizing *Pseudomonas* sp. in continuous culture. Arch Microbiol 148:63-67

Kepkay PE, Nealson KH (1982) Surface enhancement of sporulation and manganese oxidation by a marine bacillus. J Bacteriol 151:1022-1026

Kepkay PE, Nealson KH (1987b) Growth of a manganese oxidizing *Pseudomonas* sp. in continuous culture. Arch Microbiol 148:63-67

Kieber DJ, McDaniel J, Mopper K (1989) Photochemical source of biological substrates in sea water—implication for carbon cycling. Nature 341:637-639

Kinniburgh DG, Jackson ML (1981) Cation adsorption by hydrous metal oxides and clays. In: MA Anderson, AJ Rubin (eds) Adsorption of Inorganics at Solid-Liquid Interfaces p 91-160, Ann Arbor Science Publishers, Ann Arbor, Michigan

Kostka JE, Luther III GW, Nealson KH (1995) Chemical and biological reduction of Mn(III)-pyrophosphate complexes—potential importance of dissolved Mn(III) as an environmental oxidant. Geochim Cosmochim Acta 59:885-894

Krueger WB, Kolodziej WB (1976) Measurement of cellular copper levels in Bacillus megaterium during exponential growth and sporulation. Microbios 17:141-147

Laddaga RA, Chu L, Misra TK, Silver S (1987) Nucleotide sequence and expression of the mercurial resistance operon from *Staphylococcus aureus* plasmid pI258. Proc Natl Acad Sci 84:5106-5110

Laha S, Luthy RG (1990) Oxidation of aniline and other primary aromatic amines by manganese dioxide. Environ Sci Technol 24:363-373

Lee Y, Tebo BM (1994) Cobalt oxidation by the marine manganese(II)-oxidizing *Bacillus* sp. strain SG-1. Appl Environ Microbiol 60:2949-2957

Lewis BL, Landing WM (1991) The biogeochemistry of manganese and iron in the Black Sea. Deep-Sea Res 38:S773-S803

Luther III GW (1990) The frontier-molecular-orbital theory approach in geochemical processes. In: W Stumm (ed) Aquatic Chemical Kinetics, p 173-198, John Wiley & Sons, New York

Manceau A, Combes JM (1988) Structure of Mn and Fe oxides and oxyhydroxides: A topological approach by EXAFS. Phys Chem Minerals 15:283-295

Manceau A, Gorshkov AI, Drits VA (1992a) Structural chemistry of Mn, Fe, Co, and Ni in manganese hydrous oxides: Part I. Information from XANES spectroscopy. Am Mineral 77:1133-1143

Manceau A, Gorshkov AI, Drits VA (1992b) Structural chemistry of Mn, Fe, Co, and Ni in manganese hydrous oxides: Part II. Information from EXAFS spectroscopy and electron and X-ray diffraction. Am Mineral 77:1144-1157

Mandernack KW, Fogel ML, Usui A, Tebo BM (1995a) Oxygen isotope analyses of chemically and microbially produced manganese oxides and manganates. Geochim Cosmochim Acta 59:4409-4425

Mandernack KW, Post J, Tebo BM (1995b) Manganese mineral formation by bacterial spores of a marine *Bacillus*, strain SG-1: Evidence for the direct oxidation of Mn(II) to Mn(IV). Geochim Cosmochim Acta 59:4393-4408

Mandernack KW, Tebo BM (1993) Manganese scavenging and oxidation at hydrothermal vents and in vent plumes. Geochim Cosmochim Acta 57:3907-3923

Mann S, Sparks NHC, Scott GHE, de Vrind-de Jong EW (1988) Oxidation of manganese and formation of Mn_3O_4 (hausmannite) by spore coats of a marine *Bacillus* sp. Appl Environ Microbiol 54:2140-2143

Mita N, Maruyama A, Usui A, Higashihara T, Hariya Y (1994) A growing deposit of hydrous manganese oxide prodcued by microbial mediation at a hot spring, Japan. Geochem J 28:71-80

Miyajima T (1992) Biological manganese oxidation in a lake I: Occurrence and distribution of *Metallogenium* sp. and its kinetic properties. Arch Hydrobiol 124:317-335

Moffett J, Ho J (1996) Oxidation of cobalt and manganese in seawater via a common microbially catalyzed pathway. Geochim Cosmochim Acta 60:3415-3424

Moffett JW (1994a) A radiotracer study of cerium and manganese uptake onto suspended particles in Chesapeake Bay. Geochim Cosmochim Acta 58:695-703

Moffett JW (1994b) The relationship between cerium and manganese oxidation in the marine environment. Linmol Oceanogr 39:1309-1318

Morgan JJ, Stumm W (1964) Colloid-chemical properties of manganese dioxide. J Colloid Sci 19:347-359

Morgan JJ, Stumm W (1965) The role of multivalent metal oxides in limnological transformations as exemplified by iron and manganese. In: O Jaag (ed), Second Water Pollution Research Conference 1, p 103-131 Pergamon, New York

Muller J-P, Manceau A, Calas G, Allard T, Ildefonse P, Hazemann J-L (1995) Crystal chemistry of kaolinite and Fe-Mn oxides: relation with formation conditions of low temperature systems. Am J Sci 295:1115-1155

Murray JW, Dillard JG, Giovanoli R, Moers H, Stumm W (1985) Oxidation of Mn(II): Initial mineralogy, oxidation state and aging. Geochim Cosmochim Acta 49:463-470

Nealson KH (1978) The isolation and characterization of marine bacteria which catalyze manganese oxidation. In: WE Krumbein (ed) Environmental Biogeochemistry and Geomicrobiology Methods, Metals and Assessment 3:847-858 Ann Arbor Science Publishers, Ann Arbor, Michigan

Nealson KH (1983) The microbial manganese cycle. In: WE Krumbein (ed) Microbial Geochemistry, p 191-221 Blackwell Scientific, Boston

Nealson KH, Ford J (1980) Surface enhancement of bacterial manganese oxidation: Implications for aquatic environments. Geomicrobiol J 2:21-37

Nealson KH, Rosson RA, Myers CR (1989) Mechanisms of oxidation and reduction of manganese. In: TJ Beveridge, RJ Doyle (eds) Metal Ions and Bacteria, p 383-411 John Wiley & Sons, New York

Nealson KH, Tebo BM, Rosson RA (1988) Occurrence and mechanisms of microbial oxidation of manganese. Adv Appl Microbiol 33:279-318

Perez J, Jeffries TW (1992) Roles of manganese and organic acid chelators in regulating lignin degradation and biosynthesis of peroxidases by *Phanerochaete chrysosporium.* Appl Environ Microbiol 58:2401-2409

Piper DZ, Basler JR, Bischoff JL (1984) Oxidation state of marine manganese nodules. Geochim Cosmochim Acta 48:2347-2355

Popp JL, Kalyanaraman B, Kirk TK (1990) Lignin peroxidase oxidation of Mn^{+2} in the presence of veratryl alcohol, malonic or oxalic acid, and oxygen. Biochem 29:10475-10480

Post JE, Veblen DR (1990) Crystal structure determinations of synthetic sodium, magnesium, and potassium birnessite using TEM and the Rietveld method. Am Mineral 75:477-489

Potter RM, Rossman GR (1979) The tetravalent manganese oxides: identification, hydration, and structural relationships by infrared spectroscopy. Am Mineral 64:1199-1218

Richardson LL, Aguilar C, Nealson KH (1988) Manganese oxidation in pH and O_2 microenvironments produced by phytoplankton. Linmol Oceanogr 33:352-363

Rosson RA, Nealson KH (1982) Manganese binding and oxidation by spores of a marine bacillus. J Bacteriol 151:1027-1034

Ryden LG, Hunt LT (1993) Evolution of protein complexity: the blue copper-containing oxidases and related proteins. J Molec Evol 36:41-66

Schutt C, Ottow JCG (1978) Distribution and identification of manganese-precipitating bacteria from noncontaminated ferromanganese nodules. In: WE Krumbein (ed) Environmental Biogeochemistry and Geomicrobiology: Methods, Metals and Assessment 3:869-878, Ann Arbor Science, Ann Arbor, Michigan

Schweisfurth R (1973) Manganoxydierende Bakterien I. Isolierung und bestimmung einiger stamme von manganbakterien. Z Allg Mikrobiol 13:341-347

Siering PL (1996). Application of molecular approaches to investigate the role of microorganisms in manganese cycling in wetland environments. PhD Dissertation, Cornell University, Ithaca, New York

Siering PL, Ghiorse WC (1997) PCR detection of a putative manganese oxidation gene (*mofA*) in environmental samples and assessment of *mofA* homology among diverse manganese-oxidizing bacteria. Geomicrobiol J 14:109-125

Silvester E, Manceau A (1997) The structure of synthetic monoclinic Na-rich birnessite and hexagonal birnessite. Part 2. Results from chemical studies and EXAFS spectroscopy. Am Mineral (in press)

Solomon EI, Sundaram UM, Machonkin TE (1996) Multicopper oxidases and oxygenases. Chem Rev 96:2563-2605

Souren AWMG (1997) Comment on "Oxidation of cobalt and manganese in seawater via a common microbially catalyzed pathway" by J.W. Moffett and J. Ho. Geochim Cosmochim Acta (n press)

Stearman R, Yuan DS, Yamaguchi-Iwai Y, Klausner RD, Dancis A (1996) A permease-oxidase complex involved in high-affinity iron uptake in yeast. Science 271:1552-1557

Stone AT (1983). The reduction and dissolution of Mn(III) and Mn(IV) oxides by organics. PhD Dissertation, 302 p California Institute of Technology, Pasadena, CA

Stone AT (1987) Reductive dissolution of manganese(III/IV) oxides by substituted phenols. Environ Sci Technol 21:979-988

Stone AT, Morgan JJ (1984a) Reduction and dissolution of manganese(III) and manganese(IV) oxides by organics. 1. Reaction with hydroquinone. Environ Sci Technol 18:450-456

Stone AT, Morgan JJ (1984b) Reduction and dissolution of manganese(III) and manganese(IV) oxides by organics. 2. Survey of the reactivity of organics. Environ Sci Technol 18:617-624

Stone AT, Ulrich H-J (1989) Kinetics and reaction stoichiometry in the reductive dissolution of manganese(IV) dioxide and Co(III) oxide by hydroquinone. J Colloid Interface Sci 132:509-522

Stumm W, Morgan JJ (1996) Aquatic Chemistry. Chemical Equilibria and Rates in Natural Waters. John Wiley & Sons, New York

Sunda WG, Huntsman SA (1987) Microbial oxidation of manganese in a North Carolina estuary. Linmol Oceanogr 32:552-564

Sunda WG, Huntsman SA (1988) Effect of sunlight on redox cycles of manganese in the southwestern Sargasso Sea. Deep-Sea Res 35:1297-1317

Sunda WG, Huntsman SA (1990) Diel cycles in microbial manganese oxidation and manganese redox speciation in coastal waters of the Bahama Islands. Linmol Oceanogr 35:325-338

Sunda WG, Huntsman SA, Harvey GR (1983) Photoreduction of manganese oxides in seawater and its geochemical and biological implications. Nature 301:234-236

Sunda WG, Kieber DJ (1994) Oxidation of humic substances by manganese oxides yields low-molecular-weight organic substrates. Nature 367:62-64

Sung W, Morgan JJ (1981) Oxidative removal of Mn(II) from solution catalyzed by the γ-FeOOH (lepidocrocite) surface. Geochim Cosmochim Acta 45:2377-2383

Swofford DL (1990) PAUP: Phylogenetic analysis using parsimony.

Takematsu N, Kkabe H, Sato Y, Okabe S (1988) Todorokite formation in seawater by microbial mediation. J Oceanograph Soc Japan. 44:235-243

Tebo BM (1983). The ecology and ultrastructure of marine manganese oxidizing bacteria. PhD Dissertation, University of California, San Diego

Tebo BM (1991) Manganese(II) oxidation in the suboxic zone of the Black Sea. Deep-Sea Res 38:S883-S905

Tebo BM (1995) Metal precipitation by marine bacteria: potential for biotechnological applications. In: JK Setlow (ed) Genetic Engineering–Principles and Methods 17:231-263, Plenum Press, New York

Tebo BM (1997) Mn(II) oxidation in marine environments is likely bacterial. Geochim Cosmochim Acta (in press)

Tebo BM, Edwards DB, Kieber DJ, Sunda WG (1997) Bacterial Mn Oxidation Can Lead to the Degradation and Utilization of Natural Organic Matter. In: (ed), ASLO Aquatic Sciences Mtg, February, 1997, Santa Fe, New Mexico, Geochim Cosmochim Acta

Tebo BM, Emerson S (1985) The effect of oxygen tension, Mn(II) concentration and temperature on the microbially catalyzed Mn(II) oxidation rate in a marine fjord. Appl Environ Microbiol 50:1268-1273

Tebo BM, Emerson S (1986) Microbial manganese(II) oxidation in the marine environment: a quantitative study. Biogeochem 2:149-161

Tebo BM, Lee Y (1993) Microbial oxidation of cobalt. In: AE Torma, JE Wey, VL Lakshmanan (eds) Biohydrometallurgical Technologies I p 695-704, The Minerals, Metals, & Materials Society, Warrendale, Pennsylvania

Tebo BM, Mandernack K, Rosson RA (1988) Manganese oxidation by a spore coat or exosporium protein from spores of a manganese(II) oxidizing marine Bacillus, abstr. I-121. In: Abstr 88th Ann Mtg American Society of Microbiology 1988 p 201,

Tebo BM, Nealson KH, Emerson S, Jacobs L (1984) Microbial mediation of Mn(II) and Co(II) precipitation at the O_2/H_2S interfaces in two anoxic fjords. Linmol Oceanogr 29:1247-1258

Thamdrup B, Glud RN, Hansen JW (1994) Manganese oxidation and in situ fluxes from a coastal sediment. Geochim Cosmochim Acta 58:2563-2570

Tian ZR, Tong W, Wang JY, Duan NG, Krishnan VV, Suib SL (1997) Manganese oxide mesoporous structures: Mixed-valent semiconducting catalysts. Science 276:926-930

Tipping E (1984) Temperature dependence of Mn(II) oxidation in lakewaters: a test of biological involvement. Geochim Cosmochim Acta 48:1353-1356

Tipping E, Jones JG, Woof C (1985) Lacustrine manganese oxides: Mn oxidation states and relationships to "Mn depositing bacteria". Arch Hydrobiol 105:161-175

Tipping E, Thompson DW, Davison W (1984) Oxidation products of Mn(II) in lakewaters. Chem Geol 44:359-383

Todd JF, Elsinger RJ, Moore WS (1988) The distributions of uranium, radium and thorium isotopes in two anoxic fjords, Framvaren Fjord (Norway) and Saanich Inlet (British Columbia). Mar Chem 23:393-415

Turner S, Buseck PR (1981) Todorokites: A new family of naturally occurring manganese oxides. Science.212:1024-1027

Usui A, Mellin TA, Masato N, Makoto Y (1989) Structural stability of marine 10 Å manganates from the Ogasawara (Bonin) Arc: implication for low-temperature hydrothermal activity. Mar. Geol. 86:41-56 van Veen WL (1972) Factors affecting the oxidation of manganese by *Sphaerotilus discophorus*. Antonie van Leeuwenhoek 38:623-627

van Veen WL, Mulder EG, Deinema MH (1978) The *Sphaerotilus-Leptothrix* group of bacteria. Microbiol Rev 42:329-356

van Waasbergen LG, Hildebrand M, Tebo BM (1996) Identification and characterization of a gene cluster involved in manganese oxidation by spores of the marine *Bacillus* sp. strain SG-1. J Bacteriol 178:3517-3530

van Waasbergen LG, Hoch JA, Tebo BM (1993) Genetic analysis of the marine manganese-oxidizing *Bacillus* sp. strain SG-1: Protoplast transformation, Tn*917* mutagenesis and identification of chromosomal loci involved in manganese oxidation. J Bacteriol 175:7594-7603

Vandenabeele J, de Beer D, Germonpre R, Van de Sande R, Verstraete W (1995) Influence of nitrate on manganese removing microbial consortia from sand filters. Water Res 29:579-587

Vojak PWL, Edwards C, Jones MV (1984) Manganese oxidation and sporulation by an estuarine *Bacillus* species. Microbios 41:39-47

Wehrli B (1990) Redox reactions of metal ions at mineral surfaces. In: W Stumm (ed) Aquatic Chemical Kinetics p 311-336, John Wiley & Sons, New York

Wehrli B, Friedl G, Manceau A (1995) Reaction rates and products of manganese oxdiation at the sediment-water interface. In: CP Huang, CR O'Melia, JJ Morgan (eds) Aquatic Chemistry: Interfacial and Interspecies Processes. American Chemical Society, Washington, DC

Wei C-L, Murray JW (1991) [234]Th/[238]U disequilibria in the Black Sea. Deep-Sea Res 38:S855-S873

Zavarzin MA (1962) Symbiotic oxidation of manganese by two species of *Pseudomonas*. Mikrobiologiya 31:586-588

Chapter 8

ALGAL DEPOSITION OF CARBONATES AND SILICATES

Elisabeth W. de Vrind-de Jong and Johannes P.M. de Vrind

Leiden Institute of Chemistry
Gorlaeus Laboratory, Leiden University
Einsteinweg 55, P O Box 9502
2300 RA Leiden, The Netherlands

INTRODUCTION

Living organisms produce a wide variety of mineral structures known as biominerals (Lowenstam and Weiner 1989). These biominerals consist of different inorganic salts and can be amorphous or crystalline. They can have diverse functions or no apparent function at all, and may be formed under strict cellular control or in mere loose association with the living organism. The most abundant biominerals consist of calcium salts (carbonates and phosphates) or silicates. An enormous proportion of the calcium carbonate- and silicate biominerals are produced by eukaryotic, photosynthetic members of the kingdom of the Protoctysta: the algae (cf. Margulis et al. 1993). The algal calcium carbonate-biominerals are crystalline, in the crystal modification of either calcite or aragonite. The algal silicates are amorphous, hydrated polymerization products of silicic acid (opaline silica). Algal calcification and silicification occur in fresh as well as marine waters (cf. Borowitzka 1982a, 1989; Round and Crawford 1989).

Calcifying and silicifying algae: occurence and impact on element cycles

Biomineralization implies the conversion of inorganic ions or molecules from a mobile (soluble) form to an immobile (insoluble) form. These processes affect the cycling of the elements involved, especially when the biomineralizing species have a wide-spread and abundant distribution. Abundant calcifying algae are fresh-water members of the Chlorophyta (green algae, e.g. *Chara* species; Borowitzka 1982b, Lucas 1975, Okazaki and Tokita 1988), marine members of the same phylum (e.g. *Halimeda* species; Borowitzka 1982b, 1989; MacIntyre and Reid 1995), members of the phylum Rhodophyta (Corallinaceae, marine red algae; Borowitzka 1982b, 1989) and a group of species belonging to the mainly marine phylum Prymnesiophyta: the Coccolithophoridae (Green et al. 1989). Silicifying algae are predominantly found in the phyla Chrysophyta and Bacillariophyta (Preisig 1994, Round and Crawford 1989). The latter organisms are better known as diatoms. Two groups of biomineralizing algae can be hold responsible for a significant global deposition of carbonate or silicate: the Coccolithophoridae and the diatoms respectively. Although the latter occur in fresh as well as in marine habitats, the oceans are the environment in which the bulk of the algal biomineralization takes place. In the following the impact of the Coccolithophoridae and diatoms on ocean chemistry and element cycling will be shortly discussed.

Diatoms. Diatoms are after bacteria the most abundant worldwide aquatic organisms. Over 10,000 living species are known (Round and Crawford 1989). The cells are enclosed by two overlapping valves of opaline silica, which generally have intricate, beautifully ornamented structures (Fig. 1). Diatoms depend on silicon not only for valve formation, but for several other cellular processes as well (Sullivan and Volcani 1981, Hildebrand et al. 1993). Together with the radiolarians (silicifying heterotrophic protoctysts) the diatoms

0275-0279/97/0035-0008$05.00

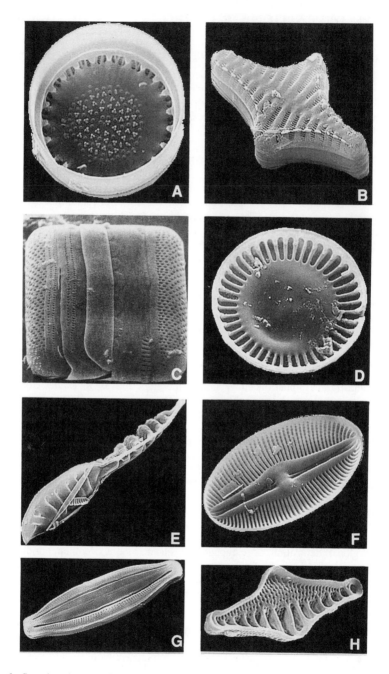

Figure 1. Scanning electron micrographs of diatoms. A: *Cyclotella*; B: *Synedra*; C: *Amphitetras*; D: *Cyclotella*; E: *Entomoneis*; F: *Diploneis*; G: *Navicula*; H: *Nitzschia* [Figs. 1A,B,D,E,F,G,H used by courtesy of Dr D.W. Johnson, Bowling State University; Fig. 1C used by permission of Jones and Bartlett Publishers, from Round and Crawford (1989), Fig 13, p. 581.]

are the major contibutors of biogenic silicon fluxes in the oceans (Reynolds 1986, Takahashi 1991). Although silicon is next to oxygen the most abundant element on Earth, the ocean water, especially the upper photic zone, is undersaturated with respect to opaline silica. This is primarily due to the removal of the element from the water by radiolarians and diatoms (Broecker 1971). The sea water probably became undersaturated with respect to opal during the Cenozoic as a result of a massive diatom expansion (Lowenstam and Weiner 1989). As a result of the undersaturation, siliceous biominerals readily dissolve after the death of the organisms. Per year about 400×10^{12} moles of silicon are recycled in the oceans by siliceous organisms and their redissolving skeletons (Reynolds 1986). About 3% of this amount settles in the sediments. At present, seasonal diatom blooms occur which are able to outcompete other photosynthetic planktonic species. These diatom blooms collapse as a result of silicon exhaustion of the water. In many cases the collapsed diatom blooms are promptly succeeded by blooms of Coccolithophoridae (Egge and Aksnes 1992).

Coccolithophoridae. The Coccolithophoridae have been abundant constituents of the world's oceans through the late Mesozoic and Cenozoic up to present. During the Cretaceous they knew an explosive radiation which resulted in the massive accumulation of lime sediments. They are characterized by the formation of crystalline calcium carbonate platelets (coccoliths) which surround the cell. These coccoliths have species-specific, delicately sculptured structures (Fig. 2). Hundreds of species are known, fossil (Tappan 1980) as well as recent (Jordan and Kleijne 1994). Several of the present species, notably *Emiliania huxleyi* (see below) and *Gephyrocapsa* species, form large blooms at mid to high lattitudes and in coastal zones (Westbroek et al. 1993). The coccolith crystals produced by these blooms strongly increase the light scattering properties of the surface water. As a result coccolithophore blooms can be detected from space with satellite imagery (Fig. 3; cf. Holligan et al. 1983, Holligan and Balch 1991). These blooms may cover thousands of square kilometers to a depth of circa sixty meters (where sufficient light can still penetrate to support photosynthesis). In such a bloom cell densities of over 10^8 cells/l have been detected (Holligan et al. 1983) It can be estimated that thousands of tonnes of calcium carbonate are precipitated by a single bloom of coccolithophorid algae. *Emiliania huxleyi* is one of the most abundant and cosmopolitan species of the Coccolithophoridae (McIntyre and Bé 1967), and regularly blooms are dominated by this species (Holligan et al. 1983, van der Wal et al. 1995). It is considered to be one of the most productive lime-secreting species on Earth (Westbroek 1991). The Coccolithophoridae, *E. huxleyi* in particular, influence the transfer of carbon from the atmosphere to the ocean and to the ocean sediments. As a result they may be an important factor in global climate forcing (Westbroek et al. 1993).

Principles of biomineralization: biologically induced or controlled

When discussing the mechanisms whereby organisms convert dissolved ions or molecules into solid minerals, we have to discriminate between the precipitation of amorphous and crystalline solids. As mentioned above, the biogenic silicates are amorphous, meaning that they do not show any detectable order over distances longer than 10 Å (Mann et al. 1983, cf. Fig. 4A). In contrast, the biogenic calcium carbonates are crystalline, for instance in the form of calcite as in the coccoliths of *E. huxleyi* (Watabe 1967, cf. Fig. 4B). The deposition of both types of minerals has to obey the same basic physico-chemical and thermodynamic principles, but the deposition of amorphous materials is more difficult to understand (cf. Simkiss and Wilbur 1989, Simkiss 1991). Consequently the formation of crystalline minerals has received considerably more attention and in the following discussion we will focuss on the latter process. Where possible, amorphous biomineralization will also be commented on.

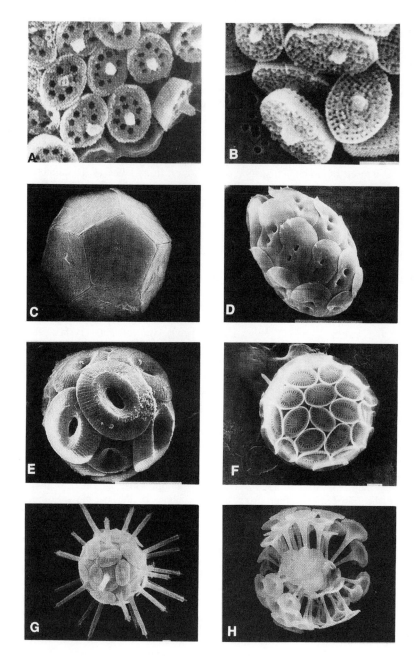

Figure 2. Scanning electron micrographs of coccoliths and Coccolithophoridae. A: coccoliths of *Corisphaera multipora*; B: coccoliths of *Laminolithus hellinicus*; C: *Braarudosphaera bigelowii*; D: *Helicosphaera carteri*; E: *Coccolithus pelagicus*; F: *Syracospaera* aff. *nodosa*; G: *Rhabdosphaera* aff. *claviger*; H: *Discosphaera tubifer*. Scale: A-C and F-H, 1 µm; D-E, 10 µm. [Used by permission of Jones and Bartlett Publishers, from Green et al. (1989), Fig. 8, p. 299.]

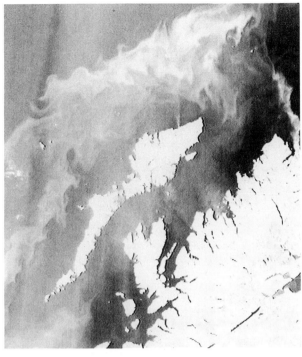

Figure 3. Satellite image of a bloom of *Emiliania huxleyi* north of Scotland. The bloom is detected by the back-scattering of light by the coccoliths and is seen as a whitish cloud in the water. [Used by permission of the editor of *J. Protozool.*, from Westbroek et al. (1989), Fig. 7, p. 371.]

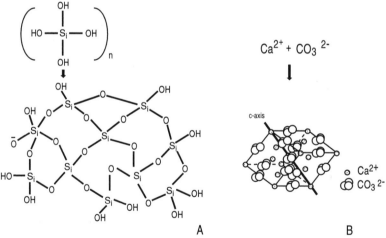

Figure 4. Amorphous and crystalline minerals. A: structure of part of an amorphous silica polymer; B: arrangement of Ca^{2+} and CO_3^{2-} ions in a calcite crystal with rhombohedral symmetry.

The relation between the standard free energy (G^0) of a substance in solution and in its solid states is schematically represented in Figure 5. The standard free energy content of a solid is lower than that of its constituents in solution; the larger the decrease in standard free energy (ΔG^0) upon precipitation of the solid, the more stable it will be. One may say that the formation of a crystal is thermodynamically favored over amorphous precipitation. For a solid to be formed a certain amount of activation energy (ΔG^*) has to be invested, for

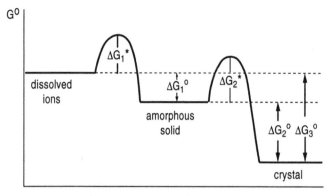

reaction coordinate

Figure 5. Relation between the standard free energy (G^0) of mineral constituents in solution and of their amorphous and crystalline solid phases. ΔG^*: activation energy for transition from one state to another.

instance to remove the water shell surrounding the constituent ions or molecules. The lower the ΔG^*, the faster the reaction proceeds; one may say that the formation of amorphous minerals is kinetically favored over crystallization. In the case of silica deposition, the energy barrier to crystallization (quartz formation) from the amorphous phase is very high because the silica polymer contains covalent bonds (cf. Fig. 4A) which have to be broken and reformed upon crystallization. The energy barrier to crystallization of crystalline calcium carbonate is much lower, especially in biomineralization where the living organisms employ several mechanisms tho lower the activation energy, as will be illustrated below.

In order to adequetely explain the mechanisms by which organisms can influence the precipitation of minerals, a descriptive summary of some basic processes of mineral precipitation and crystallization is given. For a comprehensive theoretical and mathematical survey of the subject the reader is referred to Nancollas (1982).

Supersaturation and nucleation. Before a solid can precipitate from solution, the concentration (activity) of its constituents in the solution has to exceed the solubility product. In other words, a certain degree of supersaturation has to be reached, whether the solid is amorphous or crystalline. For crystallization to proceed, ionic clusters (or nuclei) of a certain critical size which are resistant to rapid dissolution, have to be formed. Figure 6A shows the relation between the size r of these nuclei and their free energy content. Nuclei below the critical size r^* will redissolve, those above will continue to grow. The critical nuclei size depends on the degree of supersaturation S of the solution (Fig. 6B). Consequently the amount of stable nuclei in the solution depends on the degree of supersaturation. The rate at which stable nuclei are formed is called the nucleation rate J. The nucleation rate depends on the degree of supersaturation also (Fig. 6C). Figure 6C illustrates that, once a critical supersaturation has been exceeded, the nucleation rate increases exponentially (solid curve). When dealing with pure solutions containing only the crystal constituents, nucleation is said to be homogeneous. When foreign particles or surfaces are present which can bind and stabilize crystal nuclei, nucleation is heterogeneous. In the latter case the critical degree of supersaturation is decreased (dotted curve in Fig. 6C).

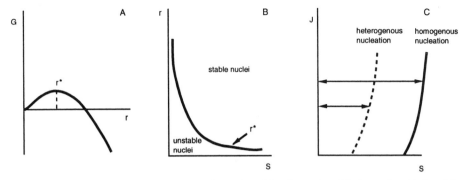

Figure 6. Relation between the free energy of crystal nuclei and crystal nuclei size r; r*: critical nuclei size (A); relation between critical nuclei size r* and degree of supersaturation S of the solution (B); relation of nucleation rate J and degree of supersaturation S, arrows denote range of supersaturation of solutions which are metastable with respect to homogeneous and heterogeneous nucleation (C).

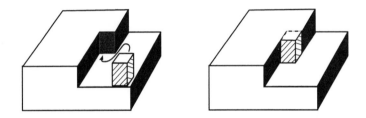

Figure 7. Crystal surfaces showing a step with a kink (growth site) and the incorporation of a new crystal element at the growth site.

Crystal growth. Crystals grow by incorporation of new crystal elements on the surface of the existing crystal lattice. This preferentially occurs at sites where the lattice is imperfect, at so-called kinks or dislocations on the crystal surface (Fig. 7). At these sites the ion activity is generally higher than in the crystal lattice resulting in a higher binding capacity for new crystal elements. New crystal elements will diffuse over the surface untill such sites, the growth sites, are reached (Fig. 7). A crystal has a limited number of growth sites. The crystal modification and morphology is determined by kinetic and thermodynamic parameters, and by the presence of other substances, like other ions, or organic molecules. At certain concentrations of Mg^{2+} in a supersaturated solution of Ca^{2+} and CO_3^{2-}, for instance, the crystallization of aragonite is favored over that of calcite (Kitano 1969). The presence of organic molecules, especially when they contain functional groups which can interact with the crystal surface, can dramatically influence crystal morphology (Addadi and Weiner 1989, 1992; Davey et al. 1991, Mann 1993a).

Termination of crystal growth. The termination of crystal growth at specific sites in specific growth stages will determine the overall morphology of the crystalline product formed. This will also apply to amorphous precipitation: termination of growth of the solid phase at specific sites will determine the final form of the product. Growth will be terminated when the concentration of the dissolved constituents drops below the critical supersaturation (cf. Fig. 6C). Alternatively, growth will be halted when the solid contacts a growth-inhibiting molecule or surface. For instance, when the mineral is formed within a delineated space, growth will proceed untill the space is filled. The final form of the solid will then be determined by the morphology of the enclosing room (cf. Lowenstam and Weiner 1989).

Biomineralization: biological interference with mineralization. Living organisms can interfere with mineralization at all levels summarized above. They can create the supersaturated conditions required for precipitation. This may be achieved by altering the chemical composition of the precipitation medium by production (and excretion) of cellular metabolites or by uptake of metabolic substrates. For instance, if CO_2 is taken up from the surrounding medium and fixed in photosynthesis, the medium becomes more alkaline which results in an increase in the concentration of CO_3^{2-} ions (cf. *"Some aspects of photosynthesis,"* below). If Ca^{2+} ions are also present, the solubility product of $CaCO_3$ can be sufficiently exceeded to allow crystallization (Borowitzka 1982b). Organisms may induce supersaturation by local concentration of the mineral constituents at binding sites of cellular surfaces (Borowitzka 1982b). Living organisms can also actively increase the concentration of the mineral constituents at certain sites by (specific) transport enzymes or intracellular transport vehicles like closed vesicles. In general, the latter processes will require an investment of energy.

Nucleation can be directed by controlling the degree of supersaturation, in other words by controlling the rate at which the mineral constituents are provided. Nucleation can be specifically initiated from a certain crystal face by supplying a substrate with stereochemical and electrochemical complementarity to the surface of the crystal face (Mann 1988, 1993b; Addadi and Weiner 1989). In a similar way direction of growth and the crystal modification can be determined by providing specific organic templates (Wada et al. 1993, Falini et al. 1996, Belcher et al. 1996). Finally, growth can be terminated by organic (macro)-molecules, synthesized by the organism, which bind to crystal growth sites and inhibit the incorporation of new crystal elements (Wheeler and Sikes 1989). Growth can also be terminated by contact with a preformed, inhibitory surface, for instance a vesicle membrane in case of intracellular mineralization.

The interference of living organisms with the mineralization process can vary from minimal, e.g. only by creating supersaturated conditions, to extensive and stringent, e.g. by creating specialized vesicles, containing specific macromolecules as templates (also called matrices) for oriented precipitation or crystallization. Lowenstam introduced the terms "biologically induced" and "biologically controlled" biomineralization for the former and latter process respectively (Lowenstam and Weiner 1989). Especially in algal calcification we can discriminate between both types of biomineralization. In the second part of this chapter representative examples will be highlighted. Algal silicification processes generally all fall in the second category, as will be shown in the third part of this chapter. In the final part new developments in the research on algal biomineralization will be discussed.

ALGAL DEPOSITION OF CALCIUM CARBONATE

Biologically induced calcification: *Chara* **and** *Halimeda*

Chara and *Halimeda* are examples of calcifying algae which induce calcification by increasing the concentration of CO_3^{2-} ions in their environment. They both occur in Ca^{2+}-rich waters: *Chara* in Ca^{2+}-rich lakes and ponds and *Halimeda* in coastal marine waters. In *Chara* the $CaCO_3$ is deposited in the form of calcite on the outer cell wall, in open contact with the surrounding medium; in *Halimeda* the crystal modification aragonite is deposited in intercellular spaces with restricted contact with the surrounding medium. In both algae the induced increase in the CO_3^{2-} concentration, and the subsequent precipitation of $CaCO_3$, is a result of photosynthetic activity. There is an intimate relationship between photosynthesis and calcification in all calcifying algae, also the species which employ biologically controlled calcification. We will shortly discuss some general aspects of algal photosynthesis which relate to the mechanisms of calcification.

Some aspects of photosynthesis. Photosynthesis in plants and algae involves the conversion of CO_2 and H_2O to carbohydrate (generally designated as $[CH_2O]$) with light as the energy source. CO_2 is reduced in the process and the reducing equivalents are ultimately withdrawn from water, resulting in the overall reaction:

$$CO_2 + H_2O \Rightarrow [CH_2O] + O_2 \tag{1}$$

The CO_2 can easily diffuse through the cell wall and cell membranes. It is fixed in the chloroplasts by coupling the molecule to an acceptor sugar molecule, ribulose-1,5-biphosphate, by the enzyme ribulose-biphosphate-carboxylase, known as RUBISCO.

Aquatic plants and algae have to acquire the CO_2 from the surrounding water in which it has a limited solubility. The following equilibria are relevant:

$$CO_{2(g)} \quad\quad \Leftrightarrow CO_{2(aq)} \tag{2}$$

$$CO_{2(aq)} + H_2O \Leftrightarrow H_2CO_3 \tag{3}$$

$$H_2CO_3 \quad\quad \Leftrightarrow H^+ + HCO_3^- \tag{4}$$

$$HCO_3^- \quad\quad \Leftrightarrow H^+ + CO_3^{2-} \tag{5}$$

The pK values of the dissociation reactions of carbonic acid vary from 6.3 to 6.0 for Reaction (4) and from 10.3 to 10.9 for Reaction (5), depending on the salinity of the environment. If CO_2 is withdrawn from solution by photosynthetic activity, Reaction 4 is drawn to the left with consumption of protons, resulting in a pH increase. A pH increase favors the production of CO_3^{2-} from HCO_3^- (Reaction 5). We may write these equations in the following overall form:

$$2\, HCO_3^- \Leftrightarrow CO_2 + CO_3^{2-} + H_2O \tag{6}$$

At a pH of around 8 (the pH of sea water and some fresh water lakes and ponds), most of the inorganic carbon will be in the form of HCO_3^- (cf. Borowitzka 1982b). Under conditions of elevated pH and of high photosynthetic activity, aquatic photosynthetic organisms may become CO_2-depleted, especially when dissolution of atmospheric CO_2 (Reaction 2) or dehydration of carbonic acid (Reaction 3) is rate-limiting. Some aquatic organisms have acquired the ability to use HCO_3^- instead of, or in addition to CO_2 for photosynthesis (Okazaki and Furaya 1985, Raven and Johnston 1991, Nimer and Merett 1992, Beer 1994). Some possess the enzyme carbonic anhydrase, which catalyzes Reaction (3) (Johnston and Raven 1986, Dixon et al. 1987, Quiroga and González 1993, Nimer et al. 1994). As will be shown below, calcification may be an additional way for concentrating CO_2 in aquatic photosynthesizers.

The example of **Chara.** The alga *Chara* is a so-called giant alga which may reach a size of several centimeters. It is a coenocytic alga, which means that the cells do not form septa after division, resulting in a multinuclear structure called the thallus (cf. definitions of protoctyst structures in Margulis et al. 1993). The thallus of *Chara* consists of a branched structure of so-called internodal cells as schematically depicted in Figure 8A. In Ca^{2+}-rich water the internodal cells precipitate calcite crystals in banded patterns which correspond to alkaline regions on the outer surface of the cells (Figs. 8B and 8C; Okazaki and Tokita 1988, McConnaughey 1995). The alkalinization of these regions is dependent on photosynthesis, since this process and the concomittant calcification is inhibited by the photosynthetic inhibitor DCMU, and the inhibitor of carbonic anhydrase Diamox (Okazaki and Tokita 1988). Borowitzka (1989) proposed a model for the uptake of HCO_3^- which subsequently is used in photosynthesis. The model is based on the operation of spatially separated proton transport enzymes (Fig. 9). Proton pumps expelling protons would be concentrated in regions with an acidic environment (Price and Whitecross 1983) and pumps

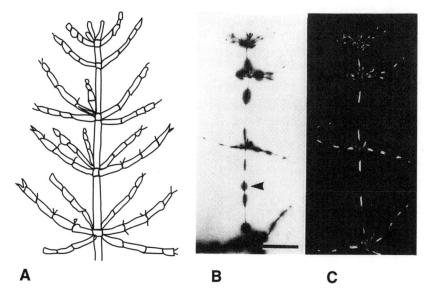

A   **B**   **C**

Figure 8. Extracellular calcification in *Chara braunii*. A: impression of the alga with internodal cells; B: cells embedded in agar containing the pH indicator phenol red; dark areas indicate alkaline regions; C: same cells as in B examined with polarized light to indicate the presence of calcite crystals. Note that calcified regions always coincide with alkaline regions; the reverse is not necessarily true (arrow). Scale: 1 cm. [Figs. 7B,C used by permission of The Japanese Society of Phycology, from Okazaki and Tokita (1988), Figs. 4A,B, p. 198.]

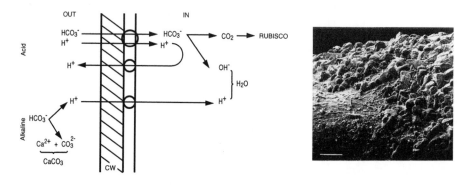

Figure 9. Model describing the generation of alkaline calcified regions on the cell walls of *Chara*. Different proton pumps are thought to operate: in the acid regions H^+ is expelled and HCO_3^- may be taken up in cotransport with H^+ using the generated proton gradient; in the alkaline regions H^+ is taken up to neutralize the OH^- ions liberated from intracellular HCO_3^- by formation of CO_2 for photosynthesis; alternatively, OH^- is transported outward. CW: cell wall; PM: plasmamembrane. The scanning electron micrograph shows the border of acid and alkaline calcified region on cell wall of *Chara corallina*; note that calcite crystals do not show preferential orientations. [Micrograph used by permission of VCH Verlagsgesellschaft mbH, from Borowitzka (1989), Fig. 3.1, p. 72.]

transporting protons into the cells (or alternatively, exporting OH^- ions) would result in alkaline regions. In this model, external Ca^{2+} and HCO_3^- ions are the source of the final $CaCO_3$ crystals. A basically different model is proposed by McConnaughey (1995), who postulates the involvement of an energy-dependent, Ca^{2+}/H^+ exchanging enzyme (a so-called Ca^{2+}-dependent ATPase) in the calcification process in *Chara*. Ion transport enzymes (HCO_3^-, H^+, or Ca^{2+} transporters) have not been biochemically identified in the alga, and the final decision which model applies most accurately to the calcification process awaits the results of the isolation and characterization of such systems.

The calcite crystals are apparently nucleated on the cell wall, but lack an organizational motif and a preferred crystal orientation (Fig. 9). They vary in size and are not associated with any organic material other than the cell wall (Borowitzka 1989). The influence of the alga on the calcification process is limited to the generation of supersaturation in the close environment of non-specific heterogeneous nucleation sites.

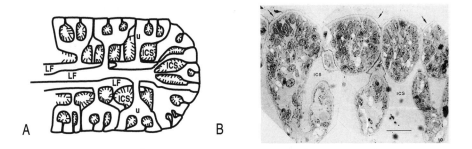

A   B

Figure 10. Site of calcification in *Halimeda*. A: impression of outer segment of the alga. LF: longitudinal filament; U: utricle; ICS: intercellular space. Intercellular spaces are filled with aragonite needles. B: electron micrograph of transverse section across peripheral utricles of a *Halimeda* species. The utricles have fused to form the ICS. Scale = 20 μm. [Fig. 9B used by permission of the editor of the *Journal of Phycology*, from Borowitzka and Larkum (1977), Fig. 10, p. 8.]

The example of **Halimeda.** The green alga *Halimeda* is a prominent constituent of tropical and subtropical marine coastal ecosystems and an important contributor to coral reefs (Hillis 1991). The organism consists of segments each of which are composed of branched, coenocytic filaments. The basic structure of such a segment, which can be several centimeters in diameter, is shown in Figure 10A. The longitudinal filaments branch into radial filaments, called utricles. The utricles are firmly apposed at the outer site of the segment, creating intercellular spaces (ICS) which are in indirect contact with the surrounding sea water (Fig. 10B). Aragonite crystals are precipitated in the ICS as a result of CO_2 uptake for photosynthesis (Fig. 11; Borowitzka 1982b, 1989). The pH increase in the ICS as a result of photosynthetic CO_2 uptake can be sufficiently large because diffusion of ions from the surrounding medium into the ICS is slow (Borowitzka 1986). In the dark the reverse of the photosynthetic reaction (cf. Reaction 1) takes place. This process is called respiration and yields energy for metabolic processes like transport of nutrients and biosynthesis. During respiration CO_2 is released which will partly diffuse into the ICS, resulting in a pH decrease. Part of the aragonite needles formed in the light probably redissolves in the dark (Borowitzka 1989).

The initial aragonite needles are heterogeneously nucleated in unspecified orientations on polysaccharide- and protein-containing fibrils of the outer layer of the cell walls facing the ICS (Borowitzka 1982a). Needles grow up to lengths of 10 μm and soon fill the entire intercellular space, but the concentration of crystals remains highest near the cell wall (Fig. 11; Flajs 1977). Like in *Chara*, there is no unambiguous evidence for the involvement of

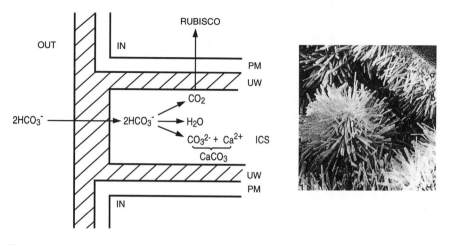

Figure 11. Model describing the generation of supersaturation with resepect to $CaCO_3$ in intercelluar space (ICS) of *Halimeda*. HCO_3^- diffuses over the utricle wall (UW) in the ICS; removal of CO_2 for photosynthesis induces an increase in the CO_3^{2-} concentration. The scanning electron micrograph shows aragonite needles in the ICS; note needles of different lenghts and orientations. Scale: 1 µm. [Micrograph used by permission of Springer-Verlag, Berlin, from Flajs (1977), Plate 1.1, p. 226.]

other organic macromolecules in calcification. One study, however, reports on the presence of an organic envelope surrounding the initial aragonite needles, suggesting a more controlled crystallization process than in *Chara* (Nakahara and Bevelander 1978). The alteration of crystals in living *Halimeda* as observed by MacIntyre and Reid (1995) may be explained by different levels of control over the crystallization process. Initial crystallization may be controlled by an organic envelope (matrix), resulting in rather uniformly sized, small crystals. Subsequent degradation of the envelope may lead to partial dissolution of the crystals (possibly in the dark, see above). Inorganic recrystallization of aragonite needles on the remnants of the initial needles would explain the distribution of differently-sized crystals in the ICS (cf. Fig. 11). Isolation of crystals in different growth stages of the alga, and careful analysis of their composition, may resolve this question. So far, calcification in *Halimeda* is considered to be an example of biologically induced mineralization.

Biologically controlled calcification: three coccolithophorid species

The coccolithophorids belong to the phylum of the Prymnesiophyta. The Prymnesiophyta are unicellular algae which may have complex life cycles, alternating between haploid and diploid stages (Billard 1994). They are motile or at least have one motile life stage. They generally produce organic body scales which are synthesized intracellularly in the Golgi apparatus (below) and exocytosed to form a multilayered cell covering (Leadbeater 1994). The haploid and diploid forms each produce scales with a characteristic microfibrillar ultrastructure (Billard 1994). In the coccolithophorids part of the body scales are calcified to form coccoliths (Faber and Preisig 1994). In general, the coccoliths constitute the outermost layer of the cell covering. *Emiliania huxleyi* is an exception in that calcifying cells do not form elaborate organic body scales.

Two types of coccoliths are distinguished: the holococcoliths, in which rhombohedral or hexagonal calcite prisms are regularly arranged on the organic scale (Figs. 2A and 2B), and heterococcoliths, in which the calcite elements can have more elaborate forms with sometimes rounded crystal faces (Figs. 2C-2H). In some species, e.g. *Coccolithus pelagicus*, different types of coccoliths are formed in different life stages. The coccolithophorids are the only known examples of algae with intracellular calcification (Borowitzka 1982b). Obviously, these biomineralization processes are strictly biologically controlled.

Coccolith formation in **Pleurochrysis carterae:** *ultrastructure of cells and coccoliths.* The coccolithophorid *Pleurochrysis carterae* is a euryhaline species with a limited distribution. Two life stages are described (Billard 1994). The non-motile, non-calcifying ("naked") form is haploid and the motile, coccolith-bearing form is diploid. Only the coccolith-bearing form is discussed here. The cells are ellipsoid with a diameter of about 15 μm. The ultrastructure of cells and coccoliths has been extensively described by Pienaar (1971), Outka and Williams (1971) and van der Wal et al. (1983c). In Figure 12 the most important cell constituents are schematically presented. The chloroplasts are the site of photosynthesis. Respiration takes place in the mitochondria, the nucleus contains the chromosomes. The nucleus is surrounded by a double membrane, of which the outer membrane extends into the cytoplasm to form a labyrinthine membrane compartment called the endoplasmatic reticulum. The endoplasmatic reticulum is also connected to the chloroplast through the so-called periplastidial endoplasmatic reticulum, a charactistic organelle of most of the Prymnesiophyta. The proteins destined for export from the cytoplasm and lipids are synthesized by the endoplasmatic reticulum. Products formed by this organelle are further processed in a stack of flattened sacs (cisternae), surrounded by rounded vesicles, the Golgi apparatus. The Golgi is also the site of oligo- and polysaccharide synthesis. The cisternae situated opposite the endoplasmatic reticulum are called the "cis face" of the Golgi, the cisternae situated near the plasmamembrane are called the "trans face". The faces of the Golgi are differentiated in such a way that different chemical and biosynthetic reactions take place at the different sites. Vesicles bud off from the cisternae and transport products between the different faces. The trans face is the site of product export from the system to the cell exterior. The cell is covered by several layers of scales. The scales consist of radial and concentric microfibrils composed of neutral and acidic polysaccharides and some protein (Brown and Romanovicz 1976). The proximal layers are formed by unmineralized scales. The distal layer consists of 150-200 coccoliths (Fig. 13). These are built by an intercalation of about thirty anvil-shaped calcite crystals on the rim of an organic scale (Okazaki et al. 1997). Each element is a single crystal. They are alternately oriented with their c-axis parallel and perpendicular to the plane of the coccolith, and called R units and V-units respectively, after the model proposed by Young et al. (1992) for the development of heterococcoliths.

Mechanisms of coccolith synthesis in **Pleurochrysis carterae:** *role of cell organelles and organic constituents.* Van der Wal et al. (1983c) presented a detailed study on the ultrastructure of coccolith morphogenesis in which they defined six stages in coccolith development. Here we will discuss a simplified version of this model, describing the three most conspicuous stages. All scales, the unmineralized and the calcified ones, are produced in the Golgi apparatus. Scales first appear in cisternae between the cis and trans face of the organelle (Fig. 12, stage 1). Special vesicles are budded off from cis cisternae. They contain electron dense particles which were first defined as coccolithosomes by Outka and Williams (1971). The vesicles containing coccolithosomes fuse with certain scale-containing cisternae, and calcite crystals begin to grow on the rim of the scales (Fig. 12, stage 2). In this stage the cisterna closely envelopes the growing coccolith. In the last stage the coccolithosomes have all been disintegrated and the crystals are completed. They are covered by an organic skin, and the Golgi cisterna becomes dilated prior to expulsion from the cell (Fig. 12, stage 3). The coccolithosomes are considered to contain precursors of the crystal elements of the coccoliths. This is supported by several observations. Van der Wal et al. (1983a) used X-ray microanalysis to demonstrate that the

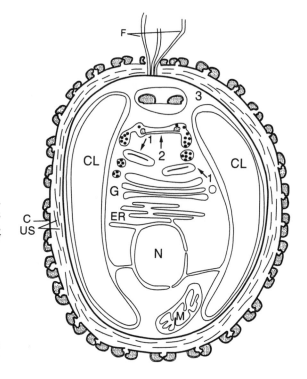

Figure 12. Schematic representation of a transverse section across a cell of *Pleurochrysis carterae*, showing three stages (1 to 3) of calcification.

Cl   chloroplast,
N   nucleus,
M   mitochondrion,
ER   endoplasmatic reticulum,
G   Golgi apparatus,
US   uncalcified scale,
C   coccolith.

The dark bodies in vesicles fusing with calcifying vesicles are coccolithosomes.

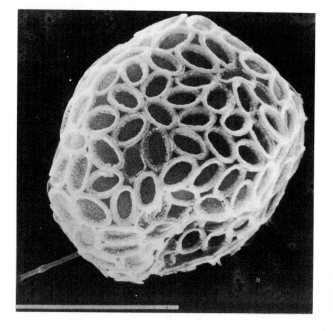

Figure 13. Scanning electron micrograph of a cell of *Pleurocrysis carterae* covered with coccoliths.

coccolithosomes contain Mg, and a high concentration of Ca, at least 6 M. Van der Wal et al. (1983c) applied a special staining specific for polysaccharides which showed that scales, coccolithosomes as well as the skin surrounding the calcite elements contain polysaccharides. Two acidic polysaccharides have been isolated from coccoliths by de Vrind-de Jong et al. (1986), and more recently by Okazaki et al. (1997). They were called PA and PB. They were only superficially characterized, but it was shown that both contained uronic acids (carboxylated sugars). Polysaccharide PB also contained sulfate groups and had a more acidic character than PA. In an elegant study Marsh et al. (1992) isolated and purified three polysaccharide fractions (PS1, PS2 and PS3) from coccoliths of *P. carterae*. They all contained uronic acids, but in addition, PS2 contained extra carboxyl groups generated by modification of part of the uronic acid moieties. PS3 contained sulfate groups. Probably PA reported by de Vrind-de Jong et al. (1986) and Okazaki et al. (1997) corresponds to PS1 and PB represents a mixture of PS2 and PS3. These polysaccharides, at least PS1 and PS2, coaggregate with Ca^{2+} ions (Marsh et al. 1992). Antibodies were raised against PS1 and PS2 and used in electron microscopical immunolocalization (Marsh 1994). Both polysaccharides were identified in the coccolithosomes and in the organic skin surrounding the crystal elements, but not in the organic scales (Fig. 14). A kinetic analysis of the intracellular turnover of coccolith constituents (the mineral ions Ca^{2+} and CO_3^{2-}, and the three acidic polysaccharides, PS1, PS2, PS3) and their incorporation in extracellular coccoliths suggests that the Ca^{2+} ions and PS1 and PS2 follow a common route, different from that of CO_3^{2-} and PS3 (Marsh 1996). Marsh (1996) proposed that the acidic polysaccharides PS1 and PS2, synthesized in the cis cisternae of the Golgi, accumulate Ca^{2+} ions and form organic calcium carriers, the coccolithosomes. Although PS3 has not been localized so far, the results of the kinetic analysis by Marsh (1996) suggest that it represents a constituent of the base scale of the coccolith.

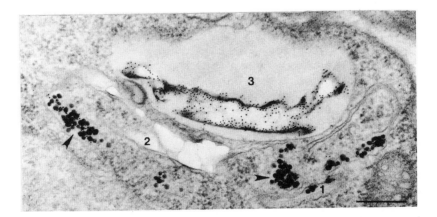

Figure 14. Transmission electron micrograph of transverse section through a cell of *Pleurochrysis carterae*. The section was treated with antibodies (IgG) raised in a rabbit against PS2 (see text). The rabbit IgG molecules were detected with antibodies raised in a goat against rabbit-IgG molecules and labeled with gold particles. Gold particles are found associated with coccolithososmes (arrows) and with the organic coat surrounding the calcite elements of coccoliths in statu nascendi. Two stages (2 and 3) of calcification are represented. White areas are holes left by dissolution of the mineral during sample preparation. Scale: 0.25 µm. [Used by permission of Springer-Verlag, Austria, from Marsh (1994), Fig. 15, p. 116.]

The role of the acidic polysaccharides in coccolith biosynthesis is most probably not limited to the transport of mineral-forming ions. The Ca^{2+}-binding capacity of PS1 and PS2 in vitro is too low to account for the amount of calcium incorporated in coccoliths relative to the amount of their organic components (Marsh 1994, 1996). In general, biological control of biomineralization is exerted via two types of organic macromolecules: a water-insoluble framework or base, in interaction with water-soluble acidic macro-molecules which, depending on the organism, consist of proteins, glycoproteins or polysaccharides, or a combination of these (Lowenstam and Weiner 1989). The acidic molecules can act as nucleation sites when fixed on the insoluble framework (Crenshaw and Ristedt 1976, Crenshaw 1991) or influence the crystal morphology by directing or directionally inhibiting crystal growth through interaction with crystal growth sites (Wheeler and Sikes 1989). Wada et al. (1993) showed that the acidic polysaccharides from $P.$ $carterae$ coccoliths as isolated by de Vrind-de Jong et al. (1986) delayed the crystallization of $CaCO_3$ from a supersaturated solution and induced the formation of calcite in the presence of Mg^{2+} ions, a condition which otherwise favors the crystallization of aragonite (Kitano et al. 1969). Okazaki et al. (1997) showed that these polysaccharides inhibit the crystallization of $CaCO_3$ in vitro (cf. Wheeler et al. 1981, Borman et al. 1982). It may thus be hypothesized that PS1 and PS2 play a role in preventing $CaCO_3$ crystallization in the coccolithosomes, and that once introduced into the coccolith-forming vesicle, they determine the crystal polymorph and the final form by specifically inhibiting crystal growth by covering certain growth sites. The way in which they release their bound Ca^{2+} ions, probably necessary for rearranging into the organic skin covering the calcite crystal, remains a matter of speculation. It is tempting to suggest that PS3 plays a role in the nucleation of the crystals, possibly by specific interaction with the rim of the base scale (the insoluble framework). The localization of PS3 in the coccoliths may be revealed using immunochemical procedures as applied for PS1 and PS2.

The calcifying vesicle has to attain supersaturation with respect to $CaCO_3$ prior to the crystallization of the calcite elements. As discussed above, the vesicles containing the coccolithosomes may be, at least in part, the carriers of coccolith calcium. It is not known how the Ca^{2+} ions are sequestered in these vesicles. In all eukaryotic cells, Ca^{2+} ions are presumed to have an important signal function (cf. Berridge 1993). Although the intracellular Ca^{2+} concentration of $P.$ $carterae$ has not been determined yet, it is likely to be strictly regulated and kept extremely low (cf. Brownlee et al. 1995). This is generally achieved by specific Ca^{2+}-transport enzymes, which remove Ca^{2+} from the cytoplasm at the expense of energy, e.g. ATP hydrolysis. These enzymes are called ATPases; they can be stimulated by Ca^{2+} ions. Ca^{2+}-stimulated ATPases have been identified in fractionated membrane preparations of $P.$ $carterae$ (Kwon and González 1994, Araki 1997). Two types were distinguished based on the effect of different inhibitors on their ATP-hydrolyzing activity: a so-called P-type ATPase, mainly associated with the plasmamembrane fraction, and a so-called V-type ATPase which was associated with membranes of the Golgi apparatus and the calcifying vesicles. The presence of a V-type ATPase in $P.$ $carterae$ was confirmed by Corstjens et al. (1996), who isolated the gene encoding a conserved subunit of the enzyme. The P-type ATPase was assumed to function in transporting Ca^{2+} ions out of the cells (Araki 1997). In the first instance the V-type ATPase was thought to be involved in the accumulation of Ca^{2+} ions in the Golgi vesicles, but Ca^{2+} transport induced by this enzyme could not be demonstrated. The V-type ATPase appeared to be involved in the Ca^{2+}-regulated transport of protons instead (Araki 1997). In analogy a proton-pumping Ca^{2+}-stimulated ATPase has been identified in a calcifying member of the Rhodophyta (Mori et al. 1996). Proton pumping may well be an important regulating factor in calcification. As demonstrated by Sikes et al. (1980), HCO_3^- is the external source of the CO_3^{2-} in $CaCO_3$ in $P.$ $carterae$ and $E.$ $huxleyi$. Consequently, calcification proceeds according to the reaction:

$$HCO_3^- + Ca^{2+} \Leftrightarrow CaCO_3 + H^+ \qquad (7)$$

The protons liberated in this reaction have to be removed from the calcifying vesicle to allow precipitation to proceed. As will be argued below, the released protons may be neutralized by OH⁻ ions liberated in photosynthetic carbon fixation.

The mechanism of HCO_3^- transport is not known. Probably specific enzymes are involved in the transport of the ions over the plasmamembrane and the membranes of the calcifying vesicle. Due to its charge, an ion does not readily diffuses over a lipid bilayer, also when the concentration gradient of the ions would support directional transport (cf. Anning et al. 1996). Direct transport of HCO_3^- in algae has hardly been studied. In a recent report evidence for the involvement of an anion-exchanging enzyme in the transport of HCO_3^- over the plasmamembrane of a giant alga has been presented (Sharkia et al. 1994). Experiments with an inhibitor of eukaryotic anion exchangers, however, do not point to a role of this transport system in HCO_3^- uptake in coccolithophorids (Anning et al. 1996).

Relation between photosynthesis and coccolith formation in **Pleuro-chrysis carterae.** The substrate for the carbon-fixing enzyme RUBISCO is CO_2. As mentioned above, the concentration of CO_2 in seawater may be limiting for photosynthesis, depending on the pH. Marine algae may benefit from mechanisms which allow them to use HCO_3^- as a source of CO_2 for photosynthesis. The chloroplasts of *P. carterae* contain carbonic anhydrase (Quiroga and González 1993), catalyzing the dehydration of HCO_3^- (Reactions 2 and 3). It may be summarized as:

$$HCO_3^- \Leftrightarrow CO_2 + OH^- \tag{8}$$

Neutralization of the hydroxyl ions liberated by photosynthetic HCO_3^- utilization by the protons released from calcification (Reaction 7), may result in mutual stimulation of the two processes according to:

$$2\,HCO_3^- + Ca^{2+} \Leftrightarrow CO_2 + CaCO_3 + H_2O \tag{9}$$

This indicates that calcifying cells are well adapted to CO_2-limiting conditions, by their ability to use HCO_3^-. This conclusion is sustained by experiments in which the net photosynthetic rate (NPS, expressed as moles O_2 evolved per number of cells per unit time, cf. Reaction 1) of *P. carterae* cells was compared with that of another *Pleurochrysis* species with a much lower rate of calcification (Israel and González 1996). At pH 7 the two species had a comparable NPS. At pH 9 however, the *P. carterae* cells still evolved O_2 at a significant rate, whereas the low-calcifying cells had an extremely low NPS. Although these low-calcifying cells also contain carbonic anhydrase (albeit in a lower concentration), they possibly are not well adapted to the use of HCO_3^- because they can not efficiently neutralize the concomittant pH rise.

The degree to which photosynthesis and calcification are interdependent appears to be influenced by conditions of inorganic carbon concentration, pH and probably by other factors as well, e.g. by the age of cultures and by nutrient concentration (cf. Nimer and Merett 1993, Nimer et al. 1996). Although calcification generally proceeds faster in the light than in the dark (Israel and González 1996), dark-calcification at rates similar as in the light have also been reported (van der Wal et al. 1987). Especially removal of extracellular coccoliths appeared to be a stimulus for coccolith formation, independent of the light- or dark condition (van der Wal et al. 1987). In the dark, calcification is probably supported by respiratory energy. In *E. huxleyi* inhibitors of respiration also inhibit dark-calcification (Sekino and Shiraiwa 1996). The dependence of calcification on light most probably also reflects its requirement for photosynthetic energy.

Coccolith formation in **Emiliania huxleyi:** *ultrastructure of cells and coccoliths.* The marine coccolithophorid *E. huxleyi* occupies a special position within the Prymnesiophyta (Green et al. 1989). The cells are coccoid and 5 to 10 µm in diameter. At least three life stages are described (Klaveness 1972b, Green et al. 1996). The coccolith-forming cells are non-motile and do not form organic scales. They are considered to be diploid (Green et al. 1996). A non-calcifying ("naked") variant of these cells exist, of which the physiological significance is still unclear (van Bleijswijk 1996). The haploid stage is probably represented by motile, scale-bearing cells, which do not produce coccoliths, but form intracellular $CaCO_3$ deposits in a special vacuole and in the space between the plasmamembrane and the cell cover (Green et al. 1996, van der Wal et al. 1985). Here we will focuss on the coccolith-forming cells, of which a schematic cross section is shown in Figure 15. A cell contains mitochondria, chloroplasts, a nucleus, an endoplasmatic and periplastidial reticulum, and a Golgi apparatus as described for *P. carterae*. In *E. huxleyi* the coccoliths are not formed in or very near the stack of Golgi cisternae, but in a specialized vesicle, the coccolith production compartment (Klaveness 1972a, de Vrind-de Jong et al. 1994). The coccolith production compartment consists of a calcifying vesicle connected to a labyrinthine membrane system called the reticular body (Klaveness 1972a). The coccolith production compartment is closely apposed to the nucleus, and probably formed by fusion of Golgi vesicles (van der Wal et al. 1983b, van Emburg et al. 1986). The cell is covered by 15-20 coccoliths (Fig. 16).

A coccolith of *E. huxleyi* consists of 30-40 single-crystal calcite elements (Watabe 1967, Mann and Sparks 1988). Placed in a radial array these unit elements form an ellipsoid proximal shield, connected by a central tube to a distal shield consisting of hammer-shaped elements (Westbroek et al. 1989, cf. Fig. 16). The shape of the calcite elements confers a specific handedness to the coccoliths (de Vrind-de Jong et al. 1994, Didymus et al. 1994). The detailed morphology of the coccoliths is strain-specific: at least two main types exist, produced by genotypically different strains A and B (Young and Westbroek 1991, van Bleijswijk et al. 1991).

Coccolith formation in **Emiliania huxleyi**: *role of cell organelles and organic constituents*. Different stages in coccolith formation have been distinguished based on the morphologies and contents of the coccolith production compartment (Westbroek et al. 1989, de Vrind-de Jong et al. 1994). The most conspicuous stages are schematically represented in Figure 15. First a flat vesicle is apposed to the nucleus, and an organic base plate is formed inside (stage 1). This base plate clearly differs from the organic scales produced by other Prymnesiophyta: they lack a distinct microstructure (van Emburg 1989). Then the reticular body is added and calcite crystals are nucleated on the rim of the base plate, probably as R- and V-units alternating in crystallographic orientation as described for *P. carterae* (Young et al. 1992). The R-units are allowed to develop to form the proximal shield elements, the central tubes and the hammer-shaped distal elements of the coccolith unit elements. They completely overgrow the V-units (Young et al. 1992, stage 2). In stage 2 the calcifying vesicle encloses the growing coccolith, and the reticular body is still elaborate. Successive growth stages of the mineral phase of the coccoliths have been recorded with electron microscopy (Fig. 17, de Vrind-de Jong et al. 1994). The organic base plate (the central area of the coccolith in statu nascendi) eventually becomes completely overgrown with calcite. In the last stage the coccolith is completed, the reticular body has disappeared (or has been consumed) and the calcifying vesicle is now dilated; the coccolith is about to be extruded and incorporated into the extracellular coccosphere (stage 3). In general, only two coccoliths in statu nascendi can be observed in one cell simultaneously. The reticular body is thought to be derived from the Golgi apparatus and to supply the coccolith precursors and membrane consituents of the calcifying vesicle. Van der Wal et al. (1983b) applied a polysaccharide-specific staining on sections of calcifying cells and demonstrated that the Golgi apparatus, the membranes and lumina of the reticular body and calcifying vesicle, and the organic base plate contain polysaccharides. The mineral phase of the coccoliths appeared to be covered by an thin polysaccharide skin. An

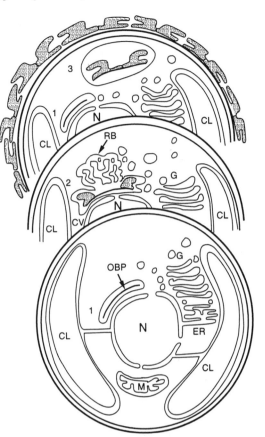

Figure 15. Schematic representation of a transverse sections across cells of *Emiliania huxleyi* showing three stages (1 to 3) of coccolith formation.

Cl   chlorolast,
N   nucleus,
M   mitochondrion,
ER   endoplasmatic reticulum,
G   Golgi apparatus,
OBP organic base plate,
CV   calcifying vesicle.
RB   reticular body.

The upper section also shows extracellular coccoliths.

Figure 16. Scanning electron micrograph of cells of *Emiliania huxleyi* covered with coccoliths. Note two detached coccoliths shown from their proximal shields and hammer-shaped elements of the distal shields.

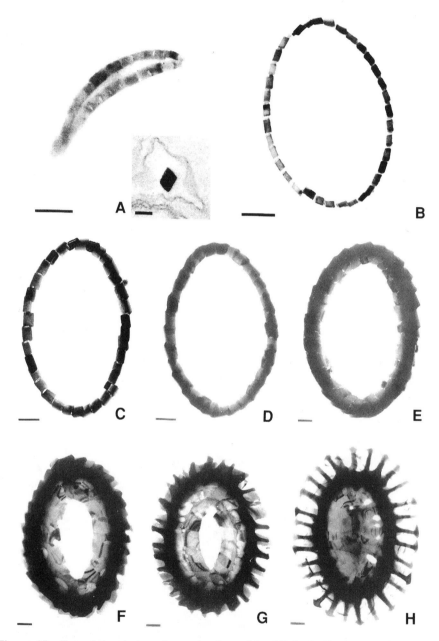

Figure 17. Transmission electron micrographs of coccoliths of *E. huxleyi* in statu nascendi. A: early growth stage of coccolith formation showing projection of protococcolith ring in its native three-dimensional shape. Inset: transverse section across initial calcite rhombohedron. B: mineral growth stage considered to be the earliest; the oblong crystals have no protrusions. C: small protrusions develop and D: crystals interlock. E-G: development of proximal shield area and distal shield elements, straight contours mark growth fronts; H: proximal and distal shield elements loose straight contours. Scales: A, B, C, D, G, H: 0,2 μm; E, F: 0,1 μm; inset A: 0,5 μm. [Used by permission of Oxford University Press, from de Vrind-de Jong et al. (1994), Fig. 8.3, p. 154.]

acidic polysaccharide has been isolated from coccoliths of *E. huxleyi* (de Jong et al. 1976). It has been characterized in more detail than the coccolith-polysaccharides of *P. carterae*. It is a complex macromolecule, containing 14 different monosaccharides, amongst which methylated, carboxylated and sulfated sugars (Fichtinger-Schepman et al. 1981). The acidic groups enable it to bind Ca^{2+}-ions (de Jong et al. 1976), but in contrast with the *P. carterae* polysaccharides, it does not precipitate with Ca^{2+}. Polysaccharide-Ca aggregates like the coccolithosomes of *P. carterae* have not been observed in *E. huxleyi*. The cellular localization of the coccolith polysaccharide was studied with antibodies (van Emburg et al. 1986). The coccolith-polysaccharide was identified in the Golgi-apparatus, the lumina of the coccolith production compartment and in an area outside the cells underlying the coccolith layer. The method used in these experiments did not have a resolution high enough to demonstrate specific association with membranes or the organic base plate. The polysaccharide was also immunolocalized on the surface of all mineral parts of the coccoliths (van Emburg et al. 1986, Fig. 18). One may conclude that the acidic polysaccharide is synthesized in the Golgi apparatus, transported via Golgi-derived vesicles which fuse to form the reticular body and calcifying vesicle, and eventually becomes associated with the mineral phase of the coccolith. Supply of Ca^{2+} ions via the Golgi-derived vesicles, in analogy with the coccolithosomes in *P. carterae*, has not been directly demonstrated. The Golgi apparatus in the motile scale-forming cell type of *E. huxleyi* contains a high concentration of Ca with respect to the cytoplasm (van der Wal et al. 1985). This may point to a Ca-storage or -transport function of the Golgi in *E. huxleyi*, possibly in association with the acidic polysaccharide (cf. Anning et al. 1996).

The organic base plate clearly functions as the insoluble substrate for crystal nucleation. Nothing more is known about its composition than that it contains polysaccharide of still undefined nature and some protein (van Emburg and de Vrind-de Jong 1989a). It is not known whether the acidic polysaccharide is associated with the base plate at any stage of calcification (cf. van Emburg and de Vrind-de Jong 1989a). The acidic polysaccharide is thought to function in the mineralization process by interaction with crystal growth sites as described for the *P. carterae* polysaccharides. It inhibits the crystallization of $CaCO_3$ in vitro (Borman et al. 1982). Indicative of a specific function of the polysaccharide in coccolith morphogenesis is the association of immunologically and electrophoretically distinct polysaccharides with different coccolith morphotypes (Borman et al. 1987, Young and Westbroek 1991, van Bleijswijk et al. 1991). The polysaccharide probably also functions as a matrix in which the extracellular coccoliths are embedded. Recently van Bleijswijk et al. (1991) and van der Kooij (unpublished) confirmed its presence outside the cells apart from its association with the mineral phase of the coccoliths.

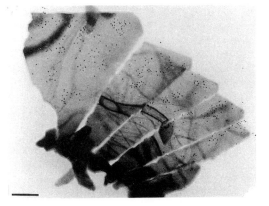

Figure 18. Transmission electron micrograph of proximal shield elements of a coccolith of *Emiliania huxleyi*. The coccoliths were treated with antibodies raised in a rabbit against the coccolith polysaccharide (see text) and subsequently with gold-anti-rabbit IgG (cf. Fig. 13). [Used by permission of Academic Press Inc., from van Emburg et al. (1986), Fig. 12, p. 257.]

Although the analysis of intracellular Ca-pools in scale-forming cells suggests that the Golgi plays a role in Ca^{2+} transport in *E. huxleyi*, the mechanisms of Ca^{2+} uptake remain obscure. The cytosolic Ca^{2+} concentration is strictly regulated and kept between 100 and 250 nM (Brownlee et al. 1995). Import of Ca^{2+} from the sea water, with a concentration of 10 mM Ca^{2+}, will not require energy and is supposed to occur passively through Ca^{2+}-selective channels (Brownlee et al. 1995). Subsequent import of Ca^{2+} from the cytosol into Ca^{2+}-sequestering organelles has to proceed against a concentration gradient and will require energy. The Golgi apparatus of plant cells is able to sequester Ca^{2+} via an enzyme which exchanges Ca^{2+} for protons (Sze et al. 1992). To make this transport go, the plant Golgi membranes contain an active proton pump (a V-type ATPase) which concentrates protons in the Golgi lumen, thus creating a proton gradient over the membrane. The gradient-driven expulsion of protons energizes the simultaneous import of Ca^{2+} into the Golgi vesicles. As a result of this transport system the lumen of the Golgi in plants is more acidic than the cytosol. It is not known whether such a Ca^{2+} transport system exists in the coccolithophorids. As stated above, a V-type ATPase was identified in the Golgi and calcifying vesicle of *P. carterae*, but Ca^{2+} transport induced by this enzyme could not be demonstrated. The coccolith production compartment of *E. huxleyi* is thought to be derived from Golgi vesicles. Calcium uptake in the coccolith production compartment via a Golgi-derived system, based on a low pH in the Golgi lumen, is less likely. The proton pump would have to be inhibited or even reversed after fusion of the Golgi vesicles to the coccolith production compartment, since the latter has to maintain a pH high enough to allow calcification to proceed, as described in Reaction (7). This is sustained by preliminary measurements of the cytosolic and vesicular pH with pH-sensitive indicators. The pH of the calcifying vesicle was determined to be at least slightly higher, certainly not lower, than that of the cytosol (Anning et al. 1996).

It is not known whether HCO_3^- is taken up actively or passively in *E. huxleyi*. Measurements and estimates of extra- and intracellular HCO_3^- concentrations (Nimer and Merett 1992, Dong et al. 1993) suggest that transport of HCO_3^- does not require energy. Transport enzymes have not been identified. The uptake appeared to be regulated by cytosolic Ca^{2+}. Inhibition of Ca^{2+} import in the cell also resulted in inhibition of HCO_3^- import (Nimer et al. 1996). The exact relation between both transport processes remains to be clarified.

Relation between photosynthesis and coccolith formation in **Emili-ania huxleyi.** The relation between calcification and photosynthesis has been extensively studied in *E. huxleyi* (e.g. Paasche 1964, Steeman-Nielsen 1966, Sikes et al. 1980, Nimer and Merett 1993, Dong et al. 1993, Brownlee et al. 1994, Brownlee et al. 1995, Nimer et al. 1996). In many aspects the connection between the two processes appears to be similar to that in *P. carterae*. The experimental evidence indicates that coccolith formation and photosynthesis are most tightly coupled under conditions of a low external CO_2 concentration. Cells with a high rate of calcification were able to grow in media equilibrated with CO_2-free air, whereas growth of cells with a low calcification rate was strongly inhibited (Dong et al. 1993). At a high pH of the medium (i.e. at a low CO_2 concentration) calcification and photosynthesis were both inhibited at low external Ca^{2+} (Nimer et al. 1996). These results indicate that calcifying cells can adapt to CO_2-limiting conditions by generating CO_2 from HCO_3^-. Chloroplasts of calcifying cells of *E. huxleyi* contain carbonic anhydrase (Nimer et al. 1994). A similar situation as in *P. carterae* can be envisaged, in which the OH^- ions liberated in photosynthetic HCO_3^- utilization are neutralized by the protons expelled from the calcifying vesicle according to Reaction (9). Under conditions of external CO_2 concentrations sufficient to support photosynthesis, the relation between photosynthesis and calcification is less evident. For instance, calcification can be inhibited under these conditions by lowering the external Ca^{2+} concentration, whereas photosynthesis still proceeds at a maximal rate (Paasche 1964, Brownlee et al. 1995). A relation between the two processes as described by Reaction (9) predicts a 1:1 ratio of inorganic carbon incorporation in calcite and organic photosynthate. This ratio can be closely approached (Paasche 1964, Sikes et al. 1980, Linschooten et al. 1991, Dong et al. 1993, Nimer and Merrett 1993), but under certain conditions, e.g. phosphorus limitation, inorganic carbon incorporation in calcite strongly exceeds that in organic material (Linschooten et al. 1991, Paasche and Brubak 1994, van Bleijswijk et al. 1994).

Emiliania huxleyi is able to form coccoliths in the dark to some extent (van Bleijswijk et al. 1994), but high rates of dark-calcification as described for *P. carterae* (van der Wal et al. 1987) have never been observed. The strong inhibition of calcification in the dark in *E. huxleyi* is most probably related to the period of cell division. Cells synchronized by a natural light-dark regime divide in the dark (Linschooten et al. 1991, Paasche and Brubak 1994). This means that in the dark period the nucleus is disintegrated, prior to cell division. The nucleus is the support of a functional coccolith production compartment (Fig. 15) and apparently the latter can not be formed in the absence of its support (van Emburg and de Vrind-de Jong 1989b). The inhibition of calcification in the dark is probably not directly related to the absence of photosynthesis. This indicates that the interdependence of photosynthesis and calcification is by no means absolute, a situation similar to that in *P. carterae*.

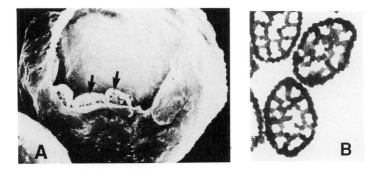

Figure 19. Scanning electron micrograph of a cell (A) and transmission electron micrograph of coccoliths (B) of the motile stage of *Coccolithus pelagicus*. Note extracellular coccoliths beneath the organic envelope surrounding the cell (arrows in A). [Used by permission of Academic Press, from Rowson et al. (1986), Fig. 1 and 4, p. 362.]

Coccolith formation in **Coccolithus pelagicus,** *an example of extracellular controlled biomineralization.* The heterococcolith-forming *Coccolithus pelagicus* (Fig 2E) and the holococcolith-forming *Crystallolithus hyalinus* (Fig. 19) were long time considered to be separate species, until Parke and Adams (1960) described that these algae represent different stages in the life cycle of one species. Since then alternating hetero- and holococcolith forming life stages have been described in other species also (Kleijne 1991). The *C. pelagicus* form is non-motile, and considered to be diploid, based on the ultrastructure of the organic scales (Billard 1994). The *C. hyalinus* form is motile, and the ultrastructure of its organic scales indicates a haploid stage. We will refer to the latter form as the motile stage of *C. pelagicus* in the following. The heterococcoliths of *C. pelagicus* are formed on an organic base scale in a special cisterna of the Golgi apparatus (Leadbeater 1994). After completion of the crystallization process the coccoliths are extruded and form a massive coccosphere (cf. Fig. 2E). The holococcoliths of the motile cells consist of a regular arrangement of similar-sized calcite rhombohedra on an organic base scale (Fig. 19B). They are situated distally from a layer of unmineralized scales, underneath a continuous skin or envelope surrounding the whole cell (Fig. 19A, Manton and Leedale 1963, Rowson et al. 1986). The exact composition and function of the envelope is not known. The scales as well as the envelope are synthesized in the Golgi apparatus, but $CaCO_3$ crystals have never been observed intracellularly (Rowson et al. 1986, Leadbeater 1994). After dissolution of the external coccoliths by lowering the pH of the medium below 7, the cells recalcify after restoration of the pH. The first calcite elements to be reformed were reported to be located on organic scales outside the plasmamembrane, near the flagellar base just underneath the envelope (Rowson et al. 1986). During the recalcification of a complete coccosphere, which took about 60 hours, no intracellular coccoliths were detected at any time. Apparently, the organic coccolith base scales are formed inside the cell and subsequently calcified extracellularly. Despite its extracellular location, the biomineralization process in the motile stage of *C. pelagicus* is clearly biologically controlled. The envelope surrounding the cell plus coccosphere probably creates a microenvironment favorable for $CaCO_3$ precipitation. The calcite crystals are not randomly arranged on the base scale, indicating directed nucleation at specific sites on the scales. Moreover, the calcite crystals appear to be covered by a skin of organic material, which may control the size of the individual crystals (Rowson et al. 1986). Possibly biomineralization in *C. pelagicus* is controlled by a coordinated excretion of the organic and inorganic constituents of the holococcoliths at a specific site (Manton and Leedale 1963, Rowson et al. 1986, de Vrind-de Jong et al. 1994).

ALGAL DEPOSITION OF SILICA

The two major algal phyla in which siliceous cell coverings are formed are the Chrysophyta and the Bacillariophyta (diatoms). Virtually all species within the Chrysophyta produce silicified resting spores, the stomatocysts (Preisig 1994). A limited number of species forms silicified scales in the vegetative state (Preisig 1994). All of the over 100,000 known diatom species produce silicified cell walls. In both phyla silicification takes place in an intracellular, membrane-bound compartment, the silica deposition vesicle (SDV). The membrane of the SDV is called the silicalemma. An immense variation in overall morphologies, detailed mineral patterns and decorations in the form of spines, bristles and other protrusions is possible (cf. Fig. 1). This morphological variability is partly inherent in the amorphous nature of the mineral. An amorphous mineral is more flexible, and can thus be more easily moulded than a crystalline mineral (Williams 1986). Contrary to a crystal it has no fracture planes. An amorphous mineral is less stable, i.e. will more readily dissolve, than crystalline material (Simkiss 1991, see also "Principles of Biomineralization," in the Introduction). The algal silica minerals are protected against dissolution by an organic coating (see below). The morphological variability of the siliceous structures, their intracellular synthesis and their association with organic components clearly show that algal silicification is a biologically controlled biominerali-zation process.

In some aspects silicification is more difficult to study than calcification, due to a number of factors. First, the enormous amount of species all forming morphologically different structures (though species-specific) frustrate a general description of the morpho-genetic process. Second, the study of silica morphogenesis can not rely on knowledge of crystal parameters of the final product as in coccolith morphogenesis. In the latter case the crystallographic orientations of the mineral in statu nascendi and of the completed crystal elements suggest specific and testable crystal-matrix relationships. Third, biochemical analysis of silicification is extremely difficult because of the inert nature of the mineral. The latter is always intimately associated with the silicalemma and the other organic components. Most of the organic constituents can only be released by treatment of the siliceous structures with hydrofluoric acid (HF). Their original structure, localization and function have not been resolved yet, with the possible exception of a special family of glycoproteins, called frustulins, which seems to be conserved in different genera of diatoms (see below). Consequently, studies on algal silica deposition have generally been devoted to ultrastructural descriptions of scale and wall development (for short reviews see Preisiger 1994 and Schmid 1994). In the following sections the present knowledge on algal silicification will be summarized. Because the data presented are compiled from studies on many different algal species, species names will be omitted in the text for reasons of clarity. They can be partly retrieved from Figure legends and from the cited literature.

Cell wall formation in diatoms

In diatoms the formation of the siliceous cell wall is tightly coupled to the cell cycle. We will shortly discuss some aspects of the eukaryotic cell cycle in general and summarize some properties of supramolecular structures involved in cytoplasmatic and organelle movement.

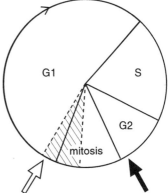

Figure 20. The eukaryotic cell cycle. Shaded area: cytoplasmatic division. Specific for diatoms: last girdle band completed (black arrow); onset of valve synthesis (white arrow). For further explanation see text.

The cell cycle. Eukaryotic cells propagating through mitotic division go through a number of stages each characterized by specific biochemical and structural alterations of the cell (Fig. 20). In the G1 stage, cell growth and maintenance requires synthesis of proteins and other cellular molecules. The end of G1 is marked by the initiation of DNA replication (which is Si-dependent in diatoms, see Sullivan and Volcani 1981). DNA replication occurs during the S stage. The following G2 stage is characterized by synthesis of proteins and other constituents needed for nuclear and cytoplasmatic division, collectively called mitosis. Nuclear division precedes cytoplasmatic separation (shaded area in Fig. 20). The synthesis of the cell wall constituents of diatoms is rather strictly limited to certain stages in the cell cycle (Li and Volcani 1985a, Schmid 1994; cf. Fig. 20 and see later discussion of *"Mechanisms of frustule formation"*). Cell division and wall synthesis involves the movement and moulding of cellular constituents and organelles. Intracellular movement is mediated by supramolecular structures called microtubules and microfilaments. These structures work separate or coordinated by attaching to cellular constituents and imposing movement by sliding, contraction and directional growth. They are composed of different proteins: tubulin polymerizes to form the hollow cylinders of the microtubules (about 25 nm in diameter), actin is the building block of the much thinner, solid fibers of the microfilaments. Both structures are implicated in cell wall formation in diatoms as well as in the Chrysophyta. For detailed information on cell cycles, mitosis, microtubules and microfilaments see Wolfe (1995).

Cell wall formation: ultrastructure of frustules. Diatoms occur in almost any habitat providing light and sometimes minute amounts of moisture and nutrients, as mobile, attached or planktonic (floating) cells (Round and Crawford 1989). They can range in size from 5 to 500 μm. They propagate through mitotic division, and only rarely form haploid gametes. Two basic types are distinguished: the circular or centric diatoms (Figs. 1A and 1D) and the elliptical or pennate diatoms (e.g. Figs. 1B and 1F). Both form silicified cell walls (called frustules) consisting of three basic elements: the epivalve, the hypovalve and one or more girdle bands. Figure 21 shows a schematic impression, including a representation of a cross section, of a hypothetical pennate diatom with two girdle bands. The epivalve is slightly larger in diameter than the hypovalve. The epivalve is loosely connected to two overlapping rings of silica, the girdle bands. The second girdle band overlaps the hypovalve. The valves of pennate diatoms generally contain a so-called raphe system, an elongated slit between two central silica ribs (Fig. 21). Radial silica ribs extend from the central ones, separated by pores (areolae, cf. Margulis et al. 1993). The pores themselves are covered by a thin plate of silica which is perforated in often complex, specific patterns (cf. Crawford 1981, Round and Crawford 1989). The pores serve as gas and nutrient exchange sites in the otherwise impermeable valves (Crawford 1981, Margulis et al. 1993). The raphe is thought to be involved in diatom sliding movement through secretion of mainly acidic polysaccharides (Hoagland et al. 1993). The valves of centric diatoms also contain pores, but a raphe system is absent. Most centric diatoms possess one or more slitted protrusions called labiate processes (Li and Volcani 1985b). Centric diatoms are evolutionary older than pennate diatoms (Round and Crawford 1989), and the raphe system is thought to be derived from the labiate process. Decorations as spines, bristles and other protrusions can add to the immense variation in diatom frustule morphology. The ultrastructure of the cells does not differ significantly from that of other unicellular algae like coccolihophorids. Characteristic for the diatom cell is the silica deposition vesicle (SDV). This will be discussed in the next section, which will focuss on morphogenesis of the pennate frustule.

Mechanisms of frustule formation. The synthesis of the valves of the frus-   tule is initiated during the last part of mitosis, the cytoplasmatic division (Fig. 20). The cell

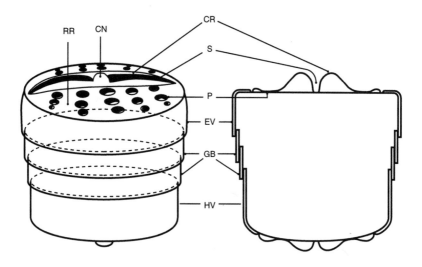

Figure 21. Impression of a frustule of a pennate diatom (A) and of a transverse section across the frustule (B). EV: epivalve; HV: hypovalve; GB: girdleband; P: pore; S: raphe slit; CR: central rib; CN: central nodule; RR: radial rib.

must have passed the S and G2 stages. DNA replication is the trigger for initiation of valve synthesis. Prior to mitosis and the formation of new valves, the cell elongates. The parent valves are rigid and can not accomodate the increased cell volume. Room is created by sliding of the girdle bands relative to the valves and to each other. The girdle bands can be considered as a flexible connection between the valves. Girdle band synthesis need not to be preceeded by DNA replication and can occur during any stage in the diatom life cycle (Volcani 1981), but generally the last girdle band to be synthesized is completed at the end of G2 (Schmid 1994). The girdle bands are formed one by one in a ring-like SDV just under the plasmamembrane and exocytosed after completion.

After mitosis one of the daughter cells is located in the parent hypovalve, the other daughter cell in the parent epivalve. Both daughter cells form new hypovalves. This results in the "promotion" of the parent hypovalve to daughter epivalve (Volcani 1981). A consequence of this division mechanism is that the cell size in most diatom populations gradually decreases over generations. Once a critical size limit is reached, the original size is restored by initiation of gamete formation and sexual reproduction (Round and Crawford 1989).

Pennate valve formation in both daughter cells is initiated by the formation of a thin elongated SDV (Fig. 22A, Volcani 1981). This SDV extends from the central nodule (Fig. 21) in both apical directions along the entire length of the cell, at the position of the future raphe system. In this early stage already the SDV is filled with polymerized silica (Fig. 22A, inset). These primary arms form the bases of one of the central ribs of the raphe system. Soon hereafter two secondary arms are formed from the central nodule colinear with the primary arms (Fig. 22D). When the primary and secondary silicified arms are completed, they form the central ribs enclosing the slit of the raphe system (Fig. 22B). During the growth of the primary and secondary arms, their SDVs also extend radially as the fingers of a glove, forming the radial ribs. At regular points the radial ribs fuse, leaving spaces for the future pores (Fig. 22B). At all stages, growth of the SDV is accompanied by simultaneous silica deposition. It seems that the shape of the SDV determines the morphology of the future valve. The shape of the SDV in its turn is supposed to be generated by an integrated interplay of the plasmamembrane, actin filaments, microtubules, cell organelles and organelle-derived vesicles. The SDV is closely apposed to the plasmamembrane, and thus copies the overall curvature of the latter. The location of the initial SDV is outlined by the endoplasmatic reticulum and mitochondria (Crawford and Schmid 1986). The SDV is thought to grow by fusion with Golgi-derived vesicles, although also the endoplasmatic reticulum has been implicated in the formation of the SDV (Crawford and Schmid 1986, Schmid 1994). Microtubules are involved in the shaping of the raphe system (Fig. 22C). Inhibitors of tubulin polymerization induce malformation of the raphe in several diatom species (Schmid 1980). Large so-called areolar or spacer vesicles (Schmid 1994) attach at regularly spaced sites to the plasmamembrane. The SDV developes around these vesicles, which thus determine the position of the future pores. How the information for the vesicle- and organelle-determined prepatterning of the SDV is translated from the DNA-level to these timed and specific interactions between different cell constituents is unknown.

The generally accepted theory of moulding and prepatterning of the SDV has been criticized by Gordon and Drum (1994). They state that diffusion-limited precipitation of silica particles within the SDV can create a certain branched pattern without recourse to a preformed mould. They modelled such a precipitation process on the basis of the following presumptions: (1) the presence, central in the SDV, of an anisotropic primary nucleation site (formed at the raphe site with mediation of microtubules), (2) the constant introduction of low-molecular weight silica particles (possibly in the form of a gel) from both sites into

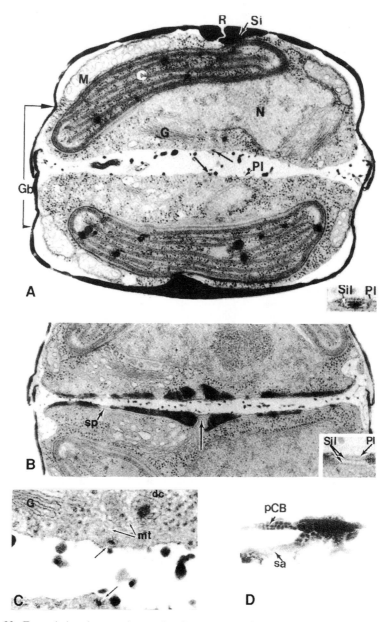

Figure 22. Transmission electron micrographs of transverse sections across dividing cells of the diatom *Navicula pelliculosa*. A: Cell containing two protoplasts in which initiation of the raphe occurs in a minute SDV (arrows). Gb: girdle bands; R: raphe; Si: silica; M: mitochondrion; C: chloroplast; N: nucleus; G: Golgi apparatus; Pl: plasmamembrane. Inset: High magnification of SDV bounded by silicallemma (Sil) containing silica (dark spot). B: Late stage of valve development showing raphe slit (arrow) and pores (sp). Inset: magnification of pore. Note continuous silicalemma beneath plasma-membrane. C: High magnification of SDV from A (arrows), showing microtubules (mt). D: Upper view of developing valve with central nodule and primary central arm or band (pCB) and secondary arm (sa). [Used by permission of Springer-Verlag, New York, from Volcani (1981), Fig. 7-4, p. 167.]

the SDV, and (3) the generation of a concentration gradient of silica from the borders of the SDV towards the centre as a result of the polymerization process. Computer-simulated precipitation of silica under these conditions resulted in patterns very similar to those of diatom frustules. Undoubtedly morphogenesis of biomineral structures is partly based on the inherent pattern-forming abilities of the precipitation processes themselves (cf. also de Vrind-de Jong et al. 1994).

The new valve is exocytosed after completion and situated outside the plasma-membrane. It is not clear what happens to the silicalemma and the plasmamebrane after frustule secretion. Different models have been proposed (Crawford and Schmid 1986). Until experimental evidence proves otherwise we support the model of Volcani (1981). Volcani proposed that a new plasmalemma is synthesized during maturation and exocytosis of the frustule and that the old plasmamembrane and the silicalemma (or their remnants) become part of the organic casing surrounding the mature valves.

The mature valves contain an organic component which can be visualized after dissolution of the silica with HF (cf. Figs. 24D and 24E). An organic ghost of the original frustule remains (Volcani 1981). Except for remnants of the plasmamembrane and the silicalemma this casing probably contains also extra organic material, added to the valves at different stages of frustule formation. The organic composition of the valves markedly changes during valve synthesis (Volcani 1981). Especially polysaccharide material appears to be added at a later stage. A high-molecular-weight glycoprotein rich in serine and glycin residues and phosphate groups was isolated from diatom valves which were thoroughly cleaned with strong oxidants prior to silica dissolution (Swift and Wheeler 1992). A very similar substance could be isolated from diatomaceous earth. This macromolecule apparently is closely asscociated with the mineral and may perform a nucleating function in silica deposition, possibly by arranging silica particles through hydrogen bonding to glycoprotein hydroxyl or phosphate groups. This molecule may also function in the transport of silica through the cell (see below).

Recently a new class of glycoproteins has been isolated from the cell wall of a lightly silicified diatom species (Kröger et al. 1994). The cell wall of this particular species consists mainly of organic material, and only the raphe area and the girdle bands are silicified. The organic wall seems to be continous with the organic casing surrounding the silicified parts and is produced within the SDV. Extraction of the cell walls with a solution of the Ca^{2+}-chelator EDTA yielded a highly acidic, Ca^{2+}-binding glycoprotein family which is characterized by repeats of Ca^{2+}-binding domains alternating with proline-rich stretches. Subsequent dissolution of the extracted cell walls with HF yielded another set of high molecular weight organic components which may be related to the glycoprotein isolated by Swift and Wheeler (1992). Antibodies against one of the acidic, EDTA-soluble glycoproteins cross-reacted with an EDTA-extract of the cell walls of highly silicified diatom species (Kröger et al. 1994). A homologous glycoprotein has been isolated from one of these highly silicified species, indicating that these macromolecules form a general component of diatom cell walls and are evolutionary conserved (Kröger et al. 1996). The glycoproteins were suggested to play a role in cell wall patterning, possibly by forming a framework of acidic glycoproteins inhibiting silica deposition at specific sites. In view of their proposed role in diatom frustule biogenesis they were called frustulins (Kröger et al. 1996). Interestingly, a protein containing a glycosylation site and rich in proline and acidic residues is thought to be involved in coccolith biogenesis in *Emiliania huxleyi* (Corstjens and de Vrind-de Jong, unpublished). It will be worth while to study the involvement of this particular family of acidic glycoproteins in the fromation of algal mineralized cell walls in general.

An unanswered question concerning silica deposition in diatoms and silicifying algae in general, is how silicon is taken up in and transported through the cell. In view of the low silicon concentrations in natural waters the uptake probably needs the investment of energy. In a particular diatom species the presence of a silicic acid transport enzyme was proposed by Sullivan and Volcani (1981), based on enzyme-kinetical and energy-dependent characteristics of the uptake system. This system may prevail in acidic environments where the primary silicon species will be silicic acid ($Si(OH)_4$). Uptake through endocytosis and subsequent intracellular transport of vesicles may prevail in alkaline environments, where silicon may be present in a more polymeric form (Gordon and Drum 1994). A cytoplasmatic silica pool can be detected prior to the onset of valve formation (Sullivan and Volcani 1981). Part of the cytoplasmatic silica pool coincides with with a relatively high phosphate concentration (Rogerson et al. 1986). It is tempting to assume that the cytoplasmatic silica pool is associated with the phosphorylated glycoprotein reported by Swift and Wheeler (1992). During transport this macromolecule may prevent premature polymerization of silica whereas it may function as a nucleator once introduced in the SDV, as suggested above. Immunolocalization of (more or less) purified organic constituents of the diatom frustule may advance our knowledge on the biochemical mechanisms involved in frustule formation. The role of organic constituents may also be clarified by in vitro studies of their inductive or inhibiting effects on silica polymerization (Swift and Wheeler 1992, cf. Harrison and Lu 1994).

Scale and stomatocyst formation in Chrysophyta

The Chrysophyceae and Synurophyceae are the two main classes in the phylum Chrysophyta in which silicified structures are formed (Mignot and Brugerolle 1982, Preisig 1994). Virtually all species within these classes form silicified stomatocysts. All of the species of the Synurophyceae form silicified body scales, which are arranged in specifically ordered arrays around the cells. In one exceptional colony-forming species, the scales do not surround individual cells, but form an investment around the colony as a whole (Pipes and Leedale 1992). A limited number of the Chrysophyceae produce silicified body scales, which generally form a less well organized cell cover. The scales are usually decorated with rims and pores, and depending on the species, may contain bristles or spines. Bristles are elongated needles which are hinged to a base plate scale. They are formed in a specialized SDV separate from that of the body scales. Their attachment to the scales is flexible. Spines are an elongation of body scales, thus rigidly attached and formed in one SDV with the scales. We will shortly discuss the morphogenesis of stomatocysts and highlight some common and specialized features of scale formation.

Stomatocyst formation. The morphogenesis of stomatocysts in both classes of the Chrysophyta has been reviewed by Sandgren (1989). Stomatocysts are spherical or oblong resistant spores which range from 3 to 35 µm in diameter. They are surrounded by a thick siliceous cell wall containing one pore plugged with organic material (Fig. 23A). Stomatocyst formation is initiated by activation of the Golgi apparatus which buds off small vesicles. These vesicles fuse to form a double-membraned sphere leaving open a small circular pore. The sphere surrounds the major part of the cytoplasm, resulting in a separation of intracystic and extracystic cytoplasm. Then silica is deposited in two phases: a thin primary silica shell is rapidly deposited followed by a more slow thickening of the wall and addition of decorations (Fig. 23B). Electron dense particles are present at the sites of active silica deposition. They may represent silica transport vesicles. The extracystic cytoplasm is contracted within the silicified wall and a siliceous collar is formed around the pore. The pore is finally filled with organic material of Golgi origin. The membrane compartment in which the silica wall is produced is analogous to the SDV in diatoms. The

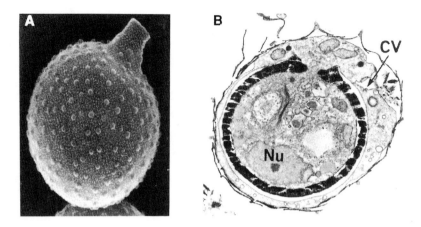

Figure 23. A: Scanning electron micrograph of stomatocyst of *Syncrypta pallida*. B: Transmission electron micrograph of a transverse section across a stomatocyst of *Paraphysomonas corynephora*; Nu: nucleus; CV: contractile vacuole. Note spherical SDV filled with silica. [Used by permission of Springer-Verlag, Austria, from Preisig (1994), Figs. 6 and 7, p. 30.]

most obvious difference with wall formation in diatoms is that the SDV in stomatocysts is completely preformed prior to silica deposition.

Scale formation. Scales of Chrysophyta are specifically ornamented structures which may be arranged in a highly ordered, species-specific manner (Figs. 24A and 24B). Scales range in size from 1 to 10 μm (Preisig 1994). The shape of the scales is often differentiated depending on their position on the cell (Lavau and Wetherbee 1994, cf. Fig. 24A). Some species contain one type of scale, others contain up to four different types, some of them decorated with bristles or spines (Beech et al. 1990). Scale morphology depends on nutrient availability like supply of silicon, phosphorus and nitrogen (Hahn et. al. 1996, Sandgren and Hall 1996). The scales are produced in SDVs which are positioned either on the periplastidial endoplasmatic reticulum on the chloroplast (most of the Synurophyceae), or in close contact with the cytoplasmatic endoplasmatic reticulum (Chrysophyceae). The SDV is thought to be of Golgi origin (cf. Pipes and Leedale 1992). The SDV is pre-shaped in the form of the mature scale prior to silica deposition (Fig. 24C), as is the case in wall formation in stomatocysts. During early stages of SDV formation, microtubules and actin filaments are closely associated with the developing SDV (Fig. 24C; cf. Brugerolle and Brichieux 1984). The microtubules are thought to be involved in moulding and migration of the SDV, the actin filaments are thought to be mainly involved in moulding (Brugerolle and Brichieux 1984). Sandgren and Barlow (1989) questioned the involvement of actin, because microfilament-disrupting drugs did not seem to affect the scale morphology whereas microtubule inhibitors resulted in malformed scales. Prior to the onset of silica deposition, fibrous material, probably organic in nature, occupies the SDV lumen (Preisig 1986). It may function as a nucleation site for silica polymerization. After dissolution of the scale with HF, an organic ghost of the scale remains (Figs. 24D and 24E). This organic material may stem from the fibres present in the SDV, but nothing is known about its nature. After completion of silicification, the SDV is released from its support, subsequently migrated to the plasmamembrane and finally the scale is incorporated in the scale case surrounding the cell. Bristles are formed in separate SDVs and secreted independently (Beech et al. 1990). They are attached to base-plate scales through a fibrillar complex at their base. This complex probably consists of adhesive glycoproteins, which

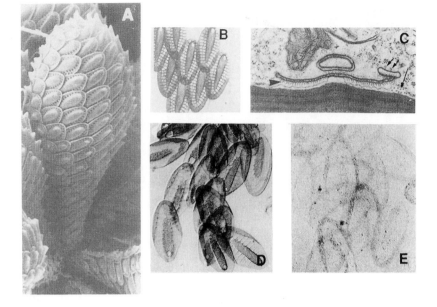

Figure 24. Scanning and transmission electron micrographs of cells, scales, and a transverse section of Chrysophyta. A: Cell of *Synura petersenii*. Note overlapping scale pattern, the pores along the rim of the scales and the central part of the scales consisting of a hollow chamber of silica B: Detached scales from *S. petersenii*. Part of the rim of the scales is curved upward. C: transverse section across SDV of *S. glabra*. The SDV is completely preformed prior to silica deposition, aided by microtubules (short arrows) and actin-like filaments (arrowhead). The SDV is positioned on the chloroplast endoplasmatic reticulum (long arrow). D and E: scales of *S. petersenii* before (D) and after (E) treatment with hydrofluoric acid. [Figs. 23A,B,C used by permission of Springer-Verlag, from Preisig (1994), Figs. 8, 9 and 10, p 32; Figs. 23D,E used by permission of Springer-Verlag, New York, from McGrory and Leadbeater (1981), Figs. 8-17, -18, p. 216.]

are also present on the rim of body scales where they overlap neighboring scales (Ludwig et al. 1996). No other macromolecular constituents of scales have been identified, nor is anything known about silicon acquisition by Chrysophyta.

The specific arrangement of scales in some scale cases has evoked discussions about the manner of insertion of new scales in the existing case. Siver and Glew (1990) proposed that new scales are secreted at one fixed site at the plasmamembrane, where they would displace the existing ones to a new position, whereas Leadbeater (1990) presented evidence for secretion of scales at varying programmed sites. When a chrysophyte culture is subjected to silicon starvation, the cells loose their scale case and become naked. Addition of silicon induces the formation of a new scale case, with secretion of scales at varying sites of the cell. After secretion of a certain number of new scales they start to assemble at one end of the cell (Sandgren and Hall 1996). These results support the mechanism proposed by Leadbeater (1990) and show that the scale case is a dynamic structure which can adept to different stages in cell development. During the production of the new scale case the cells do not divide (Sandgren and Hall 1996), indicating that scale formation can be uncoupled from cell division. Thus two main differences between silica deposition in diatoms and Chrysophyta exist: the timing of SDV moulding and silicification, and the cell cycle regulation of the biomineralization process.

CONCLUDING REMARKS

The events leading to the production of algal calcareous and siliceous biominerals are extensively documented at the ultrastructural level. At the biochemical level, some progress has been made in the field of biologically controlled calcification, i.e. coccolith formation: water-soluble, acidic polysaccharides associated with biominerals have been isolated, characterized to some extent and immunolocalized at sites of mineral deposition. These polysaccharides have been assigned functions in analogy with the macromolecules of other calcified tissues (invertebrate shells, bone, teeth, cf. Addadi and Weiner 1992, Mann 1993a,b). They are proposed to specifically promote crystal nucleation when fixed to a solid support, and to inhibit crystallization when in solution, all by interaction of their Ca^{2+}-binding groups with crystal growth sites. They are also implicated in the transport of Ca^{2+} ions to the calcification site. Although their Ca^{2+}-binding properties and in vitro interactions with $CaCO_3$ precipitation support these proposed functions, the evidence is indirect and the exact role of these polysaccharides still remains a matter of speculation. The solid supports involved in coccolith formation are base plate scales or less well structured base plates. Their compositions are poorly characterized and their role in the calcification process is also obscure. Except for one preliminary finding (cf. *"Mechanisms of frustule formation,"* above, and Corstjens and de Vrind-de Jong, unpublished, Linschooten et al. 1995) nothing is known about the possible involvement of proteins or glycoproteins in coccolith formation. Although it is now established that HCO_3^- is the carbon substrate for $CaCO_3$, it is not known how it is sequestered from the medium and transported in the cell. Nothing is known about the uptake of Ca^{2+} ions either. Some progress has been made in the isolation of cell organelles and some enzymes which may be involved in the calcification process (Ca^{2+}-dependent ATPase in *P. carterae* and carbonic anhydrase in *P. carterae* and *E. huxleyi*) have been preliminary identified.

In the field of biological silicification, biochemical information is even more scarce. Only a limited number of macromolecules have been identified: a glycoprotein probably involved in adhesion of scales and bristles to the scale case and a number of glycoproteins (most of them acidic) of which the function still has to be determined. Nothing is known about the uptake of silicon or the form in which it is transported in the cell.

The lack of information on biochemical mechanisms of calcification and silicification is inherent to the difficulties in isolating relevant macromolecules and in obtaining unambiguous evidence for their involvement in the biomineralization processes. Part of these problems may be relieved by studying biomineralization at the molecular genetical level. So far genetical information on calcifying and silicifying algae has been mainly restricted to phylogenetical data (Medlin et al. 1996 and references therein). During the last couple of years a molecular genetical approach in the study of the mechanisms of biomineralization in general is gaining increasing attention. In another chapter of this volume the molecular genetics of biomineralization are discussed in detail. Here we will summarize some recent advances and future strategies in the application of molecular genetical techniques in calcification and silicification. Finally we will briefly discuss an interdisciplinary approach to quantitatively address the impact of algal calcification on the global cycling of carbon and its effect on the global climate.

Recent developments: the molecular genetical approach

A powerful technique to demonstrate the involvement of a macromolecule in biomineralization, is to identify the gene encoding the macromolecule (or genes encoding enzymes involved in its synthesis), to interfere with the expression of the gene, and subsequently study the effects of this interference on the process under study. To follow this approach, certain conditions have to be fulfilled. First, genes possibly involved in biomineralization have to be identified. Second, vectors containing genes of interest have to be introduced into the cells under study (and transformed cells have to be selected for). Third, the introduced gene has to be expressed in the transformed cell. Suppose these conditions were fulfilled, several strategies can be followed of which one will be examplified here (Fig. 25). A relevant gene can be inserted in opposite orientation in a transforming plasmid and transformed to the biomineralizing cell. Expression of the oppositely orientated gene from the plasmid will result in so-called antisense RNA, which is complementary to the wild type messenger RNA (mRNA). The antisense RNA will hybridize to the mRNA and thus interfere with its translation to protein. As a result the wild type gene is knocked out and the effect of the lack of its product can be studied.

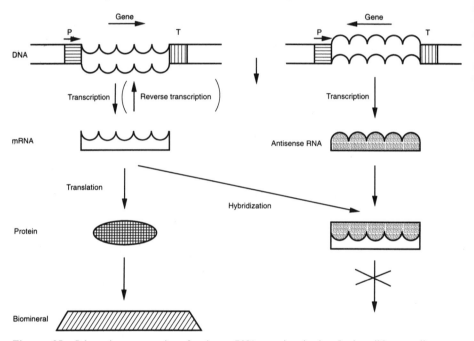

Figure 25. Schematic representation of antisense RNA gene-inactivation. In the wild type cell a gene, part of the genomic double-stranded DNA, is transcribed into messenger RNA (mRNA). This occurs by copying the nucleotide sequence of one of the two DNA strands (the coding or sense strand) into the RNA nucleotide sequence. Transcription takes place in one direction, starting from the initiation point (the promotor P) and ending at the termination point (the terminator T). The mRNA is translated into a protein, which, in this example, is involved in the formation of a biomineral. Suppose the gene has been isolated and cloned in a certain vector between a promotor and terminator, but in the opposite orientation compared to its natural situation. After the recombinant vector has been transformed to the wild type cell, the wild type gene as well as the gene lying on the vector will be transcribed into RNA. The latter RNA, however, is now copied from the other, non-coding or antisense strand of the DNA. The antisense RNA thus generated is complementary to the mRNA and will form a duplex (hybridize) with the mRNA, thus preventing translation into the "biomineralization protein." The transformed cell will be unable to produce a biomineral.

How can genes involved in biomineralization (at least preliminary) be identified? One method depends on the isolation of proteins which are suspected to play a role in the process. Antibodies raised against these proteins can be used to screen a complementary-DNA (cDNA) library for the production of homologous products. (A cDNA library consists of cloned cDNA fragments which have been reversely transcribed from isolated mRnas; cf. Fig. 25). The clones containing relevant genes can be isolated and characterized. By this approach we have isolated a gene encoding a protein putatively involved in coccolith formation (Corstjens and de Vrind-de Jong, unpublished). Genes encoding members of the frustulin family have been isolated by the same procedure (Kröger et al. 1996). Alternatively, one may apply e.g. subtractive cDNA library methodology (or related techniques). This method depends on the isolation of mRNA populations from cells which are actively biomineralizing and from cells which are inactive. Comparison of the two mRNA populations (via their respective cDNA libraries) identifies those genes which are preferentially expressed during biomineralization. By this method silicon-responsive cDNA clones have been isolated (Hildebrand et al. 1993) and genes preferentially expressed in the period of active calcification in *E. huxleyi* were preliminary identified (Corstjens et al. 1995). A third approach depends on the homology between certain genes from different organisms. If certain enzymes (e.g. transport enzymes, polysaccharide-synthesizing enzymes) are thought to be involved in biomineralization, their genes may be isolated from DNA libraries using DNA probes which are based on conserved sequences of homologous genes retrieved from data banks. A subunit from a V-type ATPase could thus be identified in *P. carterae* (Corstjens et al. 1996).

Development of procedures for transformation of especially marine algae is still in its infancy. Transformed cells have to be separated from non-transformed cells, e.g. by introducing an antibiotic resistance (normally derived from prokaryotes) on the transforming vector. Selection thus depends on the expression of heterologous genes in the algal cells. Recently two diatom species have been succesfully transformed with a plasmid containing a neomycin resistance by microprojectile bombardment (Dunahay et al. 1995). Expression of the heterologous gene was accomplished by cloning it between the transcription start (promotor) and the transcription terminator of a diatom gene. The progress made in isolation of genes relevant to the biomineralization processes and the promising start in the development of transformation procedures for marine algae will undoubtedly raise our understanding of the mechanisms of biomineralization to a higher level.

Recent developments: the interdisciplinary approach

Intuitively it is clear that calcifying and silicifying algae, especially those with a cosmopolitan distribution and blooming properties, have an enormous impact on the geochemical cycling of the elements. Although estimates have been made of the biological impact on element cycles in some cases (cf. Reynolds 1986), only recently a research programme has been initiated to quantitatively address biosphere-geosphere interactions and their relevance to global climate forcing. The Global Emiliania Modelling Initiative (GEM) uses *E. huxleyi* as a model system to study the wider implications of climate forcing by the open marine biota. The ultimate purpose of this interdiciplinary programme is to create a series of compatible, nested models, each representing different organization levels, covering the molecular, organismal, ecological and global levels of organization (Westbroek 1991, Westbroek et al. 1993, Westbroek et al. 1994). Calcifying algae like *E. huxleyi* have a distinct effect on the global climate. Sinking of coccoliths ($CaCO_3$) out of the surface waters causes the efflux of CO_2 to the atmosphere (cf. Reaction 9), whereas the drawdown of dead cells has the opposite effect (cf. Reaction 1). Moreover, blooms of *E. huxleyi* produce dimethyl sulphide (DMS), a gas which upon oxidation induces the formation of white clouds, increasing the Earth's albedo. The net effect of these organisms on the global climate will depend on the balance between the production of $CaCO_3$, photosynthate and DMS. *Emiliania huxleyi* was chosen as a model system because it is readily accessible to studies on the molecular and physiological scale in the laboratory, and on the ecological, global and geological scale in the field. At the moment, models describing the organism at the organismal level (Zonneveld 1996) and on the ecological

302   *GEOMICROBIOLOGY*

level (Tyrell and Taylor 1996) are well underway. Modelling *E. huxleyi* at the molecular level awaits the advances in biochemical and molecular genetical analysis of the relevant processes. Whether ultimately a mechanistic model of the biosphere as a whole can be proposed, remains a question to which the Global Emiliania Modelling Initiative hopes to find an answer.

ACKNOWLEDGMENTS

We thank Michiel de Kuijper for his assistance and Paul Corstjens for fruitful discussions during the preparation of this manuscript.

REFERENCES

Addadi L, Weiner S (1989) Stereochemical and structural relations between macromolecules and crystals in biomineralization. In: Mann S, Webb J, Williams RJP (eds). Biomineralization. Chemical and biochemical perspectives. VCH Verlagsgesellschaft, Weinheim, Germany, p 133-156
Addadi L, Weiner S (1992) Control and design principles in biological mineralization.Angew Chem Int'l Ed Engl 31:153-169
Anning T, Nimer N, Merrett MJ, Brownlee C (1996) Costs and benefits of calcification in coccolithophorids. J Mar Syst 9:45-56
Araki Y (1997) Calcium ion-stimulated ATPases in a calcifying coccolithophorid: Lokalization, characterization, and possible role during coccolithogenesis. PhD Dissertation, Univ of California, Los Angeles, 108 p
Beech PL, Wetherbee R, Pickett-Heaps JD (1990) Secretion and development of bristles in Mallomonas splendens (Synurophyceae). J Phycol 26:112-122
Beer S (1994) Mechanisms of inorganic carbon acquisition in marine macroalgae (with special reference to the chlorophyta). In: Chapman DJ, Round F (eds) Prog Phycol Res, Biopress Ltd, Bristol, p 179-207
Belcher AM, Wu XH, Christensen RJ, Hansma PK, Stucky GD, Morse DE (1996) Control of crystal phase switching and orientation by soluble mollusc-shell proteins. Nature 381:56-58
Berridge M (1993) Inositol triphosphate and calcium signalling. Nature 361:315-325
Billard C (1994) Life cycles. In: Green JC, Leadbeater BSC (eds) The Haptophyte Algae. The Systematics Association Spec Vol 51:167-186 Clarendon Press, Oxford
Borman AH, de Jong EW, Huizinga M, Kok DJ, Westbroek P, Bosch L (1982) The role in $CaCO_3$ crystallization of an acid Ca^{2+}-binding polysaccharide associated with coccoliths of Emiliania huxleyi. Eur J Biochem 129:179-183
Borman AH, de Jong EW, Thierry R, Westbroek P, Bosch L (1987) Coccolith-associated polysaccharides from cells of Emiliania huxleyi (Haptophyceae). J Phycol 23:118-123
Borowitzka MA (1982a) Morphological and cytological aspects of algal calcification. Int'l Rev Cytol 74:127-162.
Borowitzka MA (1982b) Mechanisms in algal calcification. Prog Phyc Res 1: 137-177
Borowitzka MA (1986) Physiology and biochemistry of calcification in the Chlorophyceae. In: Green JC, Leadbeater BSC (eds) The Haptophyte Algae. The Systematics Association Spec Vol 51:107-124 Clarendon Press, Oxford
Borowitzka MA (1989) Carbonate calcification in algae: Initiation and control. In: Mann S, Webb J, Williams RJP (eds) Biomineralization: Chemical and Biochemical Perspectives. VCH Verlagsgesellschaft, Weinheim, Germany p 63-94
Broecker WS (1971) A kinetic model for the chemical composition of sea water. Quatern Res 1:188-207
Brown RM, Romanovicz DK (1976) Biogenesis and structure of Golgi-derived cellulosic scales in Pleurochrysis. Appl Polym Symp 28:537-585
Brownlee C, Davies M, Nimer, N, Dong LF, Merett MJ (1995) Calcification, photosynthesis and intracellular regulation in Emiliania huxleyi. In: Allemand D, Cuif J-P (eds) Biomineralization 93. Bulletin de l'Institut océanographique Monaco, numéro spécial 14,2, p 19-35
Brownlee C, Nimer N, Dong LF, Merrett MJ (1994) Cellular regulation during calcification in Emiliania huxleyi. In: Green JC, Leadbeater BSC (eds) The Haptophyte Algae. The Systematics Association Spec Vol 51149-166 Clarendon Press, Oxford
Brugerolle G, Bricheux G (1984) Actin microfilaments are involved in scale formation of the Chrysomonad cell Synura. Protoplasma 123:203-212
Corstjens PLAM, Akari Y, Westbroek P, González EL (1996) A gene encoding the 16 kD proteolipid subunit of a vacuolar-type H^+-ATPase from Pleurochrysis carterae strain 136. Plant Physiol 111:652 (PGR 96-038)

Corstjens PLAM, de Vrind-de Jong EW, Westbroek P (1995) Identification of genes involved in calcification in Emiliania huxleyi. Abstr Conf Emiliania huxleyi and the Oceanic Carbon Cycle, The Natural History Museum, London, April 1995, p 13

Crawford RM (1981) The siliceous components of the diatom cell wall and their morphological variations. In: Simpson TL, Volcani BE (eds) Silicon and siliceous structures in biological systems. Springer-Verlag, New York, p 129-156

Crawford RM, Schmid A-MM (1986) Ultrastructure of silica deposition in diatoms. In: Leadbeater BSC, Riding R (eds) Biomineralization in lower plants and animals. The Systematics Association Spec Vol 30:291-314 Clarendon Press, Oxford

Crenshaw MA (1991) Mineral induction by immobilized polyanions. In: Suga S, Nakahara H (eds) Mechanisms and phylogeny of mineralization in biological systems. Springer-Verlag, Tokyo, p 101-105

Crenshaw MA, Ristedt H (1976) The histochemical localization of reactive groups in septal nacre from Nautilus pompilius L. In: Watabe N, Wilbur KM (eds) The mechanisms of mineralization in the ivertebrates and plants. Univ South Carolina Press, Columbia, p 355-367

Davey RJ, Black SN, Bromley LA, Cottier D, Dobbs B, Rout JE (1991) Molecular design based on recognition at inorganic surfaces. Nature 353:549-550

de Jong EW, Bosch L, Westbroek P (1976) Isolation and characterization of a Ca^{2+}-binding polysaccharide associated with coccoliths of Emiliania huxleyi (Lohmann) Kamptner. Eur J Biochem 70:611-621

de Vrind-de Jong EW, Borman AH, Thierry R, Westbroek P, Grüter M, Kamerling JP (1986) Calcification in the coccolithophorids Emiliania huxleyi and Pleurochrysis carterae. II. Biochemical aspects. In: Leadbeater BSC, Riding R (eds) Biomineralization in lower plants and animals. The Systematics Association Spec Vol 30:205-217 Clarendon Press, Oxford

de Vrind-de Jong EW, van Emburg PR, de Vrind JPM (1994) Mechanisms of calcification: Emiliania huxleyi as a model system. In: Green JC, Leadbeater BSC (eds) The Haptophyte Algae. The Systematics Association Spec Vol 51:149-166 Clarendon Press, Oxford

Didymus JM, Young JR, Mann S (1994) Construction and morphogenesis of the chiral ultrastructure of coccoliths from the marine alga Emiliania huxleyi. Proc Roy Soc London B 258:237-245

Dixon GK, Patel BN, Merett MJ (1987) Role of intracellular carbonic anhydrase in inorganic carbon assimilation by Polyphyridium purpureum. Planta (Berl) 172:443-449

Dong LF, Nimer NA, Okus E, Merrett MJ (1993) Dissolved inorganic carbon utilization in relation to calcite production in Emiliania huxleyi. New Phytol 123:679-684

Dunahay TG, Jarvis EE, Roessler PG (1995) Genetic transformation of the diatoms Cyclotella cryptica and Navicula saprophila. J Phycol 31:1004-1012

Egge JK, Aksnes DL (1992) Silicate as regulating nutrient in phytoplankton competition. Mar Ecol Prog Ser 83:281-289

Faber WW, Preisig HR (1994) Calcified stuctures and calcification in protists. Protoplasma 181:78-105

Falini G, Albeck S, Weiner S, Addadi L (1996) Control of aragonite or calcite polymorphism by mollusk shell macromolecules. Science 271:67-69

Fichtinger-Schepman AMJ, Kamerling JP, Versluis C, Vliegenthart JFG (1981) Structural studies of the methylated acidic polysaccharide associated with coccoliths of Emiliania huxleyi (Lohmann) Kamptner. Carbohydr Res 93:105-123

Flajs G (1977) Skeletal structures of some calcifying algae. In: Flügel E (ed) Fossil algae. Springer-Verlag, Berlin p 225-231

Gordon R, Drum RW (1994) The chemical basis of diatom morphogenesis. Int'l Rev Cyt 150:243-372

Green JC, Course PA, Tarran GA (1996) The life-cycle of Emiliania huxleyi: A briefreview and a study of relative ploidy levels analysed by flow cytometry. J Mar Syst 9:33-44

Green JC, Perch-Nielsen K, Westbroek P (1989) Phylum Prymnesiophyta. In: Margulis L, Corliss JO, Melkonian M, Chapman DJ (eds) Handbook of Protoctysta. Jones and Bartlett, Boston, p 293-317

Hahn A, Gutowski A, Geissler U (1996) Scale and bristle morphology of Mallomonas tonsurata (Synurophyceae) in cultures with varied nutrient supply. Bot Acta 109:239-247

Harrison CC, Lu Y (1994) In vivo and in vitro studies of polymer controlled silicification. In: Allemand D, Cuif J-P (eds) Biomineralization 93. Bulletin de l'Institut océanographique Monaco, numéro spécial 14,1, p 151-158

Hildebrand M, Higgins DR, Busser K, Volvcani BE (1993) Silicon-responsive cDNA clones isolated from the marine diatom Cylindrotheca fusiformis. Gene 132:213-218

Hillis L (1991) Recent calcified Halimedaceae. In: Riding R (ed.) Calcareous algae andstromatolites. Springer-Verlag, New York, p 167-187

Hoagland KL, Rosowski JR, Gretz MR, Roemer SC (1993) Diatom extracellular polymeric substances: function, fine structure, chemistry and physiology. J Phycol 29:537-566

Holligan PM, Balch WM (1991) From the ocean to cells: coccolithophore optics and biogeochemistry. In: Demers S (ed) Particle analysis in oceanography (NATO ASI ser, G27). Springer-Verlag, Berlin, p 301-324

Holligan PM, Viollier M, Harbour DS, Champagne-Philippe M (1983) Satellite and ship studies of coccolithophore production along a continental shelf edge. Nature 304:339342

Israel AA, González EL (1996) Photosynthesis and inorganic carbon utilization in Pleurochrysis sp. (Haptophyta), a coccolithophorid alga. Mar Ecol Prog Ser 137:243-250

Johnston AM, Raven JA (1986) Utilization of bicarbonate ions by the macroalga Ascophyllum nodosum (L.) Le Jolis. Plant Cell Environ 9:175-184

Jordan RW, Kleijne A (1994) A classification system for living coccolithophores. In: Winter A, Siesser WG (eds.) Coccolithophores. Cambridge Univ Press, UK, p 83-105

Kitano Y, Kanamori N, Tokuyama A (1969) Effects of organic matter on solubilitiesand crystal form of carbonates. Am Zoologist 9:681-688

Klaveness D (1972a) Coccolithus huxleyi (Lohmann) Kamptner I. Morphological observations on the vegetative cells and the process of coccolith formation. Protistologica 8:335-346

Klaveness D (1972b) Coccolithus huxleyi (Lohmann) Kamptner II. The flagellate cell, aberrant cell types, vegetative propagation and live cycles. Br Phycol J 7:309-318

Kleijne A (1991) Holococcolithophorids from the Indian Ocean, Red Sea, Mediterranean Sea and North Atlantic Ocean. Mar Micropal 17:1-76

Kröger N, Bergsdorf C, Sumper M (1994) A new calcium binding glycoprotein family constitutes a major diatom cell wall component. EMBO J 13:4676-4683

Kröger N, Bergsdorf C, Sumper M (1996) Frustulins: domain conservation in a protein family associated with diatom cell walls. Eur J Biochem 239:259-264

Kwon D-K, González EL (1994) Lokalization of Ca^{2+}-stimulated ATPase in the coccolith-producing compartment of cells of Pleurochrysis sp. (Prymnesiophyceae). J Phycol 30:689-695

Lavau S, Wetherbee R (1994) Structure and development of the scale case of Mallomonas adamas (Synurophyceae). Protoplasma 181:259-268

Leadbeater BSC (1990) Ultrastructure and assembly of the scale case in Synura (Synurophyceae Andersen). Br Phycol J 25:117-132

Leadbeater BSC (1994) Cell coverings. In: Green JC, Leadbeater BSC (eds) The Haptophyte Algae. The Systematics Association Spec Vol 51:23-46 Clarendon Press, Oxford

Li C-W, Volcani BE (1985a) Studies on the biochemistry and fine structure of silica shell formation in diatoms IX. Sequential valve formation in a centric diatom, Chaetoceros rostratum. Protoplasma 124:30-41

Li C-W, Volcani BE (1985b) Studies on the biochemistry and fine structure of silica shell formation in diatoms X. Morphogenesis of the labiate process in centric diatoms. Protoplasma 124:147-156

Linschooten C, Borman AH, Corstjens PLAM, Westbroek P, de Vrind-de Jong EW (1995) Intlracellular polysaccharide-containing fractions of Emiliania huxleyi: precursors in coccolith biosynthesis?. Abstr Conf Emiliania huxleyi and the Oceanic Carbon Cycle, The Natural History Museum, London, April 1995, p 20

Linschooten C, van Bleijswijk JDL, van Emburg PR, de Vrind JPM, Kempers ES, Westbroek P, de Vrind-de Jong EW (1991) Role of the light-dark cycle and medium composition on the production of coccoliths by Emiliania huxleyi (Haptophyceae). J Phycol 27:82-86

Lowenstam HA, Weiner S (1989) On Biomineralization. Oxford University Press, New York, Oxford, 324 p

Lucas WJ (1975) Analysis of the diffusion symmetry developped by the alkaline and acid bands which form at the surface of Chara corallina cells. J Exp Bot 26:271-286

Ludwig M, Lind JL, Miller EA, Wetherbee R (1996) High molecular mass glycoproteins associated with the siliceous scales and bristles of Mallomonas splendens (Synurophyceae) may be involved in cell surface development and maintenance. Planta 199:219-228

MacIntyre IG, Reid RP (1995) Crystal alteration in a living calcareous alga (Halimeda): implications for studies in skeletal diagenesis. J Sedimentary Res A65:143-153

Mann S (1988) Molecular recognition in biomineralization. Nature 332:119-124

Mann S, Archibald DD, Didymus JM, Douglas T, Heywood BR, Meldrum FC, Reeves NJ (1993a). Crystallization at inorganicorganic interfaces: biominerals and biomimetic synthesis. Science 261:1286-1292

Mann S (1993b) Molecular tectonics in biomineralization and biomimetic materials chemistry. Nature 365:499-505

Mann S, Perry CC, Williams RJP, Fyfe CA, Gobbi GC, Kennedy GJ (1983) The characterization of the nature of silica in biological systems. J Chem Soc Chem Commun 168:168-170

Mann S, Sparks NHC (1988) Single crystalline nature of coccolith elements of the marine alga Emiliania huxleyi as determined by electron diffraction and high-resolution transmission electron microscopy. Proc Roy Soc London B 234:441-453

Manton I, Leedale GF, (1963) Observations on the micro-anatomy of Coccolithus hyalinus Gaarder and Markali. Arch Microbiol 47:115-136

Margulis L, McKhann HI, Olendzensky L, Hiebert S (1993). Illustrated Glossary of Protoctysta, Jones and Bartlett, Boston, London, 288 p

Marsh MH (1994) Polyanion-mediated mineralization-assembly and reorganization of acidic polysaccharides in the Golgi system of a coccolithophorid alga during mineral deposition. Protoplasma 177:108-122

Marsh MH (1996) Polyanion-mediated mineralization- a kinetic analysis of the calcium-carrier hypothesis in the phytoflagellate Pleurochrysis carterae. Protoplasma 190:181-188

Marsh MH, Chang D-K, King GC (1992) Isolation and characterization of a novel acidic polysaccharide containing tartrate and glyoxylate residues from the mineralized scales of a unicellular coccolithophorid alga Pleurochrysis carterae. J Biol Chem 267:20507-20512

McConnaughey T (1995) Ion transport and the generation of biomineral supersaturation. In: Allemand D, Cuif J-P (eds) Biomineralization 93. Bulletin de l'Institut océanographique, Monaco, Numéro special 14,2, p 1-18

McGrory CB, Leadbeater BSC (1981) Ultrastructure and deposition of silica in the Chrysophyceae. In: Simpson TL, Volcani BE (eds) Silicon and siliceous structures in biological systems. Springer-Verlag, New York, p 201-230

McIntyre A, Bé A (1967) Modern Coccolithophoridae in the Atlantic ocean. I. Placoliths and crytoliths. Deep-sea Res 14:561-597

Medlin LK, Barker GLA, Campbell L, Green JC, Hayes PK, Marie D, Wrieden S, Vaulot D (1996) Genetic characterization of Emiliania huxleyi (Haptophyta). J Mar Syst 9:13-31

Mignot J-P, Brugerolle G (1982) Scale formation in Chrysomonad flagellates. J Ultrastruct Res 81:13-26

Mori IC, Sato G, Okazaki M (1996) Ca^{2+}-dependent ATPase associated with plasma membrane from a calcareous alga Serraticardia maxima (Corallinaceae, Rhodophyta). Phycol Res 44:193-202

Nakahara H, Bevelander G (1978) The formation of calcium crystals in Halimeda incrassata with special reference to the role of organic matrix. Japn J Phyc 26:9-12

Nancollas GH (1982) Biological Mineralization and Demineralization. Springer-Verlag, Berlin, Heidelberg, New York, 415 p

Nimer NA, Guan Q, Merett MJ (1994) Extra- and intracellular carbonic anhydrase in relation to culture age in a high calcifying strain of Emiliania huxleyi Lohman. New Phytol 126:601-607

Nimer NA, Merett MJ (1992) Calcification and utilization of inorganic carbon by the coccolithophorid Emiliania huxleyi Lohman. New Phytol 121:173-177

Nimer NA, Merrett MJ (1993) Calcification rate in Emiliania huxleyi Lohmann in response to light, nitrate and availability of inorganic carbon. New Phytol 123:673-677

Nimer NA, Merrett MJ, Brownlee C (1996) Inorganic carbon transport in relation toculture age and inorganic carbon concentration in a high-calcifying strain of Emiliania huxleyi (Prymnesiophyceae). J Phycol 32:813-818

Okazaki M, Furuya K (1985) Mechanisms in algal calcification. Japan J Phycol 33:328-344

Okazaki M, Tokita M (1988) Calcification of Chara braunii (Charophyta) caused by alkaline band formation coupled with photosynthesis. Japan J Phycol 36:193-201.

Okazaki M, Sato T, Mutho N, Wada N, Umegaki T (1997) Calcified scales (coccoliths) of Pleurochrysis carterae (Haptophyta): Structure, crystallography, and acid polysaccharides. J Mar Biotechnol (in press)

Outka DE, Williams DC (1971) Sequential coccolith morphogenesis in Hymenomonas carterae.   J Protozool 18:285-297

Paasche E (1964) A tracer study of the inorganic carbon uptake during coccolith formation and photosynthesis in the coccolithophorid Coccolithus huxleyi. Physiol Plant, Suppl III, p 5-82

Paasche E, Brubak S (1994) Enhanced calcification in the coccolithophorid Emiliania huxleyi (Haptophyta) under phosphorus limitation. Phycologia 33:324-330

Parke M, Adams I (1960) The motile (Crystallolithus hyalinus Gaarder and Markali) and non-motile phases in the life history of Coccolithus pelagicus (Wallich) Schiller. J Mar Biol Assoc UK 39:263274

Pienaar RN (1971) Coccolith production in Hymenomonas carterae. Protoplasma 73:217-224

Pipes LD, Leedale GF (1992) Scale formation in Tessellaria volvocina (Synurophyceae). Br phycol J 27:11-19

Preisig HR (1986) Biomineralization in the Chrysophyceae. In: Leadbeater BSC, Riding R (eds) Biomineralization in lower plants and animals. The Systematics Association Spec Vol 30:327-344 Clarendon Press, Oxford

Preisig HR (1994) Siliceous structures and silicifcation in flagellated protists. Protoplasma 181:29-42

Price GD, Whitecross MI (1983) Cytochemical localization of ATPase activity on the plasmalemma of Chara corallina. Protoplasma 116:65-74

GEOMICROBIOLOGY

Quiroga O, González EL (1993) Carbonic anhydrase in the chloroplast of a coccolithophorid (Prymnesiophyceae). J Phycol 29:321-324

Raven JA, Johnston AM (1991) Mechanisms of inorganic carbon acquisition in marine phytoplankton and their implications for the use of other resources. Limnol Oceanograph 36:1701-1714

Reynolds CS (1986) Diatoms and the geochemical cycling of silicon. In: Leadbeater BSC, Riding R (eds) Biomineralization in lower plants and animals. The Systematics Association Spec Vol 30:269-289 Clarendon Press, Oxford

Rogerson A, deFreitas ASW, McInnes AG (1986) Cytoplasmatic silicon in the centric diatom Thalassiosira pseudonana localized by electron spectroscopic imaging. Can J Microbiol 33:128-131

Round FE, Crawford RM (1989) Phylum Bacillariophyta. In: Margulis L, Corliss JO, Melkonian M, Chapman DJ (eds) Handbook of Protoctysta. Jones and Bartlett, Boston, p 574-596

Rowson JD, Leadbeater BSC, Green JC (1986) Calcium carbonate deposition in the motile (Crystallolithus) phase of Coccolithus pelagicus (Prymnesiophyceae) Br Phycol J 21:359-370

Sandgren CD (1989) SEM investigations of statospore (stomatocyst) development in diverse members of the Chrysophyceae and Synurophyceae. Beih Nova Hedwigia 95:45-69

Sandgren CD, Barlow SB (1989) Siliceous scale production in Chrysophyte algae. II. SEM observations regarding the effects of metabolic inhibitors on scale regeneration in laboratory populations of scale-free Synura petersenii cells Beih Nova Hedwigia 95:27-44

Sandgren CD, Hall SA, Barlow SB (1996) Siliceous scale production in Chrysophyte and Synurophyte algae. I. Effects of silica-limited growth on cell silica content, scale morphology, and the construction of the scale layer of Synura petersenii. J Phycol 32:675-692

Schmid A-MM (1980) Valve morphogenesis in diatoms: a pattern-related filamentous system in pennates and the effect of APM, colchicin and osmotic pressure. Beih Nova Hedwigia 33:811-848

Schmid A-MM (1994) Aspects of morphogenesis and function of diatom cell walls with implications for taxonomy. Protoplasma 181:43-60

Sekino K, Shiraiwa Y (1996) Evidence for the involvement of mitochondrial respiration in calcification in a marine coccolithophorid, Emiliania huxleyi. Plant Cell Physiol 37:1030-1033

Sharkia R, Beer S, Cabantchik ZI (1994) A membrane-located polypeptide of Ulva sp. which may be involved in HCO_3^- uptake is recognized by antibodies raised against the human red-blood-cell anion-exchange protein. CRC Crit Rev Microbiol 3:1-26

Sikes CS, Roer RD, Wilbur KM (1980) Photosynthesis and coccolith formation: inorganic carbon sources and net inorganic reaction of deposition. Limnol Oceanogr 25:248-261

Simkiss K (1991) Amorphous minerals and theories of biomineralization. In: Suga S, Nakahara H (eds) Mechanisms and phylogeny of mineralization in biological systems. Springer-Verlag, Tokyo, p 375-389.

Simkiss K, Wilbur KM (1989) Biomineralization. Cell biology and mineral deposition. Academic Press, San Diego, CA, 337 p

Siver PA, Glew JR (1990) The arrangement of scales and bristles on Mallomonas (Chrysophyceae): a proposed mechanism for the formation of the cell covering. Can J Bot 68:374-380

Steeman-Nielsen E (1966) The uptake of free CO_2 and HCO_3^- during photosynthesis of planktonic algae with special reference to the coccolithophorid Coccolithus huxleyi. Physiol Plant 19:232-240

Sullivan CW, Volcani BE (1981) Silicon in the cellular metabolism of diatoms. In: Simpson TL, Volcani BE (eds) Silicon and siliceous structures in biological systems. Springer-Verlag, New York, p 15-42

Swift DM, Wheeler AP (1992) Evidence of an organic matrix from diatom biosilica. J Phycol 28:202-209

Sze H, Ward JM, Lai S, Perera I (1992) Vacuolar-type H^+-translocating ATPases in plant endomembranes: Subunit organisation and multigene families. J Exp Biol 172:123-136

Tappan H (1980). The paleobiology of plant protists. WH Freeman, San Francisco, 1028 p

Takahashi K (1991) Mineral flux and biogeochemical cycles of marine planktonic protozoa. In: Reid PC (ed) Protozoa and their role in marine processes. NATO ASI Series G25, Springer-Verlag, Berlin, p 347-359

Tyrell T, Taylor AH (1996) A modelling study of Emiliania huxleyi in the NE Atlantic. J Mar Syst 9:83-112

van Bleijswijk JDL (1996) Ecophysiology of the calcifying marine alga Emiliania huxleyi. PhD Dissertation, Groningen Univ, Groningen, the Netherlands, 71 p

van Bleijswijk JDL, Kempers R, Veldhuis MJ, Westbroek P (1994) Cell and growth characteristics of types A and B of Emiliania huxleyi (Prymnesiophyceae) as determined by flow cytometry and chemical analysis. J Phycol 30:230-241

van Bleijswijk JDL, van der Wal P, Kempers R, Veldhuis MJ, Young JR, Muyzer G, deVrind-de Jong EW, Westbroek P (1991) Distribution of two types of Emiliania huxleyi (Prymnesiophyceae) in the northeast Atlantic region as determined by immunofluorescence and coccolith morphology. J Phycol 27:566-570 van der Wal P, de Bruijn WC, Westbroek P (1985) Cytochemical and X-ray microanalysis studies of intracellular calcium pools in scale-bearing cells of the coccolithophorid Emiliania huxleyi. Protoplasma 124:1-9

van der Wal P, de Jong L, Westbroek P, de Bruijn WC (1983a) Calcification in the coccolithophorid alga Hymenomonas carterae. Ecol Bull 35:251-258

van der Wal P, de Jong EW, Westbroek P, de Bruijn WC, Mulder-Stapel AA (1983b) Ultrastructural polysaccharide localization in calcifying and naked cells of the coccolithophorid Emiliania huxleyi. Protoplasma 118:157-168

van der Wal P, de Jong EW, Westbroek P, de Bruijn WC, Mulder-Stapel AA (1983c)Polysaccharide localization, coccolith formation, and Golgi dynamics in the coccolithophorid Hymenomonas carterae. J Ultrastruct Res 85:139-158

van der Wal P, de Vrind JPM, de Vrind-de Jong EW, Borman AH (1987) Incompleteness of the coccosphere as a possible stimulus for coccolith formation in Pleurochrysis carterae (Prymnesiophyceae). J Phycol 23:218-221

van der Wal P, Kempers ES, Veldhuis MJW (1995) Production and downward flux of organic matter and calcite in a North Sea bloom of the coccolithophorid Emiliania huxleyi. Mar Ecol Prog Ser 126:247-265

van Emburg PR, de Vrind-de Jong EW (1989a) The organic constituents of coccoliths. In: Coccolith formation in Emiliania huxleyi. van Emburg PR, PhD Dissertation, Leiden Univ, Leiden, the Netherlands, p 48-60

van Emburg PR, de Vrind-de Jong EW (1989b) Some morphological observations on the coccolith production compartment. In: Coccolith formation in Emiliania huxleyi. van Emburg PR, PhD Dissertation, Leiden Univ, Leiden, the Netherlands, p 61-74

van Emburg PR, de Vrind-de Jong EW, Daems WTh (1986) Immunochemical localization of a polysaccharide from biomineral structures (coccoliths) of Emiliania huxleyi. J Ultrastruct and Mol Struct Res 94:246-259

Volcani, BE (1981) Cell wall formation in diatoms: morphogenesis and biochemistry. In: Simpson TL, Volcani BE (eds) Silicon and siliceous structures in biological systems. Springer-Verlag, New York, p 157-200

Wada N, Okazaki M, Tachikawa S (1993) Effects of calcium-binding polysaccharides from calcareous algae on calcium carbonate polymorphs under conditions of double diffusion. J Cryst Growth 132:115-121

Watabe N (1967) Crystallographic analysis of the coccolith of Coccolithus huxleyi. Calc Tiss Res 1:114-121

Westbroek P (1991) Life as a geological force. W.W. Norton, New York, 240 p

Westbroek P, Brown ChW, van Bleijswijk J, Brownlee C, Brummer GJ, Conte M, Egge J, Fernández E, Jordan RW, Knappertbusch M, Stefels J, Veldhuis M, van der Wal P. Young J (1993) A model system approach to biological climate forcing. The example of Emiliania huxleyi. Global and Planetary Change 8:27-46

Westbroek P, Buddemeier B, Coleman M, Kok D, Fautin D, Stal L (1994) Strategies for the study of climate forcing by calcification. In: Doumenge F (ed) Biomineralization 93. Bulletin de l'Institut océanographique, Monaco, Numéro special 13, p 37-60

Westbroek P, Young JR, Linschooten K (1989) Coccolith production (biomineralization) in the marine alga Emiliania huxleyi. J Protozool 36:368-373

Wheeler AP, George JW, Evans CA (1981) Control of calcium carbonate nucleation and crystal growth by soluble matrix of oyster shell. Science 212:1397-1398

Wheeler AP, Sikes CS (1989) Matrix-crystal interactions in $CaCO_3$ biomineralization. In: Mann S, Webb J, Williams RJP (eds) Biomineralization. Chemical and biochemical perspectives. VCH Verlagsgesellschaft, Weinheim, Germany, p 95-131

Williams RJP (1986) Introduction to silicon chemistry and biochemistry. In: Silicon Biochemistry. Ciba Foundation Symp 121, Wiley, Chichester, UK, p 24-39

Wolfe SL (1995) Introduction to cell and molecular biology. Wadsworth Publishing Co, Belmont, MA

Young JR, Didymus JM, Bown PR, Prins B, Mann S (1992) Crystal assembly and phylogenetic evolution in heterococcoliths. Nature 356:516-518

Young JR, Westbroek P (1991) Genotypic variation within the coccolithophorid species Emiliania huxleyi. Mar Micropaleontol. 18:5-23

Zonneveld C (1996) Modelling the kinetics of non-limiting nutrients in microalgae. J Mar Syst 9:121-136

Chapter 9

REACTIONS OF EXTRACELLULAR ORGANIC LIGANDS WITH DISSOLVED METAL IONS AND MINERAL SURFACES

Alan T. Stone

Department of Geography and Environmental Engineering
The Johns Hopkins University
Baltimore, Maryland 21218, U.S.A.

INTRODUCTION

Uptake and release of chemicals by living organisms are necessary conditions for life, and cause perceptible geochemical changes in the surrounding environment (Lovelock 1975). The coupling of biological processes with geochemical processes has been modified by continual evolutionary pressures over the long history of life on earth. For this reason, uptake and release strategies employed by living organisms represent optimal solutions to particular physiological and biochemical design constraints.

Organisms can perturb the geochemical speciation of metallic elements by releasing inorganic and organic chemicals into the surrounding extracellular environment. A previous review (Stone et al. 1994) focused upon the reductant properties of extracellular organic chemicals. Low molecular-weight organic compounds can be selected that represent the range of functional groups and molecular structures found in extracellular organic compounds. Once the thermodynamics, kinetics, and mechanisms of reactions involving these representative compounds have been explored in the laboratory, it is often possible to draw analogies that allow the reactivities of other, more complex organic compounds to be predicted.

The present work examines the effects of extracellular organic ligands on metal ion speciation. We begin with chemical inventory, and examine how equilibrium relationships between species in aqueous systems can be quantified. Properties of metal ions and organic ligands affecting complex formation are then summarized, and physiological functions that benefit from extracellular metal ion chelation are discussed. Complexation of +II and +III metal ions, redox equilibria, and adsorption are then examined using a series of equilibrium calculations and illustrative examples. Strengths and limitations of equilibrium approaches towards speciation, and the challenges inherent in kinetic and mechanistic approaches will be discussed. We end with a brief look at functional group and structure differences among siderophores (used to acquire Fe^{III}) and other extracellular organic ligands, and a discussion of how these differences may affect biogeochemical behavior.

METAL IONS AND LIGANDS: SPECIATION AND INVENTORY

In theory, each chemical species is defined as a set of constituent atoms linked in a particular bonding and solvation arrangement, for which a partial molar Gibbs Free Energy (μ^o_i, in energy units per mole) can be defined. Activity corrections arising from long-range electrostatic interactions among chemial species are then accounted for using the Debye-Huckel limiting law (e.g. Stumm and Morgan 1996). In practice, our ability to distinguish and therefore define distinct chemical species is a function of the analytical methods available to the researcher. "Dissolved" species are distinguished from "adsorbed" and

0275-0279/97/0035-0009$05.00

"precipitated" species by filtration or by centrifugation. Distinguishing "adsorbed" from "precipitated" species is a little more subtle than it might first appear; a continuum exists that ranges from isolated adsorbed species to "islands" of coadsorbed species and finally to distinct phases. Chemical principles and wet chemical methods used in the past to distinguish adsorbed from precipitated species are giving way to more advanced surface-sensitive techniques (e.g. scanning tunnelling microscopy, and ion microprobes.) Protonation level is determined by acid-base titration, while other stoichiometric relationships are determined by measurements made as the concentrations of other chemicals are systematically changed. Bond lengths and geometries are determined using spectroscopic methods such as UV, visible, infared, Raman, NMR, EPR, and EXAFS spectroscopy.

Lewis bases (also called ligands in this chapter) are atoms or molecules that possess an electron pair that can be donated to Lewis acids (henceforth called protons and metal ions.) Ligands, protons, and metal ions can be considered a "basis set" which can be combined to form a number of possible chemical species. Each chemical species can be considered as a compartment in a mass balance equation that accounts for a certain amount of the ligand, proton, or metal ion "units" added to the system.

Water (H_2O), ever present at a concentration of 55.6 molar, is the "default" ligand in aquatic systems. Hydroxide ion (OH^-) is present at concentrations that increase by a factor of ten for every unit increase in pH; oxo ion (O^{2-}) becomes important when a metal ion is capable of inducing hydroxide ion deprotonation. Because of these solvent-derived constituents, multiple metal ion and ligand species exist in even two- or three-component systems.

Consider a system that consists of calcium (Ca^{II}, added as $Ca(OH)_2$) and oxalic acid (HOOCCOOH, abbreviated H_2L) added to pure water. The thermodynamic data base CRITICAL (Martell et al. 1995) lists a number of possible dissolved species, allowing us to write mass balance equations for Ca_T (total dissolved calcium) and L_T (total dissolved oxalate):

$$Ca_T = [Ca^{2+}] + [CaOH^+] + [CaL^o(aq)] \tag{1}$$

$$L_T = [H_2L^o] + [HL^-] + [L^{2-}] + [CaL^o(aq)] \tag{2}$$

Assume for the moment that amounts of added calcium and oxalic acid are low, and no solid phase forms. If Ca_T and L_T are fixed, the following principles apply:

(1) In order for the concentration of metal-ligand complex to increase, the concentrations of other species in the mass balance equations for metal ion and ligand must decrease.

(2) Hydroxide ions compete with the organic ligand for available metal ions; protons compete with the metal ion for available ligand.

If we additionally assume that equilibrium is attained, then concentrations of the species listed in Equations 1 and 2 are linked to one another through equilibrium constants:

$$^*K_1 = \frac{[CaOH^+][H^+]}{[Ca^{2+}]} \qquad K_L = \frac{[CaL^o(aq)]}{[Ca^{2+}][L^{2-}]} \tag{3}$$

$$K_{a1} = \frac{[HL^-][H^+]}{[H_2L]} \qquad K_{a2} = \frac{[L^{2-}][H^+]}{[HL^-]} \tag{4}$$

Substituting these equations into Equations (1) and (2) yields:

$$Ca_T = (1.0 + {^*K_1}/[H^+] + K_L[L^{2-}])[Ca^{2+}] \tag{5}$$

$$L_T = ([H^+]^2/K_{a1}K_{a2} + [H^+]/K_{a2} + K_L[Ca^{2+}])[L^{2-}] \tag{6}$$

These two equations illustrate a third principle:

(3) In the absence of solid phases, the concentration of all mononuclear metal species are directly proportional to the concentration of "free" metal ion. Hence, the concentration of "free" metal ion serves as a master variable for the system.

If we move to systems containing higher amounts of Ca^{II}, the possibility of solid phase formation must be dealt with. CRITICAL lists three possible solid phases in the system being considered: $Ca(OH)_2(s)$, $CaL(H_2O)(s)$, and $CaL(H_2O)_3(s)$. If a solid forms, then total *dissolved* calcium (Ca_T) is less than total *added* calcium (TOTCa, in moles per liter of solution). Similarly, L_T is less than TOTL:

$$TOTCa = Ca_T + m(Ca(OH)_2,s) + m(CaL(H_2O),s) + m(CaL(H_2O)_3,s) \tag{7}$$

$$TOTL = L_T + m(CaL(H_2O),s) + m(CaL(H_2O)_3,s) \tag{8}$$

In the two equations just presented, m refers to the moles of each solid that has precipitated per liter of solution). Although the algebra required for solving these equations is more complex, solutions are readily obtained using computer-based models such as HYDRAQL (Papelis et al. 1988) and FITEQL (Herberlin and Westall 1994).

When a solid phase is present, Principle 1 no longer holds. Changing the pH or adding a new ligand to the system can cause dissolution or precipitation of the solid to occur, changing the total dissolved metal ion concentration Ca_T. Principle 2 still holds, and there is an interesting new twist to Principle 3. As long as the solid phase is present, the solid helps set the free metal ion concentration through the appropriate solubility product constant:

$$^*K_{sp}(Ca(OH)_2(s)) = \frac{[Ca^{2+}]}{[H^+]^2 a_{Ca(OH)_2}} \tag{9}$$

$$K_{sp}(CaL(H_2O)(s)) = \frac{[Ca^{2+}][L^-]}{a_{CaL(H_2O)}} \tag{10}$$

$$K_{sp}(CaL(H_2O)_3(s)) = \frac{[Ca^{2+}][L^-]}{a_{CaL(H_2O)_3}} \tag{11}$$

If $Ca(OH)_2(s)$ has been identified in a system and the pH is known, for example, then $[Ca^{2+}]$ is fixed by Equation (9). If either $CaL(H_2O)(s)$ or $CaL(H_2O)_3(s)$ are present and the free oxalate concentration is known, then $[Ca^{2+}]$ is fixed by either Equation (10) or (11).

In addition to being precipitated or dissolved, calcium- and oxalate-containing species may be *adsorbed*:

Adsorption of calcium only: $>S,Ca(H)_l{}^{x+}$

Adsorption of oxalate only: $>S,L(H)_m{}^{y+}$

Calcium plus oxalate adsorption: $>S,CaL(H)_n{}^{z+}$

The symbol ">S" denotes a surface site. Acid-base titrations can be performed that are analogous in many ways to titrations of homogeneous solutions, and have provided useful insight into the proton stoichiometry of "free" surfaces and surfaces coated with adsorbate (e.g. Sjoberg and Lovgren 1993). Experiments performed at different metal ion and ligand concentrations and different surface area loadings provide additional information about stoichiometry. If metal ion and ligand adsorb to the same site, two possibilities exist: "ligand-like" adsorption where the ligand acts as a bridge between a metal atom on the surface and the metal ion adsorbate, and "metal-like" adsorption where the metal ion serves as a bridge between hydroxide ion or some other Lewis base on the surface and the ligand adsorbate (Schindler 1990). Using spectroscopy to explore surface speciation is inherently more difficult than analogous solution speciation studies.

In the system being discussed here, knowing the physical surface area of each precipitated solid is absolutely essential. For each type of site, it is desirable to measure the total concentration of sites in contact with solution (S_T, in moles/L), presumably by performing an adsorption isotherm measurement. Structural and stoichiometric differences among sites on the same solid are known to affect adsorption. When two adsorbates are added to the same suspension, the degree of competition has been observed to be far greater than predicted from a model assuming that all surface sites are chemically equivalent (e.g. Vasudevan and Stone, submitted).

If species concentrations change over time (either rapidly or slowly), then the system is kinetically-controlled. Our ability to predict how rapidly such changes take place rely heavily upon our ability to accurately quantify system composition. Hence, kinetics in solution are much better known and quantifiable than kinetics involving mineral-water interfaces.

Although speciation and chemical inventory are always in need of refinement, a basic framework is in place that allows us to explore extracellular organic ligands and to make semi-quantitative comparisons of their effects on metal ion speciation.

LEWIS BASE FUNCTIONAL GROUPS

With very few exceptions, electron pairs essential for Lewis base (i.e. ligand) activity reside on oxygen, nitrogen, and sulfur atoms within biological molecules. Table 1 lists common functional groups bearing these atoms, and provides a good basis for identifying proton- and metal ion- coordinating biological molecules.

Table 1. Oxygen-, nitrogen-, and sulfur-containing organic functional groups exhibit a very wide range of basicity, reflected in pK_a values that span 23 log units.

Compound or class	Conjugate Acid	Conjugate Base	pK_a
Carbonyl	$R-\overset{O^{..-H^+}}{\underset{}{C}}-R$	$R-\overset{O}{\underset{}{C}}-R$	approx −7
Thiol	$RCH_2-\overset{H^+}{\underset{}{S}}H$	RCH_2-SH	approx −7
Ester	$R-\overset{O^{..-H^+}}{\underset{}{C}}-OR$	$R-\overset{O}{\underset{}{C}}-OR$	approx. −6
Ether	$RCH_2-\overset{H^+}{\underset{}{O}}CH_2R$	RCH_2-OCH_2R	approx. −3·5
Alcohol	$RCH_2-\overset{H^+}{\underset{}{O}}H$	RCH_2-OH	approx. −2
Amide	$R-\overset{O^{..-H^+}}{\underset{}{C}}-NH_2$	$R-\overset{O}{\underset{}{C}}-NH_2$	approx. −0·5

Table 1, continued.

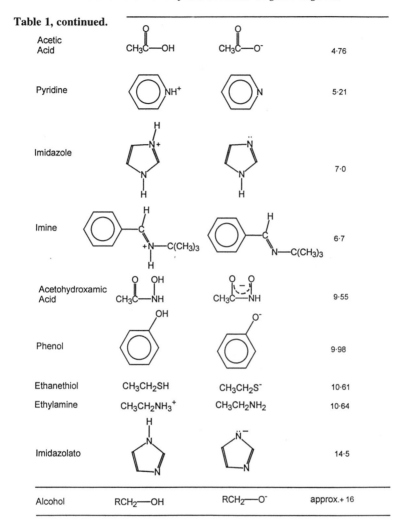

Acetic Acid	CH₃C(=O)—OH	CH₃C(=O)—O⁻	4·76
Pyridine			5·21
Imidazole			7·0
Imine			6·7
Acetohydroxamic Acid			9·55
Phenol			9·98
Ethanethiol	CH_3CH_2SH	$CH_3CH_2S^-$	10·61
Ethylamine	$CH_3CH_2NH_3^+$	$CH_3CH_2NH_2$	10·64
Imidazolato			14·5
Alcohol	RCH_2—OH	RCH_2—O⁻	approx.+ 16

Taking the minus log of the equilibrium constants for acid base reactions (e.g. Eqn. 4) is useful, since the resulting pK_a values correspond to the pH where the conjugate acid and conjugate base are present at the same concentration. Table 1 lists pK_a values for important organic functional groups. Proton-ligand bonds share many traits with metal ion-ligand bonds, and hence pK_a values are widely accepted indicators of the "basicity" of each conjugate base. The following principle must, however, be kept in mind:

4. *Electronic (inductive and resonance) and steric interactions within organic molecules can be quite large. For this reason, pK_a values and other properties of functional groups in simple molecules may be a poor indication of pK_a values and properties in larger, more complex biological molecules.*

Functional groups at the top of Table 1 are very poor Lewis bases, exhibiting pK_a values below 0.0. [In a 1.0 molar strong acid solution, less than 50% of each functional group would be protonated.] For both ester and amide functional groups, the carbonyl oxygen atom is believed to be more basic than the oxygen or nitrogen bridging atom, and hence more likely to participate in metal ion coordination (March 1985, p 222).

In catalysis, metal ion coordination by the functional groups at the top of Table 1 can be important, since complexation of only one organic molecule in 1000 or in 10^6 can provide an efficient route towards products. If we want the organic ligand to capture a significant fraction of the metal ion present, however, the functional groups at the top of Table 1 can usually be disregarded. [Nonactin and other macrocyclic ethers and enniantin and other peptides respresent exceptions to this rule; multiple ether and carbonyl functional groups are used to effectively coordinate alkali and alkaline earth metal ions (Cowan 1993).]

For the functional groups in the middle portion of Table 1, the degree of metal ion coordination depends upon the balance of two trends: (1) higher basicity yields a desirable increase in metal ion-ligand bond strength, and (2) higher basicity yields an undesirable increase in the degree of protonation. Subsequent sections will explore the coordination properties of this group of functional groups in much greater detail. When two or more of these groups are arranged for chelate ring formation, detectable metal ion coordination usually takes place.

The alcoholate anion RCH_2O^- at the bottom of Table 1 is so basic that protonation almost always wins out over metal ion coordination. In subsequent sections, we will pay close attention to factors that allow metal ions to displace the coordinated proton, and thus gain access to this unusually strong Lewis base group.

METAL ION PROPERTIES AFFECTING COMPLEX FORMATION

The metal ion properties listed in Table 2 allow us to make generalizations about the nature and strength of metal ion-ligand bonds. All bonds receive both an ionic contribution and a covalent contribution to bonding. The ionic contribution is primarily driven by short-range electrostatic attractions, which become stronger as the charge of the participating metal ion and Lewis base are increased, and as their radii decrease. Thus, ionic contributions to bonding increase dramatically in going from +II metal ions to +III and +IV metal ions. OH^- and NH_3 are often used as representative anionic and neutral ligands. OH^- appears as a focused point of negative charge, which results in a strong ionic contribution to bonding. NH_3 bears only a partial negative charge on the nitrogen atom, which results in a weak-to-moderate ionic contribution to bonding.

Covalent bonding arises from electron sharing across a metal ion-Lewis base molecular orbital, and is most effective when either the Highest Occupied Molecular Orbital (HOMO) or the Lowest Unoccupied Molecular Orbital (LUMO) is a d-orbital. For this reason, transition metal ions exhibit greater covalent bonding than lighter metals (e.g. Al^{III}) and metal ions to the left on the periodic table (the alkali metal ions and the alkaline earth metal ions.) Among homologous series of ligands, covalent contributions to bonding typically increase as the atomic weight of the Lewis base atom increases (e.g. $OH^- << SH^- < SeH^-$). The orientation of orbitals in space also has a significant effect on bond strength, especially when transition metal ions are involved. Crystal field theory and molecular orbital theory represent increasingly sophisticated attempts to explain this and related phenomena.

PHYSIOLOGICAL FUNCTIONS THAT BENEFIT FROM EXTRACELLULAR ORGANIC LIGAND RELEASE

Eight examples of physiological functions that benefit from extracellular organic ligand release are listed below:

(1) Solubilize iron and other essential metallic elements so that they can be acquired and used in the synthesis of metalloenzymes and other biomolecules.

(2) Solubilize Fe^{III} and other oxidized forms of metallic elements for use as terminal electron acceptors.

(3) Sequester toxic metal ions (or lower free toxic metal ion activity) so that they do not interfere with Functions 1 and 2.

(4) Solubilize iron and other metallic elements so that anions associated with them (e.g. phosphate) can be taken up by cells.

(5) Sequester metallic elements employed by competitor organisms.

(6) Form insoluble precipitates that possess desirable biological properties (e.g. protecting spores from desiccation, protecting spores from dissolved oxidants, adsorbing toxic metal ions from solution).

(7) Dissolve precipitates that interfere physically (e.g. impeding nutrient transport) or chemically (e.g. that adsorb nutrient elements) with physiological functions.

(8) Solubilize the oxidant Mn^{III} (or other reactants) for delivery outside the cell.

Table 2. Important properties of selected metal ions.
Radii are from Shannon (1976) as cited by Langmuir (1997).

Metal ion	Radius (Å)	Electronic configuration	logK (Rxn 1)	logK (Rxn 2)
Mg^{II}	0.72	$[Ne]3s^o$	-11.4	0.23*
Ca^{II}	1.00	$[Ar]4s^o$	-12.7	0.10
Mn^{II}	0.83	$[Ar]3d^5$	-10.6	1.0
Fe^{II}	0.78	$[Ar]3d^6$	-9.4	1.5
Co^{II}	0.75	$[Ar]3d^7$	-9.7	2.1
Ni^{II}	0.69	$[Ar]3d^8$	-9.9	2.9
Cu^{II}	0.73	$[Ar]3d^9$	-7.5	4.2
Zn^{II}	0.74	$[Ar]3d^{10}$	-9.0	2.4
Cd^{II}	0.95	$[Kr]4d^{10}$	-10.1	2.62*
Pb^{II}	1.19	$[Xe]4f^{14}5d^{10}6s^2$	-7.6	1.5
Al^{III}	0.54	$[Ne]3s^o$	-5.0	
Cr^{III}	0.62	$[Ar]3d^3$	-3.7	4.4
Mn^{III}		$[Ar]3d^4$	+0.4*	
Fe^{III}	0.65	$[Ar]3d^5$	-2.2	
Co^{III}	0.61	$[Ar]3d^6$	-0.5*	
Si^{IV}	0.26	$[Ne]3s^o$		
Ti^{IV}	0.61	$[Ar]3d^o$		
Mn^{IV}	0.53	$[Ar]3d^o$		

Rxn 1: $Me^{n+} + H_2O = MeOH^{(n-1)+} + H^+$

Rxn 2: $Me^{n+} + NH_3 = MeNH_3^{n+}$

Examples 1-3 represent the most direct coupling of complex formation and physiological function. Complex formation alters the ability of a metallic element to serve as a nutrient, toxin, or energy source (e.g. electron acceptor). Example 4 takes advantage of the fact that the speciation and inventory of all constituent chemicals are linked through mass balance equations; complex formation with extracellular organic ligand frees up other ligands within the system. Example 5 is an example of "ecological biochemistry" (Harborne 1993); organisms go to great lengths to assist the growth and metabolic activity of beneficial organisms and impede the growth and metabolic activity of competitor organisms. Examples 6 and 7 reflect that chemical, physical, and biological attributes of aquatic environments are interconnected. Dissolved, adsorbed, or precipitated metal species can have physical characteristics that either promote or impede physiological function. Example 8 is an unusual use of extracellular chelating agents. The white rot fungus uses carboxylate-containing organic ligands to transport the Mn^{III} oxidant away from the cell and into contact with lignin.

COMPLEXATION OF +II METAL IONS

Complexation of Cu^{II}

Our objective is to quantitatively explore how the functional groups and structures of simple organic ligands affect their ability to form complexes with +II metal ions. We begin with the six carboxylic acids shown in Figure 1. Equilibrium constants from the data base CRITICAL (Martell et al. 1995), combined with the equilibrium computer program HYDRAQL (Papelis et al. 1988) were used to calculate concentrations of Cu^{2+}(aq) and Cu^{II}-organic ligand complexes under the following set of conditions: 1.0 mM Cu^{II}, 1.0 mM L_T, 10 mM $NaNO_3$ constant ionic strength, and pH 5.0.

The pK_a of acetic acid is 4.76. With maleic acid, inclusion of a second carboxylic acid causes the pK_a of one carboxylic acid group to increase by 1.5 log units and the other to decrease by 2.8 log units. pK_a differences among the five dicarboxylic acids are a function of the distance between the carboxylic acid groups, their orientation in space relative to one another, and electronic interactions with the linking atoms.

Protons can only coordinate one Lewis base group at a time. (Simultaneous coordination of two Lewis base groups, called hydrogen bonding, is actually quite weak.) Metal ions can coordinate two, three, four, and more Lewis base groups simultaneously, depending upon their coordination number. For a number of reasons, metal ions have a considerable competitive advantage over protons whenever an organic ligand possesses two or more suitably placed Lewis base groups. This "chelate effect" involves both entropic and enthalpic considerations (Hancock and Martell 1989). Most if not all biological ligands employ the chelate effect to ensure capture of the target metal ion.

Figure 1 indicates that acetic acid complexes 6.6% of added Cu^{II}, and fumaric acid complexes 10.4% of added Cu^{II}. The two oxygens of the carboxylate group of acetate are too close to simultaneously coordinate a central metal ion. The two carboxylate groups of fumarate are rigidly pointing away from one another, and also cannot simultaneously coordinate a central metal ion.

The remaining four ligands shown in Figure 1 are capable of chelate ring formation. Maleic acid possesses two carboxylic acids rigidly pointed towards each other; succinic acid, malonic acid, and oxalic acid can freely rotate their carboxylate groups towards one another. As a consequence, Cu^{II} complexation rises to 27.1% for succinic acid, 37.9% for maleic acid, 98.5% for malonic acid, and nearly 100% for oxalic acid. Oxalic acid, which forms a five-membered chelate ring with Cu^{II}, performs better than the other dicarboxylic

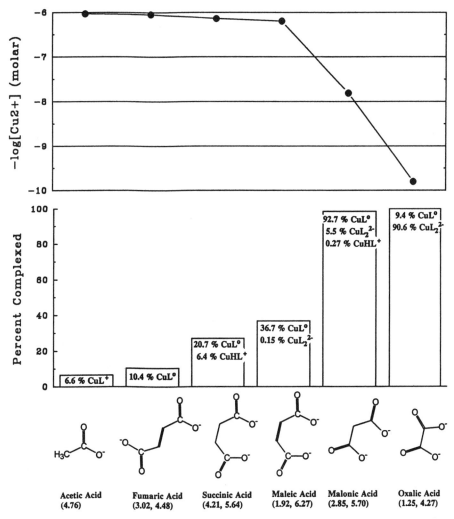

Figure 1. The logarithm of the free Cu^{2+} concentration and the percent Cu^{II} complexed in the presence of one monocarboxylic acid and five dicarboxylic acids. Reaction conditions: 1.0 μM Cu^{II}, 1.0 mM L_T, 10 mM $NaNO_3$, pH 5.0.

acids which form six- and seven-membered chelate rings. This decrease in stability with increasing chelate ring size is commonly observed.

The logK for $CuL^0(aq)$ is 5.80 for malonic acid and 6.19 for oxalic acid. Similarly, the logß for CuL_2^{2-} is 8.24 for malonic acid and 10.23 for oxalic acid. Despite differences in logK betweeen the two ligands, the percent Cu^{II} complexed by the two ligands is both nearly 100%. This similarity is deceptive, however, since raising logK values by a few log units would not cause the percent Cu^{II} complexed to change significantly. Instead, differences in $[Cu^{2+}]$, the free Cu^{II} concentration, should be noted (top of Fig. 1). $[Cu^{2+}]$ is nearly 1.5 log units lower in the oxalic acid system than in the malonic acid system. Since concentrations of all monomeric Cu^{II} species are directly proportional to $[Cu^{2+}]$, the difference in speciation between the two systems is in fact quite pronounced.

The differences in logK between the two ligands is important for another reason. If oxalic acid and malonic acid were placed in the same solution, oxalic acid would capture a greater percentage of added Cu^{II}, proportional to the differences in the equilibrium constants.

Cu^{II} speciation as a function of pH in the presence of malonic acid and salicylic acid is presented in Figure 2. Both ligands are capable of forming six-membered chelate rings with Cu^{II}. pK_{a1} for both ligands is nearly the same. pK_{a2} values are quite different: 5.70 for the carboxylic acid group of malonic acid, and 13.7 for the phenolic group of salicylic acid. Because salicylic acid has the more basic group, the logK for formation of the Cu^{II}-salicylic acid complex $CuL^o(aq)$ is 5.5 units higher than the logK for the Cu^{II}-malonic acid complex with the same stoichiometry.

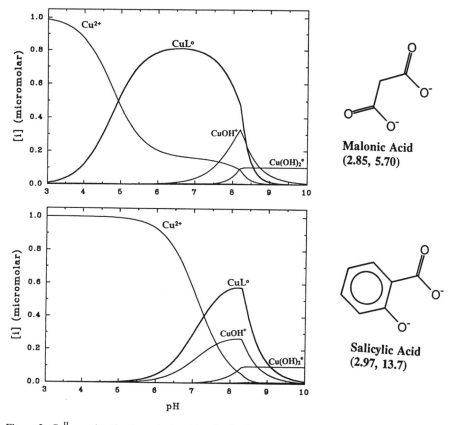

Figure 2. Cu^{II} complexation by malonic acid and salicylic acid as a function of pH. Reaction conditions: 1.0 µM Cu^{II}, 20 µM L_T, 10 mM $NaNO_3$.

As Figure 2 indicates, Cu^{II}-salicylic acid complexes reach their highest concentrations in the range $8 < pH < 8.5$. Below this pH, protons out-compete Cu^{II} for available ligand. Above this pH, OH^- out-competes salicylate ions for available Cu^{II}. (Precipitation of the solid phase $Cu(OH)_2(s)$ depresses Cu_T at pH values above 8.7).

Within the range $8 < pH < 8.5$, Cu^{II} complexation by malonic acid is comparable to complexation by salicylic acid. As the pH is decreased, malonic acid is actually a much more effective ligand for Cu^{II}. Because of its lower value for pK_{a2}, competition with protons for available ligand is much less severe, and significant complex is observed at pHs as low as 4.5.

An important conclusion from this example is that the more effective ligand for complexing a metal ion cannot be identified by comparing logK values for complex formation alone. The pK_as of the ligand must be considered as well.

Metal-to-metal comparisons

In order to compare differences in complexation behavior among +II metal ions, two ligands will be employed. Complexation by malonic acid is completely dominated by ionic bonding, owing to the two anionic carboxylate groups present; complexation by glycine arises from moderate ionic bonding and a modest covalent contribution to bonding.

Figure 3a indicates that logK values for both ligands increase as we move left to right in the periodic table, reaching a maximum in logK with Cu^{II}. This trend, called the "Irving-Williams Series" (Irving and Williams 1948) is observed with practically all oxygen- and nitrogen-bearing ligands. Alkaline earth metal ions, Mn^{II}, and Fe^{II} exhibit relatively large ionic radii and hence form relatively weak ionic bonds. Co^{II}, Ni^{II}, and Cu^{II} possess smaller radii which favor ionic bonding, and an increasing ability to form covalent bonds. The high logK for Cu^{II} cannot be explained without evoking the orientation of filled and unfilled d-orbitals in space; its shape is distorted from a true sphere in a way that favors complex formation.

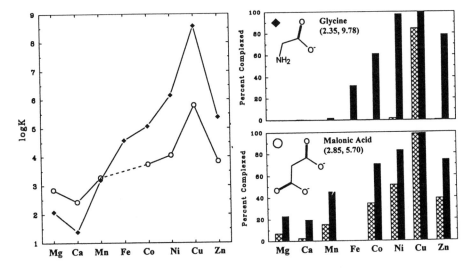

Figure 3. Complexation of +II divalent metal ions by malonic acid and by glycine. (A) logK values for the reaction $Me^{II} + L^{n-} = MeL^{(2-n)}$. Percent complexed at pH 5.0 (crosshatched bars) and at pH 8.0 (filled bars) have been calculated under the following reaction conditions: 1.0 μM Me^{II}, 1.0 mM L_T, 10 mM $NaNO_3$. (Note that logK values for Fe^{II} complexation by malonic acid are not available in CRITICAL.)

Figures 3b and 3c show the percent that each +II metal ion is complexed at pH 5.0 (crosshatched bars) and at pH 8.0 (filled bars) by the two ligands. The calculations employed 1.0 μM Me^{II} and 1.0 mM L_T in a 1.0 mM $NaNO_3$ medium.

Malonic acid is the more acidic of the two ligands, and hence competition with H^+ for free ligand is relatively unimportant. At pH 5.0, five of the seven metal ions considered are complexed to a significant extent. Raising the pH to 8.0 caused a slight-to-moderate increase in the percent of each metal ion complexed.

pK_{a2} for glycine is 4.1 log units higher that that of malonic acid. As a consequence, the monoprotonated species HL^o is predominant at pH 5.0, and only Cu^{II} is complexed to a significant extent. Raising the pH to 8.0 lowers $[H^+]$ by three orders of magnitude. Metal ions can compete more effectively for free ligand, and a dramatic increase in percent complexed is observed.

Selectivity is very much a function of the pK_as of the ligand under consideration and the pH at which metal-to-metal comparisons are made. At pH 5.0, the selectivity of glycine for Cu^{II} relative to other $+II$ metal ions is quite pronounced. At pH 8.0, much of this selectivity disappears; Co^{II}, Ni^{II}, and Zn^{II} are all complexed to a significant extent by glycine. Malonic acid, regardless of pH, is a much less selective ligand than glycine.

COMPLEXATION OF +III METAL IONS

Solubility limitations

Because of their greater charge and smaller radii, $+III$ metal ions form bonds to anionic Lewis base groups that are much stronger and more ionic in character than those formed by $+II$ metal ions. From the very start, we must contend with the solvent-derived ligands OH^- and O^{2-}. logK values for the reaction $Me^{n+} + H_2O = MeOH^{(n-1)+} + H^+$ are given in Table 2, column 4. At a pH equal to -logK, the concentration of the "free" metal ion equals the concentration of the first hydroxo species. Using pH = -logK as a benchmark, Cu^{II} and Pb^{II} are the only $+II$ metal ions that form significant concentrations of hydroxo species below pH 9. *All* the $+III$ metal ions listed, in contrast, form significant concentrations of hydroxo species at neutral pH, and most form significant concentrations well into the acidic pH range.

Once three hydroxide ions have coordinated to a $+III$ metal ion, a neutral mononuclear complex has formed, which can polymerize into solids with stoichiometry of $Me(OH)_3(s)$. $+III$ Metal ions can also induce deprotonation of coordination hydroxide ions, yielding hydroxide-oxide ($MeOOH(s)$) and oxide ($Me_2O_3(s)$) solids. From a different perspective, the three stoichiometries simply represent different hydration levels, for example:

$$Me_2O_3(s) + 3\ H_2O\ =\ 2\ MeOOH(s) + 2\ H_2O\ =\ 3\ Me(OH)_3(s)\ .$$

Two solids can possess the same stoichiometry but contain atoms that are arranged differently in three-dimensional space. $FeOOH$(goethite) and $FeOOH$(lepidocrocite), for example, are stoichiometrically equivalent but possess different crystal structures, solubilities, and other properties. Differences in particle size and defect density also affect solubility, but will be ignored here.

Concentrations of Fe^{3+}(aq) and various Fe^{III}-hydroxo complexes in equilibrium with $Fe(OH)_3$(amorphous), $FeOOH$(goethite), and Fe_2O_3(hematite) are shown as a function of pH in Figure 4. Differences in total soluble Fe^{III} (Fe^{III}_T) are enormous. Goethite and hematite are between six and seven orders-of-magnitude less soluble than amorphous $Fe(OH)_3$ at pH 7. The mineral lepidocrocite, not shown in the figure, possesses an intermediate solubility; it is three orders-of-magnitude less soluble than amorphous $Fe(OH)_3$.

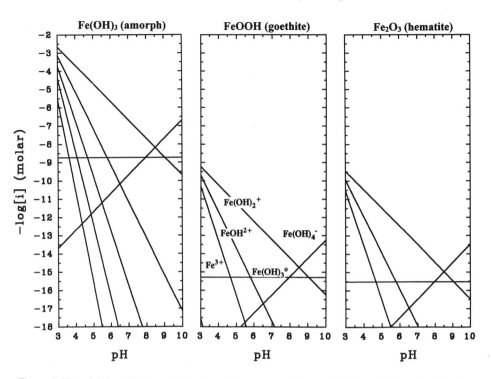

Figure 4. The solubility of Fe(OH)$_3$(amorphous), FeOOH(goethite), and Fe$_2$O$_3$(hematite) as a function of pH (in 10 mM NaNO$_3$). Calculations employed solubility product constants from Langmuir (1997) and equilibrium constants for hydroxo species from CRITICAL (Martell et al., 1995).

Fe(OH)$_3$(amorphous) typically forms first when Fe^{III} salts are added to water at 25°C and 1 atm. During timescales of days to weeks, slow conversion into crystalline solids takes place. Those with intermediate thermodynamic stability (e.g. lepidocrocite) appear first, but slowly convert into FeOOH(goethite) and Fe$_2$O$_3$(hematite). Temperature, water availability, other chemical constituents, and participation by organisms all affect the distribution and availability of solid phase Fe^{III}; Cornell and Schwertmann (1996) have presented an excellent review of this subject.

Concentrations of Al^{3+}(aq) and various Al^{III}-hydroxo complexes in equilibrium with Al(OH)$_3$(amorphous), AlOOH(diaspore), and Al$_2$O$_3$(corundum) are shown as a function of pH in Figure 5; solubilities at pH 7 differ by more than three orders-of-magnitude. When the stoichiometries and crystal structure are the same, Al^{III}-containing solids are much more soluble than Fe^{III}-containing solids. AlOOH(diaspore), for example, is nearly five orders-of-magnitude more soluble than FeOOH(goethite).

Figures 4 and 5 draw attention to the importance of identifying and characterizing the solid phases employed in biogeochemical studies. Answers to many of the issues raised in this chapter depend strongly upon the identity of the solubility-limiting phase.

Solubilization of Fe^{III}

Calculations presented in this section focus upon the solids Fe(OH)$_3$(amorphous) and FeOOH(goethite). Both solids are easily synthesized in the laboratory and hence are widely

322 GEOMICROBIOLOGY

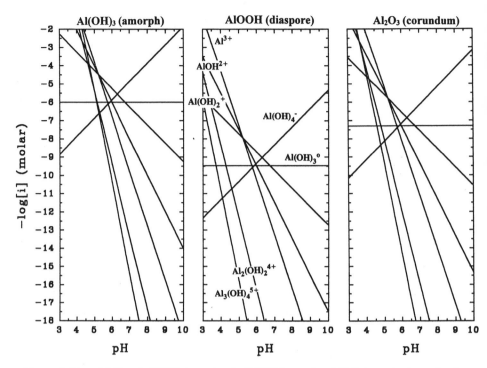

Figure 5. The solubility of Al(OH)$_3$(amorphous), AlOOH(diaspore), and Al$_2$O$_3$(corundum) as a function of pH (in 10 mM NaNO$_3$). Solubility product constants were obtained from the following sources: amorphous Al(OH)$_3$ from Langmuir (1997), diaspore from Robie et al. (1978), and corundum from Robie et al. (1978). Equilibrium constants for hydroxo species from CRITICAL (Martell et al., 1995).

used in experimental studies. In addition, they represent the broad range of Fe^{III} solubility encountered in soils.

To successfully complex Fe^{III}, organic ligands must gain a competitive advantage over OH$^-$ and O^{2-}. OH$^-$ and O^{2-} form stable bridges between metal ions, but are unable to occupy two coordinative positions around a central metal ion. Bi- and multidentate ligands enjoy a competitive advantage over OH$^-$ and O^{2-}, while monodentate ligands do not. OH$^-$ and O^{2-} take advantage of the propensity of +III metal ions to form ionic bonds. Carboxylate, phenolate, and alcoholate functional groups, which all contain anionic oxygen donor atoms, form ionic bonds of comparable bond strength.

Equilibrium Fe^{III} speciation as a function of pH in the presence of four bidentate ligands is shown in Figure 6. The following set of conditions has been used for these calculations: 1 μM Fe^{III}, 20 μM organic ligand, and 10 mM NaNO$_3$. Fe(OH)$_3$(s) serves as the solubility-limiting phase.

In the absence of organic ligands, Fe^{III} speciation is dominated by Fe(OH)$_2^+$ below pH 6.4, and by the solid phase Fe(OH)$_3$(s) at higher pHs. At a concentration of 20 μM, malonic acid perturbs Fe^{III} speciation through formation of the complexes FeL$^+$ and FeL$_2^-$. The effect of malonic acid is limited to relatively acidic pH values. Because the carboxylate groups are weak bases, there is little protonation to interfere with complex formation. logK for Fe^{III}-malonic acid complex formation is low, however, enabling hydroxo complexes and Fe(OH)$_3$(s) to grow in importance as the pH is increased. Salicylic acid possesses one

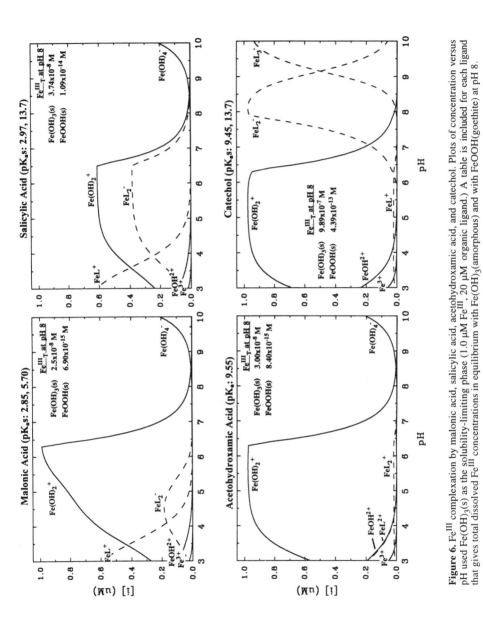

Figure 6. Fe^{III} complexation by malonic acid, salicylic acid, acetohydroxamic acid, and catechol. Plots of concentration versus pH used $Fe(OH)_3(s)$ as the solubility-limiting phase (1.0 μM Fe^{III}, 20 μM organic ligand.) A table is included for each ligand that gives total dissolved Fe^{III} concentrations in equilibrium with $Fe(OH)_3$(amorphous) and with FeOOH(goethite) at pH 8.

carboxylate and a more basic phenolate group. Greater basicity raises logK and allows salicylic acid to perturb Fe^{III} speciation throughout both the acidic and the neutral pH range. Catechol possesses two basic phenolate groups. The degree of protonation is so high that Fe^{III} complexation is prevented from occurring under acidic pH conditions. Under neutral and alkaline pH conditions, however, greater basicity and larger logK values result in effective complexation of Fe^{III} and prevention of $Fe(OH)_3(s)$ formation.

The two most common functional groups used by organisms for the acquisition of Fe^{III} are the catecholate and hydroxamate functional groups. Fe^{III} complexation by catechol has just been discussed. Fe^{III} complexation by acetohydroxamic acid is facilitated by electronic delocalization within the hydroxamate monoanion, which places enough electron density on the carbonyl oxygen for five-membered chelate ring formation to occur (van der Helm and Winkelmann 1994):

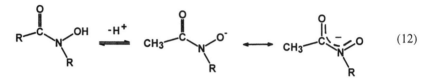

$$(12)$$

Under the conditions selected for our calculations, acetohydroxamic acid does not significantly perturb Fe^{III} speciation (Fig. 6). logK values for Fe^{III}-acetohydroxamic acid complex formation are lower than those of the other three bidentate ligands. These low logK values reflect weaker ionic bonding by monoanionic ligands in comparison to dianionic ligands.

Figure 6 also presents calculations of total dissolved Fe^{III} concentrations with respect to $Fe(OH)_3(s)$ and $FeOOH(s)$, at pH 8. Given that $Fe(OH)_3(s)$ is the most soluble Fe^{III} solid under consideration, it is important to note that catechol is the only bidentate ligand able to prevent its formation. Concentrations of Fe^{III}-malonic acid and Fe^{III}-salicylic acid complexes only surpass the concentration of $Fe(OH)_2^+$ at very low pHs. Performance of the bidentate ligands is even poorer when $FeOOH(s)$ precipitation is allowed in the calculation. At pH 8 in the presence of $FeOOH(s)$, *none* of the four bidentate ligands yields significant dissolved concentrations of Fe^{III}.

We now turn to organic ligands that possess three or more Lewis base groups suitably placed for coordinating a central metal ion. Citric acid (shown in Fig. 7) possesses three carboxylic acid groups and one alcohol group. Within cells, citric acid plays an essential role as part of the tricarboxylic acid cycle. Citric acid is used as a selective extractant in soil chemistry studies. In addition, the citrate moiety is found in a number of siderophores, including rhizoferrin (Drechsel et al. 1991, Carrano et al. 1996), staphyloferrin A (Meiwes et al. 1990, Konetschny-Rapp et al. 1990), staphyloferrin B (Drechsel et al. 1993), and vibrioferrin (Yamamoto et al. 1992).

Figure 7 shows the speciation of 1.0 μM Fe^{III} in the presence of 20 μM citric acid. When $Fe(OH)_3(s)$ is used as the solubility-limiting phase, complexation by citric acid dominates Fe^{III} speciation up to pH 8, when solid phase precipitation begins to take place. When $FeOOH(s)$ is used as the solubility-limiting phase, the greatest amount of Fe^{III}-citric acid complexation (nearly 4%) occurs at pH 4. (Note that the y-axis scale of this plot covers 0.10 μM, instead of the usual 1.0 μM). Fe^{III} complexation drops to negligible levels above pH 6.

L^{3-} is frequently used as the reference protonation level for citric acid; the three carboxylic acid groups are deprotonated, while the alcohol group is still protonated.

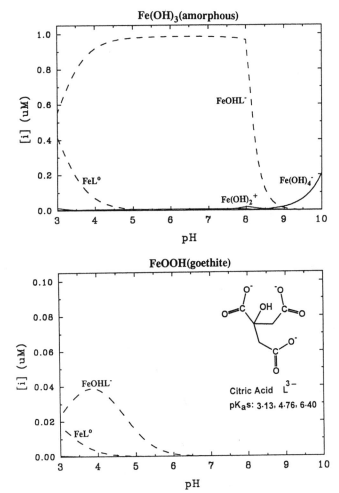

Figure 7. Fe^{III} complexation by citric acid, using $Fe(OH)_3(s)$ and $FeOOH(s)$ as possible solubility-limiting phases (1.0 μM Fe^{III}, 20 μM citric acid.)

The species FeL^0 occurs under acidic pH conditions, but loses a proton as the pH is increased:

$$FeL^0 (+ H_2O) = FeOHL^- + H^+ \tag{13}$$

The stoichiometry $FeOHL^-$ can be interpreted in two ways, both of which result from the strong Lewis acidity of the Fe^{III} cation. First, deprotonation of a coordinated water molecule could occur, yielding a ternary Fe^{III}-hydroxo-citrate complex. Second, deprotonation of the alcohol group of coordinated citrate could occur, making a direct Fe^{III}-alcoholate bond possible. In the case of Fe^{III}, the second explanation appears correct; deprotonation of the alcohol group provides a strong base for coordinating Fe^{III}, substantially contributing to the stability of the complex (Carrano et al. 1996).

Mugineic acid, shown in Figure 8, is a siderophore used by graminaceous plants (grasses) for the acquisition of extracellular Fe^{III}. In order for graminaceous plants to grow successfully in calcareous soils, mugineic acid will have to be effective at acquiring Fe^{III} under slightly alkaline pH conditions. Like citric acid, mugineic acid possesses three

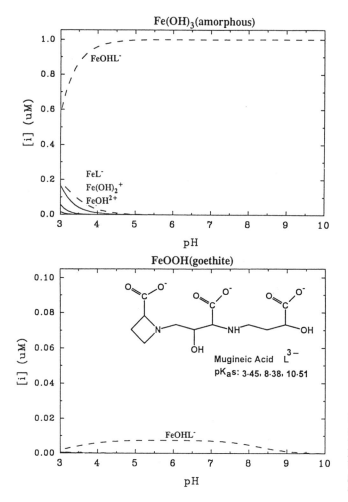

Figure 8. Fe^{III} complexation by mugineic acid, using both $Fe(OH)_3(s)$ and $FeOOH(s)$ as possible solubility-limiting phases (1.0 µM Fe^{III}, 20 µM citric acid).

carboxylic acid groups. In place of one alcohol group, mugineic acid has two alcohol groups and two amine groups. Participation of the amine groups in Fe^{III} complexation is an interesting possibility. Their intermediate basicities (pK_as of 8.38 and 10.51) help raise logK without the high degree of protonation exhibited by alcohol groups.

Figure 8 shows the speciation of 1.0 µM Fe^{III} in the presence of 20 µM mugineic acid. When $Fe(OH)_3(s)$ is used as the solubility-limiting phase, mugineic acid succeeds in solubilizing Fe^{III} throughout the pH range considered, whereas citric acid is only able to solubilize Fe^{III} up to pH 8. When $FeOOH(s)$ is used as the solubility-limiting phase, citric acid solubilizes more Fe^{III} below pH 6, but mugineic acid maintains a modest concentration of dissolved Fe^{III} up to pH 8. Thus, mugineic acid is a better candidate for Fe^{III} acquisition in calcareous soils than citric acid. As with citric acid, deprotonation of the alcohol group and formation of an Fe^{III}-alcoholate bond is believed to occur (Murakami et al. 1989).

Figure 9 shows the speciation of 1.0 µM Fe^{III} speciation in the presence of 20 µM desferriferrithiocin and aerobactin, using $FeOOH(s)$ as the solubility-limiting phase. Desferriferrithiocin complexes Fe^{III} through the phenolate oxygen, the imine nitrogen, and

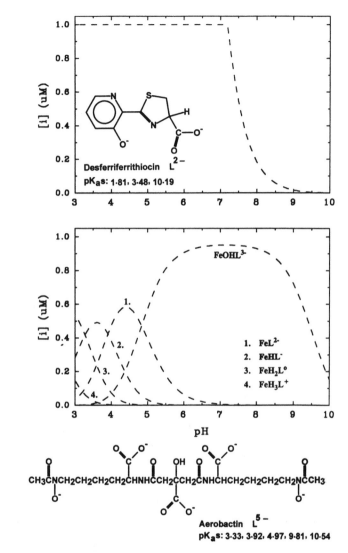

Figure 9. Fe^{III} complex-ation by desferriferrithiocin and aerobactin, using FeOOH(s) as the solubility-limiting phase (1.0 μM Fe^{III}, 20 μM organic ligand).

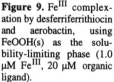

the carboxylate oxygen atoms (Langemann et al. 1996). The fact that desferriferrithiocin complexes more Fe^{III} over a wider pH range than citric acid and mugineic acid cannot be explained by the number and basicities of ligand donor groups; electronic and steric effects within the molecule are probably important. Aerobactin possesses a greater number of ligand donor groups than the other organic ligands considered: two hydroxamate groups, three carboxylate groups, and an alcohol group. This large number of anionic oxygen-donor groups readily explains the efficient complexation of Fe^{III} by aerobactin (Harris et al. 1979).

Sequestration of Al^{III}

Free Al^{3+}(aq) is an established growth-limiting factor for plants grown in acid soils (Rengel 1996). In addition, it has been established that the ability of beans to resist

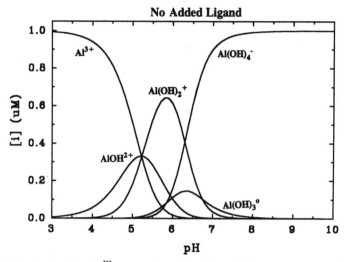

Figure 10. Speciation of 1 µM Al^{III} as a function of pH. The solid phase Al(OH)$_3$(amorphous) fails to precipitate throughout the pH range examined.

aluminum toxicity is positively correlated with citric acid release by roots (Miyaska et al. 1991). In order to explore effects of speciation on toxicity, a series of calculations have been made, which explore the effects of oxygen-donor ligands on dissolved Al^{III} speciation. Using the most soluble Al^{III}-containing phase to perform these calculations provides a worst-case scenario; 1.0 µM Al^{III} fails to precipitate within the pH range examined (Fig. 10).

Carboxylate groups are involved in the complexation of Al^{III} by oxalic acid and citric acid. Because of the low pK_as of such groups, efficient Al^{III} complexation occurs under acidic pH conditions. Competition with OH$^-$ for available Al^{III} causes an eventual decrease in the concentration of Al^{III}-organic ligand concentrations as the pH is increased (Fig. 11).

In the citric acid system, successive deprotonation of the AlL^0 complex is believed to occur:

$$AlL^0 (+ H_2O) = AlOHL^- + H^+ \qquad (14)$$

$$AlL^0 (+ 2 H_2O) = Al(OH)_2L^{2-} + 2 H^+ \qquad (15)$$

Although deprotonation of the alcohol group can be postulated, it is not believed to be as important as in the analogous Fe^{III} system (e.g. Carrano et al. 1996). Instead, the Al^{III} system is dominated by ternary Al^{III}-hydroxo-citrate complexes. The propensity of Al^{III} to form ternary complexes of this kind is extensively discussed by Ohman (1988) and Ohman and Sjoberg (1988).

Salicylic acid uses one phenolate group and catechol uses two phenolate groups to complex Al^{III}. Because phenolate groups possess high pK_as, protonation is significant at acidic pHs, and the degree of complex formation is low. Concentrations of Al^{III}-organic ligand complexes peak between pH 5 and 6. Concentrations diminish at higher pH because of the formation of Al^{III}-hydroxo species.

As shown in Figure 12, the plot of log[Al^{3+}] versus pH in the presence of 20 µM catechol is barely distinguishable from plots calculated in the absence of added ligand. Thus, catechol has a negligible effect on [Al^{3+}(aq)]. Salicylic acid depresses [Al^{3+}(aq)] near pH 5, but has a negligible effect at pHs much higher or lower than this value. Oxalic acid and citric acid, in contrast, have a greater effect on [Al^{3+}(aq)] that extends over a wider

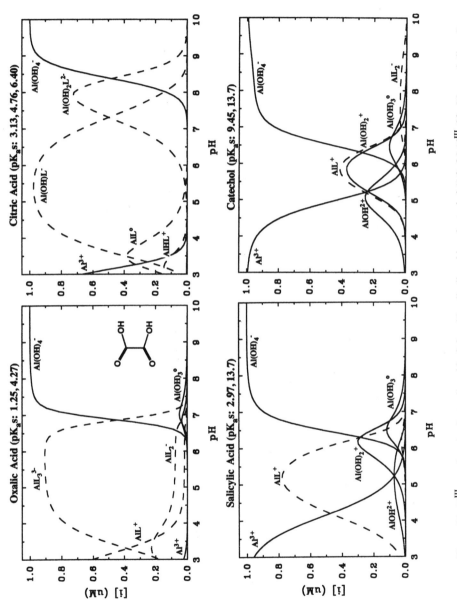

Figure 11. Al^{III} complexation by oxalic acid, citric acid, salicylic acid, and catechol (1.0 μM Al^{III}, 20 μM organic ligand).

range of pH in comparison to the other two ligands. Oxalic acid is the best at depressing [Al^{3+}(aq)] at pH 3, but it is overshadowed by citric acid above pH 5. Citric acid is the only ligand capable of depressing [Al^{3+}(aq)] by more than an order-of-magnitude at neutral pH.

These calculations could of course be repeated with a less soluble Al^{III} phase, such as AlOOH(diaspore). When the solid phase is present at equilibrium, [Al^{3+}] is fixed by solublity product constants (e.g. Eqns. 9-11); adding a new organic ligand increases Al_T (total dissolved aluminum) but does not affect [Al^{3+}] or the concentrations of Al^{III} complexes with other inorganic and organic ligands. Once enough organic ligand has been added to completely dissolve the solid, Al_T reaches its maximum value of 1.0 mM. Adding any more organic ligand would cause [Al^{3+}] to decrease, and hence lower concentrations of Al^{III} complexes with other inorganic and organic ligands.

COMPLEXATION AND REDOX EQUILIBRIA

Iron, because of its relatively high abundance and ability to undergo changes in oxidation state, has been appropriated by organisms as a reagent in intracellular redox reactions. The charge, ion size, and electronic properties of Fe^{II} and Fe^{III} differ substantially. It is thus possible for organisms to tailor organic ligands (e.g. hemes, cytochromes) that selectively stabilize the +II oxidation state or the +III oxidation state, and hence control the redox potential and reactivity of the iron system. If the redox potential is high enough, Fe^{III} can serve as a terminal electron acceptor, and hence contribute to the energy needs of cells.

Extracellular organic ligands intended to solubilize iron affect the redox potential of the Fe^{III}/Fe^{II} half-reaction, and hence may affect iron acquisition and iron respiration. The thermodynamics of these systems will now be explored using the representative organic ligands citric acid and mugineic acid.

Fe^{III} complexation by citric acid and mugineic acid has already been discussed. Complexation by citric acid reaches a maximum in the acidic range near pH 4 (Fig. 7), while complexation by mugineic acid extends upward into the neutral and slightly alkaline pH range (Fig. 8).

Fe^{II} complexation by citric and mugineic acid as a function of pH is shown in Figure 13. Fe^{II} yields logK values that are lower than those of Fe^{III}. This lowers the degree of complexation under acidic conditions, where protons out-compete Fe^{II} for available ligand. Above pH 5 (with citric acid) and pH 6 (with mugineic acid), Fe^{II}-organic ligand complexes become significant and remain high, even at relatively alkaline pH values. Precipitation of Fe^{II} as $Fe(OH)_2$(s) requires pH values greater than 9.0 to occur.

In our calculations, separate mass balance equations are written for Fe^{II} and Fe^{III}, each yielding a total system concentration of 1.0 µM. After mass balance equations and equilibrium constants have been combined to yield the speciation of each oxidation state,

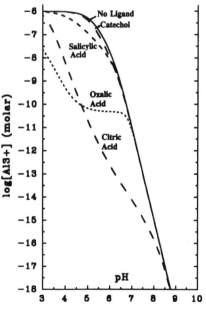

Figure 12. Citric acid, oxalic acid, salicylic acid, (and catechol) depress log[Al^{3+}] relative to ligand-free solutions (1.0 µM Al^{III}, 20 µM organic ligand).

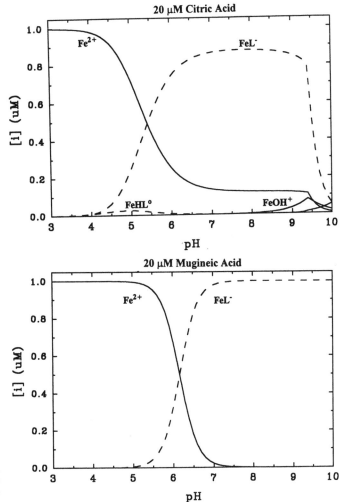

Figure 13. Fe^{II} complexation by citric acid and by mugineic acid (1.0 μM Fe^{II}, 20 μM organic ligand).

the redox potential can be calculated based upon the $E°$ for the iron half-reaction and the Nernst equation:

$$Fe^{3+}(aq) + e^- = Fe^{2+} \qquad E° = 0.770 \text{ volts} \qquad (16)$$

$$Eh = E° + (RT/F) \ln \frac{[Fe^{3+}]}{[Fe^{2+}]} \qquad (17)$$

Equation (17) indicates that any change to the system that lowers the $[Fe^{3+}]/[Fe^{2+}]$ ratio lowers the redox potential Eh and makes Fe^{III} a weaker oxidant in the thermodynamic sense. Increasing this ratio raises Eh and makes Fe^{III} a stronger oxidant.

Eh when FeOOH(s) serves as the solubility-limiting phase

Eh-pH plots presented in Figure 14 are relatively simple to interpret. The dashed curve represents Eh in the absence of organic ligand; smooth curves represent Eh in the presence of organic ligand concentrations up to 2.0 mM. The low solubility of FeOOH(s) places constraints on $[Fe^{3+}]$:

$$FeOOH(s) + 3 H^+ = Fe^{3+} + 2 H_2O \qquad {}^*K_{so} = \frac{[Fe^{3+}]}{[H^+]^3 a_{FeOOH}} \qquad (18)$$

$$[Fe^{3+}] = {}^*K_{so}[H^+]^3 a_{FeOOH} \qquad (19)$$

If the pH is known and the presence of FeOOH(s) has been established, then $[Fe^{3+}]$ is found directly from Equation (19). As long as FeOOH(s) is present, the formation of Fe^{III}-organic ligand complexes has no effect on either $[Fe^{3+}]$ or Eh.

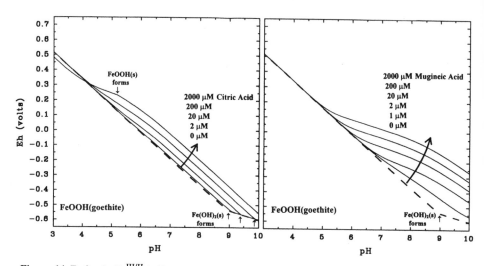

Figure 14. E_h for the $Fe^{III/II}$ half-reaction in systems using $Fe(OH)_3$(goethite) as the solubility-limiting phase: effects of citric acid and mugineic acid addition (1.0 μM Fe^{II} and 1.0 μM Fe^{III}).

A close examination of Figure 14 reveals that FeOOH(s) is present in all the calculations involving mugineic acid, and in all the calculations involving citric acid except one: FeOOH(s) does not form in the 2.0 mM citric acid system below pH 5.3. This curve behaves differently from the rest, since it is the only one where $[Fe^{3+}]$ is not governed by Equation (19). The lowering of $[Fe^{3+}]$ is enough to cause a decrease in Eh.

Fe^{II} precipitates as $Fe(OH)_2$(s) above pH 9 when no ligand has been added, and in the presence of 2.0 μM, 20 μM, and 200 μM citric acid. Arrows mark the pH where $Fe(OH)_2$(s) begins to form. These arrows move to the right as the citric acid concentration is increased. If the ligand concentration is low enough and the pH is high enough, both FeOOH(s) and $Fe(OH)_2$(s) exist at equilibrium. Whenever these two solids coexist, Eh is fixed by solubility product equations, and further decreases in the ligand concentration have no effect on Eh.

At low pH, citric acid and mugineic acid have a negligible effect on Eh. The reason for this is clear: $[Fe^{3+}]$ is set by FeOOH(s) (except in the 2.0 mM citric acid calculation), and Fe^{II}-organic ligand complex formation is poor. As the pH is increased, however, both organic ligands cause significant decreases in $[Fe^{2+}]$ and increases in Eh. At high pH, mugineic acid yields the greatest increases in Eh, as shown in Figure 13.

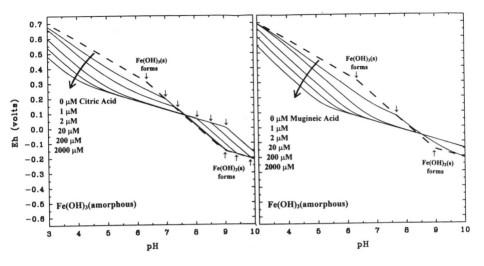

Figure 15. Eh for the $Fe^{III/II}$ half-reaction in systems using $Fe(OH)_3$(amorphous) as the solubility-limiting phase: effects of citric acid and mugineic acid addition (1.0 μM Fe^{II} and 1.0 μM Fe^{III}).

Eh when $Fe(OH)_3$(s) serves as the solubility-limiting phase

Switching to $Fe(OH)_3$(s) as the solubility-limiting phase greatly alters the appearance of the Eh-pH plots (Fig. 15). Below pH 6.4, $Fe(OH)_3$(s) does not precipitate. In this pH range, the percent Fe^{III} complexed by organic ligand is greater than the percent Fe^{II} complexed. $[Fe^{3+}]/[Fe^{2+}]$ drops substantially as organic ligand concentrations are increased, causing Eh to also decrease.

1 μM Mugineic acid shifts the pH for $Fe(OH)_3$(s) precipitation to pH 7.8. Even higher concentrations of mugineic acid prevent the formation of $Fe(OH)_3$(s) throughout the pH range considered. Each increment in citric acid concentration shifts the pH required for $Fe(OH)_3$(s) precipitation, resulting in a series of "steps" marked by discontinuities. Each increment in citric acid concentration also shifts the pH required for $Fe(OH)_2$(s) precipitation.

Overall, citric acid lowers Eh below approximately pH 7.5, but increases Eh at more alkaline pHs. Similarly, mugineic acid lowers Eh below pH 8.5, but raises Eh at higher values. $Fe(OH)_3$(s) precipitation, $Fe(OH)_2$(s) precipitation, Fe^{III} complexation, and Fe^{II} complexation all play an important role in determining Eh.

Figures 14 and 15 assume complete equilibrium between dissolved species and the solubility-limiting phases considered in the calculation. In real laboratory experiments, this equilibrium assumption can be called into question. If an Fe^{III} chloride salt has been added, for example, is the solid phase governing Eh $Fe(OH)_3$(amorphous), FeOOH(goethite), or some other phase that we have not yet considered? If well-characterized FeOOH(goethite) has been added, can we be sure that its dissolution/precipitation occurs rapidly when the pH is changed or when organic ligand has been added? If new solid phase forms, can we be sure that it is FeOOH(goethite), and not some new phase? These issues can only be resolved by chemical and structural analysis. Capillary electrophoresis and HPLC allow us to measure speciation in supernatant solution in considerable detail. X-ray diffraction, electron microscopy, probe-based microscopic techniques, and spectroscopic methods can be used to identify changes in the solid phase.

ORGANIC LIGAND ADSORPTION

Free metal ions, free ligands, and metal ion-ligand complexes are all capable of adsorbing onto mineral surfaces. In principle, quantitative descriptions of adsorption phenomena should be included in calculations whenever a mineral surface is present. Although full treatment is beyond the scope of this work, a brief look into structure-property relationships affecting organic ligand adsorption is informative. Readers interested in a more complete description should refer to reviews by Dzombak and Morel (1990), Schindler (1990), Stumm (1992), Stone et al. (1993) and others.

An interesting study of fumaric acid and maleic acid adsorption onto δ-Al$_2$O$_3$(s) (Fig. 16) was reported by Yao and Yeh (1996). Experiments were performed within relatively short timescales, so that dissolution of the underlying Al$_2$O$_3$(s) was not significant. Using the site density quoted in this work, 0.5 g/L δ-Al$_2$O$_3$ corresponds to a surface site concentration of 170 μM, 3.4-times the total dicarboxylic acid concentration employed.

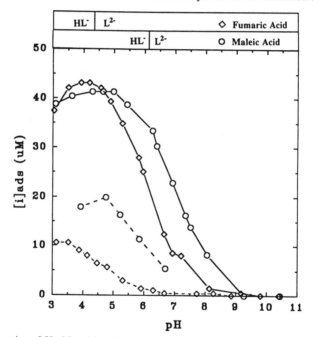

Figure 16. Adsorption of 50 μM maleic acid and fumaric acid onto 0.5 g/L δ-Al$_2$O$_3$ at low ionic strength (5 mM NaClO$_4$, smooth lines), and at high ionic strength (500 mM NaClO$_4$, dashed lines). Adapted from Yao and Yeh (1996).

Cu^{II} complexation in solution by fumaric acid and maleic acid, presented in Figure 1, provides an interesting comparison to the δ-Al$_2$O$_3$ adsorption results:

(1) Using a 1000:1 ligand-to-metal ion concentration ratio, fumaric acid complexed only 10.6% of dissolved Cu^{II}, while maleic acid complexed only 36.9%. Fumaric acid and maleic acid adsorption onto δ-Al$_2$O$_3$ at low ionic strength, in contrast, is nearly equal. Approximately 90% adsorption is achieved. At pH 5.0, adsorption is stronger than complex formation in solution, and less sensitive to organic ligand structure.

(2) Cu^{II}-fumaric acid and Cu^{II}-maleic acid complex formation in solution reaches a maximum near neutral pH (similar to the plot shown in Fig. 2 for malonic acid). Maleic acid and fumaric acid adsorption onto δ-Al$_2$O$_3$, in contrast, reaches a maximum at pHs

slightly below pK_{a2} for each ligand. This phenomenon has been discussed extensively in previous studies of dicarboxylic acid adsorption (e.g. Kummert and Stumm 1980).

(3) Cu^{II} complexation in solution by maleic acid and fumaric acid is relatively insensitive to ionic strength (not shown). In contrast, maleic acid adsorption decreases by 50% and fumaric acid adsorption decreases by 81% when the ionic strength is increased from 5 mM to 500 mM.

Benzoic acid, with a single carboxylic acid group, yields measureable adsorption onto Al^{III} hydrous oxides near its pK_a (Kummert and Stumm 1980). Adsorption of citric acid, with three carboxylic acid groups, has also been studied (Cambier and Sposito 1990). An ionic strength of 20 mM and a AlOOH(s) loading of 4.9 g/L were employed, corresponding to a surface site concentration of 1.90 mM. As shown in Figure 17, adsorption of 0.13 mM, 0.80 mM, and 1.5 mM citric acid increases with decreasing pH until essentially all the citric acid is adsorbed. When 2.1 mM citric acid is added, the organic ligand concentration exceeds the surface site concentration, and hence complete adsorption cannot occur.

It can be concluded that organic ligands possessing free carboxylic acid groups adsorb to an appreciable extent onto Al^{III} hydrous oxides, with a maximum occurring in the vicinity of the carboxylic acid group pK_a. Fe^{III} and Ti^{IV} hydrous oxides show a similar tendency to adsorb carboxylic acid-containing compounds. Compounds possessing phenolic -OH groups adsorb to hydrous oxides, but to a somewhat lesser extent (Vasudevan and Stone 1996). Replacing carboxylic acid and phenolic oxygen-donor groups with amino and pyridyl nitrogen-donor group lessens adsorption onto Fe^{III} and Ti^{IV} hydrous oxides, and causes negligible adsorption onto Al^{III} hydrous oxides (Ludwig and Schindler 1995; Vasudevan and Stone, submitted.)

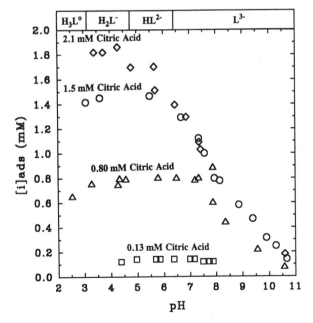

Figure 17. Adsorption of citric acid as a function of pH in suspensions containing 4.9 g/L AlOOH(pseudoboehmite) and 20 mM $NaClO_4$. Adapted from Cambier and Sposito (1991).

REMAINING CHALLENGES TOWARDS UNDERSTANDING METAL ION AND EXTRACELLULAR ORGANIC LIGAND SPECIATION

Previous sections have focused upon metal ion-ligand complexes that are relatively well-characterized, and have employed an equilibrium approach towards metal ion and organic ligand speciation. Despite recent efforts (e.g. Murakami et al. 1989, Chen et al. 1994, Carrano et al. 1996, Shenker et al. 1996) logK values for many extracellular organic ligands are either not known or provisional. In addition, kinetics, rather than equilibrium, probably controls speciation in a number of important situations. Many issues, including the ones listed below, will have to be addressed in order to fully understand metal ion-extracellular organic ligand interactions in the environment.

Slow rates of +II metal ion desorption from mineral surfaces. The longer +II metal ions are brought into contact with hydrous oxides, the longer it takes to desorb them by acidification (e.g. Ainsworth et al. 1994) and by organic ligand addition (e.g. Coughlin and Stone 1995). The three-dimensional structure and complex chemical properties of hydrous oxide surfaces must be accounted for; it is likely that metal ions diffuse deeper into the structure over time, or undergo changes in surface coordination that lower their responsiveness to changes in aqueous medium composition.

Kinetics of ligand exchange reactions. Many siderophores exhibit appreciable logK values for complex formation with +II metal ions. For this reason, siderophores may form complexes with common ions such as Ca^{II} and Mg^{II} before contact with Fe^{III} is made. Similarly, "free" Fe^{3+}(aq) is not likely to be encountered in aquatic systems. Siderophores will have to displace coordinated inorganic or organic ligands in order to acquire Fe^{III}. The kinetics and mechanisms of ligand exchange reactions and metal ion exchange reactions of this kind have been eloquently described in a series of papers by Hering and Morel (1988 1989, 1990). Many of the concepts developed for solution phase reactions also apply to surface chemical reactions. It has recently been shown, for example, that free EDTA dissolves amorphous Fe^{III} hydrous oxide much more quickly than metal ion-EDTA complexes, and that dissolution rates increase as rates of metal ion-EDTA dissociation increase (Nowack and Sigg 1997).

Ternary complex formation in solution and on mineral surfaces. As has already been mentioned, Al^{III} has an unusual ability to coordinate organic ligands and hydroxide ions simultaneously (Ohman 1988, Ohman and Sjoberg 1988). Similar ternary complexes have been postulated to occur on hydrous oxide surfaces (Schindler 1990). Metal ions that simultaneously coordinate surface hydroxo groups and solute ligands serve to enhance ligand adsorption; organic ligands that simultaneously coordinate surface metal atoms and solute metal ions serve to enhance metal ion adsorption. Dramatic effects of this kind have recently been reported in systems consisting of Al_2O_3, cobalt, and the synthetic chelating agents EDTA and NTA (Girvin et al. 1993, Girvin et al. 1994). It is likely that ternary surface complexes involving biological organic ligands are also important.

Organic ligand adsorption versus ligand-assisted dissolution. Replacing one coordinated ligand with another can change bond lengths and bond strengths between the central metal ion and its other coordinated ligands. For this reason, replacing surface-bound OH^- and H_2O with adsorbed organic ligands may weaken bonds between the metal atom within the surface site and the underlying mineral lattice, enabling ligand-assisted dissolution to occur (see Stumm 1992). When does adsorption occur without subsequent dissolution of the mineral surface? When does ligand-assisted dissolution occur? These questions can best be addressed in several stages. As discussed in previous sections, mono- and bi-dentate ligands that are not capable of forming appreciable concentrations of metal ion-ligand complexes in solution nevertheless adsorb onto hydrous oxide surfaces.

For compounds such as these (e.g. benzoic acid, fumaric acid, maleic acid), adsorption will occur without ligand-assisted dissolution.

Multidentate ligands, in contrast, have the capacity to form appreciable concentrations of metal ion-ligand complexes in solution. Our understanding of the thermodynamics of adsorption is not sufficient to evaluate whether an adsorbed multidentate ligand or the corresponding metal ion-multidentate ligand complex in solution is thermodynamically more stable. It is quite possible that the complex in solution is ultimately most stable. Adsorption versus ligand-assisted dissolution may therefore be a kinetic issue. Over short experimental timescales, adsorption may dominate, but over long experimental timescales, ligand-assisted dissolution may dominate.

Since the mid 1980s, a great deal of attention has been devoted to the kinetics and mechanisms of ligand-assisted dissolution reactions. Most of these studies have focused upon relatively simple biological ligands (e.g. oxalic acid and citric acid) and synthetic chelating agents (e.g. EDTA, DTPA, and NTA) (see Stumm (1992), Blesa et al (1994), and Cornell and Schwertmann (1996)). Recently, ligand-assisted Fe^{III} hydrous oxide dissolution by hydroxamic acid (Holmen and Casey 1996), by a bacterial siderophore (Hersman et al. 1995), and by the plant siderophore mugineic acid (Inoue et al. 1993) have been reported. Rates of ligand-assisted dissolution are undoubtably affected by (1) the functional groups and structure of the organic ligand, (2) the stoichiometry and structure of the mineral surface, and (3) aqueous phase and surface speciation. These areas are excellent subjects for future research.

Breakdown of free and metal ion-coordinated organic ligands: Extracellular organic ligands possess functional groups and structures that are subject to chemical and biological degradation (e.g. Fig. 18). Mugineic acid and related plant siderophores are readily degraded by soil microorganisms (Marschner 1995). Catecholate groups are oxidized quickly by $Mn^{III,IV}$ hydrous oxides, and more slowly by molecular oxygen. Although hydrolysis of amide linkages and imine Lewis base groups can be postulated (Fig. 18), reaction rates are expected to be low. Oxidation reactions (McBride and Wesselink 1988) and hydrolysis reactions (Stone and Torrents 1995) can be catalyzed by dissolved metal ions and by hydrous oxide surfaces. Ligand breakdown may be desirable in some circumstances, but undesirable in others. Ligand breakdown is one of a number of possible mechanisms for recovering nutrient metal ions from metal ion-ligand complexes within cells. Outside of cells, ligand breakdown must be matched by new biosynthesis and release. Organisms are known to regulate siderophore release in response to changes in the extracellular medium. It has even been speculated that organisms may even change which siderophore they produce in response to these changes.

A SURVEY OF EXTRACELLULAR ORGANIC LIGANDS

An amazing diversity of siderophores and other extracellular organic ligands have been isolated and characterized. The thermodynamics and stereochemistries of many of their metal ion-organic complexes have been characterized; mechanisms of biosynthesis and pathways of biological uptake have been extensively studied. Very little is known, however, about complex formation with non-target metal ions, adsorption onto naturally-occurring mineral surfaces, ligand-assisted dissolution, and rates of ligand exchange and breakdown. These areas are crucial towards understanding the role of extracellular organic ligands within a broader biogeochemical context.

Specific questions can be raised about a number of the siderophores and extracellular organic ligands presented in Figures 19-21. Ferrioxamine B and rhodotorulic acid (Raymond et al. 1984) are hydroxamate siderophores with an open molecular structure,

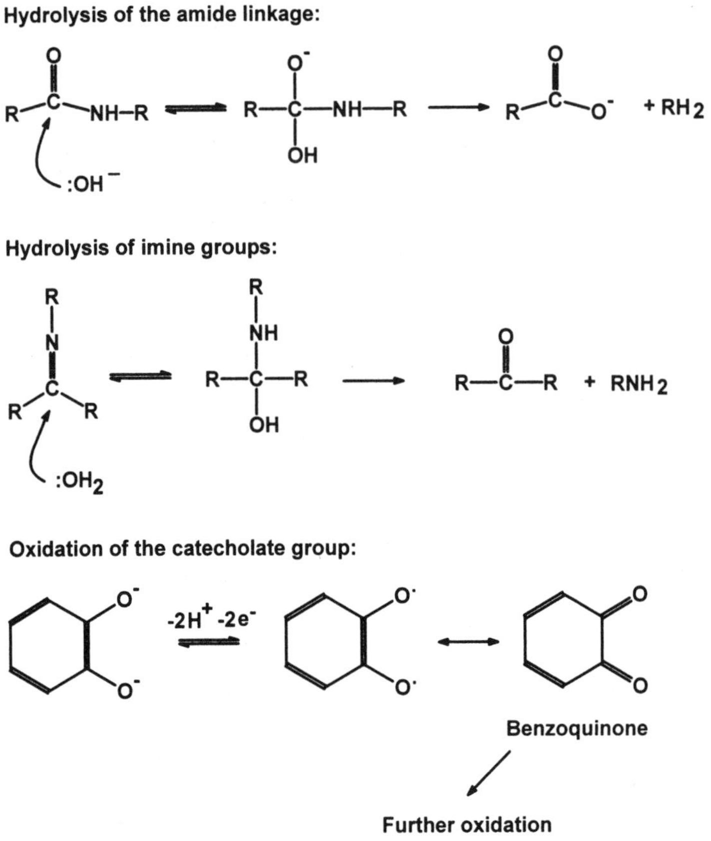

Hydrolysis of the amide linkage:

Hydrolysis of imine groups:

Oxidation of the catecholate group:

Benzoquinone

Further oxidation

Figure 18. Three chemical reactions resulting in the breakdown of extracellular chelating agents.

while alcaligin has a closed ring structure (Nishio et al. 1988). How does the closed ring affect selectivity towards Fe^{III} relative to non-target metal ions? Does the closed ring structure adsorb in ways that are similar or different to those of the open molecular structures? Enterobactin (Loomis and Raymond 1991) is a representative catecholate siderophore. To what extent does oxidation of the free siderophore take place, and does formation of the Fe^{III}-enterobactin complex protect the ligand from oxidation? All four siderophores shown in Figure 19 possess amide linkages. Does metal ion-catalyzed hydrolysis of this linkage ever occur, by the target metal ion, by non-target metal ions, or by metal-containing sites on mineral surfaces?

The six organic ligands shown in Figure 20 are synthesized by plants through closely related pathways (Ma et al. 1995). Mugineic acid, 2'-deoxymugineic acid, 3-hydroxymugineic acid, avenic acid A, and distichonic acid A are used exclusively for Fe^{III} acquisition outside plants (Sugiura and Nomoto 1984). Nicotianamine is used exclusively inside plants for the regulation of +II metal ion activity (Benes et al. 1983, Anderegg and Ripperger 1989, Scholz et al. 1992, Stephan et al. 1996). Note that nicotianamine can be converted into 2'deoxymugineic acid by simply replacing an amine group with an alcohol group. Why does this very small change in molecular structure cause plants to use the two compounds in such different ways? Among the compounds used for Fe^{III} acquisition, which form the strongest complex with Fe^{III}? How different is their selectivity for Fe^{III}

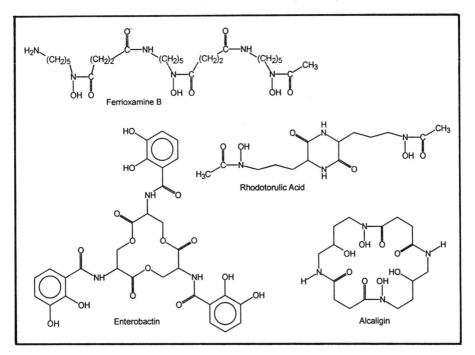

Figure 19. Three hydroxamate-based siderophores and one catecholate-based siderophores synthesized by organisms.

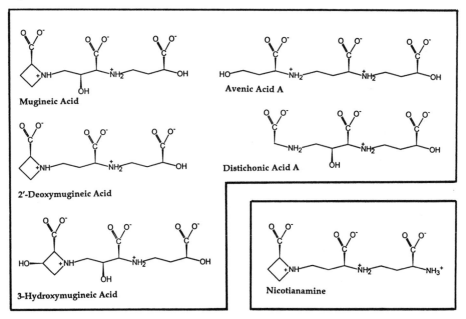

Figure 20. Nicotianamine is used inside plants for regulating +II metal ion speciation; mugineic acid and related structures are used outside of cells for the acquisition of Fe^{III}.

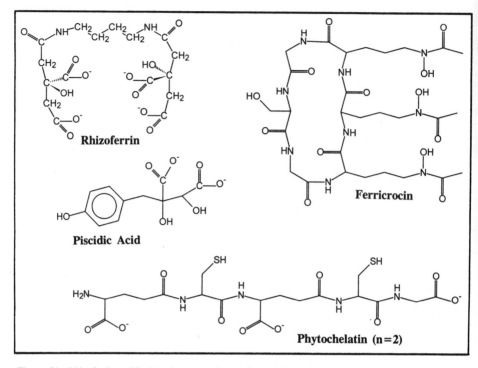

Figure 21. Rhizoferrin and ferricrocin are two fungal-derived siderophores. Piscidic acid is used by plants to release the nutrient phosphate from Fe^{III}/Al^{III} phosphate solids in soils. The phytochelatins are a class of organic ligands released organisms for the sequestration of heavy metal ions.

relative to non-target metal ions such as Ni^{II} and Al^{III}? Do they possess similar adsorption properties, and how effectively can they recover Fe^{III} from amorphous iron oxide, goethite, and hematite? Rhizoferrin and ferricrocin (Fig. 21) are both siderophores produced by fungi. Rhizoferrin consists of two citric acid molecules bound together using amide linkages to 1,4-diaminobutane (Drechsel et al. 1991, Carrano et al. 1996). Ferricrocin is also held together by amide linkages, but it uses hydroxamate groups for Fe^{III} coordination, instead of carboxylic acid groups (Prabhu et al. 1996). How does Fe^{III} acquisition by the mugineic acid family of compounds (tricarboxylic acids) compare with rhizoferrin (a tetracarboxylic acid)? Since the pK_as for rhizoferrin and ferricrocin are so different, is the pH of maximum Fe^{III} complex formation also different? Are their surface chemical properties similar to or different from those of the mugineic acid family of compounds?

It has been postulated that piscidic acid (Fig. 21) release by plants represents an attempt to recover the nutrient phosphate from Fe^{III}-phosphate and Al^{III}-phosphate solids in soils (Ae et al. 1990; Otani et al. 1996). Are logK values for this ligand high enough to accomplish this task? Do the alcoholate groups participate in complex formation with Fe^{III}, or with Al^{III}? Does adsorption of piscidic acid by hydrous oxides and clays interfere with phosphate solubilization? How does the structure of this ligand affect the kinetics of phosphate solubilization?

The phytochelatins are a class of organic ligands released by organisms for the sequestration of heavy metal ions (Grill et al. 1985, Gekeler et al. 1988, Strasdeit et al.

1991). To what extent do the thiol groups impart selectivity towards heavy metal ions? Does complex formation with lighter +II metal ions and +III metal ions ever occur? Prior to heavy metal ion complexation in soils, are phytochelatins primarily aqueous species, or do they sorb onto mineral surfaces via the carboxylic acid groups?

Along with the geochemical issues just raised, there are a number of important biological issues. How are metal ion-ligand complexes taken up by cells? Once inside, how are nutrient metal ions released from these complexes? How is the coupling of geochemical processes with biological processes regulated? Answers to these questions require interdisciplinary collaboration between geochemists and biologists.

CONCLUSIONS

Extracellular organic ligands play a crucial role in the coupling of biological processes with geochemical processes in the environment. Extracellular organic ligands that perform well in some geochemical situations perform poorly in others. Structure-property and structure-activity relationships help us understand which metal ion-organic ligand complexes are likely to be important. Equilibrium calculations allow us to explore how pH, aqueous concentrations, and the presence of mineral surfaces affect metal ion speciation. The kinetics and mechanisms of complex formation, adsorption/desorption, and precipitation/dissolution reactions are poorly understood, but nevertheless control metal ion speciation in many biogeochemical situations.

As our understanding of the chemistry of extracellular organic ligands is improved, we will be in a much better position to appreciate their amazing natural diversity, and comprehend their full biogeochemical significance.

ACKNOWLEDGMENTS

Special thanks to Ken Nealson and his former students, and to members of the U.S. Department of Energy Co-Contaminant Subprogram for many informative discussions. Financial support for this work was provided by the Ecological Research Division, Office of Health and Environmental Research, of the U.S. Department of Energy under Contract DE-FG02-90ER60946 as part of the Subsurface Science Program. The encouragement and support of Dr. Frank Wobber is greatly appreciated.

REFERENCES

Ae N, Arihara J, Okada K, Yoshihara T, Johansen C (1990) Phosphorus uptake by pigeon pea and its role in cropping systems of the Indian subcontinent. Science 248:477-480

Ainsworth CC, Pilon JL, Gassman PL, Van der Sluys WG (1994) Cobalt, cadmium and lead sorption to hydrous iron oxides: residence time effect. Soil Sci Soc Am J 58:1615-1623

Anderegg G, Ripperger H (1989) Correlation between metal complex formation and biological activity of nicotianamine analogues. J Chem Soc Chem Commun 1989:647-650

Benes I, Schreiber K, Ripperger H, Kircheiss A (1983) Metal complex formation by nicotianamine, a possible phytosiderophore. Experientia 39:261-262

Blesa M, Morando BJ, Regazzoni AE (1994) Dissolution of Metal Oxides. CRC Press, Boca Raton, Florida

Cambier P, Sposito G (1990) Adsorption of citric acid by synthetic pseudoboehmite. Clays Clay Minerals 39:369-374

Carrano CJ, Drechsel H, Kaiser D, Jung G, Matzanke B, Winkelmann GP, Rochel N, Albrecht-Gary AM (1996) Coordination chemistry of the carboxylate type siderophore rhizoferrin: the iron(III) complex and its metal analogs. Inorg Chem 35:6429-6436

Chen Y, Jurkevitch E, Bar-Ness E, Hadar Y (1994) Stability constants of pseudobactin complexes with transition metals. Soil Sci Soc Am J 58:390-396

Cornell RM, Schwertmann U (1996) The iron oxides—structure, properties, reactions, occurrence, and uses. Verlag VCH, Weinheim

Coughlin BR, Stone AT (1995) Nonreversible adsorption of divalent metal ions (Mn^{II}, Co^{II}, Ni^{II}, Cu^{II}, and Pb^{II}) onto goethite: effects of acidification, Fe^{II} addition, and picolinic acid addition. Environ Sci Technol 29:2445-2455

Cowan JA (1993) Inorganic Biochemistry: An Introduction. VCH Publishers, New York

Drechsel H, Metzger J, Freund S, Jung G, Boelaert JR, Winkelmann G (1991) Rhizoferrin—a novel siderophore from the fungus *Rhizopus microsporus* var. *rhizopodiformis*. BioMetals 4:238-243

Drechsel H, Thieken A, Reissbrodt R, Jung G, Winkelmann G (1993) δ-Keto acids are novel siderophores in the genera *proteus*, *providencia*, and *morganella* and are produced by amino acid deaminases. J Bacteriol 175:2727-2733

Dzombak DA, Morel FMM (1990) Surface Complexation Modeling: Hydrous Ferric Oxide. Wiley-Interscience, New York

Gekeler W, Grill E, Winnacker E-L, Zenk MH (1988) Algae sequester heavy metals via synthesis of phytochelatin complexes. Arch Microbiol 150:197-202

Girvin DC, Gassman PL, Bolton H (1993) Adsorption of aqueous cobalt ethylenediaminetetracetate by α-Al_2O_3. Soil Sci Soc Am J 57:47-57

Girvin DC, Gassman PL, Bolton H (1994) Adsorption of nitrilotriacetate (NTA), Co, and CoNTA by gibbsite. Clays Clay Minerals 44:757-768

Grill E, Winnacker E-L, Zenk MH (1985) Phytochelatins: the principal heavy-metal complexing peptides of higher plants. Science 230:674-676

Hancock RD, Martell AE (1989) Ligand design for selective complexation of metal ions in aqueous solution. Chem Rev 89:1875-1914

Harborne JB (1993) Introduction to Ecological Biochemistry, 4th ed. Academic Press, London

Harris WR, Carrano CJ, Raymond KN (1979) Coordination chemistry of microbial iron transport compounds, 16. isolation, characterization and formation constants of ferric aerobactin. J Am Chem Soc 101:2722-2727

Herbelin A, Westall JC (1994) FITEQL. Report 94-01, Department of Chemistry, Oregon State Univ, Corvallis, OR

Hering JG, Morel FMM (1988) Kinetics of trace metal complexation: role of alkaline earth metals. Environ Sci Technol 22:1469-1478

Hering JG, Morel FMM (1989) Slow coordination reactions in seawater. Geochim Cosmochim Acta 53:611-618

Hering, JG, Morel FMM (1990) The kinetics of trace metal complexation: implications for metal reactivity in natural waters. In: W Stumm (ed) Aquatic Chemical Kinetics, p 145-171. Wiley-Interscience, New York

Hersman L, Lloyd T, Sposito G (1995) Siderophore-promoted dissolution of hematite. Geochim Cosmochim Acta 59:3327-3330

Holmen BA, Casey WH (1996) Hydroxamate ligands, surface chemistry, and the mechanism of ligand-promoted dissolution of goethite [α-FeOOH(s)]. Geochim Cosmochim Acta 60:4403-4416

Inoue K, Hiradate S, Takagi S (1993) Interaction of mugineic acid with synthetically produced iron oxides. Soil Sci Soc Am J 57:1254-1260

Irving H, Williams RJP (1953) The stability of transition metal complexes. J Chem Soc 3192-3216

Konetschny-Rapp S, Jung G, Meiwes J, Zahner H (1990) Staphyloferrin A: a structurally new siderophore from staphylococci. Eur J Biochem 191:65-74

Kummert R, Stumm W (1980) The surface complexation of organic acids on hydrous δ-Al_2O_3. J Colloid Interface Sci 75:373-385

Langemann K, Heineke D, Rupprecht S, Raymond KN (1996) Nordesferriferrithiocin. Comparative coordination chemistry of a prospective iron chelating agent. Inorg. Chem. 35:5663-5673

Langmuir D (1997) Aqueous Environmental Geochemistry. Prentice-Hall, Englewood Cliffs, New Jersey

Loomis LD, Raymond KN (1991) Solution equilibria of enterobactin and metal-enterobactin complexes. Inorg Chem 30:906-911

Lovelock JE (1975) Thermodynamics and the recognition of alien biospheres. Proc R Soc London B 189:167-181

Ludwig L, Schindler PW (1995) Surface complexation on TiO_2, II. Ternary surface complexes: coadsorption of Cu(II) and organic ligands (2,2'-bipyridyl, 8-aminoquinoline, and o-phenylenediamine) onto TiO_2 (anatase). J Colloid Interface Sci 169:291-299

Ma JF, Shinada T, Matsuda C, Nomoto K (1995) Biosynthesis of phytosiderophores, mugineic acids, associated with methionine cycling J Biol Chem 270:16549-16554

March J (1985) Advanced Organic Chemistry, 3rd edit. Wiley-Interscience, New York

Marschner H (1995) Mineral Nutrition of Higher Plants, 2nd ed. Academic Press, New York

Martell AE, Smith RM, Motekaitis R (1995) NITS Critically Selected Stability Constants of Metal Complexes. Natl Inst Sci Tech, Gaithersburg, Maryland

Mcbride MB, Wesselink LG (1988) Chemisorption of catechol on gibbsite, boehmite, and noncrystalline alumina surfaces. Environ Sci Technol 22:703-708

Meiwes J, Fiedler H-P, Haag H, Zahner H, Kinetschny-Rapp S, Jung G (1990) Isolation and characterization of staphyloferrin A, a compound with siderophore activity from Staphylococcus hyicus DSM 20459. FEMS Microbiology Lett 67:201-206

Miyasaka SC, Buta JG, Howell RK, Foy CD (1991) Mechanism of aluminum tolerance in snapbeans, root exudation of citric acid. Plant Physiol 96:737-743

Murakami T, Ise K, Hayakama M, Kamei S, Takagi S (1989) Stabilities of metal complexes of mugineic acid and their specific affinities for iron(III). Chemistry Lett 1089:2137-2140

Nishio T, Tanaka N, Hiratake J, Katsube Y, Ishida Y, Oda J (1988) Isolation and structure of the novel dihydroxamate siderophore alcaligin. J Am Chem Soc 110:8733-8734

Nowack B, Sigg L (1997) Dissolution of iron(III) (hydr)oxides by metal-EDTA complexes. Geochim Cosmochim Acta 61:951-963

Ohman L-O (1988) Equilibrium and structural studies of silicon(IV) and aluminum(III) in aqueous solution, 17. stable and metastable complexes in the system H^+-Al^{3+}-citric acid. Inorg Chem 27:2565-2570

Ohman L-O, Sjoberg S (1988) Thermodynamic calculations with special reference to the aqueous aluminum system. In: JR Kramer, HE Allen (eds) Metal Speciation: Theory, Analysis and Application, p 1-40 Lewis Publishers, Chelsea, Michigan

Otani T, Ae N, Tanaka H (1996) Phosphorus (P) uptake mechanisms of crops grown in soils with low P status, II. Significance of organic acids in root exudates of pigeonpea. Soil Sci. Plant Nutr. 42:553-560

Papelis C, Hayes KF, Leckie JO (1988) HYDRAQL. Technical Report 306, Dept Civil Engineering, Stanford Univ, Stanford, California

Prabhu V, Biolchini PF, Boyer GL (1996) Detection and identification of ferricrocin produced by ectendomycorrhizal fungi in the genus *Wilcoxina*. BioMetals 9:229-234

Rengel Z (1996) Uptake of aluminum by plant cells. New Phytologist 134:389-406

Raymond KN, Miller G, Matzanke BF (1984) Complexation of iron by siderophores, a review of their solution and structural chemistry and biological function. Topics in Current Chem 123:49-102

Schindler PW (1990) Co-adsorption of metal ions and organic ligands: formation of ternary surface complexes. In: MF Hochella and AF White (eds) Mineral-Water Interface Geochemistry, Rev Mineral 23:281-307

Scholz G, Becker R, Pich A, Stephan UW (1992) Nicotianamine—a common constituent of strategies I and II of iron acquisition by plants: a review. J Plant Nutrition 15:1647-1665

Shannon RD (1976) Revised effective ionic radii and systematic studies of interatomic distances in halides and chalcogenides. Acta Crystallogr A32:751-767

Shenker M, Hadar Y, Chen Y (1996) Stability constants of the fungal siderophore rhizoferrin with various microelements and calcium. Soil Sci Soc Am J 60:1140-1144

Sjoberg S, Lovgren L (1993) The application of potentiometric techniques to study complexation reactions at the mineral/water interface. Aquatic Sciences 55/4:324-335

Stephan UW, Schmidke I, Stephan VW, Scholz G (1996) The nicotianamine molecule is made-to-measure for complexation of metal micronutrients in plants. BioMetals 9:84-90

Stone AT, Torrents A, Smolen J, Vasudevan D, Hadley J (1993) Adsorption of organic compounds possessing ligand donor groups at the oxide/water interface. Environ Sci Technol 27:895-909

Stone AT, Godtfredsen KL, Deng B (1994) Sources and reactivity of reductants encountered in aquatic environments. In: G Bidoglio, W Stumm (eds) Chemistry of Aquatic Systems: Local and Global Perspectives, p 337-374 ECSC, EEC, EAEC, Brussels and Luxembourg

Stone AT, Torrents A (1995) The role of dissolved metals and metal-containing surfaces in catalyzing the hydrolysis of organic pollutants. In: PM Huang, J Berthelin, J-M Bollag, WB McGill, AL Page (eds) Environmental Impact of Soil Component Interactions. Lewis Publishers, Chelsea, Michigan

Strasdeit H, Duhme A-K, Kneer R, Zenk MH, Hermes C, Nolting H-F (1991) Evidence for discrete $Cd(Scys)_4$ units in cadmium phytochelatin complexes from EXAFS spectroscopy. J Chem Soc Chem Commun 1191:1129-1130

Stumm W (1992) Chemistry of the Solid-Water Interface: Processes at the Solid-Water and Particle-Water Interface. Wiley-Interscience, New York

Stumm W, Morgan JJ (1996) Aquatic Chemistry, 3rd ed. Wiley-Interscience, New York

Sugiura Y, Nomoto K (1984) Phytosiderophores: structures and properties of mugineic acid and their metal complexes. Structure Bonding 58:107-135

van der Helm R, Winkelmann G (1994) Hydroxamates and polycarboxylates as iron transport agents (siderophores) in fungi. In: G Winkelmann, DR Winge (eds) Metal Ions in Fungi. Marcel Dekker, New York

Vasudevan D, Stone AT (1996) Adsorption of catechols, 2-aminophenols, and 1,2-phenylenediamines at the (hydr)oxide/water interface: effect of ring substituents on the adsorption onto TiO_2. Environ Sci Technol 30:1604-1613

Vasudevan DE, Stone AT (submitted) Adsorption of 4-nitrocatechol, 4-nitro-2-aminophenol, and 4-nitro-1,2-phenylenediamine at the (hydr)oxide/water interface: effect of metal (hydr)oxide properties

Yamamoto S, Fujiata Y, Okujo N, Minami C, Matsuura S, Shinoda S (1992) Isolation and partial characterization of a compound with siderophore activity from *Vibrio parahaemolyticus*. FEMS Microbiol Lett 94:181-186

Yao H-L, Yeh H-H (1996) Fumarate, maleate, and succinate adsorption on hydrous δ-Al$_2$O$_3$, 1. Comparison of the adsorption maximaand their significance. Langmuir 12:29812988

Yao H-L, Yeh H-H (1996) Fumarate, maleate, and succinate adsorption on hydrous δ-Al$_2$O$_3$, 2. Electrophoresis observations and ionic strength effects on adsorption. Langmuir 12:2989-2994

Chapter 10

THE BACTERIAL VIEW OF THE PERIODIC TABLE: SPECIFIC FUNCTIONS FOR ALL ELEMENTS

Simon Silver

Department of Microbiology and Immunology
University of Illinois
Chicago, Illinois 60612 U.S.A.

STRATEGIES FOR METAL HANDLING BY MICROORGANISMS

This chapter describes the general mechanisms by which microorganisms transport metals and other nutrients, with emphasis on the strategies they have evolved to deal with otherwise toxic concentrations of metals. The underlying theme is that for each and every inorganic cation or anion that is encountered in normal environments, there are corresponding genes and proteins that govern movements (Silver 1996a, Silver and Phung 1996). Nothing of consequence occurs uncontrolled. For required nutrients there are highly specific membrane transport systems that concentrate needed nutrients from dilute media. Once inside the cell, some inorganic nutrients are sequestered (for example, the protein metallothionein binds intracellular cadmium, copper and zinc) or enzymatically incorporated into specific proteins (for example using the enzyme ferrochelatase, which places cationic iron into heme groups). Excess ions are either stored (for example, iron in ferritin) or excreted (again by highly specific membrane transporters). Nothing is left to chance.

Cations of related elements often are associated in pairs, with one needed for intracellular nutrition and the other not used intracellularly (but sometimes coupled for co-transport or cellular signaling, and needing a highly specific efflux mechanism). For example, K and Na both serve biological functions. However, K^+ (with several parallel uptake transporter pathways) is essential for intracellular nutrition whereas Na^+ is not. There are no intracellular Na^+-specific enzymes and sodium functions basically outside of microbial cells and for many bacteria (*E. coli* is a well-known example) Na^+ is not needed at all for growth or survival. There are a few "normally" proton-driven membrane transport pumps that are in a few rare bacteria Na^+ gradient driven instead. A similar situation occurs with Mg and Ca. Mg^{2+} functions in many intracellular roles and must be transported inward and carefully regulated; Ca^{2+} is not needed within the cell and is maintained at low intracellular levels by efflux transport pathways. Ca^{2+} and Na^+ are indeed frequently used for biological processes, such as signal transduction or as co-transport substrates, but these are secondary processes, not common to all cells. Ca^{2+} and Na^+ frequently function extracellularly rather than intracellularly in biological processes. This is true of animal cells as well as free-living bacterial cells, surviving in basically distilled water or in hypertonic saline environments.

For some higher atomic number elements of the Period Table (the lanthanides and actinides, including uranium and trans-uranium elements), there appear to be no specific genes or proteins for metal ion resistances. Presumably, these were not encountered early in life on Earth in natural environments at levels that were toxic. Similarly, there are no known genes or proteins specific for Al^{3+}. Although Al is the second most abundant cation in the Earth's crust (after Si), it occurs in extremely low concentrations in most solutions near the Earth's surface (see Table 2, Banfield and Hamers, this volume). This, combined with the low toxicity of aluminum cations means that specific biological roles for Al^{3+}

0275-0279/97/0035-0010$05.00

would have low survival value and furthermore that resistance systems for Al^{3+} are not needed. There also are no resistance genes for halides, although halides are abundant in the environment and toxic in higher concentrations. This is somewhat of a surprise.

Bacteria have genes (and proteins) specific for transport of all needed nutrients and for resistances to the toxic ions of most heavy metal elements. Required inorganic nutrients include the cations NH_4^+, K^+, Mg^{2+}, Co^{2+}, Cu^{2+}, Fe^{3+}, Mn^{2+}, Ni^{2+}, Zn^{2+} and other trace cations (Silver 1996a) and the oxyanions PO_4^{3-}, SO_4^{2-} and less abundant anions. Figure 1 presents a current summary of inorganic nutrient transport systems and Figure 2 a listing of inorganic ion resistance systems and their biochemical mechanisms. Toxic inorganics with genetically-defined resistances include Ag^+, AsO_2^-, AsO_4^{3-}, Cd^{2+}, Co^{2+}, CrO_4^{2-}, Cu^{2+}, Hg^{2+}, Ni^{2+}, Pb^{2+}, Sb^{3+}, TeO_3^{2-}, Tl^+ and Zn^{2+}.

The Periodic Table of Nutrient Cation and Oxyanion Transport Systems

 1) K^+ . Four separate systems in *E. coli*. Three chemiosmotic and one ATPase.

 2) Mg^{2+}. Three separate systems in *S. typhimurium*. One chemiosmotic and two ATPase.

 3) Fe^{3+}. At least five separate systems in *E. coli*. Specificities for different siderophores.

 4) Mn^{2+}. Found in Gram positive and Gram negative bacteria. Chemiosmotic and ATPases

 5) Zn^{2+}. Newly reported ATPase in *E. coli*, preliminary evidence.

 6) Ni^{2+}. ABC ATPase in *E. coli*.

 7) PO_4^{3-}. Separate Pit (chemiosmotic) and Pst (ABC ATPase) systems in *E. coli* and *Bacillus*.

 8) SO_4^{2-}. Five component ABC-ATPase in *S. typhimurium*.

Figure 1. Nutrient cation and anion transport systems.

Heavy Metals
PLASMID HEAVY METAL RESISTANCE SYSTEMS AND MECHANISMS

☹ 1. Hg^{2+}. *mer*. Hg^{2+} and organomercurials are enzymatically detoxified.

☹ 2. AsO_4^{3-}, AsO_2^-, SbO^+. *ars*. Arsenate is enzymatically reduced to arsenite by ArsC Arsenite and antimonite are "pumped" out by the membrane protein ArsB that functions chemiosmotically alone or with the additional ArsA protein as an ATPase.

☹ 3. Cd^{2+}. *cadA*. Cd^{2+} (and Zn^{2+}) are pumped from Gram positive bacteria by a P-type ATPase with a phospho-aspartate intermediate.

☹ 4. Cd^{2+}, Zn^{2+}, Co^{2+}, and Ni^{2+}. *czc*. Cd^{2+}, Zn^{2+}, Co^{2+}, and Ni^{2+} are pumped from Gram negative bacteria by a three polypeptide membrane complex that functions as a divalent cation/2 H^+ antiporter. The complex consists of an inner membrane protein (CzcA), an outer membrane protein (CzcC) and a protein associated with both membranes (CzcB).

☹ 5. Ag^+. Ag^+ resistance results from pumping from bacteria by three polypeptide chemiosmotic exchanger plus a P-type ATPase.

☹ 6. Cu^{2+}. *cop*. Plasmid Cu^{2+} resistance results from a four polypeptide complex, consisting of an inner membrane protein, an outer membrane protein, and two periplasmic copper-binding proteins. In *Pseudomonas*, Cop results in periplasmic sequestration of Cu^{2+}. In addition, chromosomally-encoded P-type ATPases provide partial resistance by effluxing Cu^{2+} or Cu^+.

☹ 7. CrO_4^{2-}. *chr*. Chromate resistance results from a single membrane polypeptide that causes reduced net cellular uptake, but efflux has not been demonstrated.

☹ 8. TeO_3^{2-}. *tel*. Tellurite resistance results from any of several genetically-unrelated plasmid systems. Although reduction to metallic Te^0 frequently occurs, this does not seem to be the primary resistance mechanism.

☹ 9. Pb^{2+}. *pum*. Lead resistance appears to be due to an efflux ATPase in Gram negative and accumulation of intracellular $Pb_3(PO_4)_2$ in Gram positive bacteria.

Resistance systems await understanding for bismuth (Bi), boron (B), thallium (Tl) and tin (Sn).

Figure 2. Plasmid toxic metal resistance systems and mechanisms (modified from Silver 1996b).

MECHANISMS OF METAL RESISTANCE

Toxic metal resistance systems probably arose shortly after life started, in an already metal-abundant world. As with nutrient organic compounds, the environment provided strong selection pressures for the transport and accumulation of needed inorganic nutrients and for the removal or detoxification of abundant toxic cations and oxyanions. The recent activities of humans create locally polluted environments, which indeed exert a high selection pressure for metal resistance on microbial populations. Highly specific systems have evolved in response to these pressures (Silver 1996b, Silver and Phung 1996).

However, there is nothing new about toxic heavy metal resistance determinants, as microbes have been exposed to these cations and oxyanions periodically since the origin of cellular life.

Plasmid and chromosomal genes

Most frequently, toxic metal resistance systems are found on plasmids, small circular DNA molecular that can readily move from one cell to another, when there is a need for their functions. Plasmid-based resistance facilitates transfer of toxic metal resistance from cell to cell over a short time scale (weeks or months) would not allow new invention by evolution of such resistance mechanisms. In other organisms, however, resistances to toxic metals are coded for by chromosomal genes, which may indicate that most bacteria experience selection for toxic metal resistances frequently in natural environments. Some cations such as Co^{2+}, Cu^{2+}, Ni^{2+} and Zn^{2+} are essential nutrients at low levels, but toxic at higher levels. For these cations, separate transport systems for uptake are coded for by chromosomal genes, and additional genes (either on the chromosomes or on plasmids) efflux the same cations when present at toxic levels. The paired uptake and efflux systems are carefully regulated, assuring homeostasis and maintenance of stable intracellular cation levels, more or less independent of external instabilities (Silver 1996a,b; Silver and Ji 1994, Silver and Phung 1996).

Three generalizations about metal resistance mechanisms may be made:

(1) Plasmid-determined metal resistance systems are very specific, as much so as those for antibiotic resistances, or sugar or amino acid metabolism. There is no general mechanism for resistance to all heavy metal ions.

(2) Metal-ion resistance systems have been found within every bacterial group tested, from *Escherichia coli* to *Streptomyces*. These mechanisms have not been found in eukaryotic microbes (which appear to have different mechanisms) or Archaea, for which both thermophiles and halophiles live in metal-rich environments but for which such mechanisms have never been reported.

(3) The mechanisms of resistance are most often efflux "pumping" (removing toxic ions that had entered the cell by means of transport systems designed for uptake of nutrient cations or oxyanions) and less often enzymatic detoxification (generally redox chemistry) converting a more toxic to a less toxic or less available metal-ion species.

Efflux pumps

It would seem easier to keep toxic ions out (by evolving highly specific uptake transport systems), rather than to expend ATP bringing in toxic ions and then more energy pumping them out. Apparently, the design penalties in terms of chemical specificities and rates for having uptake pumps more specific is greater than the genetic cost of having plasmid genes (in part of the overall population) that can spread when needed.

The efflux pumps that are the protein basis of plasmid resistance systems can be either ATPases or membrane potential driven. Transport ATPases are enzymes that use the chemical energy from cleavage of the high-energy phospho-ester bond of adenosine triphosphate (ATP) to drive the formation of concentration gradients (a potential gradient). Membrane potential-driven pumps couple a membrane potential gradient (which often is about 200mV, internal negative, across most bacterial membranes) to establish a concentration gradient. Alternatively, "primary gradients" of protons (H^+) or Na^+ (high outside/low inside) drive the formation of secondary gradients of nutrients (for example K^+) or toxic (for example Cd^{2+} in some bacteria) metals by "co-transport" either in the same direction or opposite direction. Other examples of ATPase pumps include the Cu^{2+}

efflux ATPases of many bacteria and the arsenite-ATPase of Gram-negative bacteria. Other examples of secondary gradient-driven pumps are the arsenite efflux systems of the chromosomes of *Bacillus* and *E. coli* and of plasmids in Gram-positive bacteria.

The mechanisms are not precisely the same in all bacterial types: while the mercury-resistance and arsenic-resistance systems are highly homologous in all bacteria studied, cadmium resistance involves ATPases in Gram-positive bacteria and unrelated chemiosmotic proton/cation exchangers in Gram-negative bacteria. These divalent cation systems appear to be of independent evolutionary origin and never to have shared a common ancestral system. There is also a well-described bacterial metallothionein, found so far only in cyanobacteria, that confers resistances to Cd^{2+} and Zn^{2+} (Turner and Robinson 1995). Metallothionein protects the cellular interior by sequestration—specific binding of toxic cations within the cell but so tightly that harm is avoided.

The mechanisms of bacterial plasmid-determined resistances to many toxic inorganic cations and anions have been studied by my own and other laboratories. Because of limitations of space and the need here for only a global and general picture, I will limit references frequently only to review articles (Silver 1996a,b; Silver and Phung 1996, Silver and Ji 1994, Ji and Silver 1995) from our laboratory. These reviews have more specific references to the contributions of a large group of investigators.

Archaea

No one has reported resistance mechanisms for Archaea such as described in this report for bacteria. Whether this reflects a difference in biology or a lack of research effort is unclear. Archaea grow in environments with high levels of toxic heavy metal ions and therefore are expected to have the same or alternative mechanisms of metal resistances. A step in this direction was the publication (Bult et al. 1996) of the entire 1.66 million nucleotide DNA sequence of the methanogen *Methanococcus jannaschii*. This sequence, unlike those of bacteria that are now being released almost monthly, contains only two sequenced potential genes that are reported as homologous to those we earlier reported from bacteria. These encode homologs for ArsA (the ATPase subunit of the arsenite effluxing membrane transport system) and ChrA (the membrane protein required for chromate resistance). Why there are so few such recognizable genes in a deep-sea vent-dwelling organism (expected to have been exposed to a range of toxic heavy metals), compared with the larger number for the human pathogen *Haemophilus influenzae* is unclear.

METAL RESISTANCE IN BACTERIA: CASE STUDIES

Mercury

Mercury in the environment. Mercury is an important metal contaminant in "natural environments" such as volcanoes and vents in the bottom of the sea. There is therefore a natural mercury geocycle (Fig. 3) with microbial activity important at many levels. Bacteria oxidize Hg(0) that is deposited world-wide in rain to Hg(II), which is taken up and absorbed more effectively by living organisms. Microbial oxidation of Hg(0) is carried out by the widely-found enzyme catalase. Bacteria also reduce Hg(II) found in waters and soil to Hg(0), which is volatile in stirred environments such as sewage systems or from the surfaces of leafs. Microbes also methylate inorganic Hg(II) to methylmercury, which is both more toxic and "bio-accumulated." It is methylmercury of biological origin that is found in fish and other seafood and that represents a major human toxicity problem. However, methylmercury levels are high in tuna (for example) which live in the cleanest water in the middle of the sea (and therefore it is presumably "natural" methylmercury in

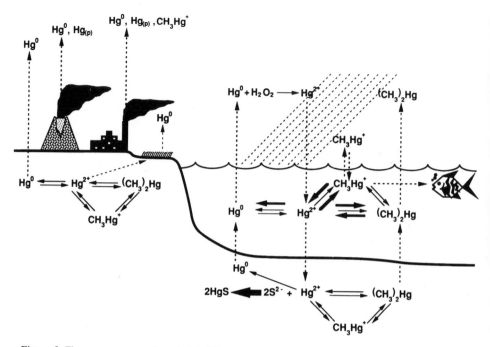

Figure 3. The mercury geocycle and global biocycle, showing both microbial activities and geophysical and industrial release of mercury compounds.

tuna) and in fish from Minamata Bay (the site of a major mercury pollution problem) in Kyushu, Japan (Silver et al. 1994) or from the gold mining region of Brazil (where liquid mercury is thought to be used by more than a hundred-thousand miners panning for gold). Again the conclusion is that toxic mercury levels have existed from the origin of life, although it is clearly correct that human activities have significantly impacted on local and global levels of mercury (and other metals). Finally, an enzyme organomercurial lyase which is found in our mercury resistance systems (see below) cleaves the mercury-carbon bond in methylmercury (and also phenylmercury), completing the cycle and yielding Hg(II).

Mercury resistance in bacteria. Closely related systems for resistance to inorganic mercury have been found on plasmids of all bacteria tested. For one example, in the collection of some 800 antibiotic-resistance plasmids from various Gram-negative bacteria, 25% carried mercury resistance (Schottel et al. 1974). In most cases, the number of genes (approximately six) and the functions of the genes are the same (Silver and Walderhaug 1995, Silver and Phung 1996) for all bacterial types. Almost all *mer* systems start with a regulatory gene, *merR*, whose product is a unique positively-acting activator protein that in the presence of Hg^{2+} twists and bends the operator DNA region, allowing RNA polymerase to synthesize messenger RNA (O'Halloran 1993, Summers 1992).

In the mercury resistance systems of Gram-negative bacteria, the regulatory gene is transcribed separately from the remaining genes. This allows tighter control of the *mer* operon than is possible with Gram-positive bacteria, where the regulatory gene is the first gene on the multi-gene *mer* operon, so that the gene that regulates and the genes for enzymes that bind and change the chemical form of mercury are always transcribed together (Silver and Walderhaug 1995, Ji and Silver 1995). Following the regulatory gene, there are one to three genes in the operon whose products are involved in transport of toxic Hg^{2+} across the cell membrane to the intracellular detoxifying enzyme, which is called mercuric reductase. In some *mer* operons, the gene for mercuric reductase is followed immediately by a small gene which encodes the enzyme organomercurial lyase. This lyase breaks the carbon-mercury bond in highly toxic organomercurials such as phenyl- and methyl-mercury.

The structure of the *Bacillus* mercuric reductase was solved by X-ray diffraction (Schiering et al. 1991). The structure is very similar to that of glutathione reductase from mammalian sources, and is a homo-dimer, with each subunit containing a highly conserved active site with two critical cysteine residues a bound FAD per subunit, and an NADPH-binding site for electron transfer from NADPH to FAD to the substrate Hg^{2+}. The active site includes the redox-active disulfide region on one subunit and the substrate-binding site at the C-terminal (including conserved vicinal Cys residues) of the other subunit (Schiering et al. 1991). The point here is that mechanistic studies have progressed and we anticipate that these studies at the atomic level will affect use of this system in "bioremediation" and understanding of functioning of the global mercury cycle.

Two lessons concerning mercury resistance genes in the environment come from the experiences with mercury resistant *Bacillus* isolates from the methylmercury polluted area at Minamata Bay Japan (Nakamura and Silver 1994, Silver et al. 1994) and the quantitating of mercuric resistance genes from Gram-negative bacteria from North American mercury polluted sites (Rochelle et al. 1991). The conclusions from both studies are (a) the genetic determinants that have been extensively studied in the laboratory are representative of what is out there in the environment and (b) what occurs in the environment is sophisticated, so that a thorough understanding from laboratory studies is essential to understand the distribution of heavy metal resistance genes and the chemical transformations in real environments.

Arsenic

Arsenic in the environment. Water supplies in many areas of many counties are extensively polluted or threatened by high concentrations of arsenic, sometimes from natural sources and sometimes from the activities of humans. Most recently, there were reports of arsenic in drinking water in India that reaches levels that cause human disease, sometimes cancers. In Taiwan and China, high arsenic in drinking water causes a condition called "black foot" where tissues die and gangrene sets in leading to the loss of toes and feet. Since microbial activities can in some cases mobilize otherwise mineral-bound arsenic or alternatively immobilize otherwise water-soluble arsenic, the transformations of arsenic by bacteria is important in geocycles. In mining, especially during bioleaching of gold from arsenopyrite ores (Rawlings and Silver 1995), soluble arsenic (mostly arsenate) can build up in large bodies of mining waste water to levels of 0.4 M (30 g/l). This clearly provides a localized environment where microbial activities will either help or harm. Another example, just north of Boston occurs where industrial activity produced a large amount of arsenic waste, which was immobilized in buried waste materials but has recently been releasing (apparently through microbial activities; Ahmann et al. 1994) troublesome amounts of arsenic into local recreational waters. The overall message is as above: it is not simply chemistry that determines the speciation of arsenic in the environment. Microbial activities play major roles. There are many less familiar and less studied environments where arsenic is significant: Many sea foods such as shrimp contain high levels of natural organoarsenicals, synthesized by living organisms. There must a marine arsenic geocycle of synthesis and degradation of natural organoarsenicals. I am told that all chicken sold in the United States has been fed industrial-made organoarsenicals as a growth stimulant. What happens to the waste arsenic is not known.

Arsenic resistance. Understanding of bacterial arsenic resistance systems is less complete and more recent than that of mercury resistance. Fundamentally the same genes (and encoded biochemical mechanism) are found on plasmids in Gram-negative and Gram-positive bacteria (Silver and Phung 1996, Silver 1996a). Closely similar gene clusters have been found to determine normal background arsenic resistance both in *Bacillus* and in *E. coli* (Silver and Phung 1996, Silver 1996a) and as yet uncharacterized systems have different bases for arsenic resistance in environmentally important bacteria (Ahmann et al. 1994, Anderson et al. 1992, Macey et al. 1996). Therefore what we think today may be only the first half of a larger picture of microbial arsenic transformations. To start, the same cluster of genes, the *ars* operon, confers resistances to As(III), As(V) and Sb(III). However, the number of genes can vary somewhat and the details of their functions can differ. Seven *ars* operons have been sequenced (Silver 1996a). There are two extra genes, *arsA* and *arsD*, on the plasmids of Gram-negative bacteria that are missing from arsenic resistance systems of plasmids of Gram-positive bacteria and the chromosomal systems of both *E. coli* and *Bacillus subtilis*. The ArsD protein is a secondary regulator of *ars* operon transcription (Chen and Rosen 1997), so its presence or absence might have little effect on resistance. Its existence in two different plasmids, however, indicates a role in the environment under conditions more subtle than we have come to understand. The ArsA protein is a membrane-associated ATPase (Kaur and Rosen 1993) attached to the ArsB inner-membrane protein (Tisa and Rosen 1990, Wu et al. 1992) and energizing the arsenite efflux pump by ATP hydrolysis (Dey and Rosen 1995, Silver and Phung 1996). It is ArsA that make this an ATPase pump and in the absence of ArsA, the remaining ArsB protein functions as a membrane potential-driven secondary pump, providing a lower level of resistance (Dey and Rosen 1995, Silver and Phung 1996). Such alternative energy coupling is unique among known bacterial uptake or efflux transport systems. To date, all other systems that have been studied are either chemiosmotic or ATP-driven transporters. The arsenite pump is the only one that can be converted from one mode of energy coupling to the other by addition of or removal of genes, which happens in natural systems (Silver and Phung 1996) and can be reconstructed also in laboratory studies (Dey and Rosen 1995).

A gene product common to all *ars* operons is ArsC, the reductase enzyme that reduces less toxic arsenate (As(V)) to more toxic arsenite (As(III)) (25,28,30). It is only As(III) and not As(V) that is pumped out from the cells by the ArsB transport protein. It seems illogical from an environmental biology or chemistry point of view to convert a less toxic compound to a more toxic form, but ArsC activity is closely coupled with efflux from the cells (Ji and Silver 1992) so that intracellular arsenite never accumulates.

Arsenate reductases from plasmids of Gram-negative bacteria and Gram-positive bacteria both reduce arsenate and both confer arsenate resistance (Gladysheva et al. 1994, Ji et al. 1994). However, their in vitro measured properties are very different and their energy coupling is different. Arsenate reductase of Gram-positives derives reducing power from a small protein called thioredoxin (Ji and Silver 1992, Ji et al. 1994), which is used in many processes of central metabolism of bacteria and higher organisms. In contrast arsenate reductase of Gram-negative bacteria uses glutaredoxin (Gladysheva et al. 1994), which is a related but different protein. The small coupling proteins are not exchangeable.

In addition to plasmid arsenic resistance that is well understood and for which clusters of genes have been isolated and sequenced, there are bacterial arsenic metabolism systems that involve oxidation of arsenite to arsenic (e.g. Anderson et al. 1992), reduction of arsenate to arsenite as part of an oxyanion-coupled anaerobic respiration (Ahman et al. 1994, May et al. 1996) or the coupled cleavage of carbon-arsenic bonds with oxidation to arsenate (Quinn and McMullan 1995). These systems appear to be of major environmental concern in arsenic-containing settings, but they have not been approached by molecular genetics as yet.

Copper resistance in bacteria

We do not know how copper in natural environments effect or select for resistant bacteria. What has been studied to date are one set of related systems for plasmid resistance in Gram-negative bacteria that have been selected by agricultural use of copper salts, both as sprays for tomato plants and walnuts, and as growth-stimulating food additives for pigs, and another totally different system found on the chromsomes of many bacteria, both Gram-positive and Gram-negative, and apparently representing a low level normal homeostatic mechanism for regulating and maintaining the intracellular copper needed for nutrition in the presence of varying high and low levels of copper in the environment.

Strong copper resistance has been described with plasmids in Gram-negative bacteria from agricultural sources (Silver and Phung 1996, Brown et al. 1994). There are two regulatory genes are called *pcoR* and *pcoS* in *E. coli* and the five structural genes are *pcoABCDE* (Brown et al. 1995, Cooksey 1994) (Fig. 4). For *Pseudomonas*, the comparable genes are called *copR, copS* and *copABCD* and there is no equivalent to *pcoE*. The Cop/PcoR and Cop/PcoS proteins were the first example among the metal resistance systems of transcriptional regulation by a "two component" regulatory system. (A second and third example will be described below.) The sensor protein PcoS is found in the membrane and probably can be labeled with radioactive ^{32}P-phosphate at a specific conserved histidine residue with from radioactive ATP. The DNA-binding responder protein PcoR is that to have the phosphate transferred from PcoS to a specific conserved apartate amino acid residue to form ^{32}P-labeled PcoR (Brown et al. 1994, Cooksey 1994).

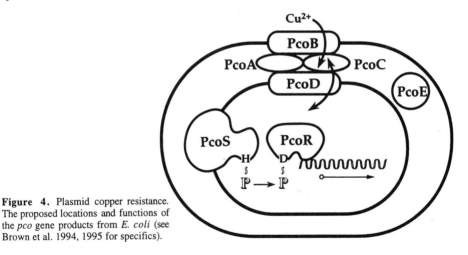

Figure 4. Plasmid copper resistance. The proposed locations and functions of the *pco* gene products from *E. coli* (see Brown et al. 1994, 1995 for specifics).

The four structural proteins determining copper resistance have been characterized in *Pseudomonas* and are the inner membrane protein CopD, the outer membrane protein CopB, and two periplasmic proteins CopA and CopC (Cooksey 1994, Cha and Cooksey 1991) (Fig. 4). CopA and CopC are blue copper-containing proteins. It is thought that storage of excess copper in the periplasmic space between the outer and inner cell membranes protects the cell from toxic copper. How CopD and CopB are involved in movement of copper across the membranes is not understood. A major problem in understanding of this system is that colonies of the copper resistant *Pseudomonas* turn blue when grown in high copper-containing media, while those of other bacteria turn brown, and show no sign of periplasmic copper storage. Furthermore, there is preliminary evidence for copper efflux (not uptake) associated with the *E. coli* copper resistance system (Brown et al. 1995). How the same genes can lead to different overall processes is unclear. However, the *E. coli* plasmid system (Brown et al. 1994 1995) includes an additional gene *pcoE*, the product of which is a periplasmic copper binding protein, and that is highly produced (Rouch and Brown 1997). Models for these gene functions have changed from year to year. However, all models of copper homeostasis and resistance involve the control of cellular uptake and efflux of copper by chromosomal genes, as well as plasmid systems for additional resistance in high copper environments (Brown et al. 1994, Cooksey 1994).

Chromosomal copper resistance in *Enterococcus hirae* is entirely different from that reported above and is indeed the best understood copper transport and resistance system (Odermatt et al. 1993, Solioz and Vulpe 1996). Two genes, *copA* and *copB*, that determine respectively uptake and efflux P-type ATPases, are found in a single operon. The system is regulated in response to both copper-starvation (when the CopA uptake ATPase is needed) and copper-excess (when the CopB efflux ATPase is needed) (Odermatt and Solioz 1995). *Enterococcus* CopA and CopB have the same names but different structures and functions from the *Pseudomonas* plasmid genes for copper resistance (Cooksey 1994). *E. hirae* mutants lacking the CopA uptake ATPase become somewhat copper-resistant and require higher levels of medium copper for growth. Bacterial mutants lacking the CopB efflux ATPase become copper-hypersensitive. The CopB copper efflux ATPase of *E. hirae* is unusual among bacterial cation efflux ATPases, in that actual subcellular ATPase, ATP-labeling and transport data are available. Solioz and Odermatt (1995) isolated inside-out, subcellular, membrane vesicles from *E. hirae* cells and the vesicles required ATP in order to accumulate $^{64}Cu^+$ and $^{110m}Ag^+$. The in vitro substrate for CopB is thought to be Cu^+ rather than Cu^{2+}. Whether copper is taken up initially as Cu^{2+} and subsequently reduced to Cu^+ or whether copper is reduced at the cell surface (before or concomitant with transport) is not established.

Bacterial metallothionein in cyanobacteria

Bacterial metallothioneins, functionally homologous to the small (approximately 60 amino acids long), thiol-rich (perhaps 20 of those 60 amino acids are cysteines) mammalian metal binding proteins, have rarely been reported and has been studied in detail only for the cyanobacterial genus *Synechococcus* (Gupta et al. 1992, 1993; Turner and Robinson 1995). The 58 amino acid polypeptide product of the *smtA* gene contains nine cysteine residues, which are clustered in groups of 4 and 5 respectively, as are the cysteines in animal metallothioneins. Metallothionein cysteines are clustered in two domains that bind divalent cations independently. The synthesis of metallothionein is regulated at three levels (Turner and Robinson 1995). Firstly, the SmtB repressor protein binds divalent cations and dissociates from the target DNA. Secondly, there is gene amplification so that tandem multiple copies of the metallothionein locus are produced in metal-stressed cells; and thirdly, a specific deletion between repeated sequences on the DNA removes most of the *smtB* gene for the repressor protein.

Cadmium resistance in Gram-positive bacteria results from a P-type ATPase

The Cd^{2+} efflux ATPase is found in Gram-positive bacteria from diverse sources, including soil bacilli and clinical *Listeria* (Silver et al. 1993, Silver and Phung 1996, Silver and Walderhaug 1995). The protein structure as diagrammed (Fig. 5a) is typical of P-type ATPases. It is presented here as a strong example of the claim that understanding only comes from the details. As in chemical sciences, to skim the surface often means missing the point. The Cd^{2+} efflux ATPase starts with a metal-binding motif, including a pair of adjoining cysteine amino acids with metal-binding thiol groups. Six metal-binding

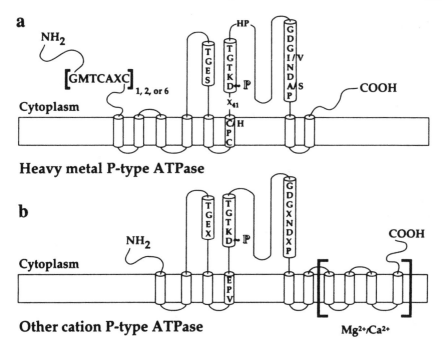

a

Heavy metal P-type ATPase

b

Other cation P-type ATPase

Figure 5. Heavy-metal cation (e.g. Cd^{2+}) (a) and nutrient cation (e.g. Ca^{2+} and Mg^{2+}) (b) P-type ATPases. The predicted motifs (cation-binding, phosphatase, membrane channel, and aspartyl kinase) regions are shown. Modified after Solioz and Vulpe (1996) and earlier models.

motifs occur in the human copper transporting efflux ATPases that are defective in the hereditary diseases Menkes and Wilson's (Silver et al. 1993, Silver and Walderhaug 1995). The remarkable similarity of these ATPases between animals, plants and bacteria is a major recent finding. There follows a membrane ATPase region closely homologous to other P-type ATPases. This includes the eight predicted membrane spanning regions shown in Figure 5a for the heavy metal translocating ATPases (Solioz and Vulpe 1996), the sixth of which is thought to be involved in the cation translocation pathway. It includes a conserved proline residue (as shown) between cysteines that are found in the Cd^{2+} ATPases and related proteins. Melchers et al. (1996) recently provided detailed data using protein fusions supporting the eight segment model for a presumed copper effluxing ATPase from *Helicobacter pylori*. Whereas the bacterial potassium ATPase KdpB had been modeled as having six membrane spanning segments in the comparable region, the Mg^{2+} ATPases and (postulated) Ca^{2+} ATPases have an addition four segments toward the carboxyl end for a total of 10 as shown in Figure 5b. Therefore, we conclude that P-type ATPases differ in membrane topology and length, depending more on cation specificity than on the difference between uptake and efflux directions.

Two intracellular domains shown in the model and common to P-type ATPases are the aspartyl kinase domain (including the site of aspartyl-phosphorylation) and the phosphatase domain, involved in removing the phosphate from the aspartate residue during the reaction and transport cycle. The name for this class of ATPases, "P-type", is used since they are the only transport ATPases that have a covalent phospho-protein intermediate.

Large plasmids of the soil chemilithotrophic autotroph *Alcaligenes* have numerous heavy metal resistance determinants, including for mercury, chromate resistance, and three related ones for divalent cations called *czc* (for Cd^{2+}, Zn^{2+} and Co^{2+} resistances), *ncc* (for Ni^{2+}, Cd^{2+} and Co^{2+} resistances), and *cnr* (for Co^{2+} and Ni^{2+} resistances) (Diels et al. 1995, Nies and Silver 1995, Silver and Phung 1996). These closely related systems contain basically the same three structural proteins. Indeed, mutations of the Cnr system give additional Zn^{2+} resistance, again showing that the two systems are fundamentally the same. Czc is an efflux pump that functions as a chemiosmotic divalent cation/proton antiporter (Nies et al. 1989, Nies 1995). The proteins involved have become the paradigm for a new family of three-component chemiosmotic exporters (Diels et al. 1995, Silver and Phung 1996), which we call CBA systems for the order of transcription of the genes and to contrast them with the ABC ("ATP-binding cassette" multicomponent ATPases). CzcC is thought to be an outer membrane protein. CzcB appears to be a "membrane fusion protein" that bridges the inner and outer cell membrane of Gram-negative bacteria (references in Diels et al. 1995, and Ji and Silver 1995). And CzcA is the basic inner membrane transport protein of over 1000 amino acids in length. Several additional regulatory genes are involved, but there is incomplete understanding of their number or functions. Two of these, *czcR* and *czcS* encode a second pair of cation-sensing sensor kinase (CzcS) and transphosphorylated responder (CzcR) proteins, homologous to PcoS and PcroR for plasmid copper resistance.

Chromate resistance and chromate reduction in Gram-negative bacteria

Chromate resistance and chromate reduction both occur, but resistance to chromate governed by bacterial plasmids appears to have nothing to do with chromate reduction. Furthermore, it is not clear whether chromate reduction ability that has been found with several bacterial isolates confers resistance to CrO_4^{2-} (Ohtake and Silver 1994). Plasmid-determined chromate resistance results from reduced uptake of CrO_4^{2-} by the resistant cells. The DNA sequences of the *P. aeruginosa* and *R. eutrophus* chromate resistance systems share homologous *chrA* genes, which encode membrane proteins. A third *chrA* gene was found on a plasmid of a cyanobacterium (Nicholson and Lauderbach 1995), so we expect more examples of chromate resistance operons will be found.

Tellurite resistance in Gram-negative bacteria

There are well-studied and indeed several sequenced determinants of plasmid-governed tellurite resistance, but in each case we do not understand the mechanism of tellurite resistance (Ji and Silver 1995, Walter and Taylor 1992). Tellurite resistance does not appear to involve reduction to black metallic tellurium—which indeed occurs, especially if resistance allows cell growth. As the British philosopher Lugwig Wittgenstein suggested, if you do not know anything then be quiet. It is enough to say that microbial tellurite resistance and metabolism exist, but are inadequately understood.

Silver resistance in enteric bacteria

We have started working on the genetic basis of plasmid-determined silver resistance, beginning with a plasmid from *Salmonella* (K. Matsui, A. Gupta and S. Silver, in preparation) and Figure 6 represents a summary of current understanding and our first presentation of as yet unpublished results. It requires 14 kb of DNA to encode the 8 genes apparently involved in bacterial silver resistance and the functions of 7 of these are diagrammed in Figure 6. Two of these, *silS* and *silR* encode another pair of cation-sensing sensor kinase (SilS) and transphosphorylated responder (SilR) proteins, homologous to PcoS/PcroR for copper resistance and CzcS/CzcR for cadmium, zinc and cobalt resistances (Fig. 6). As with the *pco* system, *silRS* is followed by *silE*, the determinant of a periplasmic (direct data are available) Ag^+-binding protein, homologous to PcoE. However, the remainder of the silver resistance determinant is transcribed in the opposite

Silver Resistance Gene Functions

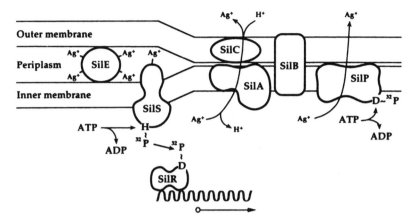

Figure 6. Model for proteins and functions of the new plasmid silver resistance determinant (from Silver et al., in preparation).

direction (unlike the situation in *pco*) and encodes both a three component CBA system (Fig. 6), weakly homologous to CzcCBA and a P-type ATPase that is generally speaking in the family of heavy metal responding enzymes diagrammed in Figure 5a, except for the absence of a CX_2C metal-binding motif. In the same position, a series of histidine residues may be functioning to recognize and function with Ag^+. After cloning and sequencing the first such silver resistance determinant, gene-specific DNA probes were used to show that similar (but not identical) systems occur in a wide range of candidate enteric bacteria from clinical sources, including burn patients, where silver salts were used as antiseptics, and metallic silver catheters (A. Gupta et al., in preparation). Whereas, silver resistance is initially being studied in clinical isolates, as had been the case for mercury and arsenic resistances in earlier years, we anticipate that this resistance system will be important for developing metal-resistant microbes for the mining industry, which is frequently becoming an applied microbiology subject (Rawlings and Silver 1995). We expect that the next few years will allow us to solve the molecular genetic and biochemical basis for highly specific Ag^+ resistance in some detail, as well as to find the comparable systems in other groups of bacteria. It already is clear, however, that reduction of Ag^+ to metallic Ag^0 is not involved.

Other toxic metal resistances

In addition to the specific resistances discussed above, several additional resistances are listed in the Figure 2. These have been studied still less and are material for future research. Lead (Pb^{2+}) resistance on plasmids appears to have a different basis in soil *Alcaligenes*, where still another cation-specific P-type ATPase has been found (D. van der Lelie, personal communication) and in *Staphylococcus*, where a process involving precipitation in intracellular lead-phosphate granules (Levinson et al. 1996) has been proposed. A plasmid resistance to tributyltin (used as anti-fouling compounds for ship hulls) has been reported (Miller et al. 1995).

FINAL COMMENT

I have presented a summary of how microbial cells "cope" with the inorganic cations and anions from elements of the Periodic Table. The approach here has been from the gene to the physiological process. Such an approach reflects the success of microbial genetics in our lifetimes, where a genetic approach is usually the most powerful. A quite different approach to understanding the biology of inorganic elements is being called "bioinorganic chemistry" (e.g. Fraústo da Silva and Williams 1991) and is generally based more on chemical principles, minimizing the powers of microbiology and genetics. Obviously, the middle ground is what is sought by all concerned—that is fundamental understanding behind problems of universal and practical importance.

ACKNOWLEDGMENTS

The references below are highly selected and frequently to review articles written by myself and many colleagues and friends (frequently listed as authors or co-authors below) over 30 years. This a limitation necessitated by space and by the needs of the intended audience only for a general view, and not for a highly technical listing of where facts were first reported. I also apologize for the excessive and undo burden placed on the lead editor of this volume to hook the author, land the manuscript, and prepare it for consumption.

REFERENCES

Ahmann D, Robert AL, Krumholz LR, Morel FMM (1994) Microbe grows by reducing arsenic. Nature 371:750

Anderson GL, Williams J, Hille R (1992) The purification and characterization of arsenite oxidase from *Alcaligenes faecalis,* a molybdenum-containing hydroxylase. J Biol Chem 267:23674-23682

Bröer S, Ji G, Bröer A, Silver S (1993) Arsenic efflux governed by the arsenic resistance determinant of *Staphylococcus aureus* plasmid pI258. J Bacteriol 175:3480-3485

Brown NL, Lee BTO, Silver S (1994) Bacterial transport of and resistance to copper. In: Sigel H, Sigel A (eds) Metal Ions in Biological Systems 30:405-434, Marcel Dekker, New York

Brown NL, Barrett SR, Camakaris J, Lee BTO, Rouch DA (1995) Molecular genetics and transport analysis of the copper-resistance determinant (*pco*) from *Escherichia coli* plasmid pRJ1004. Molec Microbiol 17:1153-1166

Bult C, White O, Olsen GJ , Zhou L, Fleischmann RD, Sutton GG, Blake JA, FitzGerald LM, Clayton RA, JD Gocayne, AR Kerlavage, BA Dougherty, J-F Tomb, MD Adams, CI Reich, R Overbeek, EF Kirkness, KG Weinstock, JM Merrick, A Glodek, JL Scott, NSM Geohagen, J Weidman, JL Fuhrmann, D Nguyen, TR Utterback, JM Kelley, JD Peterson, PW Sadow, MC Hanna, MD Cotton, KM Roberts, MA Hurst, BP Kaine, M Borodovsky, H-P Klenk, CM Fraser, HO Smith CR Woese, JC Venter (1996) Complete genome sequence of the methanogenic Archaeon, *Methanococcus jannaschii.* Science 273:1058-1073

Carlin A, Shi W, Dey S, Rosen BP (1995) The *ars* operon of *Escherichia coli* confers arsenical and antimonial resistance. J Bacteriol 177:981-986

Cervantes C, Ji G, Ramírez JL, Silver S (1994) Resistance to arsenic compounds in microorganisms. FEMS Microbiol Rev 15:355-367

Cervantes C, Ohtake H, Chu L, Misra TK, Silver S (1990) Cloning, nucleotide sequence and expression of the chromate resistance determinant of *Pseudomonas aeruginosa* plasmid pUM505. J Bacteriol 172:287-291

Cha J-S, Cooksey DA (1991) Copper resistance in *Pseudomonas syringae* mediated by periplasmic and outer membrane proteins. Proc Natl Acad Sci 88:8915-8919

Cha J-S, Cooksey DA (1993) Copper hypersensitivity and uptake in *Pseudomonas syringae* containing cloned components of the copper resistance operon. Appl Environ Microbiol 59:1671-1674

Chen Y, Rosen BP (1997) Metalloregulatory properties of the ArsD repressor. J Biol Chem 272:14257-14262

Clayton RA, White O, Ketchum KA, Venter JC (1997) The first genome from the third domain of life. Nature 387:459-462

Cooksey DA (1993) Copper uptake and resistance in bacteria. Molec Microbiol 7:1-5

Cooksey DA (1994) Molecular mechanisms of copper resistance and accumulation in bacteria. FEMS Microbiol Rev 14:381-386

Dey S, Rosen BP (1995) Dual mode of energy coupling by the oxyanion-translocating ArsB protein. J Bacteriol 177:385-389

Diels L, Dong Q, van der Lelie D, Baeyens W, Mergeay M (1995) The *czc* operon of *Alcaligenes eutrophus* CH34: from resistance mechanism to the removal of heavy metals. J Industr Microbiol 14:142-153

Diorio C, Cai J, Marmor J, Shinder R, Dubow MS (1995) An *Escherichia coli* chromosomal *ars* operon homolog is functional in arsenic detoxification and is conserved in Gram-negative bacteria. J Bacteriol 177:2050-2056

Distefano MD, Moore MJ, Walsh CT (1990) Active site of mercuric reductase resides at the subunit interface and requires Cys_{135} and Cys_{140} from one subunit and Cys_{558} and Cys_{559} from the adjacent subunit: evidence from in vivo and in vitro heterodimer formation. Biochemistry 29:2703-2713

Fraústo da Silva JJR, Williams RJP (1991) The Biological Chemistry of the Elements. The Inorganic Chemistry of Life. Clarendon Press, Oxford, 561 p

Gladysheva TB, Oden KL, Rosen BP (1994) Properties of the arsenate reductase of plasmid R773. Biochemistry 33:7288-7293

Gupta A, Whitton BA, Morby AP, Huckle JW, Robinson NJ (1992) Amplification and rearrangement of a prokaryotic metallothionein locus *smt* in *Synechococcus* PCC 6301 selected for tolerance to cadmium. Proc Roy Soc Lond B248:273-281

Gupta A, Morby AP, Turner JS, Whitton BA, Robinson NJ (1993) Deletion within the metallothionein locus of cadmium-tolerant *Synechococcus* PCC 6301 involving a highly iterated palindrome (HIP1). Molec Microbiol 7:189-195

Ji G, Garber EA, Armes LG, Chen C-M, Fuchs JA, Silver S (1994) Arsenate reductase of *Staphylococcus aureus* plasmid pI258. Biochemistry 33:7294-7299

Ji G, Silver S (1992) Regulation and expression of the arsenic resistance operon from *Staphylococcus aureus* plasmid pI258. J Bacteriol 174:3684-3694

Ji G, Silver S (1992) Reduction of arsenate to arsenite by the ArsC protein of the arsenic resistance operon of *Staphylococcus aureus* plasmid pI258. Proc Natl Acad Sci 89:7974-7978

Ji G, Silver S (1995) Plasmid resistance mechanisms for heavy metals of environmental concern. J Industr Microbiol 14:61-75

Kaur P, Rosen BP (1993) Complementation between nucleotide binding domains in an anion-translocating ATPase. J Bacteriol 175:351-357

Koonin EV (1993) A superfamily of ATPases with diverse functions containing either classical or deviant ATP-binding motif. J Molec Biol 229:1165-1174

Kuroda M, Dey S, Sanders OI, Rosen BP (1997) Alternate energy coupling of ArsB, the membrane subunit of the Ars anion-translocating ATPase. J Biol Chem 272:326-331

Levinson HS, Mahler I, Blackwelder P, Hood T (1996) Lead resistance and sensitivity in *Staphylococcus aureus*. FEMS Microbiol Letters 145:421-425

Lippard SJ, Berg JM (1994) Principles of Bioinorganic Chemistry. University Science Books, Mill Valley, California

Macy JM, Nunan K, Hagen KD, Dixon DR, Harbour PJ, Cahill M, Sly LI (1996) *Chrysiogenes arsenatis* gen. nov., sp. nov., a new arsenate-respiring bacterium isolated from gold mine wastewater. Intl J System Bacteriol 46:1153-1157

Melchers K, Weitzenegger T, Buhmann A, Steinhilber W, Sachs G, Schäfer KP (1996) Cloning and membrane topology of a P type ATPase from *Helicobacter pylori*. J Biol Chem 271:446-457

Miller CE, Wuertz S, Cooney JJ, Pfister RM (1995) Plasmids in tributyltin-resistant bacteria from fresh and estuarine waters. J Industr Microbiol 14:337-342

Nakamura K., Silver S (1994) Molecular analysis of mercury-resistant *Bacillus* isolates from sediment of Minamata Bay, Japan. Appl Environ Microbiol 60:4596-4599

Nies A, Nies DH, Silver S (1990) Nucleotide sequence and expression of a plasmid-encoded chromate resistance determinant from *Alcaligenes eutrophus*. J Biol Chem 265:5648-5653

Nies DH (1995) The cobalt, zinc, and cadmium efflux system CzcABC from *Alcaligenes eutrophus* functions as a cation-proton antiporter in *Escherichia coli*. J Bacteriol 177:2707-2712

Nies DH, Nies A, Chu L, Silver S (1989) Expression and nucleotide sequence of a plasmid-determined divalent cation efflux system from *Alcaligenes eutrophus*. Proc Natl Acad Sci 86:7351-7355

Nies DH, Silver S (1995) Ion efflux systems involved in bacterial metal resistances. J Industr Microbiol 14:186-199

Odermatt A, Suter H, Krapf R, M Solioz (1993) Primary structure of two P-type ATPases involved in copper homeostasis in *Enterococcus hirae*. J Biol Chem 268:12775-12779

Odermatt A, Suter H, Krapf R, Solioz M (1994) Induction of the putative copper ATPases, CopA and CopB, of *Enterococcus hirae* by Ag^+ and Cu^{2+}, and Ag^+ extrusion by CopB. Biochem Biophys Res Commun 202:44-48

Odermatt A, Solioz M (1995) Two *trans*-acting metalloregulatory proteins controlling expression of the copper-ATPases of *Enterococcus hirae*. J Biol Chem 270:4349-4354

Ohtake H, Cervantes C, Silver S (1987) Decreased chromate uptake in *Pseudomonas fluorescens* carrying a chromate resistance plasmid. J Bacteriol 169:3853-3856

Ohtake H, Silver S (1994) Bacterial reduction of toxic chromate. In: Chaudhry GR (ed) Biological Degradation and Bioremediation of Toxic Chemicals, p 403-415 Chapman and Hall, London

Quinn JP, McMullan G (1995) Carbon-arsenic bond cleavage by a newly isolated Gram-negative bacterium, strain ASV2. Microbiology 141:721-727

Rawlings DE, Silver S (1995) Mining with microbes. Bio/Technology 13:773-778

Rochelle PA, Wetherbee MK, Olson BH (1991) Distribution of DNA sequences encoding narrow- and broad-spectrum mercury resistance. Appl Environ Microbiol 57:1581-1589

Rosenstein R, Nikoleit K, Götz F (1994) Binding of ArsR, the repressor of the *Staphylococcus xylosus* (pSX267) arsenic resistance operon to a sequence with dyad symmetry within the *ars* promoter. Molec Gen Genet 242:566-572

Rosenstein R, Peschel A, Wieland B, Götz F (1992) Expression and regulation of the antimonite, arsenite, and arsenate resistance operon of *Staphylococcus xylosus* plasmid pSX267. J Bacteriol 174:3676-3683

Rouch DA, Brown NL (1997) Copper-inducible transcriptional regulation from two promoters in the *Escherichia coli* copper resistance determinant *pco*. Microbiology 143:1191-1202

Schiering N, Kabsch W, Moore MJ, Distefano MD, Walsh CT, Pai EF (1991) Structure of the detoxification catalyst mercuric ion reductase from *Bacillus* sp. strain RC607. Nature 352:168-171

Schottel J, Mandal A, Clark D, Silver S, Hedges RW (1974) Volatilisation of mercury and organo-mercurials determined by inducible R-factor systems in enteric bacteria. Nature 251:335-337

Silver S (1996a) Transport of inorganic cations. In: Neidhardt FC et al. (eds) *Escherichia coli* and *Salmonella typhimurium*: Cellular and Molecular Biology, 2nd ed. p 1091-1102 ASM Press, Washington, DC

Silver S (1996b) Bacterial resistances to toxic metal ions. Gene 179:9-19

Silver S, Ji G (1994) Newer systems for bacterial resistances to toxic heavy metals. Environ Health Perspect 102 (Suppl 3):107-113

Silver S, Nucifora G, Phung LT (1993) Human Menkes X chromosome disease and the staphylococcal cadmium resistance ATPase: a remarkable similarity in protein sequences. Molec Microbiol 10:7-12

Silver S, Phung LT (1996) Bacterial heavy metal resistance: new surprises. Ann Rev Microbiol 50:753-789

Silver S, Endo G, Nakamura K (1994) Mercury in the environment and laboratory. J Japan Soc Water Environ 17:235-243

Silver S, Walderhaug M (1995) Bacterial plasmid-mediated resistances to mercury, cadmium and copper. In: Aspects Goyer RA, Cherian MG (eds) Toxicology of Metals. Biochemical, p 435-458 Springer-Verlag, Berlin

Sofia HJ, Burland V, Daniels DL, Plunkett G III, Blattner FR (1994) Analysis of the *Escherichia coli* genome. V. DNA sequence of the region from 76.0 to 81.5 minutes. Nucleic Acids Res 22:2576-2586

Solioz M, Odermatt A (1995) Copper and silver transport by CopB-ATPase in membrane vesicles of *Enterococcus hirae*. J Biol Chem 270:9217-9221

Solioz M, Vulpe C (1996) CPx-type ATPases: a class of P-type ATPases that pump heavy metals. Trends Biochem Sci 21:237-241

Summers AO (1992) Untwist and shout: a heavy metal-responsive transcriptional regulator. J Bacteriol 174:3097-3101

Summers AO, Silver S (1972) Mercury resistance in a plasmid-bearing strain of *Escherichia coli*. J Bacteriol 112:1228-1236

Takemaru K-I, Mizuno M, Sato T, Takeuchi M, Kobyashi Y (1995) Complete nucleotide sequence of a *skin* element excised by DNA rearrangement during sporulation in *Bacillus subtilis*. Microbiology 141:323-327

Tisa LS, Rosen BP (1990) Molecular characterization of an anion pump: the ArsB protein is the membrane anchor for the ArsA protein. J Biol Chem 265:190-194

Turner JS, Glands PD, Sampson ACR, Robinson NJ (1996) Zn^{2+}-sensing by the cyanobacterial metallothionein repressor SmtB: different motifs mediate metal-induced protein-DNA dissociation. Nucleic Acids Res 24:3714-3721

Turner JS, Robinson NJ (1995) Cyanobacterial metallothioneins: biochemistry and molecular genetics. J Industr Microbiol 14:119-125

Walter EG, Taylor DE(1992) Plasmid-mediated resistance to tellurite: expressed and cryptic. Plasmid 27:52-64

Wu J, Rosen BP (1993) Metalloregulated expression of the *ars* operon. J Biol Chem 268:52-58

Wu J., Tisa LS, Rosen BP (1992) Membrane topology of the ArsB protein, the membrane subunit of an anion-translocating ATPase. J Biol Chem 267:12570-12576

Xu C, Rosen BP (1997) Dimerization is essential for DNA binding and repression by the ArsR metalloregulatory protein of *Escherichia coli*. J Biol Chem 272:15734-15738

Chapter 11

GEOMICROBIOLOGY OF
SULFIDE MINERAL OXIDATION

D. Kirk Nordstrom

U.S. Geological Survey
3215 Marine Street
Boulder, Colorado 80303 U.S.A.

Gordon Southam

Department of Biological Sciences
Northern Arizona University
Flagstaff, Arizona 86011 U.S.A.

INTRODUCTION

Current estimates on the global flux of sulfur (Berner and Berner 1996) indicate that the amount of sulfate from pyrite weathering, contributing to the average river concentration of sulfate, is about 11% of the total and that this amount is increasing because of increased mining activities. Of the rock-derived sulfate contributions to river sulfate, about two-thirds comes from evaporite bed dissolution (mostly gypsum) and about one-third comes from pyrite oxidation. The occurrence of acid mine drainage, caused by pyrite oxidation, is a major water quality problem throughout the world. Acid mine waters typically have pH values in the range of 2 to 4 and high concentrations of metals known to be toxic to aquatic organisms (Ash et al. 1951, Barton 1978, Kelly 1988, Martin and Mills 1976, Nordstrom and Ball 1985). Acid sulfate soils, most often occurring in lagoonal and estuarine environments affected by tidal cycles, cause agricultural problems from the periodic oxidation of pyrite (Van Breemen 1976, Pons et al. 1982). Occasionally the construction of highways or building structures involves excavation into pyritiferous rock that requires specially-designed engineering features to avoid rapid deterioration from the acid waters that can rapidly develop (Byerly 1996). The same process is also very important in the weathering and enrichment of sulfide ore deposits (Lindgren 1928). Biohydrometallurgists capitalize on the microbial catalysis of sulfide mineral oxidation to increase the efficiency of metal extraction (Murr et al. 1978, Lawrence et al. 1986, Barrett et al. 1993). Hence, the geomicrobiology of sulfide mineral oxidation has important ramifications from the global cycling of elements to the industrial extraction of metals.

The association of microorganisms with pyrite oxidation and the formation of acid mine drainage has a long history that begins with the discovery that microbes can utilize inorganic compounds as well as organic compounds such as metabolites. S.N. Winogradsky (1888) recognized that some microbes could derive their metabolic energy from the oxidation of inorganic compounds, such as iron and sulfur (Sokolova and Karavaiko 1968). This property affords one of the major divisions among microorganisms: the lithotrophs, e.g. *Thiobacillus* spp., that gain energy from the oxidation of inorganic compounds, and the heterotrophs (like ourselves) who gain energy from the oxidation of organic compounds. Autotrophs obtain their carbon requirements for growth through CO_2 fixation. Thus, *Thiobacillus* spp. have also been called lithoautotrophs. Occasionally the word chemoautotroph is used for lithoautotroph. Organisms that receive solar radiation for

0275-0279/97/0035-0011$05.00

their energy source and obtain cellular carbon from CO_2 fixation are known as photoautotrophs.

Nathanson (1902) first isolated a member of the bacterial genus *Thiobacillus*, noted for its ability to oxidize sulfur. The acidophilic bacterium, *Thiobacillus thiooxidans*, was isolated and identified by Waksman and Jåffe (1921, 1922) from soils containing free sulfur and phosphate. Rudolfs and Helbronner (1922) observed that bacteria could attack zinc sulfide. Twenty-five years later, the acidophilic autotroph *Thiobacillus ferrooxidans*, was isolated by Colmer and Hinkle (1947), implicated in the formation of acid mine drainage, and was found to oxidize both iron and sulfur. Numerous members of the *Thiobacillus* genus have now been identified and the sulfur compounds they utilize are shown in Table 1. Other acidophilic iron- and sulfur-oxidizers that have been found or implicated in the formation of acid mine waters are also shown in Table 1.

Acid mine waters have been known to contain abundant microbial life; indeed, they are often the only form of life under these conditions. Powell and Parr (1919) and later Carpentor and Herndon (1933) suggested that pyrite oxidation and the consequent acid

Table 1. Members of the Bacteria genera *Thiobacillus* (arranged alphabetically by species name), *Leptospirillum*, and *Sulfobacillus*; the inorganic substances they utilize (adapted from Kelley and Harrison 1984 and Barrett et al. 1993, except where otherwise indicated), and four additional Archaea spp. that are known to be associated with acid mine waters and pyrite oxidation. Acidophilic species are shown in bold type.

Thiobacillus albertis	H_2S, $S_2O_3^{2-}$
***Thiobacillus acidophilus*[1]**	$S°$, $S_2O_3^{2-}$, $S_3O_6^{2-}$, $S_4O_6^{2-}$
Thiobacillus denitrificans	H_2S, $S°$, $S_2O_3^{2-}$, $S_4O_6^{2-}$
Thiobacillus delicatus	$S°$, $S_2O_3^{2-}$, $S_4O_6^{2-}$
Thiobacillus ferrooxidans	H_2S, sulfide minerals, $S°$, $S_2O_3^{2-}$, $S_4O_6^{2-}$, Fe^{2+}
Thiobacillus halophilus[2]	$S°$
Thiobacillus intermedius	$S°$, $S_2O_3^{2-}$, $S_4O_6^{2-}$
Thiobacillus neapolitanus	H_2S, sulfide minerals, $S°$, $S_2O_3^{2-}$, $S_3O_6^{2-}$, $S_4O_6^{2-}$
Thiobacillus novellus	$S_2O_3^{2-}$, $S_4O_6^{2-}$
Thiobacillus perometabolis	$S°$, $S_2O_3^{2-}$, $S_4O_6^{2-}$
Thiobacillus tepidarius	H_2S, $S°$, $S_2O_3^{2-}$, $S_3O_6^{2-}$, $S_4O_6^{2-}$
Thiobacillus thermophilica[3]	H_2S, sulfide minerals, $S°$
Thiobacillus thiooxidans	$S°$, $S_2O_3^{2-}$, $S_4O_6^{2-}$
Thiobacillus thioparus[4]	H_2S, sulfide minerals, $S°$, $S_2O_3^{2-}$, $S_3O_6^{2-}$, $S_4O_6^{2-}$
Thiobacillus versutus	H_2S, $S_2O_3^{2-}$
Leptospirillum ferrooxidans	Fe^{2+}, sulfide minerals
Leptospirillum thermoferrooxidans	Fe^{2+}, sulfide minerals
Sulfobacillus thermosulfidooxidans	Fe^{2+}, $S°$, sulfide minerals
Archaea spp.	
Acidianus brierleyi	Fe^{2+}, $S°$, sulfide minerals
Sulfolobus solfataricus	$S°$
Sulfolobus ambivalens	$S°$
Sulfolobus acidocaldarius	Fe^{2+}, $S°$

[1] Also known as *T. organoparus*
[2] Wood and Kelly, 1991
[3] Egorova and Deryugina, 1963 (not a *Thiobacillus* sp.)
[4] Range of pH = 3 to 10

mine drainage from coal deposits may be catalyzed by bacteria. Lackey (1938) investigated 62 West Virginia streams affected by acid mine drainage, described acid slime streamers, and found flagellates, rhizopods, ciliates, and green algae. In his investigation of acidified surface waters and soils in West Virginia and Pennsylvania, Joseph (1953) observed Gram-positive and Gram-negative bacilli and cocci, actinomycetes, fungi, green algae, diatoms. In acid mine waters from a copper mine in the southwestern United States, Ehrlich (1963a) found yeasts, flagellates, and amoebas. Acid slime streamers of the type described by Lackey (1938) are a common occurrence (Temple and Koehler 1954, Nordstrom 1977, Johnson et al. 1979) and have been described in further detail by Dugan et al. (1970).

One of the most widely studied autotrophs and a key bacterium in the catalysis of pyrite oxidation is *Thiobacillus ferrooxidans*. It was isolated and identified as Gram-negative, acidophilic, and rod-shaped by Colmer and Hinkle (1947), Colmer et al. (1950), Temple and Colmer (1951), and Temple and Delchamps (1953) and shown to be essential to the production of acid mine waters (Leathen et al. 1953). Following these discoveries, there has been considerable research on both the abiotic and the microbially-catalyzed reactions for sulfide mineral oxidation. By the 1960s, the essential role of bacteria in the oxidation of pyrite, especially in coal, had been well-established (Lyalikova 1960, Silverman et al. 1961).

Two principal activities have motivated studies on microbial catalysis of sulfide minerals: the need to eliminate or reduce the deleterious effects of acid mine drainage and the advantages gained in biohydrometallurgy when bacteria are used to improve the extraction of metals from ores and mine waste materials. The reactions and the microbial ecology are complicated, rates and mechanisms are difficult to resolve, and the interactions of bacteria at sulfide surfaces are poorly understood. In this chapter we attempt to summarize the current state of knowledge of sulfide mineral oxidation with a focus on the role of bacteria in this important process.

SUMMARY OF SULFIDE MINERAL OXIDATION REACTIONS

The three basic ingredients responsible for the formation of acid mine waters are pyrite, water, and oxygen. The overall reaction is commonly written as

$$FeS_2 + 15/4 \ O_2 + 7/2 \ H_2O \Rightarrow Fe(OH)_3 + 2 \ H_2SO_4 \tag{1}$$

where one mole of ferric hydroxide and two moles of sulfuric acid are produced for every mole of pyrite oxidized. This reaction, however, is extraordinarily simplistic compared to the realities of chemistry and the actual processes occurring in the environment. Some of the complicating features are tabulated in the following list:

(1) Pyrite occurs over a large range of grain size and surface area. It also occurs in several different crystal forms and with large variations in defect structure, crystallinity, and trace element content. All of these properties of pyrite can affect the rate of reaction.

(2) For each mole of pyrite oxidized in Reaction (1), 1 electron is lost by oxidation of iron, 14 electrons are lost by oxidation of disulfide, and 15 electrons are gained by the reduction of oxygen. In addition, the iron is hydrolyzed and precipitated. These reactions cannot take place in a single step. Only one or two electrons are commonly transferred at a time (Basolo and Pearson 1967). Hence, there could be 15 or more reactions with as many possible rate-determining steps to consider. To further complicate matters, other oxidizing agents have been implicated, chiefly ferric iron.

Fortunately all the intermediate reactions need not be determined to delineate the rate-controlling mechanisms involved with pyrite oxidation.

(3) The product of pyrite oxidation is not a pure ferric hydroxide phase but a mixture of phases with variable stoichiometry, including:

goethite (α-FeOOH),   ferrihydrite   ($Fe_5OH_8 \cdot 4H_2O$),
jarosite ($KFe_3(SO_4)_2(OH)_6$),   schwertmannite ($Fe_8O_8SO_4(OH)_6$).

(4) Products, such as elemental sulfur, have been found to form from pyrite oxidation.

Important reviews on pyrite oxidation have been published by Lowson (1982), Nordstrom (1982), and Evangelou (1995). Reviews on the oxidation of pyrite and other mineral sulfides can be found in Dutrizac and MacDonald (1974) and Nordstrom and Alpers (1997). This section highlights the major findings from these reviews and related current research.

Stokes (1901) recognized that the oxidation of pyrite by ferric iron had been known for a long time. The following balanced chemical reaction

$$FeS_2 + 14\ Fe^{3+} + 8\ H_2O \Rightarrow 15\ Fe^{2+} + 2\ SO_4^{2-} + 16\ H^+ \tag{2}$$

has been confirmed by several experimental studies (Garrels and Thompson 1960, Wiersma and Rimstidt 1984, McKibben and Barnes 1986). Since ferric iron is insoluble at circumneutral pH values, Reaction (2) requires acidic conditions. Oxygen is still needed to oxidize more ferric iron from ferrous:

$$Fe^{2+} + 1/4\ O_2 + H^+ \Rightarrow Fe^{3+} + 1/2\ H_2O \tag{3}$$

but the oxygen does not have to diffuse to the pyrite surfaces. Pyrite can oxidize in the absence of oxygen. Nevertheless, the overall rate of pyrite oxidation in a tailings pile, in a mine, or in a waste rock pile will largely be determined by the overall rate of oxygen transport (advection, convection, and diffusion).

Elemental sulfur may form during pyrite oxidation, but the mechanism is not well understood. At ambient environmental temperatures, little or no sulfur forms. The formation of sulfur during pyrite oxidation increases with temperature. Bergholm (1955) found that detectable sulfur appeared at 60°C and Lowson (1982) notes that sulfur formation reaches a maximum at 100°-150°C. Both authors also found that low pH, longer run times, increased ferric iron concentration, and increased oxygen concentration increase the production of sulfur.

When pyrite is attacked by acid solutions, the iron is easily leached from the surface layer, leaving a sulfur-rich surface. Buckley and Woods (1987) reviewed the X-ray photoelectron spectroscopy (XPS) literature on the surface oxidation of pyrite and discussed the effect of experimental techniques on the results. They also concluded from their own work that iron tends to be easily leached from the surface producing an iron-deficient (or sulfide-rich) surface that does not have the properties of elemental sulfur unless prolonged strong acid attack was used. Several other studies, using XPS and Raman spectroscopy, have confirmed the initial dissolution under acidic conditions first releases iron and forms a layer containing disulfide, monosulfide, and polysulfides (Sasaki 1994, Nesbitt and Muir 1994, Sasaki et al. 1995). With continued oxidation monosulfide decreases, disulfide increases proportionally, polysulfides increase, and eventually thiosulfate and sulfate form (Nesbitt and Muir 1994). The formation of the disulfide species corroborates the early work of Sato (1960) on the mechanism of oxidation of sulfide ore bodies. The degradation of this surface by either ferric iron or microorganisms, or both, is a critical and poorly understood aspect of pyrite oxidation. As Norris (1990) and Sasaki et al. (1995) mention, this sulfur-rich surface should be energetically favorable for *Thiobacillus thiooxidans* and may explain their association with *T. ferrooxidans*.

The initial oxidation and acid leaching of other sulfide minerals have also been subject to surface spectroscopy. Several studies on both air and air-saturated aqueous solutions of pyrrhotite oxidation have shown the formation of a Fe(III) oxyhydroxide by diffusion of structural iron to the surface and oxidation (Buckley and Woods 1985, Pratt et al. 1994a,b; Mycroft et al. 1995). Underneath this layer is a sulfur-enriched layer that will eventually form marcasite (Mycroft et al. 1995). The oxidation of chalcopyrite in air also leads to iron migration to the surface and formation of an Fe(III) oxyhydroxide leaving a residual CuS_2 layer (Buckley and Woods 1984). Chalcopyrite oxidation in acid solution leads to the migration and dissolution of iron and a residual CuS_2 surface (Yin et al. 1995). At temperatures of 80° to 100°C, most of the sulfur from the oxidation of chalcopyrite appears as elemental sulfur (Dutrizac 1990). Arsenopyrite oxidation also forms Fe(III) oxy-hydroxides at the surface but arsenic also migrates to the surface and rapidly oxidizes and solubilizes when in contact with water (Nesbitt et al. 1995).

The formation of unstable sulfoxyanions of intermediate oxidation state between elemental sulfur and sulfate further complicates the chemical interpretation. One of the earliest reports was from Steger and Desjardins (1978) who found a thiosulfate compound on the surface of pyrite undergoing oxidation. Recently, a copper thiosulfate mineral has been reported from the oxidation of copper sulfides. Sulfoxyanion formation from pyrite oxidation at pH values of 6 to 9 was studied by Goldhaber (1983) who found thiosulfate, polythionate, and sulfite but only at high stirring rates. The proportions of sulfoxyanions were sensitive to the solution pH. Polythionates were dominant at low pH, thiosulfate was dominant at high pH, and some sulfite formed at the highest pH values. Moses et al. (1987) corroborated Goldhaber's results with improved analytical methods, but they found that when ferric iron replaced oxygen as the oxidant, sulfoxyanions were no longer detectable. These results implicate ferric iron as a powerful oxidant for all types of reduced sulfur species. Williamson and Rimstidt (1993) have demonstrated the rapid oxidation of thiosulfate by ferric iron.

The formation of oxidation products from sulfide minerals depends on the type and composition of the mineral. The first simple distinction is between the metal monosulfides (such as sphalerite, ZnS, greenockite, CdS, covellite, CuS, pyrrhotite (var. troilite), FeS, galena, PbS, and millerite, NiS) and the disulfides (such as pyrite and marcasite, FeS_2, and arsenopyrite, $FeAsS^1$). Minerals such as molybdenite, MoS_2, and chalcopyrite, $CuFeS_2$ are not true disulfides because the sulfur atoms are bonded to metal atoms and not to other sulfur atoms. The monosulfides react with acid to form H_2S (Parsons and Ingraham 1970), the reactivity increasing with increasing solubility. Upon contact with air, H_2S will rapidly oxidize to form sulfur, thiosulfate, sulfite, and, ultimately but more slowly, sulfate. The disulfides, however, do not form H_2S, upon reaction with acids. Disulfides contain sulfur with an oxidation state of S(-I) intermediate between S(-II) and S(0) and a tendency to form both elemental sulfur and thiosulfate upon oxidative attack in acid solutions, depending on pH, concentration of oxidants, and temperature. Ahonen et al. (1986) have shown that pyrrhotite forms sulfur during acid dissolution at 28°C and pyrite does not. The sulfur formed from oxidation of H_2S. This distinction is fundamental to the nature of acid weathering reactions for sulfide minerals. The occurrence of hydrogen sulfide, sulfur, and thiosulfate in mine waste environments indicate either sulfate reduction or acid dissolution of monosulfides. Viable, dissimilatory sulfate-reducing bacteria have been found within the saturated zone of acidic mine tailings (Fortin et al. 1995). Stable isotope compositions of sulfur should be definitive in distinguishing these sources. Thiosulfate, sulfur, sulfite, and polythionates are not likely to be found from disulfide oxidation because the autocatalytic

---

[1] An isostructural solid solution exists between pyrite and löllingite ($FeAs_2$) with arsenopyrite as a stable intermediate at exactly 0.5 mole fraction of arsenic substituted for sulfur.

oxidation on a semi-conducting surface such as pyrite is much too rapid. The results from Goldhaber's experiments cannot be reproduced without using very fast stirring rates and high shearing force at the mineral surface. The results of thiosulfate formation from pyrite oxidation in the column leaching experiments of Granger and Warren (1969) were faulty (Nordstrom and Alpers 1997) because they contaminated their columns with sodium sulfide solutions. Dissolved sulfide rapidly forms thiosulfate when exposed to the air.

Galvanic protection has been postulated to affect sulfide mineral oxidation rates during the weathering of mineral deposits (Sveshnikov and Dobychin 1956, Sveshnikov and Ryss 1964). The more electroconductive the metal sulfide, the slower its rate of oxidation whereas the less conductive sulfides exhibit increased reaction rates as long as they are in electronic contact with the more conductive sulfide. Kwong (1995) has demonstrated this process and has shown that relative rates can be predicted according to the sequence of standard electrode potentials as compiled by Sato (1992). Furthermore, Kwong et al. (1995) have shown that although bacterially-mediated oxidation increases the oxidation rates, the electrochemical sequence remains the same.

Compilation of sulfide mineral oxidation studies

We have compiled 29 minerals for which oxidation studies have been done at temperatures below 50°C either with or without microbial participation. This list is based primarily on the reviews by Dutrizac and MacDonald (1974a), Ehrlich (1996), and Nordstrom and Alpers (1997). More information and references can be found in these reviews. The following compilation has been subdivided into four tables (Tables 2a-d) in which the iron sulfides are found in Table 2a, copper sulfides and selenides in Table 2b, arsenic and antimony sulfides in Table 2c, and the remaining sulfides in Table 2d. They have also been placed in alphabetical order within each table. A few comments are needed here. Of course, many more studies have been done on minerals like pyrite than can be referenced in this paper. The studies with valuable rate information have been preferentially included. The references themselves are given in shorthand notation and decoded at the end of all the tables. In most of these studies the surface areas were not measured. Furthermore, the effect of ferric iron concentrations were not always obtained nor distinguished from microbial action. All the necessary reaction products were usually not determined so that the stoichiometry is not always known. In that regard, Ehrlich (1964) found the pH to decrease during enargite oxidation by *T. ferrooxidans* but it increased in the abiotic control. The chemical interpretation would suggest that copper was solubilized and oxidized in both solutions but that the sulfur and arsenic were oxidized only in the presence of bacteria.

Most of the microbial studies included a sterile control but were not necessarily cited in the abiotic references. Qualitative to quantitative data is commonly available and show that microbes increase the rate of reaction when compared to the sterile control but the cause and mechanism are unclear. Many of the studies used impure minerals. These qualitative experiments were, of course, done to improve metal extraction from minerals, not to provide quantitative rate information.

Role of bacteria

Although the association of microbes with sulfide mineral oxidation has been known for many decades, it has taken a considerable amount of research to determine the rates, mechanisms, and metabolic pathways. Discussion has focused on whether bacteria, primarily the genus *Thiobacillus*, degrade sulfides by a *direct* or an *indirect oxidation* mechanism (Silverman 1967). During direct oxidation, the bacteria attach themselves onto the sulfide mineral surface and directly solubilize the surface through hypothesized enzymatic oxidation reactions. Indirect oxidation occurs through microbial catalysis of aqueous ferrous to ferric ion oxidation (Reaction 3) and then direct oxidation of the sulfide by the ferric ion (Reaction 2). Adsorption of either *Thiobacillus ferrooxidans* or *Sulfolobus acidocaldarius* to pyrite surfaces is rapid (Bagdigian and Myerson 1986, Chen and Skidmore 1987, 1988). Surface etch patterns that reflect bacterial attachment have been

Table 2a. Oxidation studies on iron sulfide minerals.

Mineral / Oxidant	Formula	Abiotic reference	Microbial reference
marcasite/Fe³⁺	FeS_2	WR84, R94	--
pyrite/O₂	FeS_2	B54, BJ58, G83, MB86, M87, MH91, N94, O91	B54, BJ58, SE64, O91, P91
pyrite/Fe³⁺	FeS_2	MB86, BJ89, M87, MH91, R94	--
pyrrhotite/O₂	$Fe_{1-x}S$	TC76, NS94	P91, B93

Table 2b. Oxidation studies of copper sulfides and selenides.

Mineral	Formula	Abiotic reference	Microbial reference
bornite	Cu_5FeS_4	S31, KO63, L70, D70a, KO69, D70b, U67	B54, L66
carrollite	Co_2CuS_4	DC64	DC64
chalcopyrite	$CuFeS_2$	S33, BS34, I62, R94	BA57, I61, D64, T76
chalcocite	Cu_2S	S30a, T67, KO69, M69,	B54, RT63, NB72, S76, B77
covellite	CuS	S30b, TI67, M71, DM74b, WR86	S76, RV78
cubanite	$CuFe_2S_3$	D70c	--
digenite	Cu_9S_5	T67	--
klockmannite	$CuSe$	--	TH72
copper-selenide	Cu_2Se	K66	--
copper-telluride	Cu_2Te	K66	--

Table 2c. Oxidation studies of arsenic, antimony, and gallium sulfides.

Mineral	Formula	Abiotic reference	Microbial reference
arsenopyrite	FeAsS	R94	E64
cobaltite	CoAsS	--	SC61
enargite	Cu_3AsS_4	S33, BS34, KG52	E64
gallium sulfide	Ga_2S_3	--	T78
orpiment	As_2S_3	--	E63b
realgar	AsS	--	E63b
stibnite	Sb_2S_3	TT63	T74, TG77
tennantite	$Cu_{12}As_4S_{13}$	S33, BS34	--
tetrahedrite	$Cu_{12}Sb_4S_{13}$	BS34, Y80	B54

Table 2d. Oxidation studies of other metal sulfides.

Mineral	Formula	Abiotic reference	Microbial reference
cinnabar	HgS	P39, B75	B89
cobalt sulfide	CoS	--	T71, T74
galena	PbS	H70, R94	ST74, TS74
greenockite	CdS	--	B71, T74
millerite	NiS	--	RT63, DT64, T71, T74
molybdenite	MoS$_2$	U52,	BA57, BJ58, B65, BM73
pentlandite	(Fe, Ni)$_9$S$_8$	--	DM74c
sphalerite	ZnS	R94	RH22, I61, MP61, T72, KR77

Table 2e. References on oxidation studies of sulfide minerals.

B54	Bryner et al (1954)	I62	Ichikuni (1962)	S30	Sullivan (1930b)
B65	Bhappu et al. (1965)	K66	Kholmanskikh et al. (1966)	S31	Sullivan (1931)
B71	Brissette et al. (1971)	KG52	Koch and Grasselly (1952)	S33	Sullivan (1933)
B75	Burkstaller et al. (1975)	KO63	Kopylov and Orlov (1963)	S76	Sakaguchi et al. (1976)
B77	Beck (1977)	KO69	Kopylov and Orlov (1969)	SC61	Sutton and Corrick (1961)
B89	Baldi et al. (1989)	KR77	Khalid and Ralph (1977)	SE64	Silverman and Ehrlich (1964)
B93	Bhatti et al. (1993)	L66	Landesman et al. (1966)	ST74	Silver and Torma (1974)
BA57	Bryner and Anderson (1975)	L70	Lowe (1970)	T67	Thomas et al. (1967)
BJ58	Bryner and Jamieson (1958)	M69	Mulak (1969)	T71	Torma (1971)
BJ89	Brown and Jurinak (1989)	M71	Mulak (1971)	T72	Torma et al. (1972)
BL77	Brierley and Le Roux (1977)	M87	Moses et al. (1987)	T74	Torma et al. (1974)
BM73	Brierley and Murr (1973)	MB86	McKibben and Barnes (1986)	T76	Torma et al. (1976)
BS34	Brown and Sullivan (1934)	MH91	Moses and Herman (1991)	T78	Torma (1978)
D64	Duncan et al. (1964)	MP61	Malouf and Prater (1961)	TC76	Tervari and Campbell (1976)
D70a	Dutrizac et al. (1970a)	N94	Nicholson (1994)	TG77	Torma and Gabra (1977)
D70b	Dutrizac et al. (1970b)	NB72	Nielson and Beck (1972)	TH72	Torma and Habashi (1972)
DC64	De Cuyper (1964)	NS94	Nicholson and Scharer (1994)	TI67	Thomas and Ingraham (1967)
DM74a	Dutrizac and MacDonald (1974b)	O91	Olson (1991)	TS74	Torma and Subramanian(1974)
DM74b	Dutrizac and MacDonald(1974c)	P39	Pande (1939)	TT63	Tugov and Tsyganov (1963)
DT64	Duncan and Trussell (1964)	P91	Pinka (1991)	U52	Usataya (1952)
E63b	Ehrlich (1963b)	R94	Rimstidt et al. (1994)	U67	Uchida (1967)
E64	Ehrlich (1964)	RH22	Rudolfs and Helbronner (1922)	WR84	Wiersma and Rimstidt (1984)
G83	Goldhaber (1983)	RT63	Razzell and Trussell (1963)	WR86	Walsh and Rimstidt (1986)
H70	Haver et al. (1970)	RV78	Rickard and Vanselow (1978)	Y80	Yakhontova et al. (1980)
I61	Ivanov et al. (1961)	S30a	Sullivan (1930a)		

observed (Bennett and Tributsch 1978). The actual mechanism of enzymatic oxidation is not entirely clear and is discussed further in the section on "Microbial Oxidation of Sulfide Minerals," below. Furthermore, it seems unnecessary to explain the rate data. We shall explore this controversy and the general interaction of bacteria with oxidizing sulfides by reviewing the empirical rate data for both ferrous to ferric oxidation and pyrite oxidation, other sulfide mineral oxidation studies, the microbial physiology of *Thiobacillus ferrooxidans*, and the microbial ecology of mine waste environments.

AQUEOUS Fe(II) OXIDATION KINETICS

The aqueous oxidation of ferrous to ferric ion has been firmly established for acid solutions where the rate is relatively slow. The rate increases with increasing pH but also becomes sensitive to oxygen and anionic ligand concentrations. At circumneutral pH values, the abiotic rate is so fast that bacterial catalysis has not been clearly demonstrated and it is not needed to explain the rate of aqueous iron oxidation. Singer and Stumm (1968, 1970a) showed that the rate increased rapidly with increasing pH above a value of about 4. Below this value the rate leveled out and became nearly independent of pH at about 3×10^{-12} mol L^{-1} s^{-1}. The presence of *T. ferrooxidans* increased this rate by five orders of magnitude to about 3×10^{-7} mol L^{-1} s^{-1} (Singer and Stumm 1970a,b). This catalysis is most remarkable in the enhancement of inorganic reactions and has a major effect on pyrite weathering. Measurements from microbial oxidation experiments, to study optimal growth of *T. ferrooxidans* on culture media, typically gave rates of 2.8-8.3 $\times$ 10^{-7} mol L^{-1} s^{-1} (Silverman and Lundgren 1959, Lundgren et al. 1964, Lacey and Lawson 1977, Noike et al. 1983). This agreement is quite good considering differences in experimental procedure and conditions.

Field determinations of the ferrous oxidation rate are very similar to those derived from laboratory experiments. Wakao et al. (1977) estimated a rate of 3×10^{-6} mol L^{-1} s^{-1} for an acid mine water in Japan but the stream velocities were not directly measured. Nordstrom (1985) measured stream velocities in some acid mine waters in California and found the oxidation rate to vary 2-8 $\times$ 10^{-7} mol L^{-1} s^{-1}. He found the rate in the same drainage to decrease after a rainstorm, presumably due to the flushing out and dilution of the bacterial population. For pH values of 2-4, the microbially-catalysed oxidation of ferrous iron generally has an average optimal rate of 5×10^{-7} mol L^{-1} s^{-1} whether in the field or in the lab using culture media (usually 9K medium, Silverman and Lundgren 1959) and where temperatures are about 10°-30°C. In the environment, however, ferrous iron oxidation rates are likely to be lower because of numerous environmental factors including temperature fluctuations, hydrologic conditions, nutrient limitations, and UV radiation for surface waters.

A brief note is needed here regarding the units and formalism for microbial reaction rates. In chemical kinetics, reaction rates are generally zero-order, first-order, second-order, or sometimes non-integral order. The rate equations for living entities such as bacteria are more complex than for the simpler entities of molecules and atoms. Growth cycles of microbes, expressed as population densities over time, usually follow a sigmoidal type of curve. During the early phase of growth at low cell counts, there is little change with time. The doubling process, however, is exponential (like radioactive decay in reverse) so that after a period of time with little apparent growth, known as the lag phase, there is rapid growth that can only be stopped by lack of an energy source, an essential nutrient, build-up of toxic products, or by a metabolic inhibitor. Nutrient concentrations can also become high enough that the growth rate becomes constant and independent of concentration. For a lithoautotroph like *T. ferrooxidans*, this means the ferrous iron oxidation rate can become independent of the ferrous iron concentration and the rate becomes zero order. Bacterial cell division by binary fission, is the separation of a parent cell into two daughter cells. During the lag phase and exponential growth phase, the growth is dependent on ferrous concentration and a first-order or pseudo-first-order reaction rate can been observed. The convention in microbiology is to treat this changing growth curve with the Michelis-Menten equation that contains both a first order and a zero-order term.

The rates we are using assume nutrient-saturated conditions and zero-order rates. Most lab studies on the microbial rate of ferrous iron oxidation have shown the rate to be zero-order for a wide range of ferrous concentrations (Schnaitman et al. 1969, Okereke and Stevens 1991). The rate of sulfide mineral oxidation will now be considered.

SULFIDE MINERAL OXIDATION KINETICS

The conventional approach to distinguish between the direct and indirect oxidation mechanisms would be to compare the rate of pyrite oxidation by iron-oxidizing bacteria with that obtained abiotically but in the presence of ferric ion. This analysis, however, is more complicated. It is known that acidophilic iron-oxidizing bacteria will generate Fe(III) from Fe(II). The question that should be addressed is: what are the relative rates among (1) oxidation of pyrite by Fe(III), (2) oxidation of Fe(II) by bacteria, and (3) oxidation of pyrite by bacteria independent of their oxidation of Fe(II)? If there is a direct mechanism of bacterial pyrite oxidation other than by regeneration of Fe(III) then for that rate to be significant, it would have to be as fast or faster than the rate of Fe(II) oxidation and there should be a plausible mechanism to explain it. Under those circumstances, the bacterial pyrite oxidation rate would have to be faster than the abiotic oxidation rate by Fe(III). With these concepts in hand, let's look at the evidence.

Nordstrom and Alpers (1997) compiled several reports on pyrite oxidation rates and summary results are shown in Table 3. The oxidation rate with ferric ion as the oxidant is faster than the rate with oxygen as the oxidant by an order of magnitude or more. Although these rates overlap when comparing different investigations, the relative rates within any one investigation show a consistently faster rate with Fe(III). Most investigators get a rate (at pH = 2, T = 25°C) close to 1×10^{-8} mol m^{-2} s^{-1} which would be less than 9×10^{-8} mol m^{-2} s^{-1} for the bacterial rate. These results would indicate an enhancement induced by microbial attachment. Unfortunately, part of this difference could be explained by sample preparation effects and experimental design problems. Furthermore, the microbial rate was obtained by assuming a surface area based on grain size which could be low by a factor of at least 2 or 3.

Table 3. Comparison of iron and pyrite oxidation rates.

Reaction	*Abiotic Rate*	*Microbial Rate*
Oxidation of Fe(II)	3×10^{-12} mol L^{-1} s^{-1}	5×10^{-7} mol L^{-1} s^{-1}
Oxidation of pyrite by O_2	$0.3 - 3 \times 10^{-9}$ mol m^{-2} s^{-1}	8.8×10^{-8} mol m^{-2} s^{-1}
Oxidation of pyrite by Fe(III)	$1 - 2 \times 10^{-8}$ mol m^{-2} s^{-1}	

The source of the bacterially-mediated pyrite oxidation rate study (Olson 1991) was an interlaboratory comparison implemented with sterile controls. About a 34-fold increase in rate was observed with *T. ferrooxidans* present compared to the sterile control and the control gave similar (but higher) rates to those from other studies of pyrite oxidation by oxygen. Paciorek et al. (1981) similarly found a 25-fold increase in the microbially-catalyzed pyrite oxidation rate relative to the abiotic control. Lizama and Suzuki (1989) found that the pyrite oxidation rate with *T. ferrooxidans* and Fe(III) was notably faster than the abiotic rate with Fe(III). These results, however, could be an artifact of the experimental design. When iron-oxidizing bacteria are oxidizing pyrite, they are also regenerating ferric from aqueous ferrous iron whereas the sterile control has a fixed initial ferric concentration that eventually gets used up so that the reaction rate slows down. This complication is difficult to avoid in experimental studies.

Olson (1991) reported his results in terms of the rate of formation of aqueous iron which was 12.4 mg Fe L^{-1} h^{-1} or about 6×10^{-8} mol L^{-1} s^{-1}. This rate is about one order of magnitude less than the rate of bacterial oxidation of Fe(II) suggesting an inhibition of pyrite oxidation by the bacteria. The inhibiting effect of adsorbed bacterial cells on the pyrite surface was reported by Wakao et al. (1984) who concluded that pyrite oxidation proceeded by the indirect mechanism through the growth of free-floating ferrous-iron-oxidizing bacteria. In the experiments of Wakao et al. (1984), a surfactant was used to desorb the cells from the pyrite surface and the oxidation rate then increased again to near its former level. Contradicting the experiments of Wakao et al. (1984), Arkestyn (1979) found the rate of pyrite oxidation to decrease when he separated *T. ferrooxidans* from the pyrite surfaces by a dialysis bag. The observation that iron-oxidizing bacteria can oxidize aqueous Fe(II) faster than they can oxidize pyrite may indicate inhibition by bacterial adsorption on surfaces or, more likely, it may indicate the rate of pyrite oxidation by Fe(III) is slower than the oxidation rate of Fe(II) by bacteria. This conclusion would be consistent with the general observation that aqueous chemical reactions, especially when catalyzed, are faster than heterogeneous chemical reactions.

A review by Sand et al. (1995) summarized recent literature and provided further evidence for the indirect oxidation mechanism. They pointed out that subculturing *T. ferrooxidans* in an iron-free salt solution resulted in complete loss of measurable substrate degradation. The bacterial cells could not function in the absence of iron. Further unpublished work of theirs showed that the iron in the bacterial cells could not be removed from their host, i.e. exopolymers, by washing. The extracellular polymeric substances (exopolymers) facilitate the attachment of the bacteria to the pyrite surface. Sand et al. (1995) also emphasized the importance of thiosulfate as the first soluble oxidized sulfur species and a key intermediate during the oxidation of pyrite. Thiosulfate is oxidized to tetrathionate and hydrolysis of tetrathionate produces disulfane sulfonate which further decomposes to form a variety of products. Further elucidation of sulfoxyanion formation during bioleaching of pyrite was reported by Schippers et al. (1996). They monitored the formation of thiosulfate, elemental sulfur, polythionates and sulfate during pyrite oxidation by both *T. ferrooxidans* and *L. ferrooxidans* in solutions with pH just less than 2. The appearance of polythionates is to be expected since both Fe(III) and pyrite can catalyze the decomposition of thiosulfate to polythionates abiotically. These two studies find no evidence for the direct mechanism and substantial evidence for the indirect mechanism.

The consistency and precision of the rate data indicates the abiotic pyrite oxidation rate by Fe(III) is slower than the conversion of Fe(II) to Fe(III) by bacteria and slower than the oxidation of pyrite with bacteria and Fe(III). However, several caveats should be noted. The cited aqueous Fe(II) oxidation rate is the maximum possible rate achievable with no limiting nutrients and at optimal temperatures. Microbes in the environment are nearly always growth-limited by some necessary nutrient, predatory relationships, hydrologic conditions, or by other factors. Hence, their environmental growth rate will be slower than the values quoted above and they could be less than and probably not greater than the rate of pyrite oxidation by Fe(III). Furthermore, the oxidation rate of pyrite in the environment will be controlled by bacterial growth rates and these, in turn, will be controlled by environmental factors such as temperature, hydrologic variables (supply of water and oxygen), and nutrient limitations. The search for the rate-determining step and mechanism in the formation of acid mine drainage may be over. What we have learned from all of the relevant experimental rate studies is that both the rates of aqueous Fe(II) oxidation and the oxidation of pyrite are fast and comparable in the presence of iron- and sulfur-oxidizing bacteria. The overall rate is governed by environmental factors that affect the growth rate of the bacteria.

The link between the chemistry and the biology of sulfide mineral oxidation can be expressed as follows:

(1) Bacteria want to be attached, or as close as possible, to sulfide mineral surfaces to maximize their efficiency of substrate utilization. The close adherence reduces the time needed for diffusion of iron between the mineral and the bacterium, it does not necessarily require a separate mechanism for sulfide mineral degradation. Attachment also changes the surface chemistry of the solid-microbe interface (see section below).

(2) All available evidence is consistent with a chemical oxidation of sulfide minerals with aqueous catalysis by iron- and sulfur-oxidizing bacteria.

(3) No other processes need to be invoked to explain the available data. This statement is another expression of Occam's razor.

MICROBIAL OXIDATION OF SULFIDE MINERALS

General physiology of *Thiobacillus ferrooxidans*

Thiobacillus ferrooxidans is typically known for its ability to oxidize reduced sulfur compounds and ferrous iron to produce sulfuric acid and ferric iron as by-products of its metabolism (Ingledew 1982, Harrison 1984). The general nutritional requirements of this bacterium are provided by carbon dioxide fixation (providing cellular carbon; Leathen et al. 1956), pyritic minerals as sources of energy (see Table 1) for ATP, ammonia, atmospheric nitrogen, or nitrate as its source of nitrogen (Temple and Colmer 1951, Lundgren et al. 1964, Mackintosh 1978, Stevens et al. 1986), and phosphate from acid solubilization of available phosphate minerals (Lipman and McLean 1916). Previous reports on the heterotrophic ability of *T. ferrooxidans* are actually a consequence of the presence of acidophilic heterotrophs found to co-exist with *T. ferrooxidans* (Harrison et al. 1980).

The formation of ATP occurs via a chemiosmostic mechanism (see Fig. 1). The impermeable nature of the cytoplasmic membrane to protons maintains a pH (pH ~ 2 vs. pH 6) based proton motive force, allowing protons to be transported only at sites where ATPase is anchored in the membrane (Ingledew et al. 1977). In addition to its role in general cell energetics, the ATP that is formed by this reaction is used in reverse electron transport to produce NADPH (Aleem et al. 1963) which is in turn used in the reductive assimilation (fixation) of CO_2 by the Calvin-Bensen cycle (Maciag and Lundgren 1964).

In natural systems that contain iron, *T. ferrooxidans* may exclusively oxidize $Fe^{2+} \Rightarrow Fe^{3+}$ to gain energy. This oxidation, yields a chemically reactive cation which can scavenge electrons from less electronegative metal species (e.g. pyrite, Eqn. 2; Mehta and Murr 1982) forming ferrous iron once again. The formation of reduced iron from the latter provides an effective bacterial electron carrier for sustained lithotrophy. Mineral dissolution then, is a chemical process enhanced by bio-catalysis (Singer and Stumm 1970, Keller and Murr 1982, Hutchins et al. 1986, Baldi et al. 1992).

The oxidation of iron yields low levels of energy. Silverman and Lundgren (1959) calculated 18.5 moles of iron need to be oxidized to assimilate 1 mole of carbon. These calculations assume 100% metabolic efficiency which is not possible in biological systems. For *Thiobacillus ferrooxidans*, metabolic efficiency has been found to be highly variable (3.2%, Temple and Colmer 1951; 4.8-10.6%, Beck and Elsden 1958; 20.5±4.3% Silverman and Lundgren 1959; 30%, Lyalikova 1958). The poor energy yield for carbon utilization from iron oxidation remains a puzzle.

In our model (Fig. 1) of iron oxidation in *T. ferrooxidans* adapted from Ingledew et al. (1977) and Blake et al. (1992) the biological oxidation of Fe(II) to Fe(III) and the

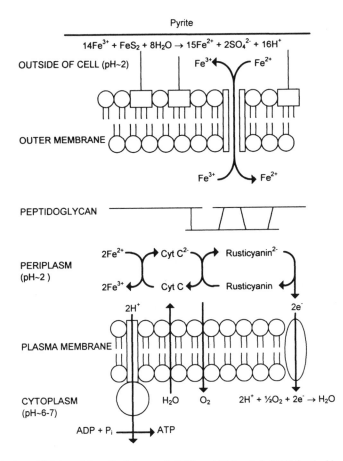

Figure 1. A model (adapted from Ingledew et al. 1977, and Blake et al. 1992) for the bioenergetics of iron oxidation by *T. ferrooxidans* based on a bidirectional diffusion gradient for Fe(II/III).

geochemical oxidation of pyrite by Fe(III) creates a bidirectional diffusion gradient which continuously drives Fe(II) into the periplasm and Fe(III) out of the periplasm. Iron will bind weakly (if at all) to the organic constituents on the cell surface and in the periplasm because they will be outcompeted by protons at pH 2. Other aspects of the model are the same as those described in the model of Blake et al. (1992).

Direct vs. indirect oxidation of non-ferrous sulfide minerals

The biooxidation of sulfide minerals presents a special problem when iron is not a component of the mineral(s). In these systems, the direct, enzymatic of susceptible mineral sulfides has been proposed in which each bacterial cell acts as a conductor of electrons from the crystal structure of chalcocite to oxygen (Ehrlich 1996). The transfer of constituent electrons from pyrite under acidic conditions has also been described by Wiersma and Rimstidt (1984). While these models are supported by an extremely close bacterial-mineral interaction (Bagdigian and Myerson 1986, Duncan and Drummond 1973, Bennet and Tributsch 1978, Norman and Snyman 1987, Southam and Beveridge 1992), the transfer of free electrons from the crystal structure only represents a partial reaction and cannot occur without an appropriate bacterial electron acceptor. The outer membrane (OM) of Gram-negative Bacteria, e.g. *T. ferrooxidans*, functions as a passive diffusion barrier (Beveridge 1981, 1988) and does not possess electron acceptors (Haddock and Jones 1977). Therefore, mineral sulfide oxidation must proceed via diffusion of reduced compounds across the outer membrane into the periplasm. Soluble ferrous iron will be present in iron-containing systems (described above; Fig. 1), acid dissolution of chalcocite releases soluble Cu^+ (Nielsen and Beck 1972) and nonbiological, autooxidation of covellite under acid conditions produces elemental sulfur (Rickard and Vanselow 1978) most likely from oxidation of H_2S formed during acid dissolution of a monosulfide,

$$CuS + 1/2\ O_2 + 2\ H^+ \Rightarrow S^0 + Cu^{2+} + H_2O \tag{4}$$

Uptake of this elemental sufur may be promoted by the production of surface active agents such as those described by Jones and Starkey (1961). The low Reynolds number of bacterial-sized particles dictates that diffusion of these soluble constituents away from the bacterial cell surface will be limited (Purcell 1977), promoting continued bacterial activity.

Thiobacillus spp.–mineral interaction in natural systems

Thiobacillus spp. can colonize naturally occurring sulfides (outcrops), low grade sulfide-containing ore (to promote acid leaching of base metals) and sulfide-bearing mine wastes (tailings) disposed of, at the earth's surface. The high surface:volume ratio created by grinding sulfides to extract base metals, creates an optimum environment for growth of *Thiobacillus* spp. For this reason, the following discussion focuses on the colonization of minerals surfaces in mine tailings environments.

The occurrence of metabolically diverse bacteria in the subsurface (Amy et al. 1996) down to at least 2.7 km (Boone et al. 1995), suggests that mineral sulfides may contain low populations of thiobacilli prior to mining. Whether or not this is true, mine tailings become quickly colonized with thiobacilli once discharged from the mill (Southam and Beveridge 1993). However, the mechanism of its colonization of tailings is not well understood (Dispirito et al. 1983).

When tailings are deposited as an aqueous slurry and allowed to drain, water is retained at the grain boundaries and in the pores between grains because of the capillary forces counteracting gravity (Nicholson et al. 1990). This area of partial water saturation, the vadose zone, provides all of the essential physical and chemical requirements for growth of *Thiobacillus* spp. In the vadose zone, the capillary border on the mineral surface supplies water to support life (i.e. protection against dessication which is deleterious to thiobacilli; Brock 1975), the pore spaces allow for the influx of gaseous oxygen (as terminal electron acceptor) and carbon dioxide (for carbon fixation) and the sulfide minerals serve as the substrate for lithotrophy (see Table 2). Drying at the surface of a sulfide tailings and freezing conditions reduce thiobacilli populations (Southam and Beveridge 1992) although near freezing conditions do not (Ahonen and Tuovinen 1992). Survival of thiobacilli during periodic wetting when microaerophilic to anaerobic environments may form (due to limited diffusion of oxygen into water saturated material) is presumably due to anaerobic metabolism (Pronk et al. 1992, Das and Mishra 1996).

Metallogenium has been described as an organism potentially responsible for the transition of mineral sulfide-rich environments from neutral to acid pH conditions (Walsh and Mitchell 1972). The proper identification of this bacterium is somewhat questionable and it is not required for this transition to occur. At the Copper Rand mine tailings, an individual *Thiobacillus ferrooxidans* subspecies (an LPS chemotype) was able to transform an alkaline pH (pH 8) tailings environment down to pH 3 (Southam and Beveridge 1993). At the Kidd Creek tailings pile, Blowes et al. (1995) demonstrated the predominant colonization with *Thiobacillus thioparus* under circumneutral pH conditions down to a pH of about 5 with *T. ferrooxidans* and *T. thiooxidans* becoming predominant at pH values of 4 and less. Measured pH is not, however, a unique or accurate indicator of viable populations of thiobacilli. Low populations of *T. ferrooxidans* can be recovered from the surface of acidic (pH < 3) mine tailings where evaporation produces low water activity which can inhibit or kill thiobacilli (Brock 1975, Southam and Beveridge 1992). Although the optimal pH for growth of *Thiobacillus* spp. is pH < 3 (Trafford et al. 1973, Amaro et al. 1991), viable thiobacilli can be cultured from tailings possessing neutral environmental pH values (see Fig. 2; Southam and Beveridge 1992). These thiobacilli must be employing acidic nanoenvironments because growth at neutral pH is not possible.

As mentioned previously and shown in Figure 3, growth on minerals is facilitated by close bacteria-mineral interaction (also see Sand et al. 1995). This close mineral interaction is important because the electron carriers responsible for iron oxidation and energy uptake are found in a gel-like (Hobot et al. 1984) periplasm (Blake et al. 1992, Harrison 1984, Ingledew et al. 1977). A close juncture will also promote the diffusion of soluble electron carriers e.g. soluble iron (Fig. 1) or Cu^+ (Nielsen and Beck 1972) or atomic % S (Rickard and Vanselow 1978). Colonization of sulfide minerals and the resulting, lithoautotrophy begins under bulk 'neutral' pH conditions and probably occurs on the nanoscale through the development of an acidic interface between the bacteria and the mineral surfaces (compare Figs. 1, 2 and 3).

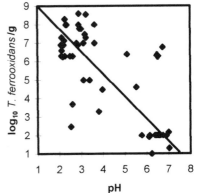

Figure 2. *Thiobacillus ferrooxidans* plotted against sample pH for samples from the Lemoine mine tailings (Southam and Beveridge 1993). The two clusters along the straight line are typical of what is expected for *Thiobacillus* spp. The outliers show (1) low populations at low pH where drying has caused death or decreased activity and (2) high populations at high pH indicating acidic nanoenvironments.

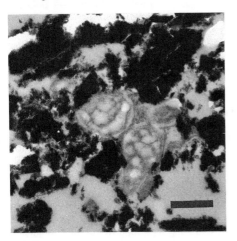

Figure 3. An unstained thin section of a tailings sample from the Lemoine mine which demonstrates the close association between *T. ferrooxidans* and a chalcopyrite mineral surface. Bar equals 0.5 µm. (Published with permission from Southam and Beveridge 1992).

Thiobacillus ferrooxidans binds preferentially to pyrite dispersed through a coal slurry containing abundant organic surfaces (Bagdigian and Myerson 1986, Mustin et al. 1992). Bennet and Tributsch (1978) and Norman and Snyman (1987) have demonstrated that *T. ferrooxidans* chooses to colonize fracture lines and dislocations on pyritic mineral surfaces. These fine grooves (0.2 to 0.5 μm) in the mineral surface eventually develop into corrosion pits, widening and enlarging until the mineral grain is destroyed (Tributsch 1976, Norman and Snyman 1987). Mineral dislocations may provide convenient physical recesses for bacterial colonization and a unique surface charge promoting the attachment of *T. ferrooxidans*. This is one way in which certain strains may exhibit mineral selectivity.

The thiobacillus surface component responsible for preferential binding to sulfide minerals and the resulting close bacteria-mineral interaction is lipopolysaccharide (LPS) because this molecule is situated on the outer membrane surface of Gram-negative Bacteria and its side chains extend beyond the usual bilayer structure of a membrane. *Thiobacillus* spp. do not possess capsular material (Shively et al. 1970, Wang *et al* 1970, Vestal et al. 1973, Hirt and Vestal 1975, Rodriguez et al. 1986, Yokota et al. 1988). LPS heterogeneity (Southam and Beveridge 1993), i.e. differences in LPS O-antigen side-chains, will control the cell surface charge character and hydrophobicity which could affect the ability of different strains to colonize different sulfide mineral surfaces. Because charge character is also pH dependent (Chakrabarti and Banerjee 1991), the reduction in pH as thiobacilli colonize mine tailings material would also affect, presumably enhance, their ability to colonize tailings as acid mine drainage develops.

Lizama and Suzuki (1991) characterized a *T. ferrooxidans* strain which could oxidize either pyrite or chalcopyrite but not sphalerite and a second strain which could oxidize either pyrite or sphalerite but not chalcopyrite. The ability to differentiate between chalcopyrite and sphalerite suggests that a form of recognition, presumably reflected in LPS chemistry (Southam and Beveridge 1993), exists towards these minerals. *Thiobacillus* spp. are also known for their ability to adapt to various types of sulfide ores prior to the initiation of active microbial leaching (Suzuki et al. 1990). This adaptation mechanism is not well understood although it might have something to do with "activation" by chemical oxidation (Moses and Herman 1991). Although phenotypic switching has been demonstrated in *T. ferrooxidans*, a relationship between LPS chemistry and mineral adapatation has not been demonstrated (Schrader and Holms 1988, Southam and Beveridge 1993).

Acid mine waters from many different types of oxidizing sulfide mineral deposits typically have high concentrations of both *T. ferrooxidans* and *T. thiooxidans*, and in roughly equal proportions (Scala et al. 1982). This observation raises the question of what role *T. thiooxidans* plays in the oxidation process because *T. ferrooxidans* can oxidize both iron and sulfur. The difference between *T. ferrooxidans* and *T. thiooxidans* is not simply that *T. thiooxidans* cannot oxidize Fe(II). At Brimstone Basin, an elemental sulfur deposit inYellowstone National Park, *T. thiooxidans* populations were 2 to 3 orders of magnitude greater than *T. ferrooxidans* (Southam, Donald, and Nordstrom, pers. comm.). Nature has selected for the most efficient sulfur-oxidizing bacterium.

Cells grown on ore are difficult to dissociate from the ore particles (Gormley and Duncan 1974, Suzuki et al. 1990, Southam and Beveridge 1992) demonstrating that a tight association must occur between *Thiobacillus* spp. and sulfide mineral surfaces. The development of a tight association between the bacteria (LPS) and sulfide minerals may proceed via a hypothetical two phase mechanism (Southam et al. 1995). First, either ionic, salt-bridging or hydrophobic interactions would reversibly attach the bacterium to the mineral surface so that an acidic interface could be established to sustain lithoautotrophy. Phase two encompasses the subsequent cementing of the bacterium to the mineral surface with iron oxy-hydroxides (Bigham et al. 1990, Bhatti et al. 1993) forming an even more acidic nanoenvironment (i.e. precipitation of $Fe(OH)_3$ yields H^+) to support bacterial growth and multiplication. The role of ferric oxy-hydroxides in glueing thiobacilli to mineral surfaces is supported by their release after solubilization of these precipitates (Ramsay et al. 1988, Southam and Beveridge 1993). Strong adherence of *T. ferrooxidans* to minerals via iron precipitates (Southam and Beveridge 1992, 1993) may have an important ecological role in reducing the diffusion of metabolic products (e.g. Fe(III) and sulfuric acid) away from the cell-mineral interface. This would help maintain an acidic nanoenvironment at the mineral surface and provide a potential source of soluble ferrous iron through repeated chemical oxidation of the sulfide (Reaction 2) thereby promoting the growth of *T. ferrooxidans*.

Effects of temperature

Environments where iron, sulfur, and sulfide minerals are oxidizing can have a large range of temperature from 0°C in some waste rock piles (Strömberg and Banwart 1994), to 50-60°C in some underground mines (Nordstrom and Potter 1977, Nordstrom and Alpers 1995) and waste rock piles (Hendy 1987), to boiling in hot spring waters (Bott and Brock 1969). Temperature is one of the distinctive and most important environmental parameters that influences the activity of microorganisms. It not only influences their rate of growth but it can also determine the predominant genus and species. Bacteria are thermally divided into psychrophiles, mesophiles, thermophiles and hyperthermophiles. Psychrophiles can grow at 0° with an optimum of 10°-15° and a range up to 30°C. Mesophiles tend to grow at temperatures below 40°C with an optimum temperature around 30°C. Thermophiles have an optimum temperature of about 50°-60°C and hyperthermophiles have an optimum temperature of >80°C. Both obligate and facultative thermophiles have also been observed.

Very little information is available on psychrophilic activity in mine waste environments but Strömberg (1997) has provided evidence that for a waste rock pile at Aitik, Sweden with a temperature range of 0° to 12°C and averaging 5°C, microbial catalysis is affecting pyrite oxidation. Ferroni et al. (1986) and Berthelot et al. (1993) found that species such as *T. ferrooxidans* can be psychrotrophic. They enriched mine water samples from Elliot Lake, Ontario and found that growth continued at temperatures as low as 2°C, with faster growth for the enriched mine water than that for *T. ferrooxidans* ATCC 33020. Both cultures had the same growth rates and the same optimum temperature over the range of 12° to 35°C.

Acidophilic iron- and sulfur-oxidizing bacteria are either mesophiles or thermophiles. Figure 4 shows the temperature range for growth of several important mesophiles and thermophiles (Norris 1990). The growth rates shown are not necessarily independent of pH. Norris (1990) has noted the increase in rate with an increase in pH for *T. ferrooxidans*. However, Nordstrom (1977, 1985) found that the ferrous iron oxidation rate for acid mine waters of pH 1-2 was virtually the same as that for 9K culture medium of pH 2-3. The influence of pH on microbial kinetics has not been clearly defined for sulfide mineral and ferrous iron oxidation. *Thiobacillus* spp. and *Leptospirillum* spp. are mesophiles with typical temperature optima in the range of 30° to 35°C, although the same species can have a different range and optimum as shown in Figure 4. These species have a temperature range that coincides well with the typical temperature range of most environments where pyrite is oxidizing. The new species, *Leptospirillum thermoferrooxidans*, is a moderate thermophile with a temperature optimum of 45°-50°C (Golovacheva et al. 1992). *Sulfobacillus thermosulfidooxidans* and strain TH1 are Gram-positive, thermophiles, and facultative lithotrophs, unlike *T. ferrooxidans*. They show greater morphological variation and are more elongated (Karavaiko et al. 1988).

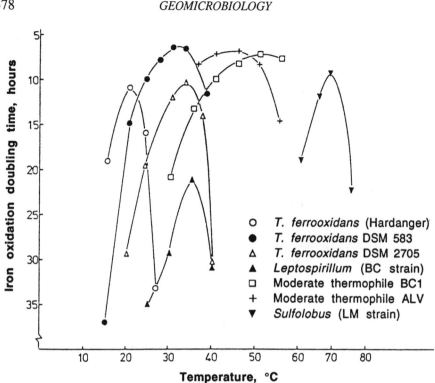

Figure 4. Temperature ranges for several acidophilic iron- and sulfur-oxidizing mesophiles and thermophiles (reproduced with permission from Norris 1990).

There are four genera of Archaea that are aerobic, acidophilic, coccoid in form, and moderate to extreme thermophiles: *Sulfolobus, Acidanus, Metallosphaera,* and *Sulfurococcus*. They oxidize reduced sulfur compounds and some oxidize Fe^{2+} and sulfide minerals. Thermophiles that oxidize iron and sulfur are *Sulfolobus* spp. and *Acidianus* spp. Since the original classification of *Sulfolobus acidocaldarius* by Brock et al. (1972), three species have been recognized: *S. acidicaldarius, S. solfataricus,* and *S. brierleyi* (Zillig et al. 1980). *S. brierleyi* has subsequently been reclassified as *Acidianus brierleyi*. *Sulfolobus* oxidizes arsenopyrite as well as pyrite (Ngubane and Baecker 1990). Norris and Parrot (1986) have demonstrated the catalytic effect of *Sulfolobus* on the oxidation of concentrates of pyrite, chalcopyrite, pentlandite, and nickeliferous pyrrhotite at a pH of 2 and 70°C.

As previously described, *Thiobacillus* spp. are Gram-negative Bacteria (Fig. 5) possessing LPS as their outermost cell surface component (Fig. 1, Berry and Murr 1980) while *Sulfolobus* spp. only possess an S-layer external to the cell membrane (Taylor et al. 1982; Fig. 6). An S-layer is a two-dimensional paracrystalline array of protein that functions as a diffusion barrier in cells on which they occur (see by Fortin et al., this volume). For *Sulfolobus* spp. the formation of a nanoenvironment scale diffusion gradient, which is responsible for the growth of *Thiobacillus* spp. on mineral surfaces (Fig. 1), must be accomplished without a periplasm. The same must be true for the mycoplasma, *Thermoplasma acidophilium* (see next section).

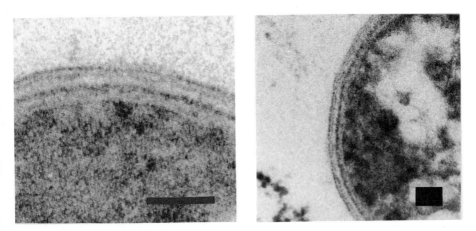

Figure 5 (left). A stained thin-section of *Thiobacillus ferrooxidans* ATCC 13661 demonstrating the usual gram-negative cell envelope structure in which lipopolysaccharides represent the outermost envelop component. Bar equals 50 nm.

Figure 6 (right). A negatively stained *Sulfolobus* spp. demonstrating the crystalline nature of the cell envelope. Bar equals 200 nm.

Based on its ability to produce bacterial endospores, *Thiobacillus thermohilica* (Egorova and Deryugina 1963) is likely a *Sulfobacillus* spp. (Golovacheva and Karavaiko 1978). Gram-positive, acidophilic iron-, sulfur- and mineral-oxidizing bacteria have not received very much attention. The fact that they are not common, or at least rarely described, may relate to the difficulty with which a Gram-positive bacterium can establish an efficient diffusion gradient.

In this review of the literature on bacterial physiology and bacterial interactions with sulfide minerals we have tried to show how *Thiobacillus* spp. function at low pH and regenerate ferric iron from ferrous. At the physiological level, the promotion of sulfide mineral oxidation by bacteria does not require their attachment to surfaces but the close proximity would certainly decrease the diffusion distance to the substrate source thereby maintaining higher concentrations of iron for growth. The overall effect is to enhance the oxidation rate. High temperature (40° to 100°C) oxidation is also enhanced by both Bacteria and Archaea iron- and sulfur-oxidizers but dominated by the Archaea at the higher temperatures.

MICROBIAL ECOLOGY IN MINE WASTE ENVIRONMENTS

In the introduction to this chapter we mentioned that mine waste environments and acid mine waters support a wide diversity of microbial life. Autotrophic bacteria, heterotrophic bacteria, green algae, fungi, yeasts, mycoplasmas, and amoebae have all been found in acid mine waters. A study by Wichlacz and Unz (1981) reported 37 acidophilic heterotrophs that were isolated and partly characterized from acid mine drainage. Belly and Brock (1974) found numerous heterotrophic bacteria in association with coal refuse piles. In the Rum Jungle mine site, Australia, a mine dump was found to have relatively low numbers of *T. ferrooxidans* and high numbers of acidophilic heterotrophs with no apparent seasonal variations, suggesting a stable population had been established (Goodman et al. 1981). The occurrence of this rich and complex microbial ecology deserves a short discussion.

As Dugan et al. (1970), Johnson et al. (1979), and Ledin and Pedersen (1996) have astutely pointed out, autotrophic thiobacilli will leak or excrete organic compounds that may be utilized by heterotrophic bacteria. Dead autotrophic cells will provide more organic sustenance for heterotrophs. Furthermore, a commensal relationship, possibly even a symbiosis (Battacharyya et al. 1990), exists between heterotrophic and autotrophic bacteria. *T. ferrooxidans* excretes pyruvate during its growth (Schnaitman and Lundgren 1965) and pyruvate can inhibit growth. The heterotroph *Acidiphilium cryptum* uses pyruvate as a carbon and energy source and its presence can enhance the growth of *T. ferrooxidans* (Harrison 1984, Wichlacz and Thompson 1988). *Acidiphilium* spp. are common inhabitants of acid mine drainage (Harrison 1981, Wichlacz et al. 1986, Kishimoto and Tano 1987) and copper dump leaching operations (Groudev and Groudeva 1993) and they are always found in association with *T. ferrooxidans* (Lobos et al. 1986, Bhattacharyya et al. 1991).

Ehrlich (1996) and Ledin and Pedersen (1996) have noted that several satellite microorganisms live in close association with *T. ferrooxidans*. Some of these are other species of thiobacilli noted previously (see Table 1). Other heterotrophic microorganisms, beside *Acidiphilium* spp., include *Pseudomonas, Bacillus, Micrococcus, Sarcina, Crenothrix, Microsporium, Aerobacter, Caulobacter,* and the fungi and yeasts *Acontium, Cladosporium, Penecillium, Trichosporon,* and *Rhodotorula*. Brock (1973) demonstrated that the lower pH limit for the growth of cyanobacteria (formerly "blue-green algae") is about 4 and only green algae can tolerate lower pH conditions. Green algae are common in acid mine waters where they have access to sufficient sunlight and continuous and steady-flowing water. Nordstrom (1977) found several types of green algae at Iron Mountain, California. The most common algae occurring in acid mine waters are *Chlamydomonas, Chlorella, Ulothrix, Chroomonas,* and *Euglena*. Additional acidophilic or acid-tolerant algae, diatoms, and higher plants are described in Kelly (1988).

One of the most unusual microorganisms found in environments where pyrite oxidizes is the Mycoplasma, *Thermoplasma acidophilum* (Darland et al. 1970). Mycoplasma are like bacteria without cell walls. They are Gram-negative, highly pleomorphic procaryotic organisms, and yet *T. acidophilum* has a pH optimum of 1-2 and a temperature optimum of about 60°C.

The microbial ecology of environments where acid waters have developed from the oxidation of pyrite and other sulfide minerals is poorly understood. Investigations have shown the possibility for a microbial succession from autotrophs to heterotrophs to more complex organisms in a food chain. If the hydrology and climate do not have sudden and large variations (e.g. steady-state flow most of the time) there are better opportunities for growth of a stable and more mature microbial community. Those environments with highly variable and dynamic conditions (high erosion rates, onset of sudden extreme storms) should discourage the development of large and stable populations of microbes.

SUMMARY AND FUTURE WORK

Any attempts to understand the degradation of pyrite and other sulfide minerals or to ameliorate the water quality hazards associated with mining of sulfide mineral deposits without recognizing the role of microorganisms are likely to fail. Literature on the subject is voluminous. More than a century ago, microbial interactions with sulfur and sulfide minerals have been scientifically investigated. Lithoautotrophs, such as *Thiobacillus ferrooxidans, Thiobacillus thiooxidans, Leptospirillum ferrooxidans,* and some of their co-habitants such as *Acidiphilium, Thermoplasma, Bacillus, Micrococcus, Chlamydo-*

monas, Chlorella, Ulothrix, and *Euglena* are an integral part of the environment where sulfide minerals are oxidizing.

Investigations on the geomicrobiology of sulfide mineral oxidation have clarified the mechanism of microbial catalysis, physiological features that allow them to respire and reproduce in acidic waters, and their microbial ecology. The main role of iron- and sulfur-oxidizing bacteria is in the oxidation of aqueous ferrous to ferric iron. The ferric iron can then rapidly attack the sulfide surface. These rates are comparable so that pyrite will oxidize from the ferric iron as rapidly as the microbes can regenerate the ferric iron. The bacterial catalysis of ferrous iron, however, will be a function of environmental conditions including (but not only) temperature, pH, density or total dissolved solids, chemical composition of the solution, ecology of commensal or predatory species, oxygen concentrations, hydrologic conditions, and mineralogy.

The initiation of pyrite oxidation does not require an elaborate sequence of different geochemical reactions that dominate at different pH ranges (e.g. Kleinmann et al. 1981). *Thiobacillus* spp. employ nanoenvironments to grow on sulfide mineral surfaces, usually as iron-cemented biofilms (Southam et al. 1995). These nanoenvironments can develop thin layers of acidic water that, at first, do not affect the bulk water chemistry. Eventually, with progressive oxidation, the nanoenvironment (nanometer scale) becomes a microenvironment (micrometer scale). Evidence for acidic microenvironments in the presence of circumneutral pH for the bulk water in the subsurface can be found in the occurrence of jarosite in certain soil horizons where the current soil water is neutral (Carson et al.1982). Jarosite can only form under acidic conditions. The occurrence of isolated cubes of goethite pseudomorphs after pyrite also suggests that pH gradients probably existed near the surface of the oxidizing pyrite but may not have affected the bulk water pH. Of course, the bulk water may have been acidic in the past and currently it is neutral, however, experiments such as those of Goldhaber (1983) and Moses et al. (1987) indicate that a lot of oxidative chemistry with large chemical gradients occurs within nanoscale to microscale dimensions at the mineral-water interface.

Some of the contradictory studies regarding bacterial adsorption on sulfide surfaces and their effect on oxidation rates need to be resolved. The mechanism of *T. ferrooxidans* mineral selectivity is not known, however, it must be conferred by the nature of its lipopolysaccharide. The *Sulfolobus* mineral interface has not been studied, nor has the distribution and importance of *Sulfolobus* and *Sulfobacillus* in mine waste environments been studied. The various factors that can inhibit autotrophic growth in sulfide mineral environments and their relative importance should be clarified.

The oxidation of elemental sulfur by *Thiobacillus* spp. requires the formation of a wetting agent that reduces the surface tension. Jones and Starkey (1961) describe a proteinaceous compound which promotes growth of *T. thiooxidans* on elemental sulfur. This wetting agent is also important for dissolution of the sulfur-rich layer formed by acid leaching of base metals from pyrite and chalcopyrite. In his review of the biochemistry of inorganic sulfur oxidation by chemolithotrophs, Kelly (1982) emphasized the oxidation of thiosulfate and polythionates with almost no information on solid sulfur oxidation. Clearly more studies are needed to elucidate the mechanism of growth on elemental sulfur.

Microbial ecology and phylogeny have undergone a revolution with the application of 16S rRNA sequence analysis. However, defining phylogenetic relationships based on sequence diversity raises new problems that have yet to be addressed, such as how to reconcile conventional taxonomy based on physiology and function with differences based on sequence types. Recent work on thiobacilli have highlighted some of these difficulties (Goebel and Stackebrandt 1994). Although strain-specific molecular probes may not be possible, the phylogenetic diversity of *Thiobacillus* spp. has been revealed with rRNA sequence analysis (Lane et al. 1985, Goebel and Stackebrandt 1994). This molecular diversity has enabled the use of 16S rRNA probe (Goebel and Stackebrandt 1994) and PCR-mediated 16S rDNA detection (Wulf-Durand et al. 1997) of acidophilic microorganisms from natural environments and bio-leaching operations. Future developments in microbial ecology of acidophilic microorganisms will likely rely on these molecular techniques.

REFERENCES

Ahonen L, Tuovinen OH (1992) Bacterial oxidation of sulfide minerals in column leaching experiments at suboptimal temperatures. Appl Environ Microbiol 58:600-606

Ahonen L, Hiltunen P, Tuovinen OH (1986) The role of pyrrhotite and pyrite in the bacterial leaching of chalcopyrite ores. In: RMR Branion, HG Ebner (eds) Fundamental and Applied Biohydrometallurgy, p 13-22, 6th Intl Symp Biohydrometall, Vancouver, BC, Canada, August 1985. Elsevier, Amsterdam

Aleem MIH, Lees H, Nicholas DJD (1963) Adenosine triphosphate-dependent reduction of nicotinamide adenine dinucleotide by ferro-cytochrome c in chemoautotrophic bacteria. Nature 200:759-761

Amy PS, Haldeman DL, eds (1997) The microbiology of the terrestrial deep subsurface. CRC Press, Boca Raton, Florida

Amaro AM, Chamorro D, Seeger M, Arrendondo R, Perrano I, Jerez CA (1991) Effect of external pH perturbations on *in vivo* protein synthesis by the acidophilic bacterium *Thiobacillus ferrooxidans*. J Bacteriol 173:910-915

Ash SH, Felegy EW, Kennedy DO, Miller PS (1951) Acid mine drainage problems—Anthracite region of Pennsylvania. US Bureau of Mines Bulletin 508

Bagdigian RM, Myerson AS (1986) The adsorption of *Thiobacillus ferrooxidans* on coal surfaces. Biotech Bioeng 28:467-469

Baldi F, Clark T, Pollack, SS Olson GJ (1992) Leaching of pyrites of various reactivities by *Thiobacillus ferrooxidans*. Appl Environ Microbiol 58:1853-1856

Baldi F, Filippelli M, Olson GJ (1989) Biotransformation of mercury by bacteria isolated from a river collecting cinnabar mine waters. Microb Ecol 17:263-274

Barrett J, Hughes MN, Karavaiko GI, Spencer PA (1993) Metal Extraction by Bacterial Oxidation of Minerals. Ellis Horwood, West Sussex, UK

Barton P (1978) The acid mine drainage. In: JO Nriagu (ed) Sulfur in the Environment, Part II: Ecological Impacts, p 313-358, Wiley-Interscience, New York

Basolo F, Pearson RG (1967) Mechanisms of Inorganic Reactions—A study of metal complexes in Solution, 2nd edition, John Wiley & Sons, New York

Beck JV (1977) Chalcocite oxidation by concentrated cell suspensions of *Thiobacillus ferrooxidans*. In: W Schwartz (ed) Conf Bacterial Leaching 1977, p 119-128 GBF, Verlag Chemie, Weinheim, Germany

Beck JV, Elsden SR (1958) Isolation and some characteristics of an iron-oxidizing bacterium. J Gen Microbiol 19:1

Belly RT, Brock TD (1974) Ecology of iron-oxidizing bacteria in pyritic materials associated with coal. J Bacteriol 117:726-732

Bennet JC, Tributsch H (1978) Bacterial leaching patterns on pyrite crystal surfaces. J Bacteriol 134:310-317

Bergholm A (1955) Oxidation av pyrit. Jernkontorets Annaler 139: 531-549. English translation: D Baxter (translator) DK Nordstrom, JM DeMonge, L Lövgren (eds) (1995) Oxidation of Pyrite. US Geol Survey Open-File Report 95-389

Berner EK, Berner RA (1996) Global Environment: Water, Air and Geochemical Cycles. Prentice-Hall, Engelwood Cliffs, New Jersey

Berry VK, Murr LE (1980) Morphological and ultrastructural study of the cell envelope of thermophilic and acidophilic microorganisms as compared to *Thiobacillus ferrooxidans*. Biotech Bioeng 22:2543-2555

Berthelot D, Leduc LG, Ferroni GD (1993) Temperature studies of iron-oxidizing autotrophs and acidophilic heterotrophs isolated from uranium mines. Can J Microbiol 39:384-388

Beveridge TJ (1981) Ultrastructure, chemistry, and function of the bacterial wall. Intl Rev Cytol 72:229-317

Beveridge TJ (1988) The bacterial surface: general considerations towards design and function. Can J Microbiol 34:363-372

Bhappu RB, Reynolds DH, Roman RJ (1965) Molybdenum recovery from sulphide and oxide ores. J Metals New York 17:1199-1205

Bhattacharyya S, Chakrabarty BK, Das A, Kundu PN, Banerjee PC (1990) *Acidiphilium symbioticum* sp. nov., an acidophilic heterotrophic bacterium from *Thiobacillus ferrooxidans* cultures isolated from Indian mines. Can J Microbiol 37:78-85

Bhatti TM, Bigham JM, Carlson L, Tuovinen OH (1993) Mineral products of pyrrhotite oxidation by *Thiobacillus ferrooxidans*. Appl Environ Microbiol 59:1984-1990

Bigham JM, Schuertmann U, Carlson L, Murad E (1990) A poorly crystallized oxyhydroxy sulfate of iron formed by bacterial oxidation of Fe (II) in acid mine waters. Geochim Cosmochim Acta 54:2743-2758

Blake R, Waskovsky J, Harrison AP (1992) Respiratory components in acidophilic bacteria that respire on iron. Geomicrobiol J 10:173-192

Blowes DW, Al T, Lortie L, Gould WD, Jambor JL (1995) Microbiological, chemical, and mineralogical characterization of the Kidd Creek Mine Tailings Impoundment, Timmins Area, Ontario. Geomicrobiol J 13:13-31

Boone DR, Liu Y, Zhao Z, Balkwill DL, Drake GR, Stevens TO Aldrich HC (1995) *Bacillus infernus* sp. nov. Fe(III)- and Mn(IV)-reducing anaerobe from the terrestrial subsurface. Intl J Sys Bacteriol 45: 441-448

Bott TL, Brock TD (1969) Bacterial growth rates above 90°C in Yellowstone hot springs. Science 164:1411-1412

Brierley JA, Le Roux NW (1977) A facultative thermophilic *Thiobacillus*-like bacterium—Oxidation of iron and pyrite. In: W Schwartz (ed) Conf Bacterial Leaching 1977, p 55-66, GBF Verlag Chemie, Weinheim, Germany

Brierley CL, Murr LE (1973) Leaching—Use of a thermophilic and chemoautotrophic microbe. Science 179:488-489

Brissette C, Champagne J, Jutras JR (1971) Bacterial leaching of cadmium sulphide. Can Inst Metall Bull 64:85-88

Brock TD (1973) Lower pH limit for the existence of blue-green algae: evolutionary and ecological implications. Science 179:480-483

Brock TD (1975) Effect of water potential on growth and iron oxidation by *Thiobacillus ferrooxidans*. Appl Microbiol 29:495-501

Brock TD, Brock KM, Belly RT, Weiss RL (1972) *Sulfolobus*: a new genus of sulfur-oxidizing bacteria living at low pH and high temperature. Arch Microbiol 84:54-68

Brock TD, Cook S, Peterson S, Mosser JL (1976) Biogeochemistry and bacteriology of ferrous iron oxidation in geothermal habitats. Geochim Cosmochim Acta 40:493-500

Brown AD, Jurinak JJ (1989) Mechanisms of pyrite oxidation in aqueous mixtures. J Environ Quality 18: 545-550

Brown SL, Sullivan JD (1934) Dissolution of various copper minerals. US Bureau Mines Report Invest RI-3228

Bryner LC, Anderson R (1957) Microorganisms in leaching sulfide minerals. Ind Eng Chem 49:1721-1724

Bryner LC, Jamieson AK (1958) Microorganisms in leaching sulfide minerals. Appl Microbiol 6:281-287

Bryner LC, Beck JV, Davis BB, Wilson DG (1954) Microorganisms in leaching sulfide minerals. Ind Eng Chem 46:2587-2592

Buckley AN, Woods RW (1984) An X-ray photoelectron scpectroscopic study of the oxidation of chalcopyrite. Aust J Chem 37:2403-2413

Buckley AN, Woods RW (1985) X-ray photoelectron spectroscopy of oxidized pyrrhotite surfaces. Appl Surf Sci 22/23:280-287

Buckley AN, Woods RW (1987) The surface oxidation of pyrite. Appl Surf Sci 27:437-452

Burkstaller JE, McCarty P, Parks GA (1975) Oxidation of cinnabar by Fe(III) in acid mine waters. Environ Sci Tech 9:676-678

Byerly DW (1996) Handling acid-producing material during construction. Environ Eng Geosci II:49-57

Carpentor LV, Herndon LK (1933) Acid Mine Drainage from Bituminous Coal Mines. West Virginia Univ Eng Expl Stn Res Bull No 19, Morgantown, WV

Carson CD, Fanning DS, Dixon JB (1982) Alfisols and ultisols with acid sulfate weathering features in Texas. In: JA Kittrick, DS Fanning, LR Hossner (eds) Acid Sulfate Weathering, p 127-146, Soil Sci Soc Am Pub No 10, Madison, Wisconsin

Chakrabarti BK, Banerjee PC (1991) Surface hydrophobicity of acidophilic heterotrophic bacterial cells in relation to their adhesion on minerals. Can J Microbiol 37:692-696

Chen C-Y, Skidmore DR (1987) Langmuir adsorption isotherm for *Sulfolobus acidocaldarius* on coal particles. Biotech Lett 9:191-194

Chen C-Y, Skidmore DR (1988) Attachment of *Sulfolobus acidocaldarius* cells in coal particles. Biotechnol Progr 4:25-30

Colmer AR, Hinkle ME (1947) The role of microorganisms in acid mine drainage. Science 106:253-256

Colmer AR, Temple KL, Hinkle ME (1950) An iron-oxidizing bacterium from the acid drainage of some bituminous coal mines. J Bacteriol 59:317-328

Darland G, Brock TD, Samsonoff W, Conti SF (1970) A thermophilic, acidophilic mycoplasma isolated from a coal refuse pile. Science 170:1416-1418

De Cuyper JA (1963) Bacterial leaching of low grade copper and cobalt ores. In: ME Wadsworth, FT Davis (eds) Intl Symp Unit Procedures in Hydrometallurgy, p 126-142 Gordon and Breach, New York

Dispirito AA, Dugan PR, Tuovinen OH (1983) Sorption of *Thiobacillus ferrooxidans* to particulate material. Biotechnol Bioeng 25:1163-1168

Dugan PR, MacMillan CB, Pfister RM (1970) Aerobic heterotrophic bacteria indigenous to pH 2.8 acid mine water: I. Microscopic examination of acid streamers. J Bacteriol 101:973-981; II. Predominant slime-producing bacteria in acid streamers. J Bacteriol 101:982-988

Duncan DW, Drummond AD (1973) Microbiological leaching of porphyry copper type mineralization: post-leaching observations. Can J Earth Sci 10:476-484

Duncan DW, Trussell PC (1964) Advances in microbiological leaching of sulphide ores. Can Metall Q 3:43-55

Duncan DW, Trussel PC, Walden PC (1964) Leaching of chalcopyrite with *Thiobacillus ferrooxidans*: effect of surfactants and shaking. Appl Microbiol 12:122-126

Dutrizac JE (1990) Elemental sulphur formation during the ferric chloride leaching of chalcopyrite. Hydrometall 23:153-176

Dutrizac JE, MacDonald RJC (1974a) Ferric ion as a leaching medium. Minerals Sci Eng 6:59-104

Dutrizac JE, MacDonald RJC (1974b) Kinetics of dissolution of covellite in acidified ferric sulphate solution. Can Metall Q 13:423-433

Dutrizac JE, MacDonald RJC (1974c) Percolation leaching of pentlandite ore. Can Inst Min Met Bull 67:169-175

Dutrizac JE, MacDonald RJC, Ingraham TR (1970a) The kinetics of dissolution of bornite in acidified ferric sulphate solutions. Metall Trans 1:225-231

Dutrizac JE, MacDonald RJC, Ingraham TR (1970b) The kinetics of dissolution of cubanite in aqueous acidic ferric sulphate solutions. Metall Trans 1:3083-3088

Egorova AA, Deryugina ZP (1963) The spore forming thermophilic thiobacterium: *Thiobacillus thermophilica* Imschenetskii nov. spec. Mikrobiologiya 32:439-446

Ehrlich HL (1963a) Microorganisms in acid drainage from a copper mine. J Bacteriol 86:350-352

Ehrlich HL (1963b) Bacterial action on orpiment. Econ Geol 58:991-994

Ehrlich HL (1964) Bacterial oxidation of arsenopyrite and enargite. Econ Geol 59:1306-1308

Ehrlich HL (1996) Geomicrobiology. 3rd edition, Marcel Dekker, New York

Evangelou VP (Bill) (1995) Pyrite Oxidation and Its Control. CRC Press, Boca Raton, Florida

Ferroni GD, Leduc LG, Todd M (1986) Isolation and temperature characterization of psychotrophic strains of *Thiobacillus ferrooxidans* from the environment of a uranium mine. J Gen Appl Microbiol 32: 169-175

Fortin D, Davis B, Southam G, Beveridge TJ (1995) Biogeochemical phenomena induced by bacteria within sulfidic mine tailings. J Ind Microbiol 14:178-185

Garrels RM, Thompson ME (1960) Oxidation of pyrite by ferric sulfate solutions. Am J Sci 258A:57-67

Goebel BM, Stackerbrandt E (1994) Cultural and phylogenetic analysis of mixed microbial populations found in natural and commercial bioleaching environments. Appl Environ Microbiol 60:1614-1621

Goldhaber MB (1983) Experimental study of metastable sulfur oxyanion formation during pyrite oxidation at pH 6-9 and 30°C. Am J Sci 283:193-217

Golovacheva RS, Karavaiko GI (1978) A new genus of thermophilic spore-forming bacteria, Sulfobacillus. Mikrobiologiya 47:815-822

Goodman AE, Khalid AM, Ralph BJ (1981) Microbial ecology of Rum Jungle. Part 1. Environmental study of sulphidic overburden dumps, environmental heap-leach piles and tailings dam area. Australian AEC AAEC/E531

Gormley LS, Duncan DW (1974) Estimation of *Thiobacillus ferrooxidans* concentrations. Can J Microbiol 20:1454-1455

Granger HC, Warren CG (1969) Unstable sulfur compounds and the origin of roll-type uranium deposits. Econ Geol 64:160-171

Groudev SN, Groudeva VI (1993) Microbial communities in four industrial copper dump leaching operations in Bulgaria. FEMS Microbiol Rev 11:261-268

Haddock BA, Jones CW (1977) Bacterial respiration. Bacteriol Rev 41:47-99

Harrison AP, Jr (1981) *Acidiphilium cryptum* gen. nov., sp. nov., heterotrophic bacterium from acidic mineral environments. Intl J Syst Bacteriol 31:327-332

Harrison AP, Jr (1984) The acidophilic thiobacilli and other acidophilic bacteria that share their habitat. Ann Rev Microbiol 38:265-292

Harrison AP Jr, Jarvis BW, Johnson JL (1980) Heterotrophic bacteria from cultures of autotrophic *Thiobacillus ferrooxidans*: relationships as studied by means of deoxyribonucleic acid homology J Bacteriol 143:448-454

Haver FP, Uchida K, Wong MM (1970) Recovery of lead and sulphur from galena concentrate, using a ferric sulphate leach. US Bureau Mines Report Invest RI-7360

Hendy NA (1987) Isolation of thermophilic iron-oxidizing bacteria from sulfidic waste rock. J Ind Microbiol 1:389-392

Hirt WE, Vestal JR (1975) Physical and chemical studies of *Thiobacillus ferrooxidans* lipopolysaccharides. J Bacteriol 123:642-650

Hobot JA, Carlemam E, Villiger W, Kellenberger E (1984) The periplasmic gel: a new concept resulting from the re-investigation of bacterial cell envelope ultrastructure by new methods. J Bacteriol 160: 143-152

Hutchins SR, Davidson MS, Brierly JA, Brierly CL (1986) Microorganisms in reclamation of metals. Ann Rev Microbiol 40:311-336

Ichikuni M (1962) The action of ferric ions on chalcopyrite. Bull Chem Soc Japan 35:1765-1768

Ingledew WJ, Cox JC, Halling PJ (1977) A proposed mechanism for energy conservation during Fe^{2+} oxidation by *Thiobacillus ferrooxidans*: Chemiosmostic coupling to net H^+ influx. FEMS Microbiol Lett 2:193-197

Ingledew WJ (1982) *Thiobacillus ferrooxidans*: The bioenergetics of an acidophilic chemolithotroph. Biochim Biophys Acta 683:89-117

Ivanov VI, Nagirynyak FI, Stepanov BA (1961) Bacterial oxidation of sulfide ores. I. Role of *Thiobacillus ferrooxidans* in the oxidation of chalcopyrite and sphalerite. Mikrobiologiya 30:688-692

Johnson DB, Kelso WI, Jenkins DA (1979) Bacterial streamer growth in a disused pyrite mine. Environ Pollut 18:107-118

Jones JE, Starkey RL (1961) Surface active substances produced by *Thiobacillus thiooxidans*. J Bacteriol 82:788-789

Joseph JM (1953) Microbiological study of acid mine waters—Preliminary report. Ohio J Sci 53:123-127

Karavaiko GI, Golovacheva RS, Pivovarova TA, Tzaplina IA, Vartanjan NS (1988) Thermophilic bacteria of the genus *Sulfobacillus*. In: PR Norris, DP Kelly (eds) Biohydrometallurgy, p 29-41, Proc Intl Symp, Science and Technology Letters, Kew, UK

Keller L, Murr LE (1982) Acid-bacterial and ferric sulfate leaching of pyrite single crystals. Biotechnol Bioeng 24:83-96

Kelly M (1988) Mining and the Freshwater Environment. Elsevier Applied Science, Amsterdam

Kelly DP, Harrison AP (1984) Genus *Thiobacillus* Beijerinck. In: JT Staley (ed) Bergeys Manual of Systematic Bacteriology III:1842-1858, Williams and Wilkins, New York

Kelly, DP (1982) Biochemistry of the chemolithotrophic oxidation of inorganic sulphur. Phil Trans R Soc London B 298:499-528

Khalid AM, Ralph BJ (1977) The leaching behavior of various zinc sulphide minerals with three *Thiobacillus* species. In: W Schwartz (ed) Conference Bacterial Leaching 1977, p 165-173, GBF Verlag Chemie, Weinheim, Germany

Kishimoto N, Tano T (1987) Acidophilic heterotrophic bacteria isolated from acidic mine drainage, sewage, and soils. J Gen Appl Microbiol 33:11-25

Kholmanskikh YB, Ilyashevich II, Kosnareva IA (1966) Kinetics of dissolution of copper and its compounds in aqueous solutions. Trudy ural nauch-issled. Proekt Inst Mednoi Prom 9:289-293

Kleinmann RLP, Crerar DA, Pacelli RR (1981) Biogeochemistry of acid mine drainage and a method to control acid formation. Min Eng March:300-305

Koch S, Grasselly G (1952) Data on the oxidation of sulphide ore deposits. Acta Miner Petrogv Szeged 6:23-29

Kopylov GA, Orlov AI (1963) Kinetics of dissolving bornite. Izv vyssk ucheb Zaved, Tsvet Metall 6:68-74

Kopylov GA, Orlov AI (1969) Rates of bornite and chalcocite dissolution in ferric sulphate. Jr Irkutsh Politckh Inst 46:127-132

Kwong YTJ (1995) Influence of galvanic sulfide oxidation on mine water chemistry. In: TP Hynes, MC Blanchette (eds) Proc Sudbury '95—Mining and the Environment 2:477-484, May 28-June 1, 1995, Sudbury, Ontario, CANMET, Ottawa, Canada

Kwong YTJ, Lawrence JR, Swerhone GDW (1995) Interplay of geochemical, electrochemical, and microbial controls in the oxidation of common sulfide minerals. Proc Intl Land Reclam Mine Drainage Conf and 3rd Intl Conf Abatement Acidic Drainage 2:419

Lackey JB (1938) The flora and fauna of surface waters polluted by acid mine drainage. Public Health Report 53:1499-1507

Landesman J, Duncan DW, Walden CC (1966) Oxidation of inorganic sulfur compounds by washed cell suspensions of *Thiobacillus ferrooxidans*. Can J Microbiol 12:957-964

Lane DJ, Stahl DA, Olsen GJ, Heller DJ, Pace NR (1985) A phylogenetic analysis of the genera *Thiobacillus* and *Thiomicrospira* by 5S ribosomal RNA sequences. J Bacteriol 163:75-81

Lawrence RW, Branion RMR, Ebner HG (1986) Fundamental and Applied Biohydrometallurgy. Proc. 6th Intl. Symp. Biohydrometallurgy, Vancouver, BC, Canada. Elsevier, Amsterdam

Leathen WW, Braley SA, McIntyre LE (1953) The role of bacteria in the formation of acid from certain sulfuritic constituents associated with bituminous coal. II. Ferrous iron-oxidizing bacteria. Appl Microbiol 1:65-68

Leathen WW, Kinsel NA, Braley SA, Sr (1956) *Ferrobacillus ferrooxidans*: a chemosynthetic autotrophic bacterium. J Bacteriol 72:700-704

Ledin M, Pedersen K (1996) The environmental impact of mine wastes—roles of microorganisms and their significance in treatment of mine wastes. Earth-Science Reviews 41:67-108

Lindgren W (1928) Mineral Deposits. McGraw-Hill, New York

Lipman JG, McLean H (1916) The oxidation of sulfur soils as a means of increasing the availability of mineral phosphates. Soil Sci 1:533-539

Lizama HM, Suzuki I (1989) Rate equations and kinetic parameters of the reactions involved in pyrite oxidation by *Thiobacillus ferrooxidans*. Appl Environ Microbiol 55:2918-2923

Lizama HM, Suzuki I (1991) Interaction of chalcopyrite and sphalerite with pyrite during leaching by *Thiobacillus thiooxidans*. Can J Microbiol 37:304-311

Lobos JH, Chisholm TE, Bopp LH, Holmes DS (1986) *Acidiphilium organovorum* sp. nov., an acidophilic heterotroph isolated from a *Thiobacillus ferrooxidans* culture. Intl J Syst Bacteriol 36:139-144

Lowe DF (1970) The kinetics of the dissolution reaction of copper and copper-iron sulphide minerals using ferric sulphate solutions. PhD Dissertation, University of Arizona, Tucson, AZ

Lowson RT (1982) Aqueous oxidation of pyrite by molecular oxygen. Chem Rev 82:461-497

Lundgren DG, Andersen KJ, Remsen CC, Mahoney RP (1964) Culture, structure, and physiology of the chemoautotroph *Ferrobacillus ferrooxidans*. Dev Indust Microbiol 6:250-259

Lyalikova NN (1958) A study of chemosynthesis in *Thiobacillus ferrooxidans*. Mikrobiologiya 27:556-559

Lyalikova NN (1960) Participation of *Thiobacillus ferrooxidans* in the oxidation of sulfide ores in pyrite beds of the Middle Ural. Mikrobiologiya 29:382-387

Maciag WJ, Lundgren DG (1964) Carbon dioxide fixation in the chemoautotroph, *Ferrobacillus ferrooxidans*. Biochem Biophys Res Comm 17:603-607

Mackintosh ME (1978) Nitrogen fixation by *Thiobacillus ferrooxidans*. J Gen Microbiol 105:215-218

Malouf EE, Prater JD (1961) Role of bacteria in the alteration of sulphide minerals, J Metals New York 13:353-356

Martin HW, Mills WR, Jr (1976) Water pollution caused by inactive ore and mineral mines. US Environ Protection Agency contract report 68-03-2212

McKibben MA, Barnes HL (1986) Oxidation of pyrite in low temperature acidic solutions—Rate laws and surface textures. Geochim Cosmochim Acta 50:1509-1520

Mehta AP, Murr LE (1982) Kinetic study of sulfide leaching by galvanic interaction between chalcopyrite, pyrite and sphalerite in the presence of *T. ferrooxidans* (30% C) and a thermophilic microorganism (55% C). Biotechnol Bioeng 24:919-940

Moses CO, Herman JS (1991) Pyrite oxidation at circumneutral pH. Geochim Cosmochim Acta 55:471-482

Moses CO, Nordstrom DK, Herman JS, Mills AL (1987) Aqueous pyrite oxidation by dissolved oxygen and by ferric iron. Geochim Cosmochim Acta 55:471-482

Mulak W (1969) Kinetics of cuprous sulphide dissolution in acidic solutions of ferric sulphate. Roczn Chem 43:1387-1394

Mulak W (1971) Kinetics of dissolving polydispersed covellite in acidic solutions of ferric sulphate. Roczn Chem 45:1417-1424

Murr LE, Torma AE, Brierley JA (eds) (1978) Metallurgical Applications of Bacterial Leaching and Related Microbiological Phenomena. Academic Press, London

Mustin C, Berthelin J, Marion P, de Donato P (1992) Corrosion and electrochemical oxidation of pyrite by *Thiobacillus ferrooxidans*. Appl Environ Microbiol 58:1175-1182

Mycroft JR, Nesbitt HW, Pratt AR (1995) X-ray photoelectron and Auger electron spectroscopy of air-oxidized pyrrhotite: Distribution of oxidized species with depth. Geochim Cosmochim Acta 59:721-733

Nathanson A (1902) Über eine neue Gruppe von Schwefelbakterien und ihren Stoffwechsel. Mitt Zool Sta Neapel 15:655-680

Nesbitt HW, Muir IJ (1994) X-ray photoelectron spectroscopic study of a pristine pyrite surface reacted with water vapor and air. Geochim Cosmochim Acta 58:4667-4679

Nesbitt HW, Muir IJ, Pratt, AR (1995) Oxidation of arsenopyrite by air and air-saturated, distilled water, and implications for mechanism of oxidation. Geochim Cosmochim Acta 59:1773-1786

Nicholson RV (1994) Iron-sulfide oxidation mechanisms: Laboratory studies. In: JL Jambor, DW Blowes (eds) Environmental Geochemistry of Sulfide Mine-Wastes, p 163-183, Mineral Assoc Canada, Short Course Handbook 22

Nicholson RV, Scharer JM (1994) Laboratory studies of pyrrhotite oxidation kinetics. In: CN Alpers, DW Blowes (eds) Environmental Geochemistry of Sulfide Oxidation. Am Chem Soc Symp Series 550: 14-30

Nielson AM, Beck JV (1972) Chalcocite oxidation and coupled carbon dioxide fixation by *Thiobacillus ferrooxidans*. Science 175:1124-1126

Nicholson RV, Gillham RW, Cherry JA, Reardon EJ (1990) Reduction of acid generation in mine tailings through the use of moisture-retaining cover layers as oxygen barriers: Reply. Can Geotechnol 27: 402-403

Ngubane WT, Baecker AAW (1990) Oxidation of gold-bearing pyrite and arsenopyrite by *Sulfolobus acidocaldarius* and *Sulfolobus* BC in airlift bioreactors. Biorecovery 1:255-269

Nordstrom DK (1977) Hydrogeochemical and microbiological aspects of the heavy metal chemistry of an acid mine drainage system. PhD Dissertation, Stanford University, Stanford, California

Nordstrom DK (1982) Aqueous pyrite oxidation and the consequent formation of secondary iron minerals. In: JA Kittrick, DS Fanning, LR Hossner (eds) Acid Sulfate Weathering, p 37-56, Soil Sci Soc Am Spec Publ 10:37-56

Nordstrom DK (1985) The rate of ferrous iron oxidation in a stream receiving acid mine effluent. In: Selected Papers in the Hydrologic Sciences. US Geol Survey Water-Supply Paper 2270:113-119

Nordstrom DK, Alpers CN (1995) Remedial investigations, decisions, and geochemical consequences at Iron Mountain Mine, California. Proc Sudbury '95—Mining and the Environment, p 633-642 Sudbury, Ontario, May 28-June 1, 1995.

Nordstrom DK, Alpers CN (1997) Geochemistry of acid mine waters. In: G Plumlee, M Logsdon (eds) Environmental Geochemistry of Mineral Deposits. Rev Econ Geol, Soc Econ Geol (in press).

Nordstrom DK, Ball JW (1985) Toxic element composition of acid mine waters from sulfide ore deposits. 2nd Intl Mine Water Symp, p 749-758 Tallares Graficos ARTE, Granada, Spain

Nordstrom DK, Potter R W, II (1977) The interactions between acid mine waters and rhyolite. 2nd Intl Symp Water-Rock Interactions. Strasbourg, France, I-15 – I-26

Norman PF, Snyman CP (1987) The biological and chemical leaching of an auriferous pyrite/arsenopyrite flotation concentrate: a microscopic examination. Geomicrobiol 6:1-10

Norris PR (1990) Acidophilic bacteria and their activity in mineral sulfide oxidation. In: HL Ehrlich, CL Brierley (eds) Microbial Mineral Recovery, p 3-27, McGraw-Hill, New York

Norris PR, Parrott L (1986) High temperature, mineral concentrate dissolution with *Sulfolobus*. In: RW Lawrence, RMR Branion, HG Ebner, (eds) Fundamental and Applied Biohydrometallurgy, p 355-365 6th Intl Symp Biohydrometall. Vancouver, BC, Canada, August 1985

Okereke A, Stevens SE, Jr (1991) Kinetics of iron oxidation by *Thiobacillus ferrooxidans*. Appl Environ Microbiol 57:1052-1056

Olson GJ (1991) Rate of pyrite bioleaching by *Thiobacillus ferrooxidans*—Results of an interlaboratory comparison. Appl Environ Microbiol 57:642-644

Paciorek KJL, Kratzer RH, Kimble PF, Toben WA, Vatasescu AL (1981) Degradation of massive pyrite: physical, chemical, and bacterial effects. Geomicrobiol J 2:363-374

Pande J (1939) Leaching sulfidic ores. German patent 680 518

Parsons HW, Ingraham TR (1970) The hydrogen sulphide route to sulfur recovery from base metal suphides. Part I: The generation of H_2S from base metal sulphides. Ottawa Dept Natural Resources, Mines Branch Information Circular 242

Pinka J (1991) Bacterial oxidation of pyrite and pyrrhotite. Erzmetall 44:571-573

Pons LJ, Van Breemen N, Driessen PM (1982) Physiography of coastal sediments and development of potential soil acidity. In: JA Kittrick, DS Fanning, LR Hossner, (eds) Acid Sulfate Weathering. Soil Sci Soc Am Special Publ 10:1-18

Powell AW, Parr SW (1919) Forms in which sulfur occurs in coal. Bull AIME:2041-2049

Pratt AR, Muir IJ, Nesbitt HW (1994a) X-ray photoelectron and auger electron spectroscopic studies of pyrrhotite and mechanism of air oxidation. Geochim Cosmochim Acta 58:827-841

Pratt AR, Nesbitt HW, Muir IJ (1994b) Generation of acids from mine waste: Oxidative leaching in dilute H_2SO_4 solutions at pH 3.0. Geochim Cosmochim Acta 58:5147-5159

Pronk JT, deBruyn JC, Bos P, Kuenen JG (1992) Anaerobic growth of *Thiobacillus ferrooxidans*. Appl Environ Microbiol 58:2227-2230

Purcell E (1977) Life at low Reynold's number. Am J Phys 45:3-11

Ramsay B, Ramsay J, de Tremblay M, Chavarie C (1988) A method for the quantification of bacterial protein in the presence of jarosite. Geomicrobiol J 6:171-177

Razzell WE, Trussell PC (1963) Microbiological leaching of metallic sulphides. Appl Microbiol 11: 105-110

Rickard PAD, Vanselow DG (1978) Investigations into the kinetics and stoichiometry of bacterial oxidation of covellite (CuS) using a polarographic probe. Can J Microbiol 24:998-1003

Rimstidt JD, Chermak JA, Gagen PM (1994) Rates of reaction of galena, sphalerite, chalcopyrite, and arsenopyrite with Fe(III) in acidic solutions. In: CN Alpers, DW Blowes (eds) Environmental Geochemistry of Sulfide Oxidation, Am Chem Soc Symp Series 550:2-13, American Chemical Society, Washington, DC

Rodriguez M, Campos S, Gomez-Silva B (1986) Studies on native strains of *Thiobacillus ferrooxidans* III. Studies on the outer membrane of *Thiobacillus ferrooxidans*. Characterization of the lipopolysaccharide and some proteins. Biotechnol Appl Biochem 8:292-299

Rudolfs A, Helbronner A (1922) Oxidation of zinc sulphide by microorganisms. Soil Sci 14:459-464

Sakaguchi H, Torma AE, Silver M (1976) Microbiological oxidation of synthetic chalcocite and covellite by *Thiobacillus ferrooxidans*. Appl Environ Microbiol 31:7-10

Sand W, Gerke T, Hallmann R, Schippers A (1995) Sulfur chemistry, biofilm, and the (in)direct attack mechanism—a critical evaluation of bacterial leaching. Appl Microbiol Biotechnol 43:961-966

Sasaki K (1994) Effect of grinding on the rate of oxidation by oxygen in acid solutions. Geochim Cosmochim Acta 58:4649-4655

Sasaki K, Tsunekawa M, Ohtsuka T, Konno H (1995) Confirmation of a sulfur-rich layer on pyrite after oxidative dissolution by Fe(III) ions around pH 2. Geochim Cosmochim Acta 59:3155-3158

Sato M (1960) Oxidation of sulfide ore bodies. II. Oxidation mechanisms of sulfide minerals at 25°C. Econ Geol 55:1202-1234

Scala G, Mills AL, Moses CO, Nordstrom DK (1982) Distribution of autotrophic Fe and sulfur-oxidizing bacteria in mine drainage from sulfide deposits measured with the FAINT assay [abstr]. Am Soc Microbiol Ann Mtg

Schippers A, Peter-Georg J, Sand W (1996) Sulfur chemistry in bacterial leaching of pyrite. Appl Environ Microbiol 62:3424-3431

Schnaitman CA, Lundgren DG (1965) Organic compounds in the spent medium of *Ferrobacillus ferrooxidans*. Can J Microbiol 11:23-27

Schnaitman CA, Korczynski MS, Lundgren DG (1969) Kinetic studies of iron oxidation by whole cells of *Ferrobacillus ferrooxidans*. J Bacteriol 99:552-557

Schrader JA, Holmes DS (1988) Phenotypic switching of *Thiobacillus ferrooxidans*. J Bacteriol 170:3915-3923

Shively JM, Decker GL, Greenwalt JW (1970) Comparative ultrastructure of the thiobacilli. J Bacteriol 101:618-627

Silver M, Torma AE (1974) Oxidation of metal sulphides by *Thiobacillus ferrooxidans* grown on different substrates. Can J Microbiol 20:141-147

Silverman MP (1967) Mechanism of bacterial pyrite oxidation. J Bacteriol 94:1046-1051

Silverman MP, Ehrlich HL (1964) Microbial formation and degradation of minerals. Adv Appl Microbiol 6:153-206

Silverman MP, Lundgren DG (1959) Studies on the chemoautotrophic iron bacterium *Ferrobacillus ferrooxidans*. II. Manometric studies. J Bacteriol 78:326-331

Silverman MP, Rogoff MH, Wender I (1961) Bacterial oxidation of pyritic materials in coal. Appl Microbiol 9:491-496

Singer PC, Stumm W (1970) Acidic mine drainage: the rate determining step. Science 167:1121-1123

Sokolova GA, Karavaiko GI (1964) Physiology and Geochemical Activity of Thiobacilli. Izdatel'stvo Nauka, Moscow, USSR. English translation: Y Halpern (translator) E Rabinovitz (ed) (1968) US Dept Commerce, Springfield, Virginia

Southam G, Beveridge TJ (1992) Enumeration of thiobacilli with pH-neutral and acidic mine tailings and their role in the development of secondary mineral soil. Appl Environ Microbiol 58:1904-1912

Southam G, Beveridge TJ (1993) Examination of lipopolysaccharide (O-antigen) populations of *Thiobacillus ferrooxidans* from two mine tailings. Appl Environ Microbiol 59:1283-1288

Southam G, Firtel M, Blackford BL, Jericho MH, Xu W, Mulhern PJ, Beveridge TJ (1993) Transmission electron microscopy, scanning tunneling microscopy and atomic force microscopy of the cell envelope layers of the archaeobacterium *Methanospirillum hungatei* GP1. J Bacteriol 175:1946-1955

Southam G, Ferris FG, Beveridge TJ (1995) Mineralized bacterial biofilms in sulphide tailings and in acid mine drainage systems. In: HM Lappin-Scott, JW Costerton (eds) Microbial Biofilms, p 148-170 Cambridge University Press, Cambridge, UK

Steger HF, Desjardins LE (1978) Oxidation of sulfide minerals. IV. Pyrite, chalcopyrite, and pyrrhotite. Chem Geol 23:225-237

Stevens CJ, Dugan PR, Tuovinen OH (1986) Acetylene reduction (nitrogen fixation) by *Thiobacillus ferrooxidans*. Biotechnol Appl Biochem 8:351-359

Stokes HN (1901) On Pyrite and Marcasite. US Geol Survey Bull 186

Strömberg B (1997) Weathering kinetics of sulphidic mining waste: An assessment of geochemical processes in the Aitik mining waste rock deposits. PhD Dissertation, Royal Institute of Technology, Stockholm. AFR-Report 159

Strömberg B, Banwart S (1994) Kinetic modelling of geochemical processes at the Aitik mining waste rock site in northern Sweden. Appl Geochem 9:583-595

Sullivan JD (1930a) Chemistry of leaching chalcocite. US Bureau Mines TP-473

Sullivan JD (1930b) Chemistry of leaching covellite. US Bureau Mines TP 487

Sullivan JD (1931) Chemistry of leaching bornite. US Bureau Mines TP-486

Sullivan JD (1933) Chemical and physical features of copper leaching. Trans Am Inst Min Metall 106: 515-546

Sutton JA, Corrick JD (1961) Bacteria in mining and metallurgy: Leaching selected ores and minerals, experiments with *Thiobacillus ferrooxidans*. US Bureau Mines RI-5839

Suzuki I, Takeuchi TL, Yuthasastrakosol TD, Oh JK (1990) Ferrous iron and sulfur oxidation and ferric iron reduction activities of *Thiobacillus ferrooxidans* are affected by growth on ferrous iron, sulfur, or a sulfide ore. Appl Environ Microbiol 56:1620-1626

Sveshnikov GB, Dubychin SL (1956) Electrochemical solution of sulfides and dispersion aureoles of heavy metals. Geokhimya 4:413-419

Sveshnikov GB, Ryss YuS (1964) Electrochemical processes in sulfide deposits and their geochemical significance. Geokhimya 3:208-218

Taylor KA, Deatherage JF, Amos LA (1982) Structure of the S-layer of *Sulfolobus acidocaldarius*. Nature 299:840-842

Temple KL, Colmer AR (1951) The autotrophic oxidation of iron by a new bacterium—*Thiobacillus ferrooxidans*. J Bacteriol 62:605-611

Temple KL, Delchamps EW (1953) Autotrophic bacteria and the formation of acid in bituminous coal mines. Appl Microbiol 1:255-258

Temple KL, Koehler WA (1954) Drainage from Bituminous Cave Mines, West Virginia Univ Bull 25, Ser 54, Morgantown, WV

Tervari PH, Campbell AB (1978) Dissolution of iron sulfide (troilite) in aqueous sulfuric acid. J Phys Chem 80:1844-1848

Thomas G, Ingraham TR (1967) Kinetics of dissolution of synthetic covellite in aqueous acidic ferric sulphate solutions. Can Metall Q 6:153-165

Thomas G, Ingraham TR, MacDonald RJC (1967) Kinetics of dissolution of synthetic digenite and chalcocite in aqueous acidic ferric sulphate solutions. Can Metall Q 6:281-291

Torma AE (1971) Microbiological oxidation of synthetic cobalt, nickel, and zinc sulphides by *Thiobacillus ferrooxidans*. Rev Can Biol 30:209-216

Torma AE (1978) Oxidation of gallium sulphides by *Thiobacillus ferrooxidans*. Can J Microbiol 24: 888-891

Torma AE, Gabra GG (1977) Oxidation of stibnite by *Thiobacillus ferrooxidans*. Antonie v Leeuwenhoek 43:1-6

Torma AE, Habashi F (1972) Oxidation of copper (II) selenide by by *Thiobacillus ferrooxidans*. Can J Microbiol 18:1780-1781

Torma AE, Subramanian KN (1974) Selective bacterial leaching of a lead sulphide concentrate. Intl J Mineral Processing 1:125-134

Torma AE, Gabra GG, Guay R, Silver M (1976) Effects of surface active agents on the oxidation of chalcopyrite by *Thiobacillus ferrooxidans*. Hydrometall 1:301-309

Torma AE, Legault G, Kougiomoutzakis D, Oullet R (1974) Kinetics of bio-oxidation of metal sulphides. Can J Chem Eng 52:515-517

Torma AE, Walden CC, Duncan DW, Branion RM (1972) The effect of carbon dioxide and particle surface area on the microbiological leaching of a zinc sulphide concentrate. Biotech Bioeng 15:777-786

Trafford BD, Bloomfield C, Kelso WI, Pruden G (1973) Ochre formation in field drains in pyritic soils. J Soil Sci 24:453-460

Tributsch H (1976) The oxidative disintegration of sulfide crystals by *Thiobacillus ferrooxidans*. Naturwiss 63:88

Tugov NO, Tsyganov GA (1963) Hydrometallurgical method of obtaining metallic antimony from concentrates. Uzbek khim Zh 7:17-21

Uchida T, Watanabe A, Furuya S (1967) Leaching of copper from copper-bearing ores with a dilute solution of sulphuric acid and ferric sulphate. Hakko Kyokaishi 25:168-172

Usataya ES (1952) Oxidation of molybdenite by water solutions. Zap Vses Miner Obshch 8:298-303

Van Breeman N (1976) Genesis and solution chemistry of acid sulfate soils in Thailand. PhD Dissertation, Agricultural University Wageningen

Vestal JR, Lundgren DG, Milner KC (1973) Toxic and immunological differences among lipopolysaccharides from *Thiobacillus ferrooxidans* grown autotrophically and heterotrophically. Can J Microbiol 19:1335-1339

Wang WS, Korczynski MS, Lundgren DG (1970) Cell envelope of an iron-oxidizing bacterium: Studies of lipopolysaccharide and peptidoglycan. J Bacteriol 104:556-565

Waksman SA, Jåffe JS (1921) Acid production by a new sulfur-oxidizing bacterium. Science 53:216

Waksman SA, Jåffe JS (1922) Microorganisms concerned in the oxidation of sulphur in the soil II. *Thiobacillus thiooxidans*, a new sulphur-oxidizing organism isolated from the soil. J Bacteriol 7:239-256

Walsh AW, Rimstidt JD (1986) Rates of reaction of covellite and blaubleibender covellite with ferric iron at pH 2.0. Can Mineral 24:35-44

Walsh F, Mitchell R (1972) An acid-tolerant iron-oxidizing Metallogenium. J Gen Microbiol 72:369-376

Wichlacz PL, Thompson DL (1988) The effect of acidophilic bacteria on the leaching of cobalt by *Thiobacillus ferrooxidans*. In: PR Norris, DP Kelly (eds) Biohydrometallurgy, Proc Intl Symp 1986, Warwick, UK. Science Tech Letters, p 77-86, Kew, UK

Wichlacz PL, Unz RF (1981) Acidophilic, heterotrophic bacteria of acidic mine waters. Appl Environ Microbiol 41:1254-1261

Wichlacz PL, Unz RF, Langworthy TA (1986) *Acidiphilium angustum* sp. nov., *Acidiphilium facilis* sp. nov., *Acidiphilium rubrum* sp. nov.: acidophilic heterotrophic bacteria isolated from acidic coal mine drainage. Intl J Syst Bacteriol 36:197-201

Wiersma CL, Rimstidt J D (1984) Rates of reaction of pyrite and marcasite with ferric iron at pH 2, Geochim Cosmochim Acta 48:85-92

Williamson MA, Rimstidt JD (1993) The rate of decomposition of the ferric-thiosulfate complex in acidic aqueous solutions. Geochim Cosmochim Acta 57:3555-3561

Winogradsky SN (1888) Über Eisenbakterien. Bot Ztg 46:261-276

Wood AP, Kelly DP (1991) Isolation and characterization of *Thiobacillus halophilus* sp. nov., a sulfur-oxidizing autotrophic eubacterium from a Western Australian hypersaline lake. Arch Microbiol 156:277-280

Yakhontova LK, Zeman I, Nesterovich LG (1980) Oxidation of tetrahedrite. Doklady, Earth Sci Sect, Akad Nauk SSSR 253:461-464

Yin Q, Kelsall GH, Vaughan DJ, Englund KER (1995) Atmospheric and electrochemical oxidation of the surface of chalcopyrite ($CuFeS_2$). Geochim Cosmochim Acta 59:1091-1100

Yokota A, Yamada Y, Imai K (1988) Lipopolysaccharides of iron-oxidizing *Leptospirillum ferrooxidans* and *Thiobacillus ferrooxidans*. J Gen Appl Microbiol 4:27-37

Zillig W, Stetter KO, Wunderl A, Schulz W, Priess H, Scholz I (1980) The *Sulfolobus "Caldariella"* group: Taxonomy on the basis of the structure of DNA-dependent RNA polymerases. Arch Microbiol 125:259-269

Chapter 12

BIOGEOCHEMICAL WEATHERING OF SILICATE MINERALS

William W. Barker, Susan A. Welch and Jillian F. Banfield

Department of Geology and Geophysics
Weeks Hall, 1215 West Dayton Street
The University of Wisconsin-Madison
Madison, Wisconsin 53706 U.S.A.

INTRODUCTION

An enormous body of increasingly interdisciplinary scientific literature reflects the profound and pervasive entwinement of biological processes and silicate mineral weathering. While the geomicrobiology of silicate mineral weathering might seem an extremely arcane subject, consider the following short list of processes affected by biogeochemically mediated cycles:

- Soil formation and plant nutrition
- Stability of architectural materials
- Geochemistry of groundwater and movement of contaminants
- Long term stability of geological nuclear repositories
- Effects of mineral weathering on climate on a geological time scale

Soil scientists are concerned with pedogenesis, the process of soil formation, and recognize the central role of organisms in mineral weathering (see, for example, Huang and Schnitzer 1986, Jones 1988). Weathering of minerals in soils releases major nutrients such as K, P, Fe, Ca, Mg, Si, as well as trace ions which are necessary for microbial and plant growth. Plant nutritionists, primarily working with model rhizospheres, have documented the dramatic and astonishingly rapid biomobilization of essential nutrients from phyllosilicates (Hinsinger et al. 1992, Leyval and Berthelin 1991).

Engineers and conservationists, concerned with the long term stability of architectural materials, have contributed greatly to defining communities of organisms and the damage they cause to natural and synthetic building materials. Several investigators (De la Torre et al. 1993, Eckhardt 1978, Krumbein et al. 1991, Palmer et al. 1991, Sand and Bock 1991, Urzi et al. 1991) have focused on the biological weathering of stone building material and glass, and have demonstrated enhanced corrosion and weathering of these materials associated with attached organisms, presumably due to the production of acidic, alkaline, or complexing compounds which would enhance the dissolution of the solid substrata. Piervittori et al. (1994) published an excellent review on the role of lichens in rock and mineral weathering, both in a natural and architectural context. Easton (1994) reviewed the literature from a geologist's perspective.

While microbial weathering of minerals in soils and structures at the Earth's surface is widely recognized, similar processes also occur in subsurface environments. Recent surveys of deep subsurface aquifers (hundreds of meters), show relatively high microbial numbers (approximately 10^5 to 10^7 cells/cm^3). Microbial abundance and diversity is variable but does not decrease systematically with depth (Balkwill 1989). While some bacteria are free living in solution, most occur attached to mineral surfaces (Hazen et al. 1991, Holm et al. 1992) where they can directly impact mineral surface chemistry, water

0275-0279/97/0035-0012$05.00

rock interaction, and groundwater geochemistry (Bennett et al. 1996, Chapelle 1993, Chapelle et al. 1987, Chapelle and Lovely 1990, Fyfe 1996, Ghiorse 1997, Gurevich 1962, Hiebert and Bennett 1992, Lovely and Chapelle 1993, McMahon et al. 1991, 1992). Attached microorganisms can etch mineral surfaces, releasing ions to solution (Bennett et al. 1996, Hiebert and Bennett 1992), and also serve as nucleation sites for a variety of secondary mineral phases (Ehrlich 1996, Ferris et al. 1988, 1989, 1994; Konhauser et al. 1994, Konhauser and Ferris 1996, Schultze-Lam et al. 1992, 1994, 1996). Microbial redox processes in aquifers can directly affect the distribution of elements such as C, O, N, Fe, Mn, S, As, U (Herring and Stumm 1990). These in turn can have indirect effects on other geochemical reactions. For an extensive review of groundwater biogeochemistry, see Chapelle (1993).

Environmentally acceptable sequestration of nuclear waste materials over periods of time approaching geological scales forms one of the most pressing environmental problems of modern society. Currently, vitrification, encapsulation of radioactive materials in a glass matrix, and storage in underground repositories appears to be the most politically expedient course of action. In these environments, particularly over the long periods of time needed to safely ensure storage, groundwater and its associated microorganisms may interact with the glass surfaces, releasing ions (possible radioactive) to the environment. Bacteria colonizing mineral surfaces in aquifers, as mentioned above, preferentially etch attachment surfaces and are well known to have the ability to solubilize glass (Staudigel et al. 1995). At some point in the future, if and when water and subsurface lithotrophic microbes interact with the vitrified waste, release of radioactive contaminants will occur.

Global warming due to accumulation of CO_2 in the atmosphere constitutes another serious environmental problem. A major feedback mechanism controlling atmospheric pCO_2 is the weathering of Ca-Mg silicates and subsequent precipitation of Ca-Mg carbonates (Berner et al. 1983). Because microorganisms may have affected biogeo-chemical weathering reactions for at least 3.8 Ga, the biota could have moderated the CO_2-Si cycle, and could have therefore affected pCO_2 and global climate (Gwiazda and Broecker 1994, Schwartzman and Volk 1991).

METHODS

Lab studies

Experimental studies of the effects of microorganisms on mineral weathering rates and reactions are commonly performed in either closed (batch) or open (flow-through) reaction vessels. In batch experiments, minerals and solutions are placed in a container, solution composition is monitored over time, and reaction rates are calculated from the evolution of solution chemistry. In flow through experiments, solution passes through reaction vessels containing mineral samples and reaction rates are calculated from effluent concentration and flow rate. Mineral weathering rates determined *in vitro* typically exceed those derived from field-based measurements by one to three orders of magnitude, because these controlled laboratory experiments cannot simulate the more complicated water-rock interactions in natural environments (Brantley 1992, Swoboda-Colberg and Drever 1992, Velbel 1986, 1990). While difficult to predict the absolute magnitude of a reaction, one may detect effects of microbial metabolites on weathering reactions.

When these types of experiments include microorganisms, unrealistically high concentrations of organic carbon and nutrients are added to solution (Avakyan et al. 1981, 1985, Berthelin and Dommergues 1972, Berthelin 1971, Karavaiko et al. 1984, 1979; Kutuzova 1969, Malinovskaya et al. 1990, Vandevivere et al. 1994). Microorganisms often significantly enhance mineral dissolution rates *in vitro* by producing high concentrations of organic and inorganic acids. It is difficult to extrapolate from these high nutrient experiments to estimate the magnitude of this effect in nature, where mineral weathering rates, organic carbon, and nutrient concentrations, are typically orders of magnitude lower. As previously mentioned, bacterial distributions in natural environments are typically heterogeneous and most are attached to mineral surfaces. In contrast, most bacteria in laboratory weathering experiments are in suspension, where any dissolution enhancing metabolites produced would be diluted by the bulk solution. Laboratory mineral dissolution experiments typically use a single strain of bacteria or else an assemblage of microorganisms which can grow in culture, and therefore may not accurately represent the distribution of microorganisms and complex biogeochemical processes occurring in nature. Additional natural variables which are difficult to simulate *in vitro* are fluctuations in water potential, temperature and light levels. Based on the results of these types of laboratory experiments we can determine whether or not microorganisms can affect geochemical reactions. It is unclear whether one can extrapolate those results to natural systems and predict the magnitude of any observed effects under field conditions.

Field studies

Extensive field evidence exists for microbially-mediated mineral weathering on rock surfaces and in soils, aquifers, and the marine environment. Natural microbially mediated mineral weathering reactions have been investigated by direct analysis of mineral surfaces and attached microorganisms using light and electron microscopy. Microbially enhanced mineral weathering has also been determined indirectly by analyzing natural waters for mineral solutes and microbial metabolites.

Bacteria are ubiquitous in the Earth's surface and subsurface environments and play an important role in mineral weathering and geochemical cycling of elements. *In situ* microcosm studies (Bennett and Hiebert 1992, Hiebert and Bennett 1992, Bennett et al. 1996) showed that feldspar and quartz weathering in a shallow aquifer increased when bacteria colonized mineral surfaces, even when solutions were supersaturated with respect to the dissolving phase. Many etch pits seen on mineral surfaces approximated the size and shape of bacteria colonizing mineral surfaces, and the authors hypothesized that attached bacteria locally increased mineral dissolution rates by creating a micro-reaction zone of organic acids and other metabolites concentrated at the mineral surface. Thorseth et al. (1992) found similar etch structures associated with colonizing cyanobacteria on palagonite surfaces. In this case, however, they attributed enhanced weathering to production of highly alkaline microenvironments. Bacteria can also play a role in the alteration of silicate minerals in marine environments. Thorseth et al. (1995a,b) examined glass rims of pillow lavas from oceanic crust. Once again they found microbe sized etch features, in this case containing DNA. The basalt chemistry had been altered in zones immediately surrounding the etch features and potassium was enriched within the microbial etch pits, indicating that viable bacteria actively altered glass surfaces.

While evidence for bacterially mediated weathering of silicate minerals is not always as compelling as finding a bacterium in a bacteria-shaped etch pit, there is often good correlation between mineral solutes and microbial metabolites in solution. Berthelin et al. (1985) found a nearly 1:1 correlation between cations released to solution and nitrate concentration in soil, indicating that bacterial nitrification releases mineral cations to solution by ion exchange and proton promoted mineral dissolution. Bacterial production of organic acids in aquifers has been cited as a mechanism for the release of solutes from minerals to solutions (Bennett and Siegel 1987, McMahon and Chapelle 1991, McMahon et al. 1992).

Abundant field evidence for fungal mediation of mineral weathering also exists, though most of this comes from literature on weathering of architectural materials (De la Torre et al. 1993, Hirsch et al. 1995, Krumbein et al. 1991, Palmer et al. 1991, Palmer and Hirsch 1991, Sand and Bock 1991, Urzi et al. 1991). Bacteria and fungi commonly coexist and it is difficult to distinguish between bacterial and fungal weathering of silicates. Fungi may be more important than bacteria in microbially-mediated weathering as they produce and excrete higher concentrations of more effective dissolution enhancing ligands (Palmer et al. 1991). But fungi are not as abundant as bacteria in most environments and do not occur in some extreme environments where bacteria flourish. Algae are also common on rock surfaces and can influence mineral weathering reactions though they are usually considered to be less effective than bacteria or fungi (Ortega-Calvo et al. 1995). While algae may not significantly impact mineral weathering reactions, they do provide a carbon rich substrate for both bacteria and fungi.

In lichens, green algae and/or cyanobacteria (phycobiont) form a symbiotic (Nash 1996) or "controlled parasitic" (Ahmadjian 1995) relationship with fungi (mycobiont). While a host of other organisms such as rotifers, protozoa, diatoms, bacteria, unlichenized cyanobacteria and fungi, tardigrades, arthropods, and gastropods inhabit these lithobiontic (Golubic et al. 1981) communities, most biogeochemical weathering research has focused on the lichen component.

Lichens aggressively attack mineral surfaces, resulting in both physical (Fry 1927) and chemical weathering of rock surfaces (see reviews by Easton 1994, Jones 1988, Piervittori et al. 1994). Early researchers used light microscopy to document extensive disaggregation of minerals and binding of the resultant particulate by lichens to form "organomineral dust" (Fry 1927). With the application of X-ray diffraction (XRD) and scanning electron microscopes (SEM) equipped with energy-dispersive X-ray microanalysis (EDS), researchers documented strongly etched mineral surfaces beneath lichens (Jones et al. 1980, Jones et al. 1981) as well as metal oxalates whose cationic character clearly mirrored that of the mineral substratum (Chisolm et al. 1987, Purvis 1984, Wilson et al. 1980, Wilson and Jones 1984). Recent research applied back-scattered electron (BSE) imaging to study biodeterioration of minerals by lithobiontic microorganisms (Ascaso and Wierzchos 1994, Ascaso et al. 1995, Barker and Banfield 1996, Sanders et al. 1994, Wierzchos and Ascaso 1994, 1996).

While these studies are important in understanding the relationships between lichens and minerals, the reactions which characterize mineral transformations take place at a scale beyond the resolving power of SEM. A high-resolution transmission electron microscope (HRTEM) provides simultaneous structural information via selected area electron diffraction (SAED) and chemical data through EDS. Thus, it is an ideal instrument with which to study biogeochemical mineral transformation reactions at the subnanometer scale. Difficulties inherent in producing samples sufficiently thin for TEM analysis while preserving the intact organic-mineral interface have limited its application. Recent developments in energy filtered transmission electron microscopy (EFTEM) have made it possible to not only collect zero loss images of exceptional clarity (chromatic aberration is eliminated by the energy filter) and contrast (without using contrast-enhancing heavy metal stains), but also electron energy loss (EELS) images that detail elemental distributions at the nanometer scale.

While TEM has been used to characterize minerals extracted from lichens, most preparative methods employed to date involve scraping the mineral grains accumulated at the lichen-rock interface free from the thallus, and removing the organic component with oxidation in H_2O_2 and ultrasonication. These techniques destroy all textural information and drastically alter the mineralogical component (Birrell 1971, Cook 1992, Douglas and

Fiessinger 1971, Lakvulich and Wiens 1970, Perez-Rodriguez and Wilson 1969). Recently, Ascaso and Ollacarizqueta (1991) reported limited success in an attempt to apply diamond knife ultramicrotomy to study the interface of lichens and a marble building stone. Barker and Banfield (1996) and Wierzchos and Ascaso (1996) used a combination of biological and physical sample preparation techniques to examine the intact organic-mineral interface by a variety of electron microscopic techniques. Such direct comparisons of physiochemical and biogeochemical weathering of silicate minerals by microorganisms comprising lithobiontic communities on outcrops of known exposure age allows both mechanisms and rates to be determined from field observations.

MECHANISMS

The following biological processes are commonly cited as explanations for increased silicate mineral weathering.

(1) Growing roots and fungal hyphae physically disrupt minerals, exposing new fresh mineral surfaces and increasing surface area available for reaction.

(2) Soil stabilization increases water retention, lengthening time for weathering reactions to occur.

(3) Acid production, primarily carbonic acid from CO_2, but also other inorganic and organic acids, accelerates weathering rates.

(4) Organic ligands directly attack mineral surfaces or form complexes with ions in solution, changing saturation state.

(5) Complex extracellular polymers moderate water potential, maintain diffusion channels, act as ligands or chelators, and serve as nucleation sites for authigenic mineral formation.

(6) Nutrient absorption, primarily K, Fe and P, decreases solution saturation state and enhances weathering.

Physical disaggregation

The ability of microorganisms, particularly lichen communities, to physically disaggregate rock surfaces, is perhaps the most strikingly obvious biological process involved in silicate mineral weathering (Fig. 1). This physical disaggregation of minerals increases mineral surface area and exposes new fresh mineral surfaces for biogeochemical weathering. When one views a thin section across the intact mineral-lithobiontic community interface, for example the sheer physical disruption required to produce such shattered mineral grains is the most immediately impressive aspect of these organomineral interfaces (Barker and Banfield 1996, Wierzchos and Ascaso 1996). The true extent of biophysical disaggregation is only fully appreciated when one views an optical thin section across the lichen-mineral interface using backscattered electron imaging. Microorganisms aggressively invade and colonize mineral substrata as soon as the space becomes available. The mechanisms involved in biophysical disaggregation of minerals are not well understood. For rock surfaces, freeze-thaw and subsequent colonization of freshly exposed mineral surfaces along grain boundaries and cleavages is probably an important mechanism. While forces generated by the alternate wetting and drying of lichen thalli are often cited as an explanation for shattered rock surfaces, apparently no one has ever measured forces which might be generated by these hydration/dehydration cycles. Griffin (1981) points out that the availability of water works in concert with the physical pore size to control the distribution of microorganisms in nonsaturated environments, most commonly soils and on rock surfaces. While bacterial movement is much more limited by falling water potential (Papendick and Campbell 1981) than that of fungi (Griffin 1981), the latter require much larger passageways in rocks and soils. Rhizosphere studies comparing mineral weathering

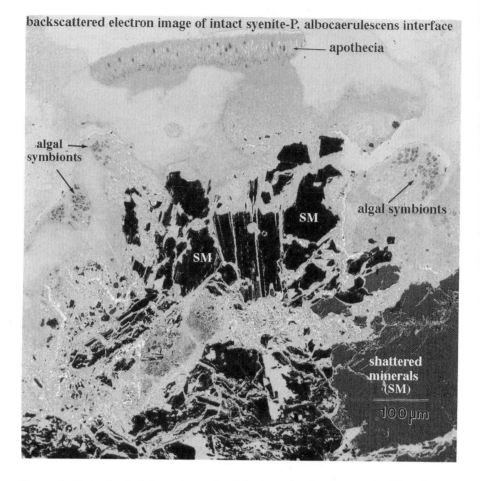

Figure 1. Backscattered electron micrograph of the intact interface between the lichen *Porpidia albocaerulescens* and metasyenite. Fungal hyphae have penetrated along grain boundaries and cleavages to extensively disaggregate the rock surface (Barker and Banfield 1996).

by roots, roots plus associated symbiotic microorganisms and roots plus symbionts plus other rhizosphere organisms demonstrate a strong correlation between K release from biotite and phlogopite for the latter group, attributable to "better soil exploration" (Berthelin and Leyval 1982, Leyval and Berthelin 1991).

Soil stabilization

It has been suggested that the most critical factor affecting the biotic enhancement of weathering is the stabilization of soils (Drever 1994, Schwartzman et al. 1993). There is a strong positive correlation between soil stabilization and organic content, although this organic content is comprised of both plant and microbial carbon compounds (Oades 1993). Roots of higher plants stabilize larger soil aggregates, and provide channels for water and oxygen to reach soil particles. Fungal hyphae physically bind soil particles together as well (Forster and Nicolson 1981a,b). These ectomycorrhyzal fungal hyphae dramatically

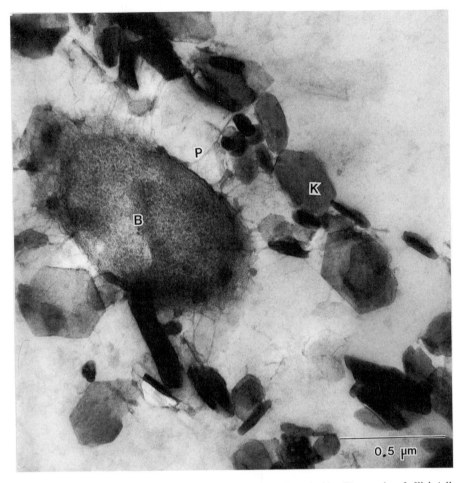

Figure 2. Kaolinite (K) enmeshed in the anionic mucopolysaccharide (P) capsule of *Klebsiella pneumoniae* (B). Ruthenium red-stained diamond knife thin section, bright-field transmission electron micrograph (Barker 1988).

increase contact area between biological and inorganic mineral surfaces, enhancing exposure of minerals to corrosive substances (water and metabolites). Interparticulate bridging of soil minerals by extracellular polysaccharides (Fig. 2) is a major component of soil aggregation (Tisdall 1994). Polymers that have high concentration of acid functional groups, primarily uronic acids, are most effective for enhancing sediment stability (Dade et al. 1990). While the aggregation of soil particles by plants and microorganisms decrease physical weathering rates, increased soil surface area and water residence time increase chemical weathering. Microbial colonization and stabilization of soils is a critical preliminary step for colonization of surfaces by higher plants (Forster and Nicolson 1981b).

Inorganic acids

Silicate mineral dissolution is a complex function of pH (Fig. 3) (Blum and Lasaga 1988, Brady and Walther 1989, Chou and Wollast 1985, Hellmann 1994, Welch and

Ullman 1993, 1996). Acidic or alkaline compounds produced by microbial metabolic processes may mediate silicate weathering by changing solution pH. At the pH of most natural environments, 5-8, experimentally determined mineral dissolution rates are lowest and essentially independent of pH (Brady and Walther 1989, Chou and Wollast 1985, Welch and Ullman 1993, 1996). As acidity increases, below pH 5, mineral dissolution rates increase by a factor of $a_{H^+}^n$ where n is the fractional dependence of mineral dissolution rate on proton activity and is related to proton adsorption on the mineral surface (Blum and Lasaga 1988). Similarly, in basic solution above pH 8-9, mineral dissolution rates increase by a factor proportional to $a_{OH^-}^m$, where m is related to the adsorption of hydroxyl ions (Blum and Lasaga 1988, Brady and Walther 1989). The pH dependence of mineral dissolution varies with mineral composition, oxidation and coordination state of ions at the mineral surface, and zero point of charge.

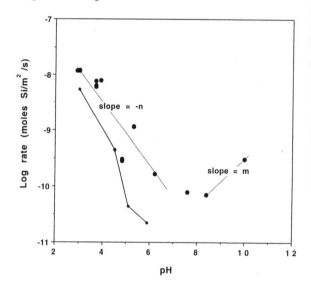

Figure 3. Si release rate from feldspar as a function of pH in inorganic ($\blacklozenge$) and organic ($\bullet$) solutions. Organic acids enhance feldspar dissolution from acidic to neutral pH (data from Welch 1991).

The combination of respired CO_2 with water to form carbonic acid constitutes the most basic biochemical species thought to accelerate mineral weathering. Carbonic acid is a relatively weak inorganic species but can affect silicate mineral weathering rates by interacting with mineral surfaces. The pH of low ionic strength solutions in equilibrium with atmospheric pCO_2 (350 ppm) is approximately 5.7. However, dissolution rates of most silicate minerals are pH independent at circumneutral pH, so carbonic acid in rain or surface run off may have little enhancement effect over hydrolysis reactions. Respiration by plant roots and microbial degradation of organic matter can elevate carbonic acid concentration in soils, sediments, and groundwater by several orders of magnitude, presumably leading to an increase in mineral weathering rates (Chapelle et al. 1987, Schwartzman and Volk 1989, 1991). Acidity in natural waters rarely decreases below pH 4.5 due to carbonic acid, however. This is approximately the pH where mineral dissolution rates start to increase with acidity (Drever 1994), and increased weathering due to carbonic acid production by organisms in soils and sediments will depend on how sensitive reactions are to changes in pH in this range (~4.5 to 6). Drever (1994) estimates that it will be at most a factor of 3. At very high total CO_2 concentrations and high ionic strengths, silicate mineral dissolution may be inhibited as carbonate complexes with metal sites on the mineral surface, or insoluble carbonates precipitate on mineral surfaces, limiting hydrolysis reactions (Ferris et al. 1994, Wogelius and Walther 1991).

Because carbonic acid is a weak acid, microbial respiration may not significantly impact mineral weathering. However, lithotrophic microorganisms also produce other strong inorganic acids, primarily sulfuric and nitric from oxidation of reduced sulfur and nitrogen compounds. Reactions involving reduced sulfur compounds generally occur at interfaces between oxygenated and anoxic environments (e.g. in biofilms, microbial mats, soils or sediment microenvironments), or due to anthropogenic activity (e.g. wells, mining, etc.) and can cause severe but localized pH changes and increased weathering of minerals (Berthelin 1983, Sand and Bock 1991; Nordstrom and Southam, this volume). Nitrifying bacteria, those that oxidize ammonia to nitrate, have been implicated in leaching of cations from rocks and soils by acidolysis and ion exchange (Berthelin et al. 1985, Lebedeva et al. 1979).

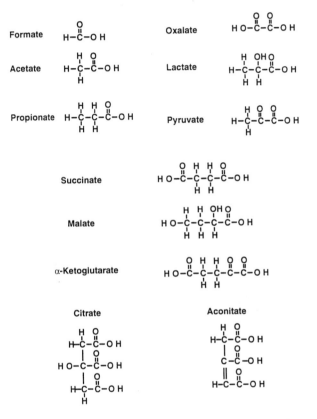

Figure 4. Structures of common organic compounds produced by microorganisms.

Organic acids

Low molecular weight organic acids, in particular oxalic, are most often cited as the main component in biogeochemical weathering of silicate minerals. Microorganisms produce and excrete a variety of low molecular weight organic ligands including acetate, oxalate, formate, propionate, lactate, pyruvate, citrate, succinate and α-ketoglutarate (Berthelin 1983, Johnston and Vestal 1993, McMahon and Chapelle 1991, Palmer et al. 1991). The simplified form of these compounds is illustrated in Figure 4. As is the case with any acid species, numerous studies have demonstrated increased mineral dissolution rates *in vitro* . Many of these ligands do occur in soils, sediment pore water, and aquifers in sufficient concentrations to significantly impact silicate mineral weathering. Several tens to hundreds of micromolar concentrations of oxalate, formate, citrate, malate, and aconitate have been found in rhizosphere soils (Fox and Comerford 1990, Grierson 1992) and in lesser amounts in non-rhizosphere soils. Organic acid concentrations in oxic uncontaminated groundwater are generally very low, on the order of a few to few tens micromolar concentration. These organic acids are often not produced *in situ* by microbes but migrate from organic rich or reducing zones (McMahon and Chapelle 1991, Thurman 1985). Organic acids, primarily fermentation products, have been measured in shallow anoxic groundwater with concentrations ranging from a few to few tens micromolar (Welch, unpub) though much higher concentrations, micromolar to millimolar, have been reported in marine sediments (Peltzer and Bada 1981, Sansone 1986).

Naturally-occurring organic ligands increase dissolution rates and affect the dissolution stoichiometry of rocks and minerals. Although most organic acids are weak, they can increase mineral weathering by proton attack at the mineral surface. Ligands can attack minerals directly by complexing with ions at the surface, weakening metal-oxygen bonds, and catalyzing dissolution reactions. Ligands also indirectly affect rates, by complexing with ions in solution, lowering solution saturation state. The enhancement of mineral dissolution by organic acids increases with organic acid concentration (Stillings et al. 1996, Vandevivere et al. 1994, Welch and Ullman 1992) or increasing reactive sites at the mineral surface (Huang and Kiang 1972, Manley and Evans 1986, Stillings et al. 1996, Welch and Ullman 1993, 1996). Bidentate ligands, such as oxalate and succinate, increase overall mineral dissolution rate much more than monodentate ligands such as acetate. Of all these acids, acetate and oxalate are the most abundant organic acids in natural systems (Fisher 1987, Kawamura and Nissenbaum 1992, Martens 1990, Sansone 1986, Sansone and Martens 1982, Surdam et al. 1984). Acetate can be produced by both abiotic and biological processes. Concentrations greater than 10,000 ppm have been measured in sedimentary basins and acetate has been detected at temperatures up to 200°C (Kharaka et al. 1986, Surdam et al. 1984). Experimental mineral dissolution studies have demonstrated that acetate enhances silicate mineral dissolution and may be important in mobilizing ions in oilfield formations waters. Acetate, however, is much less effective than oxalate in enhancing mineral weathering.

Oxalate. High concentrations of oxalate have also been measured in natural systems (Fox and Comerford 1990, Johnston and Vestal 1993, Kawamura and Nissenbaum 1992, MacGowan and Surdam 1988, Peltzer and Bada 1981). It is produced and released by a variety of plants and microorganisms, particularly fungi. In some microenvironments concentrations are high enough to precipitate oxalate salts (Ascaso et al. 1990, Graustein et al. 1977, Johnston and Vestal 1993). Because oxalate is one of the most abundant ligands detected in solutions interacting with mineral surfaces, many laboratory studies have focused on oxalate promoted mineral weathering.

Most laboratory mineral weathering studies using oxalate have demonstrated a significant increase in the release of major ions from rocks and minerals compared to controls. Manley and Evans (1986), Huang and Kiang (1970) Huang and Keller (1972) measured a one to two order of magnitude increase in release of ions from minerals compared to controls in batch dissolution experiments using high concentrations of oxalate. However, they attribute enhanced dissolution to the acidity of oxalate, and not to ligand promoted dissolution. More detailed mineral dissolution experiments showed that oxalate enhances dissolution of the feldspar framework compared to inorganic controls over a range of pH (Bevan and Savage 1989, Stillings et al. 1996, Welch and Ullman 1993,

1996). Silica release rates in 1 mM oxalate solutions were 2 to 5 times higher in acid (pH 3) and 2 to 15 times higher at near neutral pH. The effect of oxalate on Al release is even more dramatic. At near neutral pH, Al release from plagioclase feldspars was 10 to 100 times greater than in the inorganic controls. However, oxalate also enhanced the dissolution rate of quartz, olivine, and a cemented sandstone (Bennett et al. 1988, Grandstaff 1986, Johnston and Vestal, 1993) suggesting that it can bind to Si, Fe, and Mg sites as well. This implies that ligand promoted dissolution is particularly important for increasing release rates of less soluble ions (primarily Al and Fe) from minerals at the pH of natural waters (Antweiler and Drever 1983, Stillings et al. 1996, Welch and Ullman 1993).

In addition to enhancing the overall mineral dissolution rate, oxalate can also affect the nature of dissolution products. For example, oxalate increased Al release to solution from feldspars, leaving a residual solid phase enriched in Si (Mast and Drever 1987, Welch and Ullman 1992, 1993). However, in inorganic solutions at neutral pH, Si is preferentially removed, leaving an aluminum rich residual phase (Manley and Evans 1986, Stillings et al. 1996, Welch and Ullman 1993). Therefore, oxalate production by microorganisms should preferentially solubilize Al from minerals due to both its acidity and complexing ability. Occasionally, net calcium release to oxalate solutions were less than the stoichiometric ratio in the mineral phase, presumably due to the formation of insoluble calcium oxalate phases (Welch 1991). Sandstone leaching experiments showed a decrease in Ca release in oxalate solutions compared to inorganic controls (Johnston and Vestal 1993). This change in dissolution stoichiometry should result in different secondary phases forming as minerals weather in organic rich (soils, lichens) verses organic poor (groundwater) environments.

Varadachari et al. (1994) used SEM and light microscopy to examine silicate minerals (olivine, epidote, tourmaline, hornblende, biotite, microcline) reacted with oxalic acid. Minerals containing high concentrations of Al^{+3} and Fe^{+3} were extensively etched, presumably due to the formation of soluble metal-oxalate complexes. However, minerals with divalent cations (Ca^{+2}, Mg^{+2}, Fe^{+2}) tended to have these ions in the residual solid material, presumably from precipitation of insoluble divalent-metal oxalate complexes. Other unidentified amorphous and crystalline phases were also observed. Oxalate also affected the morphology of mineral etching. Garnets dissolved in oxalate were deeply etched with euhedral prismatic etch pits while garnets dissolved in inorganic solutions were not as extensively etched and had more irregular pit morphologies (Hansley 1987).

As always, extreme caution in applying laboratory derived data to natural environments must be used. In the case of oxalates, however, an insoluble crystalline oxalate forms by combination with appropriate cations. Many fungi are known to synthesize oxalic acid and their hyphae are frequently found encrusted with crystalline oxalates. Calcium oxalates also accumulate to high levels in some forest soils (Graustein et al. 1977). Additionally the presence of insoluble oxalates whose chemistry reflects that of the substratum (Wilson and Jones 1984) provides a relatively unambiguous correlation between laboratory and field studies. Reports by Purvis (1984) of copper oxalates within lichens colonizing copper sulfide bearing rocks, and by Wilson et al. (1980) of glushkinite (a rare manganese oxalate) are particularly compelling evidence of the involvement of oxalic acid in mineral weathering in natural systems.

Lichen acids

Each lichen association produces a unique and relatively diagnostic suite of compounds called lichen acids or lichen substances (Huneck and Yoshimura 1996). As a general rule, neither isolated symbiont produces lichen acids, (Jones 1988) which apparently result from specific interactions between mycobiont and phycobiont. In lichen weathering of minerals, lichen substances are often cited as dissolution agents (Williams and Rudolph 1974).

Despite limited solubility in water, saturated solutions of lichen substances have been shown to increase mineral dissolution rates in laboratory experiments (Ascaso and Galvan 1976, Iskandar and Syers 1972) via chelation. Nonetheless, reports of naturally occurring lichen acid-metal complexes are rare. Purvis et al. (1987) reported copper-norstictic acid and copper-psoromic acid (Purvis et al. 1990) complexes in oddly pigmented green lichens growing on cupriferous rocks. Usnic acid did not form a copper complex. One species of lichen growing on these cupriferous rocks contained norstictic acid that was not complexed with copper. Thus, Purvis et al. (1990) concluded that copper only complexed with lichen substances when their locations within the thallus coincidentally overlapped.

Lichen substances are usually concentrated in the upper thallus near the algal phycobiont, not at the mineral interface (Thomson, pers. comm. 1994). In other cases these substances crystallize on the upper surface as a pruinose covering. Therefore, a more plausible interpretation for the true role of lichen substances may be to deter predation (Lawrey 1980). Additionally, the lichen may use these compounds to regulate the opacity of the upper thallus to optimize light levels for the phycobiont (Lawrey 1986, Rundel 1978). In any case, the lack of reported metal-lichen acid complexes, limited solubility in water and their location in the upper thallus suggest they do not play an important role in mineral weathering.

Siderophores

Under conditions of iron limitation, microorganisms produce special iron binding ligands called siderophores which complex with and transport iron to the cell (Stone and Nealson, this volume). Siderophores are soluble, intermediate-molecular-weight (100s to 1000s dalton) compounds that have a very high affinity for Fe^{+3}. There are two major classes of these compounds, hydroxamates and catechols. Both form bidentate 5-member chelates with Fe^{+3} with stability constants on the order of 10^{30} to 10^{40} (Hersman et al. 1995, 1996; Holman and Casey, 1996). Hydroxamate siderophores produced by bacteria and fungi significantly enhanced dissolution of iron hydroxides and Fe^{+3} containing silicate minerals (Hersman et al. 1995, Watteau and Berthelin 1994). Catechols forms very strong bonds with mineral and metal surfaces (Kummert and Stumm 1980) and increase the dissolution rate of silicate minerals *in vitro* (Welch and Ullman 1996). Compounds containing catechols and hydroxamates may bind with other major ions in minerals, especially Al^{+3}, accelerating mineral dissolution (Ochs et al. 1993, Watteau and Berthelin 1994).

Polysaccharides

Bacteria and other microorganisms synthesize extracellular polymeric layers which are thought to be a complex mixture composed primarily of extracellular polysaccharides (EPS). Despite numerous studies demonstrating bacteria commonly attached to surfaces in natural environments (see Little et al., this volume), the mechanisms and biochemistry involved in microbial attachment to surfaces has received comparatively little attention in mineral weathering studies. In the initial stages of attachment, microorganisms are only weakly adsorbed to the surface substrate (reversibly attached). Bacteria seem to "sense" when they are in contact with a surface, and with time produce extracellular polysaccharides (EPS) and other compounds which enable them to irreversibly attach to surfaces (see micrograph on cover of this volume) (Fletcher and Floodgate 1973). EPS production is greater for attached verses free-living bacteria (Davies et al. 1993, Vandevivere and Kirchman 1993).

Polysaccharides impact mineral weathering in five ways: binding soil (see section on soil stabilization), antidesicant (increasing the residence time of water), maintaining diffusion pathways as water potential decreases (Chenu and Roberson 1996), acidic functional groups (chelation and ligand formation) and by acting as nucleation sites for authigenic mineral formation (Barker and Banfield 1996, Barker and Hurst 1992, Fortrin et al., this volume; Urritia and Beveridge 1993, 1994).

Schwartzman and Volk (1991) emphasized the importance of soil binding by microorganisms as a major factor in increased mineral weathering by microorganisms. Polysaccharides are widely recognized as the chief organic component involved in sediment binding by microorganisms (Dade et al. 1990, Forster and Nichols 1981a,b). Perhaps the most important barrier to microbial colonization of terrestrial surfaces would have been evolution of mechanisms with which to guard against dehydration. From a biological point of view, the large investment of a microorganism in constructing a capsule, sheath or slime layer is explained as creation of a controlled microenvironment, as well as a defense against predation and desiccation (Ophir and Gutnick 1994). From a mineral weathering perspective, the ability of these polymers to retain water is especially important as they increase the residence time for water to fuel hydrolysis reactions which characterize low temperature silicate mineral weathering. Another contribution EPS make to silicate mineral weathering is to maintain diffusion pathways as water levels decrease (Chenu and Roberson 1996). Increased residence times and interconnected diffusion pathways in surficial weathering environments characterized by irregular water supplies would increase weathering rates from this standpoint alone.

In addition to their antidessication properties, these compounds contain functional groups, primarily uronic and teichoic acids (carboxylated and phosphorylated carbohydrates, respectively), that can react with ions in solution and on surfaces. Naturally weathered minerals show extensively etched surfaces surrounded by EPS (Barker and Banfield 1996). Experimental weathering studies with slime forming microorganisms and microbial EPS showed that the polymers could indeed increase mineral dissolution, though they were not as effective as the low molecular ligands (Malinovskaya et al. 1990, Welch and Vandevivere 1994). Malinovskaya et al. (1990) determined that mixtures of polymers and low molecular weight ligands had a synergistic effect on weathering. In a similar study however, mineral dissolution was inhibited in solutions with microbial EPS compared to organic acids alone (Welch and Vandevivere 1994), indicating the higher molecular weight compounds are irreversibly binding to reactive sites on the surface, making them unavailable for further reaction. The apparent inhibition of silicate mineral dissolution could have been due to the precipitation of a secondary phase because microbial cell surfaces and extracellular polymers can also serve as nucleation sites for secondary mineral phases (Barker and Banfield 1996, Konhauser and Ferris 1996, Urrutia and Beveridge 1993, 1994), thereby moderating mineral precipitation reactions.

Proteins

The involvement of microorganisms in human disease has generated an extensive literature on microbial interaction with surfaces. Microbial pathogenicity is often intricately related to specific surface recognition and attachment mechanisms. For example, the pathogenicity of *Neisseria gonnorrheae*, is known to be directly related to its ability to synthesize proteinaceous adhesive appendages (Brinton et al. 1978). Mutated strains incapable of manufacturing these proteins lose their ability to adhere to cell surfaces in humans and their ability to cause venereal disease. The astonishing aspect of this story is that this single species is known to synthesize 50 antigenically distinct adhesion proteins (Buchanan 1975)! Is there any reason not to suspect similar complexity and specificity in the adhesive mechanisms between mineral surfaces and microorganisms? The dental literature also contains a wealth of literature concerning microbial attachment and cavity development.

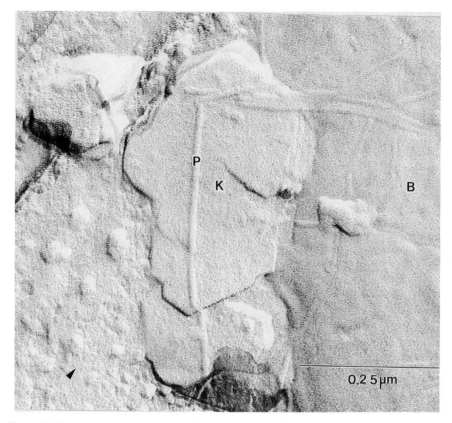

Figure 5. F-type sex pilus (P) of *Escherischia coli* (B) adsorbed onto the basal surface of kaolinite (K). Arrow indicates metal deposition direction. Propane jet cryofixed, freeze-etch Pt-C replica, bright field transmission electron micrograph (Barker 1988).

   While several excellent reports detail the involvement of proteins in mineral nucleation (Schultze-Lam et al. 1992, 1994, 1996), almost nothing is known regarding the role of these biological molecules in silicate mineral weathering. Microbial cell surfaces are comprised of approximately 40% protein, and these compounds should be able form complexes with ions very similar to ones formed by polysaccharides (Brock et al. 1994, Hughes and Poole 1989). Amino acids and polypeptides are known to adhere strongly to silicate mineral surfaces (Fig. 5) (Barker 1988, Barker and Hurst 1993, Dashman and Stotzky 1984). Microorganisms also produce enzymes that are specific for binding metal ions or that require metal ions as part of their structure, however its not known if these enzymes play a significant role in mineral weathering. It is well known that organisms such as lichens and fungi grow on durable, organic materials (e.g. polysaccharides, chitin, cellulase) that may have crystalline structures with periodicities not dissimilar to those found in silicate minerals (e.g. Chanzy et al. 1996). Microorganisms produce enzymes (e.g. chitinase and cellulase, often bound to the cell wall) specifically to degrade these substrates. Do microorganisms living on rocks (e.g. lichens) or in soils produce enzymes (or other high molecular weight molecules) that function in ways analogous to chitinase and cellulase ("mineralases"?) to break down minerals that they grow on in order to extract elements required for metabolism or structural purposes? Speculation about the existence of such "mineralases" has not lead to a sustained research effort, but that is only a matter of time. Now that advanced molecular techniques such as two dimensional gel electrophoresis have been developed and scientists trained in molecular techniques are turning their attention to geomicrobiology, a level of specificity undreamed of today in the ancient relationship between microorganisms and mineral surfaces may well emerge.

Nutrient adsorption

Microorganisms can affect mineral reactions by serving as a sink for elements released by weathering. Mineral dissolution rates are a function of solution saturation state. At far from equilibrium conditions, mineral weathering rates are independent of solution saturation state, then as solution approaches equilibrium with respect to the dissolving phase, rates decrease. Several biotic weathering studies have demonstrated that plant roots, fungi and bacteria dramatically increased K, Fe, Mg, and Al release from micas, and that these organisms served as a partial sink for these elements (Hinsinger et al. 1993, Hinsinger and Jaillard 1993, Leyval et al. 1990, Mojallali and Weed 1978). These organisms also increase the transformation of biotite to vermiculite by taking up K^+ that was released from the biotite structure (Hinsinger and Jaillaird 1993). However, none of the biotic mineral weathering studies have demonstrated unequivocally that nutrient adsorption by microorganisms increases weathering rates, because other factors, such as acid and ligand production, could also account for the observed changes in mineral weathering. Abiotic weathering experiments using micas and ion exchange resins have demonstrated that weathering rates increased substantially when ions were removed from solution (Lin and Clemency 1981).

MINERAL WEATHERING STUDIES

Quartz

Silica is an essential nutrient for many organisms. Some microorganisms, such as diatoms (Gordon and Drum 1994), radiolarians, and frustules actively precipitate intricate siliceous tests, even when bulk solutions are extremely undersaturated with respect to amorphous silica (de Vrind-de Jong and de Vrind, this volume; Konhauser et al. 1992). Microbial cell walls and EPS also serve as a substrate for amorphous silica precipitation (Fortin et al., this volume; Jones et al. 1997, Urritia and Beveridge 1993). Many higher organisms, such as plants and sponges, form siliceous support structures (phytoliths and spicules, Alexandre et al. 1997, Bavestrello et al. 1995). Trace amounts of silica are also required for other structural organic compounds. Since silica is an essential nutrient required by many organisms it seems plausible that microbes have developed mechanisms to actively extract silica from silicate minerals.

Quartz, a framework silicate with the composition SiO_2 (see details in Banfield and Hamers, this volume), is extremely resistant to chemical weathering. Unlike most other silicate minerals, quartz dissolution is unaffected by acidity except at extremely low pH, less than 2 (Brady and Walther 1990). Dissolution rates and quartz solubility are enhanced substantially as pH increases above approximately pH 8 due to deprotonation of Si-O-H bonds. Although quartz dissolution is unaffected by production of inorganic acids, both field and experimental data demonstrate that quartz is susceptible to organic ligand attack. Therefore, microbial production of organic ligands potentially can moderate geochemical reactions involving silica. Hallbauer and Jahns (1977) reported that fungal hyphae etched quartz grains as well as other silicate minerals. While the mechanism for this attack was unknown, presumably it was due in part to the production of low molecular weight organic acids. In petroleum contaminated aquifers, quartz sand was extensively etched compared to grains outside the contaminated zone (Bennett and Siegal 1987, McMahon et al. 1995). There was a strong correlation between DOC and dissolved Si, indicating that microbial degradation of oil and production of organic ligands catalyzed dissolution. Quartz dissolved even though solutions were nearly saturated with respect to amorphous silica (Bennett and Seigel 1987).

In abiotic laboratory experiments, quartz dissolution was enhanced by organic ligands compared to inorganic controls. The effect was less pronounced than that observed in the field, indicating high organic content in bulk solution is not solely responsible for the Si released from quartz (Bennett et al. 1988, Bennett and Siegel 1987). Microorganisms in culture were also able to increase dissolution of quartz and biogenic or amorphous silica by producing organic ligands or alkaline compounds (Avakyan et al. 1984, Lauwers and Heinen 1974).

Higher organisms also have the ability to etch quartz (Bavestrello et al. 1995). The Demospongiae have siliceous skeletons usually composed of opaline spicules, although some species incorporate other allocthonous particles as well. One species, *Chondrosia renformis*, forms its skeleton from a wide range of foreign particles. Bavestrello et al. (1995) examined silicate minerals from the ectosome of these sponges using SEM. Most of the silicate minerals and opaline silicate particles appeared unaltered by the sponge, though crystalline quartz was highly etched. When quartz and opaline silica were placed in the ectosomes of sponges the quartz grain size decreased and surfaces were extensively etched within a matter of days, while the opaline silica appeared unaltered. The sponges significantly increased dissolved Si released to solution from quartz sand compared to abiotic control. The authors implicated ascorbic acid because it is the reducing agent involved in producing collagen at the site of quartz dissolution, and quartz dissolved in solutions of ascorbate. However, ascorbic acid should have acted on other silicate minerals as well, particularly Al- and Fe-silicates. Natural samples of these minerals found in the sponge ectosomes had well shaped crystalline features as opposed to the spongy corrosion texture on quartz surfaces. It is likely that additional mechanisms may have acted in concert with ascorbic acid to produce the observed extensive quartz dissolution.

Feldspars

Feldspars are the most abundant aluminosilicate mineral at the Earth's surface. Their structure is described by Banfield and Hamers (this volume). Feldspars are more suseptible to microbial attack than quartz because they contain ions that react strongly with microbially produced compounds. Dissolution and weathering reactions involving feldspars have been studied extensively in the last few decades because weathering of minerals is important in neutralizing acid rain. Over geologic time scales, weathering of Ca-silicates (such as plagioclase) regulates atmospheric CO_2 concentration. Weathering of alkali feldspars also releases K^+, a major nutrient necessary for plant growth. A general feldspar weathering reaction can be described as dissolution of the primary feldspar mineral releasing ions to solution, followed by precipitation of a secondary clay phase. [For an excellent review on feldspar dissolution studies see Blum and Stillings (1995)].

Laboratory studies demonstrate that microorganisms substantially enhance feldspar dissolution rates by producing acids (Fig. 6). Vandevivere et al. (1994) measured a 200-fold increase in Si release rate from a Ca rich plagioclase in an acid producing bacterial culture compared to an abiotic control. Similarly, Berthelin (1971) measured a 200-fold increase in Al concentration from a granitic sand reacting in an acid producing microbial culture (pH ~ 3) compared to an abiotic control. Many other laboratory experiments (in nutrient rich solutions) on microbial dissolution of minerals have demonstrated a significant enhancement in concentration or release rate of major ions from feldspar. Microbes, primarily bacteria and fungi, produced acids (both inorganic and organic) and decreased bulk solution pH (Avakyan et al. 1981, Vandevivere et al. 1994, Williams and Rudolf

1974, Kutuzova 1969, Webley et al. 1963). Microorganisms can also produce alkaline compounds and increase feldspar weathering by hydroxyl promoted dissolution (Kutozova 1969).

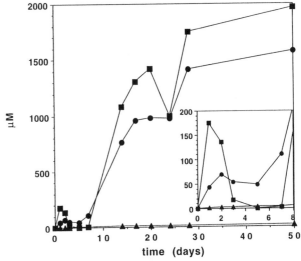

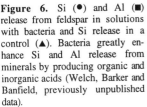

Figure 6. Si (●) and Al (■) release from feldspar in solutions with bacteria and Si release in a control (▲). Bacteria greatly enhance Si and Al release from minerals by producing organic and inorganic acids (Welch, Barker and Banfield, previously unpublished data).

In most studies of microbial weathering of feldspars, it is very difficult to decouple the synergistic effects of production of complexing organic ligands and decreasing solution pH on mineral dissolution. Vandevivere et al. (1994) demonstrated that several strains of subsurface bacteria produced organic ligands and increased Si release from feldspar compared to controls without a significant change in bulk solution pH. This enhancement of Si release was correlated with production of complexing organic acids which catalyzed Al release to solution (Welch 1996, Welch and Ullman 1995). Therefore, both biotic and abiotic experiments indicate that microbial production of organic ligands might be particularly important in enhancing feldspar weathering at near neutral pH.

As noted above, it is difficult to extrapolate results of these types of experiments to more realistic field conditions where microbial cell numbers and nutrient concentrations are much lower and bulk solution pH ranges from 5 to 9. However, there is field based evidence that microorganisms and microbial metabolites are important in weathering of feldspars in natural environments. Much of this work has focused on mineral surfaces under lichens. SEM examinations of feldspar surfaces under lichens showed highly etched mineral surfaces. Formation of etch pits appeared to be crystallographically controlled because the pits were euhedral and often occurred in linear patterns, indication that they were forming along twin planes, exsolution lamellae or other crystal defects. The etching pattern depended on the type of feldspar and the mineral face (Adamo et al. 1993, Jones et al. 1980, Wilson et al. 1980, 1981). Similar etch features were observed on feldspars weathered in different environments (Aldahan and Morad 1987, Berner and Holdren 1979) or in laboratory studies simulating natural conditions (e.g. Jones et al. 1980, Holdren and Berner 1979, Welch 1991).

The question of stoichiometry is central to understanding the effect of biological processes on silicate mineral weathering reactions. Several studies concluded that nonstoichiometric weathering characterized mineral reactions at the lichen-mineral interface. Production of organic ligands resulted in feldspar surfaces depleted in cations and enriched in Si, with amorphous or poorly crystalline material forming at the interface. In the absence of organic ligands, surfaces were enriched in Al and secondary phases, primarily clays, were formed at the lichen mineral interface (Ascaso et al. 1990, Jones et al. 1980). These observations agree with laboratory studies which show an increase in Al release to solutions with complexing organic ligands or decreased pH (Welch and Ullman 1993, 1996, Stillings et al. 1996). It is important to remember that these studies of natural material all employed intense digestion of organic materials with boiling hydrogen peroxide. As noted above, such intense chemical treatments destroy all textural information and drastically alter the mineralogical component. Barker and Banfield (1996) examined the intact organic-mineral interface in a lithobiontic lichen community using HRTEM and found no evidence of leached layers or siliceous relicts on amphibole, feldspar, or biotite.

Bacteria are also important in increasing feldspar weathering rates. *In situ* microcosm studies (Bennett et al. 1996, Bennett and Hiebert 1992, Hiebert and Bennett 1992) in an aquifer contaminated by an oil spill show that feldspar and quartz weathering was increased when bacteria colonized mineral surfaces. Many etch pits on mineral surfaces approximated the size and shape of bacteria. The authors hypothesized attached bacteria created a micro-reaction zone where organic acids and other metabolites were concentrated at the mineral surface, locally increasing the mineral dissolution rate. Microbial colonization and extent of etching depended on feldspar composition (Bennett et al. 1996). Two plagioclase feldspars, albite and oligoclase, were sparsely colonized by bacteria and only oligoclase appeared to be etched by the colonies. Two alkali feldspars, microcline and anorthoclase, had much higher surface colonization and were more deeply weathered. The authors hypothesized that bacteria were potassium limited, and were attacking the more potassium-rich feldspars preferentially. Estimates of weathering rates based on Si mass balance along a flow path and Si release from the mineral surface due to microbial etching shows that nearly all the weathering can be attributed to the bacteria. Further experimentation is needed to determine the role that irregularities in crystal structures such as dislocations and crystal strain in domain boundaries play in microbe-mineral interactions (Hochella and Banfield 1995, Lee and Parson 1995).

Micas

Micas are layer silicates that commonly contain K as the interlayer cation (see Banfield and Hamers, this volume, for structural details). Bioavailability of inorganic nutrients, such as K^+, to plants long has been of interest to soil and plant scientists. Biogeochemical weathering of micas, in particular the biotite to vermiculite transformation, has received a great deal of attention over the years. Mortland et al. (1956) demonstrated that wheat, given biotite as a sole K^+ source, converted biotite to vermiculite in a period of one year. Spyridakis et al. (1967) confirmed the rapid conversion of biotite to vermiculite and also reported epitaxial crystallization of kaolinite overgrowths. While neither of these studies mentioned rhizosphere microorganisms, no particular precautions against microbial growth are evident in the experimental designs. Weed et al. (1969) documented the ability of soil fungi to concentrate K^+ when given biotite, phlogopite and muscovite as a sole K^+ source, resulting in the production of vermiculite. While bioavailability of K^+ varied among different micas (biotite > phlogopite > muscovite), biotite functioned as well as soluble salt (KCl) as a source of K^+. The authors noted a positive correlation between levels of Na^+ and K^+. While they suggested a possible Na^+ ion exchange mechanism, they were careful to note that a portion of the experiment took place under reduced sucrose conditions. Reduced fungal metabolism and organic acid production is an alternative explanation. Mojallali and Weed (1979) specifically investigated the contribution of a soybean rhizosphere community to mica weathering and found increased production of interstratified biotite/vermiculite and phlogopite/vermiculite by rhizosphere microorganisms than by soybeans alone. Muscovite was not affected by biological weathering. The authors documented K and Al depletion from flake edges using an electron microprobe. Berthelin and Belgy (1979) reported the complete removal of K and Ti from biotite in a granitic sand by a community of bacteria and fungi after a 22-week perfusion experiment, resulting in brittle, white micaceous particles. Berthelin and Leyval (1982) found that symbiotic microorganisms in the rhizosphere increased the availability of K^+ from biotite to *Zea mays*, and further noted that addition of non-symbiotic microorganisms promoted much more biotite weathering and K uptake, an effect they attributed to "better soil exploration". Frankel (1977) attributed observed weathering of biotite in estuarine sands to interlamellar penetration by microorganisms. More recently Leyval and Berthelin (1991) confirmed these results by investigating the effects of rhizosphere microorganisms of pine to extract K^+ from phlogopite. A far more interesting result was their report of a spatial correlation between rhizosphere microorganisms and areas of highest K depletion. Fritz et al. (1994) documented a similar K^+ gradient in soil solutions from pine rhizospheres communities. Many field studies, particularly investigations of the effects of lichens on rock weathering, have documented biophysical disaggregation and mineralogical transformation of biotite to vermiculite and smectite (Barker and Banfield 1996, Wierzchos and Ascaso 1996).

Chain silicates

Natural and laboratory studies of weathering of pyroxene and amphibole (see Banfield and Hamers, this volume, for structural details) demonstrate that the cleavage-controlled reaction proceeds along cleavages and grain boundaries to produce a variety of secondary minerals. Most frequently, smectite, Al and Fe oxides and oxyhydroxides or complicated mixtures (Fig. 7) of these minerals form (e.g. Banfield et al. 1991, Banfield and Barker 1994), although kaolinite (Velbel 1989) and talc (Eggleton and Boland 1982) also have been reported. Velbel (1989) explained the transformation of hornblende to ferruginous and aluminous weathering products and the ultimate formation of relict cleavages known as microboxworks by a dissolution-reprecipitation mechanism. In the case of amphibole and pyroxene transforming to smectite, the reaction is isovolumetric (Fig. 8) (Colin et al. 1985, Banfield et al. 1991, Banfield and Barker 1994, Singh and Gilkes 1993). Due to the close structural relationship between the I beam of chain silicates and the T-O-T layer of smectites, topotactic relationships between the primary and secondary phases are often found (Eggleton 1975, Banfield et al. 1991, Banfield and Barker 1994), although this is not always the case (Singh and Gilkes 1993). Nahon and Colin (1982) found that orthopyroxene initially weathered under lateritic conditions to an amorphous product whose chemistry closely approximated the parent phase. Studies of naturally weathered coexisting amphiboles (Proust 1985, Banfield and Barker 1994) also demonstrated a close correlation between the chemistry of the parent amphibole and smectite, indicating very limited transport of dissolved constituents.

Efforts to reproduce natural weathering conditions for chain silicates in the laboratory have been partially successful. While Berner et al. (1979, 1980) were able to produce etch pits and denticulations which resembled those seen in ultrasonically cleaned, naturally weathered soil grains (Berner and Schott 1982), Schott et al. (1981) produced cation depleted silica enriched surface layers in iron-free pyroxenes and amphiboles with acid treatments. Banfield and Barker (1994) and Barker and Banfield (1996) found no evidence for surface depleted layers in naturally weathered amphiboles.

Biogeochemical and physiochemical weathering of ferrohastingsite amphibole was directly compared by Barker and Banfield (1996) in a lichen-colonized syenite boulder of known exposure age. In contrast to topotactically oriented smectite and Fe-oxyhydroxides produced at distance from the lichen thallus, biogeochemical weathering involved complete

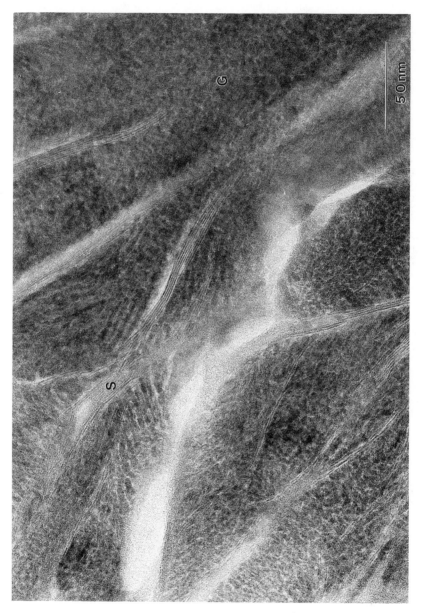

Figure 7. Alteration assemblage composed of smectite (S) and nanocrystalline goethite (G) which remains following complete transformation of coexisting riebeckite and acmite. HRTEM bright field micrograph, Ar-ion milled sample. (Barker and Banfield, previously unpublished image).

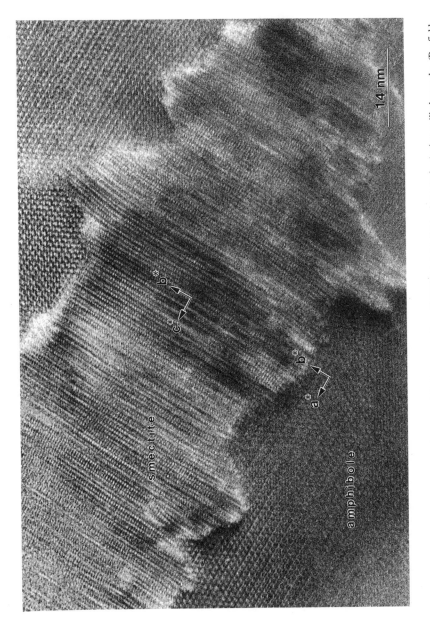

Figure 8. Isovolumetric transformation of amphibole to smectite. HRTEM bright field micrograph, Ar ion milled sample. (Banfield and Barker 1994).

dissolution and reprecipitation of a complex mixture of non-topotactically oriented smectite and Fe-oxyhydroxides bound by extracellular polymers (Fig. 9).

BIOGEOCHEMICAL WEATHERING OVER TIME

Conditions on Earth have changed substantially in the last 4.6 Ga, ostensibly from a very hot, high pressure CO_2 rich atmosphere to the hospitable climate we know today. Evidence exists, albeit controversial, that biological processes have operated throughout most of the Earth's history, even in the most extreme conditions. One major factor affecting the Earth's climate is the concentration of CO_2 in the atmosphere. Over geologic time scales ($>10^5$ years) the major sink of atmospheric CO_2 is weathering of Ca-Mg silicates and subsequent precipitation of Ca-Mg carbonates (Berner et al. 1983, Volk 1987, Walker et al. 1981). The overall reaction can be described by (Berner 1991):

$$CO_2 + 2H_2O + CaAl_2Si_2O_8 \Rightarrow Al_2Si_2O_5(OH)_4 + CaCO_3$$

At higher temperatures (due to higher atmospheric CO_2), mineral dissolution rates increase, leading to an increase in the flux of Ca^{+2} and Mg^{+2} to solution and precipitation of carbonate minerals. This in turn lowered atmospheric CO_2 concentration and global temperatures. The magnitude of this negative feedback between weathering and climate depends on the temperature dependence (or apparent activation energy) of the silicate weathering reaction.

Two early geochemical models, BLAG (Berner et al. 1983) and WHAK (Walker et al. 1981) attempted to model pCO_2 and temperature over time by considering various abiotic geological and geochemical processes. Neither model considered the effects of biological processes on silicate weathering. More recent models by Schwartzman and coauthors (Schwartzman and Volk 1989, 1991; Schwartzman et al. 1993, Schwartzman 1995, Schwartzman and Shore 1996) integrated geomicrobiological parameters in an attempt to quantify the biological acceleration of silicate mineral weathering rates, the subsequent increase in removal of CO_2 from the atmosphere, and resultant lower temperatures.

Biotic enhancement of weathering should change with temperature and pCO_2. More importantly, biological acceleration of mineral weathering should progress as life evolved and colonized the Earth's surface, from the earliest prokaryotes, to eukaryotes to the more complex metazoans. If the biotic enhancement of silicate weathering is substantial, organisms can affect the feedback between pCO_2, temperature and mineral weathering, thereby moderating the Earth's climate (Schwartzman and Volk 1989, 1991; Schwartzman et al. 1993, Schwartzman 1995, Schwartzman and Shore 1996).

Biogeochemical weathering models of Schwartzman and coauthors assume that early Earth atmosphere was a $>100°C$ pressure cooker with high pCO_2 and low O_2. The temperature of Archean Earth formed a major obstacle limiting biological evolution because more complex life forms (those with mitochondria) could not exist above 60°C (Schwartzman et al. 1993). First evidence of life occurs at 3.8 Ga (Mojzsis et al. 1996). These organisms might have been thermophilic bacteria and Archaea which colonized the Earth's surface and enhanced weathering rates, thereby cooling Earth's surface. Cyanobacteria are thought to have first appeared approximately 3.5 Ga when surface temperatures had decreased to less than 70°C. At approximately 2.1 to 2.8 Ga, eukaryotes evolved with an upper temperature limit of 60°C. At temperatures < 50°C, higher plants and animals evolved and colonized the Earth's surface. The evolution of all these organisms was constrained by the upper temperature limits, and the successive appearance of each new life form lead to increases in the biotic enhancement of weathering and a decrease in global temperatures.

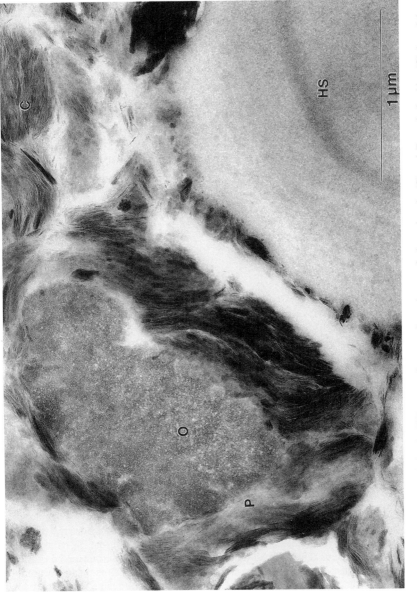

Figure 9. Fungal hyphal sheath (HS) penetrating complex mixture of clay minerals (C) and Fe-oxyhydroxides (O) bound by extracellular polymers (P). Zero loss EFTEM micrograph, diamond knife ultrathin section. (Barker, Welch and Banfield, previously unpublished image).

In attempting to evaluate the importance of biogeochemical weathering over geologic time, we must of necessity attempt to apply modern analogs, for example, weathering of minerals in the rhizosphere as a model of continental weathering since the evolution of vascular plants, weathering of rock surfaces under lichen communities as a model of biotic weathering by cryptogamic organisms, and even lithotrophic bacteria and Archaea that live in extreme environments (e.g. anoxic, high temperature, high pCO_2, low pH, low nutrient) as a model for possible biogeochemical reactions between rocks and primitive microorganisms during Precambrian times.

There is little doubt that in modern systems, the presence of soils and associated biota are important in weathering of minerals (Drever 1994, Schwartzman and Volk 1991). The colonization of the Earth's surface by vascular plants has been important in enhancing mineral weathering rates since the Silurian (Knoll and James 1987). However, the magnitude of this enhancement and the most important factors are still being debated. The enhancement in weathering has often been attributed to the activity of higher plants, though these studies often overlook the effect of associated microbes in the rhizosphere soil because it is difficult (if not impossible) to decouple the effects of vascular plants and their associated microorganisms on mineral weathering. Not all higher plants have the same enhancement effect. Silicate mineral weathering in ecosystems dominated by deciduous forests is three to four times greater than in conifer-evergreen dominated systems (Knoll and James 1987). Therefore, changes in the relative abundances of these ecosystems should affect net global biogeochemical weathering rates and CO_2 uptake.

Schwartzman and Volk (1989) argue for a present day biotic enhancement of weathering of two to three orders of magnitude times the abiotic rate. Contrary to this, however, Drever and Zobrist (1992) estimate a much smaller enhancement in biotic weathering of granitic rocks, a factor of ~7, by comparing fluxes of elements from small catchments at different elevations in the Swiss Alps, where vegetation decreases with elevation, from deciduous forests with 1 meter thick soils to bare rock with no significant vegetation. Cawley et al. (1969) estimate a factor of ~3 increase in weathering of vegetated basalt compared to abiotic rates by comparing alkalinity in runoff. Based on laboratory experiments, Drever (1994) argues that the biological production of acids and complexing organic ligands in soils should have little effect on mineral weathering (not more than a factor of two). A geochemical model of weathering in soils also indicates that biological processes (primarily acid production) should have little effect, but that physical factors, such as increasing temperature, should substantially increase mineral weathering rates (Gwiazda and Broecker 1994). The main effect of the biota on mineral weathering in soils is to stabilize soil structure, increase particle binding and interaction of mineral surfaces with solutions (Drever 1994, Schwartzman and Volk 1991). The rise of vascular plants and formation of soils may have less of a direct effect on weathering but more of an indirect effect , because this should have dramatically changed regional precipitation compared to the pre-Silurian landscape (Drever 1994).

Before the advent of vascular plants, the landscape probably was not barren, but more likely colonized by cryptogamic organisms, such as algae, fungi, and bacteria possibly similar to modern lichens (Golubic and Campbell 1979). In order to estimate the effects of primitive cryptogamic organisms on rock weathering, investigators have focused on extreme environments where vascular plants are absent but cryptogams flourish. Another ideal place to investigate rock weathering by cryptogams is Hawaiian lava flows, because ages are well know and changes in elevation and precipitation yield a range of weathering regimes. Early work on these basalts by Jackson and Keller (1970) show a thick weathering crust forming below the lichen and approximately an order of magnitude increase in weathering rate by lichens compared to bare rock surfaces. Cochran and Berner

(1992) however, found no noticeable enhancement of weathering by the lichen *Stereocaulon vulcani* , attributing Jackson and Keller's (1970) observations to trapping of aeolian debris by lichen thalli. This spawned sharp debate in the literature between Jackson, Berner, Schwartzman, and Drever (Cochran and Berner 1993, Drever 1996, Jackson 1993, 1996; Schwartzman 1993). Cochran and Berner (1996) maintain the contribution of higher plants is much greater than that of the *S. vulcani*. In a more extensive survey of these basalts, Brady et al. (1997) examined plagioclase and olivine grains in basalts that had been collected in areas that had a range of precipitation and temperature (elevation) to estimate the biogeochemical feedback between weathering and climate. The lichen *Stereocaulon vulcani* increased weathering by a factor of 2 to 18 compared to uncolonized surfaces. Though weathering increased with temperature and precipitation, these effects were different for colonized verses uncolonized surfaces. Weathering under lichens was much more sensitive to increases in precipitation than bare rock surfaces, because lichens are poikilohydric and trap water at the mineral surface, increasing time for reaction. Although overall mineral weathering rates under lichens were greater, the temperature-dependent weathering reactions (E_a) are only approximately half the abiotic values. Similar results were obtained by Welch and Ullman (1996) in laboratory mineral weathering experiments. Bacteria and microbial metabolites enhanced mineral weathering rates but decreased the apparent activation energy of the reaction. Therefore, early colonization of the Earth's surface by cryptogamic organisms probably resulted in lower pCO_2 levels due to enhanced weathering, but also resulted in a decrease in the temperature dependent feedback between weathering and climate.

In order to understand possible biogeochemical reactions in the Archean, microbes in an ecological niche that is as isolated as possible from the rest of the biosphere must be studied. Stevens and McKinley (1995) examined an assemblage of microorganisms living in deep anoxic aquifers where lithoautotrophs derive metabolic energy from reactions with basalt and water. In these extremely reducing environments, autotrophic microorganisms grow on $H_2 + CO_2$ producing methane and biomass. They speculate that microbially enhanced weathering exposes new fresh mineral surfaces for reaction.

Paleontological evidence from organic rich cherts in Australia (Awramik et al. 1983, Knoll et al. 1988, Schopf 1993, Schopf and Packer 1987) and South Africa (Walsh 1992) indicates single celled organisms may have existed on the Earth at least 3500 Ma. There is of course no small amount of controversy regarding these putative microfossils (see Buick 1991 for the most rigorously exclusive list of characteristics which must be satisfied to prove a biogenic origin). By far most of the research performed on Archean microfossils has been descriptive optical microscopy. Heaney and Veblen (1991) examined *Eosphaera tyleri* which had been interpreted as a fossilized cyanobacterium, using TEM and concluded the structure was not a microfossil. One interesting result of the intense scientific interest generated by Martian meteorite ALH84001 is that much more is known concerning the mineralogy and ultrastructure of putative Martian microfossils than their terrestrial counterparts (Mckay et al. 1996). Recent isotopic data (Mojzsis 1996) from apatites in the Isua Fm. in Greenland has been interpreted to show biologically fractionated C isotopes as far back as 3850 Ma. Allowing for the rarity of rocks older than 3800 Ma, vagaries of the fossil record and a certain period of time required to develop some sort of relatively advanced creature capable of isotopic fractionation, it seems reasonable to think microorganisms have probably coexisted with mineral surfaces for four billion years and must have exerted a profound impact on mineral weathering prior to the evolution of vascular plants. Most of these models assume terrestrial colonization by some sort of cryptogamic crust or lichen-like association. An interesting side aspect is that for a long time scientists assumed the lichen association, whether one considered it a true symbiosis (Ahmadjian 1993) or a case of controlled parasitism (Ahmadjian 1995), to be a very ancient pairing indeed. Recent genetic analyses on several lichen associations indicate that many fungal partners (mycobionts) in lichen associations are not closely related (Gargas et al. 1995). The implications are twofold: lichen associations have arisen multiple times (i.e. there are no ancestral forms) and perhaps more importantly for biology in general, parasitism may not be a necessary precondition for symbioses to develop. Another problem for extrapolating modern lichens to Archean landscapes is that of approximately 8500 described lichens, only a handful (less than 10) are cyanolichens, a formal association of cyanobacteria and fungi. One might expect this to be reversed as cyanobacteria are of a more ancient lineage than green algae. If an association between fungi (mycobiont) and a cyanobacteria or green alga (phycobiont) is so easy to form that lichenization arose multiple times, perhaps this strengthens the argument for symbioses and cryptogamic crusts arising in the Archean. Additionally, the fossil record for lichens is extremely poor (as one might expect) and the hypothesis by Retallack (1994) that the Ediacaran fauna (600 Ma) represent fossil lichens has not been widely accepted. We only raise these points to urge caution in applying knowledge about contemporary organisms and biological processes to Archean microorganisms. Similar morphologies might well be where the similarity ends.

Actual evidence for terrestrial Archean organisms is best described as weak. While evidence for Archean paleosols is clear, none are reported to contain any microfossils. Martini (1994) cites the presence of graphite as evidence of a microbial mat associated with an Archean paleosol. Hallbauer (1977) published a remarkable series of observations on coal-like seams of carbon (thucolite) which occurs in the Wiwatersrand gold reefs. Ashed thucolite was examined by a variety of methods and Hallbauer interpreted the structures as a fossil lichen, which he named *Thucomyces lichenoides*. Additionally, the report described another filamentous fungus-like microorganism and evidence for considerable biological remobilization of Au and U. While Hallbauer (1986) reported that these putative filamentous mats occurred on ventifacts, he was careful to ascribe periodic flooding of previously windswept terrain as the most likely depositional environment. A very recent paper (Barnicoat et al. 1997) dismisses this report and attributes thucolite to injection of liquid pyrobitumen along fractures in a hydrothermal event. Regardless of the nature of the earliest terrestrial microorganisms, arguably the biggest obstacle to overcome would have been the ability to withstand dehydration. In this context, perhaps a case could be made that the evolutionary progression onto land was preceded by aquatic microorganisms adapted to epilithic niches in the vadose zone of permeable rock units. Even today, the highest biomass is found at the interface between the saturated and unsaturated zone in aquifers (Madsen and Ghiorse 1993, Rudnick et al. 1992). Fluctuating water table levels would have provided as efficient a dehydration selective pressure as an intertidal zone. Regardless of the crucible in which terrestrial adaptations originated, the most effective method microorganisms apparently use today is to surround themselves with layers of extracellular polysaccharides (Ophir and Gutnick 1994).

SUGGESTIONS FOR FURTHER RESEARCH

In order to understand the impact of organisms and biological processes on low temperature silicate mineral transformations, ion concentration data, both in the mineralogical and organic phases which make up the organism mineral interface are needed. Owing to the enormous mineralogical and biological complexity, sample preparation techniques which preserve the ultrastructure of the *intact* organic mineral interface as well as elemental distributions must be applied to these complex biomineralogical samples. While recent advances in examining intact organic-mineral interfaces with a variety of high resolution electron imaging and spectroscopic techniques have contributed to our understanding of the complex processes occurring in these areas, fundamental questions remain. Microbial exopolysaccharides, hydrous polymers which contain up to 90% water, are extremely difficult to prepare for electron microscopy. Dehydration artifacts render a hydrated gel into a cobwebbed mass. Additionally, organic solvents used to dehydrate these materials prior to resin embedment effectively elutes metals from polysaccharides. Different types of polysaccharide can be distinguished with the use of selective heavy metal cytochemical staining techniques (Foster 1981). Apart from establishing the existence, location, and general type of extracellular polysaccharide with respect to mineral surfaces in rocks and soils, conventional electron microscopy is of limited usefulness in studying these polymers. For example, what is the elemental distribution of Fe in extracellular polysaccharides coating the surface of a weathering pyroxene within a crustose lichen thallus? As of now no biogeochemical study published to date has employed cryopreservation, cryoultramicrotomy, or cryomicroscopy (Fig. 10) to investigate these types of questions. Stelzer and Lehmann (1993) reviewed recent developments in cryo-sample preparation and analysis which, if applied creatively, will generate a wealth of chemical data which will greatly clarify the intimate interrelationship between biological and mineralogical processes. Of course, the sample preparation steps required to prepare a living community for examination under vacuum will always be subject to interpretational questions and the possibility of introducing artifacts is a very

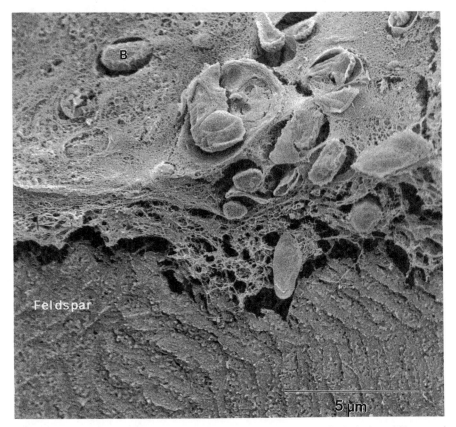

Figure 10. Bacterial biofilm (DOE SMCC 693) coating an etched plagioclase feldspar grain. Cryomicroscopy reveals the complicated structure of extracellular polysaccharide in which the cells (B) reside. High pressure, cryofixed, field emission gun; high resolution-low voltage cryo-scanning electron micrograph. (Barker, Welch and Banfield, previously unpublished image).

serious drawback, even the most sophisticated high pressure cryofixation techniques. Clearly, the ideal situation would be to quantitatively observe the living community in its natural state.

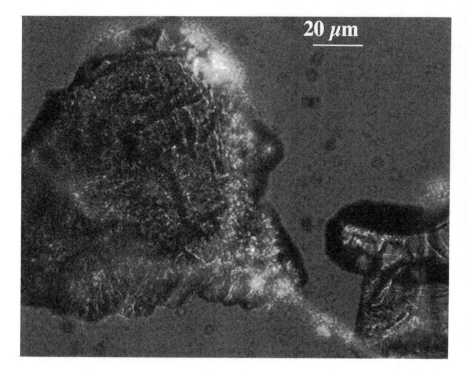

Figure 11. Epifluorescence photomicrograph of bacteria (bright spots) on feldspar grains. The bacteria (B0428 from the DOE SMCC) produce extracellular polysaccharides that "glue" mineral grains together. (Welch, Barker and Banfield, previously unpublished image).

Confocal scanning laser microscopy (CSLM) and epifluorescence (Fig. 11) offers unprecedented opportunities to study the structure and chemistry of geomicrobiological systems *in vivo* (Rautureau et al. 1993). Researchers have successfully studied the structure of live microbial biofilms (Wolfaardt et al. 1994), as well as liquid flow within these communities (Stoodley et al. 1994). In addition, CSLM imaging of fluorescent epoxy has revealed the three dimensional pore structure of a porous sandstone aquifer (Fredrich et al. 1995). By use of the proper fluorescent dyes, one can study pH gradients on a submicron scale (Chu et al. 1995, Chu and Montrose 1995), as well as monovalent, divalent and heavy metal ion concentrations. One can also explore community structure with "live-dead" bacterial stains and fluorescent probes.

The identity of members of communities of microorganisms in natural environments, their distribution, and the combined effects of their metabolisms on mineral dissolution are currently poorly understood. As noted elsewhere in this volume, extraordinarily powerful molecular techniques are now available to identify microbes (e.g. 16S rDNA analysis) and these can be applied to sediment, rock, and soil samples without culturing. Molecular information can be related to individual cells through the use of oligonucleotide probes, allowing microbial ecological information to be added to the mineralogical and geochemical perspectives. This combination of high-resolution structural, compositional, and biological approaches should provide key insights necessary for understanding natural geomicrobiological weathering of silicate minerals.

While both laboratory and field studies have demonstrated that microorganisms can play a role in mineral weathering, the magnitude of this effect is still unknown. Biological processes are extremely important in mineral weathering on rock surfaces and in well developed soils, but it is difficult to even estimate a net weathering rate, let alone discerning biotic and abiotic components. Estimates of biological enhancement of weathering in soils range from a factor of two to 3 orders of magnitude. Further work is needed to estimate the magnitude of biotic enhancement of weathering in different environments. One approach has been to place well characterized samples in different environments for several years (i.e. as in Bennett et al. 1996, Ullman et al. 1996) and then analyze samples comparing sites where organisms are attached versus 'bare sites' and attempt to estimate the rate enhancement due to biological weathering. An alternative approach is to analyze samples such as building stones, quarries, or lava flows whose exposure age is known in order to quantify field rates of biogeochemical weathering.

CONCLUSION

These are good times to be a geomicrobiologist! A remarkable confluence of interdisciplinary interest and techniques promises to exponentially increase our understanding of the timeless dance between the physical and the biological world in which we live. Consider the following short list of very recent discoveries :

• Bacteria have been found living up to 4 kilometers beneath the Earth's surface (Fredrickson and Onstott 1996).

• A new kingdom, Archaea is discovered and found to occur commonly (Barnes et al. 1994, Zimmer 1995).

• Molecular biologists find that perhaps 99.9% of soil microbes are uncultureable and unknown to science (Amann et al. 1996).

• Microorganisms capable of withstanding temperatures in excess of 100°C form the basis of chemosynthetic-based food chains around hydrothermal vents at abyssal ocean depths (Brock et al. 1994).

Of course, we have saved perhaps the most astonishing thing for last—the tantalizing possibility that Martian meteorite ALH84001 contains microbial microfossils, the first evidence that life exists elsewhere in the universe other than Earth (Mckay et al. 1996). This single electrifying report has riveted the attention of non-scientists and scientists alike, and will stimulate a tsunami of interdisciplinary geomicrobiological research. One day we will look back on this time, a time when instrumentation and creative analytical techniques converged with crumbling barriers between physical and biological sciences, as the golden age of geomicrobiology. It falls to us, the new generation of geomicrobiologists, to integrate knowledge from such diverse areas as the molecular aspects of microbial surface recognition, data on microbial adhesion and tooth decay from dental literature, and mineral dissolution data from the geochemistry literature to produce a unified picture.

ACKNOWLEDGMENTS

The authors thank Paul Ribbe for editorial support. Susan M. Paskewitz and R. Lee Penn ferretted out all the glitches. WWB is deeply indebted to Vernon Hurst for pointing him down the weathering/organoclay path. SAW thanks Bill Ullman for many years of help and support. We also acknowledge Colleen Lavin and Ya Chen for cryomicroscopy support, provided in part by the IMR, Madison, WI, through NIH Biomedical Research Technology Grant RR00570. Additional research support from the National Science Foundation (Grants EAR-9508171, EAR-9317082, CHE-9521731) and the Department of Energy (DE-FG02-93ER14328) is very gratefully acknowledged.

REFERENCES

Adamo P, Marchetiello A, Violante P (1993) The weathering of mafic rocks by lichens. Lichenologist 25:285-297

Ahmadjian V (1993) The Lichen Symbiosis. John Wiley, New York

Ahmadjian V (1995) Lichens-Specialized groups of parasitic fungi. In: K Kohmoto, US Singh, RP Singh (eds) Pathogenic and Host Specificity in Plant Diseases: Histopathological, Biochemical, Genetic, and Molecular Bases, Volume II: Eukaryotes, p 277-288 Pergamon, New York

AlDahan AA, Morad S (1987) A SEM study of dissolution textures of detrital feldspars in proterozoic sandstones, Sweden. Am J Sci 287:460-514

Alexandre A, Meunier J-M, Colin F, Koud J-M (1997) Plant impact on the biogeochemical cycle of silicon and related weathering processes. Geochim Cosmochim Acta 61:677-682

Amann R, Snaidr J, Wagner M (1996) *In situ* visualization of high genetic diversity in a natural microbial community. J Bacteriol 178:3496-500

Antweiler RC, Drever JI (1983) The weathering of a late tertiary volcanic ash: importance of organic solutes. Geochim Cosmochim Acta 47:623-629

Ascaso C, Galvan J (1976) Studies on the pedogenic action of lichen acids. Pedobiologia 16:321-331

Ascaso C, Galvan J, Ortega C (1976) The pedogenic action *of Parmelia conspersa, Rhizocarpon geographicum,* and *Umbilicaria pustulata.* Lichenologist 8:151-171

Ascaso C, Ollacarizqueta MA (1991) Structural relationship between lichen and carved stonework of Silos Monastery, Burgos, Spain. Intl Biodeterioration 27:337-349

Ascaso C, Sancho LG, Rodriguez-Pascual C. (1990) The weathering action of saxicolous lichens in maritime Antarctica. Polar Biol. 11:33-39

Ascaso C, Wierzchos J (1994) Structural aspects of the lichen-rock interface using back-scattered electron imaging. Botan Acta 107:251-256

Ascaso C, Wierzchos J, de los Rios A (1995) Cytological investigations of lithobiontic microorganisms in granitic rocks. Botan Acta 108:474-481

Avakyan ZA, Karavaiko GI, Mel'nikova EO, Krutsko VS, OstroushkoYI (1981) Role of microscopic fungi in weathering of rocks and minerals from a pegmatite deposit. Mikrobiologia 50:115-120

Avakyan ZA, Belkanova NP, Karavaiko GI, Piskunov PV (1985) Silicon compounds in solution during bacterial quartz degradation. Mikrobiologiya 54:250-256

Awramik SM, Schopf JW, Walter MR (1983) Filamentous fossil bacteria from the Archean of Western Australia. PreCambrian Res 20:357-374

Balkwill DJ (1989) Numbers, diversity, and morphological characteristics of aerobic, chemoheterotrophic bacteria in deep subsurface sediments from a site in South Carolina. Geomicrobiol J 7:33-52

Banfield JF, Jones BF, Veblen DR (1991) An AEM-TEM study of weathering and diagenesis, Abert Lake, Oregon: I. Weathering reactions in the volcanics. Geochim Cosmochim Acta 55:2781-2793

Banfield JF, Barker WW (1994) Direct observation of reactant-product interfaces formed in natural weathering of exsolved, defective amphibole to smectite: evidence for episodic, isovolumetric reactions involving structural inheritance. Geochim Cosmochim Acta 58:1419-1429

Barker WW (1988) An electron microscopic study of extralamellar organoclay complexes. PhD Dissertation, University of Georgia, Athens, Georgia

Barker WW, Banfield JF (1996) Biologically versus inorganically-mediated weathering reactions: Relationships between minerals and extracellular microbial polymers in lithobiontic communities. Chem Geol 132:55-69

Barker WW, Hurst VJ (1992) Bacterial trace fossils in Eocene kaolin of the Huber Formation of Georgia: *Phylloderma microsphaeroides,* n. ichnogen., n. ichnosp. Ichnos 2:55-60

Barker, W. W. and Hurst, V. J. (1993) Freeze etch replication of extracellular bacterial polymers adsorbed onto kaolinite. Proc 51st Ann Meet Microscopy Soc Am, p 52-53

Barns SM, Fundyga RE, Jeffries MW (1994) Remarkable archaeal diversity detected in a Yellowstone National Park hot spring environment. Proc Natl Acad Sci 91:1609-13

Barnicoat AC, Henderson IHC, Knipe RJ, Yardley BWD, Napier RW, Fox NPC, Kenyon AK, Muntingh DJ, Strydom D, Winkler KS, Lawrence SR, Cornford C (1997) Hydrothermal gold mineralization in the Witwatersrand basin. Nature 386:820-824

Bavestrello G, Arillo A, Benatti U, Cerrano C, Cattaneo-Vietti R, Cortesogno L, Gaggero L, Giovine M, Tonetti M, Sara M (1995) Quartz dissolution by the sponge *Chondrosia reniformis* (Porifera, Demospongiae). Nature 378:374-376

Bennett PC, Hiebert FK (1992) Microbial mediation of silicate diagenesis in organic-rich natural waters. In: Kharaka YK, Maest AS (eds) Water-Rock Interaction-7, Vol 1,AA Balkema, Rotterdam, p 267-270

Bennett PC, Hiebert FK, Choi WJ (1996) Microbial colonization and weathering of silicates in a petroleum-contaminated groundwater. Chem Geol 132:45-53

Bennett PC, Melcer ME, Siegel DI, Hassett JP (1988) The dissolution of quartz in dilute aqueous solutions of organic acids at 25°C. Geochim Cosmochim Acta 52:1521-1530

Bennett PC, Siegel DI (1987) Increased solubility of quartz in water due to complexation by dissolved organic compounds. Nature 326:684-687

Berner RA (1991) Model for atmospheric CO_2 over Phanerozoic time. Am J Sci 291:339-376

Berner RA, Holdren GR Jr (1979) Mechanism of feldspar weathering-II. Observations of feldspars from soils. Geochim Cosmochim Acta 43:1173-1186

Berner RA, Lasaga AC, Garrels RM (1983) The carbon-silicate geochemical cycle and its effect on atmospheric carbon dioxide over the past 100 million years. Am J Sci 283:641-683

Berner RA, Schott J (1982) Mechanism of pyroxene and amphibole weathering II. Observations of soil grains. Am J Sci 282:1214-1231

Berner RA, Sjoberg EL, Velbel, MA, Krom MD (1980) Dissolution of amphiboles and pyroxenes during weathering. Science 207:1205-1206

Berthelin J (1971) Alteration microbienne d'une arene granitique: Note preliminaire. Science du Sol 1:11-29

Berthelin J, Dommergues Y (1972) Role de produits du metabolisme microbien dans la solubilisation des mineraux d'une arene granitique. Rev Ecol Biol Sol IX, 3:397-406

Berthelin J, Belgy G (1979) Microbial degradation of phyllosilicates during simulated podzolization. Geoderma 21:297-310

Berthelin J, Bonne M, Belgy G, Wedraogo FX (1985) A major role for nitrification in weathering of minerals of brown acid forest soils. Geomicrobiology J 4:175-190

Berthelin J, Leyval C (1982) Ability of symbiotic and non-symbiotic rhizospheric microflora of maize (*Zea mays*) to weather micas and to promote plant growth and plant nutrition. Plant and Soil 68:369-377

Berthelin J (1983) Microbial weathering processes. In: Krumbein WE (ed), Microbial Geochemistry, p 223-262 Blackwell Scientific Publications, Oxford, UK

Bevan J, Savage D (1989) The effect of organic acids on the dissolution of K-feldspar under conditions relevant to burial diagenesis. Mineral Mag 53:415-425

Birrell KS (1971) Effect of pretreatment with hydrogen peroxide on composition of clay fractions extracted from organic horizons of andosols. Trans Intl Cong Soil Sci 7:179-189

Blum A, Lasaga AC (1988) Role of surface speciation in the low temperature dissolution of minerals. Nature 331:431-433

Blum AE, Stillings LL (1995) Feldspar dissolution kinetics. In: White AF, Brantley SL (eds) Chemical Weathering Rates of Silicate Minerals. Rev Mineral 31:290-351

Brady PV, Dorn RI, Brazel AJ, Clark J, Glidewell T (1997) Silicate weathering and the biologic takeover of global climate. In prep

Brady PV, Walther JV (1989) Controls on silicate dissolution rates in neutral and basic pH solutions at 25°C. Geochim Cosmochim Acta. 53:2823-2830

Brady PV, Walther JV (1990) Kinetics of quartz dissolution at low temperatures. Chem Geol 82:253-264

Brantley SL (1992) Kinetics of dissolution and precipitation: Experimental and field results. In: Kharaka YK Maest AS (eds) Water-Rock Interaction-7, 1:3-6 AA Balkema, Rotterdam

Brinton CC, Bryan J, Dillon JA Guering N, Jacobson J (1978) Uses of pili in gonorrhoeal control: Role of bacterial pili in disease, purification and properties of gonococcal pili, and progress in the development of a gonococcal pilus vaccine for gonnorhoeae In: GF Brooks, KK Holmes, WD Sawyer, FE Young (eds) Immunobiology of Neisseria Gonorrhoeae, p 155-178 American Society for Microbiology, Washington, DC

Brock TD, Madigan MT, Martinko JM, Parker J (1988) Biology of Microorganisms. Prentice Hall, Englewood Cliffs, New Jersey

Buchanan TM (1975) Antigenic heterogeneity of gonococcal pili. J Experimental Med 141:1470-1475

Buick R (1991) Microfossil recognition in Archean rocks: An appraisal of spheroids and filaments from 3500 m.y. old chert-barite unit at North Pole, Western Australia. Palaios 5:441-459

Chanzy H, Andre I, Taravel FR (1996) Molecular and crystal structures of inulin from electron diffraction data. Macromolecules 29:4626-4635

Chapelle FH (1993) Ground-Water Microbiology and Geochemistry. John Wiley & Sons, New York.

Chapelle FH, Lovely DR (1990) Rates of microbial metabolism in deep coastal plain aquifers. Appl Environ Microbiol 56:1865-1874

422 *GEOMICROBIOLOGY*

Chapelle FH, Zelibor JL Jr, Grimes DY, Knobel LL (1987) Bacteria in deep coastal plain sediments of Maryland: A possible source of CO_2 to groundwater. Water Resources Res 23:1625-1632

Cawley JL, Burruss RC, Holland HD (1969) Chemical weathering in Central Iceland: An analog of pre-Silurian weathering. Science 165:391-392

Chenu C, Roberson EB (1996) Diffusion of glucose in microbial extracellular polysaccharide as affected by water potential. Soil Biol Biochem 28:877-884

Chisolm JE, Jones GC, Purvis OW (1987) Hydrated copper oxalate, moolooite, in lichens. Mineral Mag 51:715-718

Chou L, Wollast R (1985) Steady-state kinetics and dissolution mechanisms of albite. Am J Sci 285:963-993

Chu S, Montrose MH (1995) Extracellular pH regulation in microdomains of colonic crypts: Effects of short chain fatty acids. Proc Natl Acad Sci 92:3303-3307

Chu S, Brownell WE, Montrose MH (1995) Quantitative confocal imaging along the crypt-to-surface axis of colonic crypts. Am J Physiol 269:C1557-C1564

Cochran MF, Berner RA (1992) The quantitative role of plants in weathering. In: Kharaka YK, Maest AS (eds) Water-Rock Interaction-7, 1:473-476 AA Balkema, Rotterdam

Cochran MF, Berner RA (1993) Reply to Comments on "Weathering, plants, and the long-term carbon cycle." Geochim Cosmochim Acta 57:2147-2148

Cochran MF, Berner RA (1996) Promotion of chemical weathering by higher plants: field observations on Hawaiian basalts. Chem Geol 132:71-78

Colin F, Noack Y, Trescases JJ, Nahon D (1985) L'alteration lateritique debutante des pyroxenites de Jacubs, Niquelandia, Bresil. Clay Mineral 20:93-113

Cook RJ (1992) A comparison of methods for the extraction of smectites from calcareous rocks by acid dissolution techniques. Clay Mineral 27:73-80

Dade WB, Davis JD, Nichols PD, Nowell ARM, Thistle D, Trexler MB, White DC (1990) Effects of bacterial exopolymer adhesion on the entrainment of sand. Geomicrobiology J 8:1-16

Dashman T, Stotzky G (1984) Adsorption and binding of peptides on homoionic montmorillonite and kaolinite. Soil Biol Biochem 16:51-55

Davies DG, Chakrabarty AM, Geesey GG (1993) Exopolysaccharide production in Biofilms: Substratum activation of alginate gene expression by Pseudomonas aeruginosa. Appl Environ Microbiol 59:1181-1186

De La Torre M, Gomez-Alarcon G, Vizcaino C, Garcia T (1993) Biochemical mechanisms of stone alteration carried out by filamentous fungi living in stone monuments. Biogeochem 19:129-147

Douglas LA, Fiessinger F (1971) Degradation of clay minerals by H_2O_2 treatments to oxidize organic matter. Clays Clay Minerals 19:67-68

Drever JI (1994) The effect of land plants on the weathering rate of silicate minerals. Geochim Cosmochim Acta 58: 2325-2332

Drever JI, Zobrist J (1992) Chemical weathering of silicate rocks as a function of elevation. Geochim Cosmochim Acta 56:3209-3216

Drever JI (1996) reply to the Comment by T.A. Jackson on "The effect of land plants on weathering rates of silicate minerals." Geocim Cosmochim Acta 60:725

Easton RM (1994) Lichens and rocks: a review. Geoscience Canada 21:59-76

Eckhardt FEW (1978) Microorganisms and weathering of a sandstone monument. In: Krumbein WEK (ed) Environmental Biogeochemistry and Geomicrobiology 2: 675-686 Ann Arbor Science Publ, Ann Arbor, Michigan

Eggleton RA (1975)Nontronite topotaxial after hedenbergite. Am Mineral 60:1063-1068

Eggleton RA, Boland JN (1982) Weathering of enstatite to talc through a series of transitional phases. Clays Clay Mineral 30:11-20

Ehrlich HL (1996) How microbes influence mineral growth and dissolution. Chem Geol 132:1-3

Ferris FG, Fyfe WS, Beveridge TJ (1988) Metallic ion binding by *Bacillus subtilis*: Implications for the fossilization of microorganisms. Geology 16:149-152

Ferris FG, Schultze S, Witten TC, Fyfe WS, Beveridge TJ (1989) Metal interactions with microbial biofilms in acidic and neutral pH environments. Appl Environ Microbiol 55:1249-1257

Ferris FG, Wiese RG, Fyfe WS (1994) Precipitation of carbonate minerals by microorganisms: Implications for silicate weathering and the global carbon dioxide budget. Geomicrobiology J 12:1-13

Fisher JB (1987) Distribution and occurrence of aliphatic acid anions in deep subsurface waters. Geochim Cosmochim Acta 51:2459-2468

Fletcher M, Floodgate GD (1973) An electron-microscopic demonstration of acidic polysaccharide involved in the adhesion of a marine bacterium to solid surfaces. J Gen Microbiology 74:325-334

Forster SM, Nicolson TH (1981a) Aggregation of sand from a maritime embryo sand dune by microorganisms and higher plants. Soil Biol Biochem 13:199-203

Forster SM, Nicholson TH (1981b) Microbial aggregation of sand in a maritime dune succession. Soil Biol Biochem 13:205-208

Foster RC (1981) Polysaccharides in soil fabrics. Science 214: 665-667

Fox TR, Comerford NB (1990) Low-molecular-weight organics in selected forest soils of the southeastern USA. Soil Sci Soc Am J, 54:1139-1144

Frankel L (1977) Microorganism induced weathering of biotite and hornblende grains in estuarine sands. J Sed Pet 47:849-854

Fredrich JT, Menendez B, Wong TF (1995) Imaging the pore structure of geomaterials. Science 268:276-279

Fredrickson JK, Onstott TC (1996) Microbes deep inside the Earth. Scientific Am 275:68-73

Fritz E, Knoche D, Meyer D (1994) A new approach for rhizosphere research by X-ray microanalysis of micro-liter soils solutions. Plant and Soil 161:219-223

Fry, EJ (1927) The mechanical action of crustaceous lichens on substrata of shale, schist, gneiss, limestone, and obsidian. Ann Botany 41:437-460

Fyfe WS (1996) The biosphere is going deep. Science 273:448

Gargas A, DePriest PT, Grube M, Tehler A (1995) Multiple origins of lichen symbioses in fungi suggested by SSU rDNA phylogeny. Science 268:1492-1495

Ghiorse WC (1997) Subterranean Life. Science 275:789-790

Golubic S, Friedmann EI, Schneider J (1981) The lithobiontic ecological niche, with special reference to microorganisms. J Sedimentary Petrol 51:475-478

Golubic S, Campbell SE (1979) Analogous microbial forms in recent subaerial habitats and in Precambrian cherts: Gloeothece coerulea Geitler and Eosynechococcus moorei Hofman. Precambr Res 8:201-217

Gordon R, Drum RW (1994) The chemical basis of diatom morphogenesis. Intl Rev Cytol 150:243-372

Grandstaff DE (1986) The dissolution rate of forsteritic olivine from Hawaiian beach sand In: Colman SM, Dethier DP (eds) Rates of Chemical Weathering of Rocks and Minerals, p 41-59 Academia Press, New York

Graustein WC, Cromack K Jr, Sollins P (1977) Calcium oxalate: Occurrence in soils and effect on nutrient and geochemical cycles. Science 198:1252-1254

Grierson PF (1992) Organic acids in the rhizosphere of *Banksia integrifolia* L.f. Plant and Soil 144:259-265

Griffin DM (1981) Water potential as a selective factor in the microbial ecology of soils. In: Parr JF, Gardner RI, Elliott LF (eds) Water Potential Relations in Soil Microbiology, Soil Sci Soc Am Spec Publ No. 9, SSSA, Madison, WI, p 141-151

Gurevich (1962) The role of microorganisms in producing the chemical composition of groundwater. In: Kuznetsov SI (ed) Geologic Activity of Microorganisms, p 65-75 Trans Institute of Microbiology No. IX, Consultants Bureau, New York

Gwiazda RH and Broecker WS (1994) The separate and combined effects of temperature, soil pCO_2, and organic acidity on silicate weathering in the soil environment: Formulation of a model and results. Global Biogeochem Cycles 8:141-155

Hallbauer DK (1986) The mineralogy and geochemistry of Witwatersrand pyrite, gold, uranium, and carbonaceous matter. In: Anhaeusser CR, Maske S (eds) Mineral Deposits of South Africa, p 731-752 Geol Soc of South Africa, Johannesburg

Hallbauer DK, Jahns HM (1977) Attack of lichens on quartzitic rock surfaces. Lichenologist 9:119-122

Hallbauer DK, Jahns HM, Beltmann HA (1977) Morphological and anatomical observations on some Precambrian plants from the Witwatersrand, South Africa. Geologische Rundschau. 66:477-491

Hansley PL (1987) petrologic and experimental evidence for etching of garnets by organic acids in the upper Jurassic Morrison formation, northwest New Mexico. J Sed Petrol 57:666-681

Hazen TC, Jimenez L, de Victoria GL (1991) Comparison of bacteria from deep subsurface sediment and adjacent groundwater. Microb. Ecol. 22:293-304

Heaney PJ, Veblen DR (1991) An examination of spherulitic dubiomicrofossils in Precambrian banded iron formations using transmission electron microscope. Precambrian Res 49:355-372

Hellmann R (1994) The albite-water system: Part I. The kinetics of dissolution as a function of pH at 100, 200, and 300°C. Geochim Cosmochim Acta 58:595-611

Herring JG, Stumm W (1990) Oxidative and reductive dissolution of minerals. In: Hochella MF, White AF (eds) Mineral-Water Interface Geochemistry. Rev Mineral 23:427-466

Hersman L, Maurice P, Sposito G (1996) Iron acquisition from hydrous Fe(III)-oxides by an aerobic Pseudomonas sp. Chemical Geology 132:25-31

Hersman L, Lloyd T, Sposito G (1995) Siderophore-promoted dissolution of hematite. Geochim Cosmochim Acta 59:3327-3330

Hiebert FK, Bennett PC (1992) Microbial control of silicate weathering in organic-rich ground water. Science 258:278-281

Hinsinger P, Elsass F, Jaillard B, Robert M (1993) Root-induced irreversible transformation of a trioctahedral mica in the rhizosphere of rape. J Soil Sci 44:535-545

Hinsinger P, Jaillard B (1993) Root induced release of interlayer potassium and vermiculitization of phlogopite as related to potassium depletion in the rhizosphere of rye grass. J Soil Sci 44:525-534

Hinsinger P, Jaillard B, Dufey JE (1992) Rapid weathering of a trioctahedral mica by the roots of ryegrass. Soil Soc Am J 56:977-982

Hirsch P, Eckhardt FEW, Palmer RJ (1995) Methods for the study of rock-inhabiting microorganisms- A mini review. J Microbiol Methods 23:143-167

Hochella MF Jr, Banfield JF (1995) Chemical weathering of silicates in nature: a microscopic perspective with theoretical considerations. In: White AF, Brantley SL (eds) Chemical Weathering Rates of Silicate Minerals. Rev Mineral 31:354-406

Holdren GR Jr, Berner RA (1979) Mechanism of feldspar weathering -1. Experimental studies. Geochim Cosmochim Acta 43:1161-1171

Holm PE, Nielsen PH, Albrechtsen H-J, Christensen TA (1992) Importance of unattached bacteria and bacteria attached to sediment in determining potentials for the degradation of xenobiotic organic contaminants in an aerobic aquifer. Appl Environ Microbiol 58:3020-3026

Huang WH, Keller WD (1970) Dissolution of rock-forming silicate minerals in organic acids: Simulated first stage weathering of fresh mineral surfaces. Am Mineral 55:2076-2094

Huang WH, Kiang WC (1972) Laboratory dissolution of plagioclase in water and organic acids at room temperature. Am Mineral 57:1849-1859

Huang PM, Schnitzer M (1986) Interactions of Soil Microbes with Natural Organics and Microbes. Soil Sci Soc Am Spec Publ No. 17, SSSA, Madison, Wisconsin

Hughes MN, Poole RK (1989) Metals and Microorganisms. Chapman and Hall, London.

Huneck S, Yoshimura I (1996) Identification of Lichen Substances. Springer-Verlag, New York

Iskander IK, Syers JK (1972) Metal-complex formation by lichen compounds. J Soil Sci 23:255-265

Jackson TA (1993) Comment on "Weathering, plants, and the long-term carbon cycle" by Robert A. Berner. Geochim Cosmochim Acta 57:2141-2144

Jackson TA (1996) Comment on "The effect of land plants on weathering rates of silicate minerals" by James I. Drever. Geochim Cosmochim Acta 60:723-724

Jackson TA, Keller WD (1970) Comparative study of the role of lichens and inorganic processes in the chemical weathering of recent Hawaiian lava flows. Am J Sci 269:446-466

Johnston CG, Vestal JR (1993) Biogeochemistry of oxalate in Antarctic cryptoendolithic lichen-dominated community. Microb Ecol 25:305-319

Jones B, Renaut RW, Rosen MR (1997) Biogenicity of silica precipitation around geysers and hot-spring vents, North Island, New Zealand. J Sedimentary Res 67:88-104

Jones D (1988) Lichens and pedogenesis. In: Galun M (ed) Handbook of Lichenology, p 109-124 CRC Press, Boca Raton, Florida

Jones D, Wilson MJ, Tait JM (1980) Weathering of basalt by Pertusaria corallina. Lichenologist 12:277-289

Jones D, Wilson MJ, McHardy WJ (1981) Lichen weathering of rock-forming minerals: application of scanning electron microscopy and microprobe analysis. J Micro 124:95-104

Karavaiko GI, Avakyan ZA, Krutsko VS, Mel'nikova EO, Zhdanov AV, Piskunov VP (1979) Microbiological investigations on a spodumene deposit. Mikrobiologiya 48:393-398

Karavaiko GN, Belkanova NP, Eroshchev-Shak VA, Avakyan ZA (1984) Role of microorganisms and some physiochemical factors of the medium in quartz destruction. Mikrobiologiya 53:795-800

Kawamura K and Nissenbaum A (1992) High abundance of low molecular weight organic acids in hypersaline spring water associated with a salt diapir. Org Geochem 18:469-476

Kharaka YK, Law LL, Carothers WW, and Goerlitz DF (1986) Role of organic species dissolved in formation waters from sedimentary basins in mineral diagenesis. In: Gautier DL (ed). Roles of Organic Matter in Sediment Diagenesis. Soc Econ Paleontologists Mineralogists Spec Publ 38:111-122

Knoll AH, Strother PK, Rossi S (1988) Distribution and diagenesis of microfossils from the lower Proterozoic Duck Creek Dolomite, Western Australia. Precambrian Res 38:257-279

Knoll MA, James WC (1987) Effect of the advent and diversification of vascular land plants on mineral weathering through geologic time. Geology 15:1099-1102

Konhauser KO Ferris FG (1996) Diversity of iron and silica precipitation by microbial mats in hydrothermal waters, Iceland: Implications for Precambrian iron formations. Geology 24:323-326

Konhauser KO, Mann H, Fyfe WS (1992) Prolific organic SiO_2 precipitation in a solute-deficient river: Rio Negro, Brazil. Geology 20:227-230

Konhauser KO, Schultze-lam S, Ferris FG, Fyfe SF, Longstaffe FJ, Beveridge TJ (1994) Mineral precipitation by epilithic biofilms in the Speed River, Ontario, Canada. Appl Environ Microbiol 60:549-553

Krumbein WE, Urzi CE, Gehrmann C (1991) Biocorrosion and biodeterioration of antique and medieval glass. Geomicrobiology J 9:139-160

Kummert R, Stumm W (1980) The surface complexation of organic acids on hydrous -Al_2O_3. J Colloid Interface Sci 75:373-385

Kutuzova RS (1969) Release of silica from minerals as a result of microbial activity. Mikrobiologiya 38:596-602

Lakvulich LM, Wiens JH (1970) Comparison of organic matter destruction by hydrogen peroxide and sodium hypochlorite and its effects on selected mineral constituents. Soil Sci Soc Am Proc 34:755-758

Lauwers AM, Heinen W (1974) Biodegradation and utilization of silica and quartz. Arch Microbiol 95:67-78

Lawrey JD (1980) Correlations between lichen secondary chemistry and grazing activity by *Pallifera varia*. Bryologist 83: 328-334

Lawrey JD (1986) Biological role of lichen substances. Bryologist 89:111-122

Lebedeva EV, Lyalikova NN, Bugel'skii YY (1979) Participation of nitrifying bacteria in the weathering of serpentized ultrabasic rock. Mikrobiologia 47:898-904

Lee Mr, Parsons I (1995) Microtextural controls of weathering of perthitic alkali feldspars. Geochim Cosmochim Acta 59:4465-4488

Leyval C, Berthelin J (1991) Weathering of a mica by roots and rhizospheric microorganisms of pine. Soil Sci Soc Am J 55:1009-1016

Leyval C, Laheurte F, Belgy G, Bethelin J (1990) Weathering of micas in the rhizospheres of maize, pine, and beech seedlings influenced by mycorrhizal and bacterial inoculation. Symbiosis 9:105-109

Lin FC, Clemency CV (1981) Dissolution kinetics of phlogopite: II, Open system using an ion-exchange resin. Clays Clay Minerals 29:107-112.

Lovely DR, Chapelle FH (1995) Deep subsurface microbial processes. Rev Geophys 33:365-381

Madsen EL, Ghiorse WC (1993) Groundwater microbiology: susbsurface ecosystem processes. In: Ford TE (ed.) Aquatic Microbiology, p 167-214 Blackwell Scientific, London

Malinovskaya IM, Kosenko LV, Votselko SK, Podgorskii VS (1990) Role of Bacillus mucilagosus polysaccharide in degradation of silicate minerals. Mikrobiologiya 59:49-55

Manley EP, Evans LJ (1986) Dissolution of feldspars by low-molecular-weight aliphatic and aromatic acids. Soil Sci 141:106-112

Marshall HG, JC Walker, WR Kuhn (1988) Long-term climate change and the geochemical cycle of carbon. J Geophys Res 93:791-801

Martens CS (1990) Generation of short chain organic acid anions in hydrothermally altered sediments of the Guaymas Basin, Gulf of California. Appl Geochem 5:71-76

Martini JEJ (1994) A late Archaean-Palaeoproterozoic (2.6 Ga) Palaeosol on ultramafics in the eastern Transvaal, South Africa. Precambrian Res 67:159-180.

Mast MA, Drever JI (1987) The effect of oxalate on the dissolution rates of oligoclase and tremolite. Geochim Cosmochim Acta 51:2559-2568

MacGowan DB, Surdam RC (1988) Difunctional carboxylic acid anions in oilfield waters. Org Geochem 12:245-259

McKay DS, Gibson EK Jr., Thomas-Keprta KL, Vali H, Romanek CS, Clemett SJ, Chillier XDF, Maechling CR, Zare RN (1996) Search for past life on Mars: Possible relic biogenic activity in martian meteorite ALH84001. Science 273:924-930

McMahon PB, Chapelle FH (1991) Microbial organic acid production in aquitard sediments and its role in aquifer geochemistry. Nature 349:233-235

McMahon PB, Chapelle FH, Falls WF, Bradley PM (1992) Role of microbial processes in linking sandstone diagenesis with organic rich clays. J Sediment Petrol 62:1-10

McMahon PB, Vroblesky DA, Bradley PM, Chapelle FH, Gullett CD (1995) Evidence for enhanced mineral dissolution in organic acid-rich shallow ground water. Ground Water 33:207-216

Mojallali H, Weed SB (1982) Weathering of micas by mycorrhizal soybean plants. Soil Sci Soc Am J 42:367-372

Mojzsis SJ, Arrhenius G, McKeegan KD, Harrison TM, Nutman AP, Friend CRL (1996) Evidence for life on Earth before 3,800 million years ago. Nature 384:55-59

Mortland MM, Lawton K, Uehara G (1956) Alteration of biotite to vermiculite by plant growth. Soil Sci 82:477-481

Nash TH (1996) Lichen Biology. Cambridge University Press, New York, 303 p

Oades JM (1993) The role of biology in the formation, stabilization and degradation of soil structure. Geoderma 56:377-400

Ochs M, Brunner I, Stumm W, Cosovic B (1993) Effects of root exudates and humic substances on weathering kinetics. Water Air Soil Pollution 68:213-229

Ophir T, Gutnick DL (1994) A role for exopolysaccharides in the protection of microorganisms from desiccation. Appl Environ Micobiol 60:740-745.

Ortega-Calvo JJ, Arino X, Hernadez-Marine M, Saiz-Jimenez C (1995) Factors affecting the weathering and colonization of monuments by phototrophic microorganisms. Sci Tot Environ 167:329-341

Palmer RJ Jr, Hirsch P (1991) Photosynthesis-based microbial communities on two churches in Northern Germany: Weathering of granite and glazed brick. Geomicrobiology J 9:103-118

Palmer RJ Jr., Siebert J, Hirsch P (1991) Biomass and organic acids in sandstone of a weathered building: Production by bacterial and fungal isolates. Microb Ecol 21:253-266

Papendick RI, Campbell GS (1981) Theory and measurement of water potential. In: Parr JF, Gardner RI, Elliott LF (eds) Water Potential Relations in Soil Microbiology, Soil Sci Soc Am Spec Publ No. 9:1-22 SSSA, Madison, Wisconsin

Peltzer ET, Bada JL (1981) Low molecular weight a-hydroxy carboxylic and dicarboxylic acids in reducing marine sediments. Geochim Cosmochim Acta 45:1847-1854

Perez-Rodriguez JL, Wilson MJ (1969) Effects of pretreatment on a 14-Å swelling mineral from Gartly, Aberdeenshire. Clay Mineral 8:39-45

Piervittori R, Salvadori O, Laccisaglia A (1994) Literature on lichens and biodeterioration of stonework. I. Lichenologist 26:171-192

Proust D (1985) Amphibole weathering in a glaucophane-schist (Ile De Groix, Morbihan, France). Clay Mineral 20:161-170

Purvis OW (1984) The occurrence of copper oxalate in lichens growing on copper sulphide-bearing rocks in Scandinavia. Lichenologist 16:197-204

Purvis OW, Elix JA, Broomhead JA, Jones GC (1987) The occurrence of copper-norstictic acid in lichens from cupriferous substrata. Lichenologist 19:193-203

Purvis OW, Elix JA, Gaul KA (1990) The occurrence of copper-psoromic acid from cupriferous substrata. Lichenologist 22:345-354

Rautureau M, Cooke RU, Boyde A (1993) The application of confocal microscopy to the study of stone weathering. Earth Surface Proc Landforms 18:769-775

Retallack GJ (1994) Were the Ediacaran fossils lichens? Paleobiology 20:523-544

Rudnick DT, Levine SN, Ghiorse WC (1992) A flow through microcosm system for testing microbial nutrient limitation in subsurface sediments. In: Stanford JA, Simmons JJ (eds) Proc 1st Intl Conf Ground Water Ecology. American Water Resources Association, Bethesda, MD, p 47-58

Rundel PW (1978) The ecological role of secondary lichen substances. Biochem Systematics Ecol 6:157-170

Sand W and Bock E (1991) Biodeterioration of mineral materials by microorganisms- Biogenic sulfuric and nitric acid corrosion of concrete and natural stone. Geomicrobiology J 9:129-138

Sanders WB, Ascaso C, Wierzchos J (1994) Physical interactions of two rhizomorph-forming lichens with their rock substrate. Botan Acta 107:432-439

Sansone FJ (1986) Depth distribution of short-chain organic acids turnover in Cape Lookout Bight sediments. Geochim Cosmochim Acta 50, 99-105

Sansone FJ, Martens CS (1982) Volatile fatty acid cycling in organic-rich marine sediments. Geochim Cosmochim Acta 46:1575-1589

Schopf JW (1993) Microfossils of the early Archean Apex chert: New evidence of the antiquity of life. Science 260:640-646

Schopf JW, Packer BM (1987) Early Archean microfossils (2.2 billion to 3.5 billion-year-old) microfossils from Warrawoona Group, Australia. Science 237:70-73

Schott J, Berner RA, Sjoberg EL (1981) Mechanism of pyroxene and amphibole weathering—I. Experimental studies of iron-free minerals. Geochim Cosmochim Acta 45:2123-2135

Schultze-Lam S, Harauz G, Beveridge TJ (1992) Participation of a cyanobacterial S-layer in fine grain mineral formation. J Bacteriology 174:7971-7981

Schultze-Lam S, Beveridge TJ (1994) Nucleation of celestite and strontianite on a cyanobacterial S-layer. Appl Environ Mic 60:447-453

Schultze-Lam S, Fortin D, Davis BS, Beveridge TJ (1996) Mineralization of bacterial surfaces. Chem Geol 132:171-181

Schwartzman D (1993) Comment on "Weathering, plants, and the long-term carbon cycle" by Robert A. Berner. Geochim Cosmochim Acta 57:2145-2146

Schwartzman DW (1995) Temperature and the evolution of the Earth's biosphere. In: Shostak GS (ed) Progress in the Search for Extraterrestrial Life, Astronomical Society of the Pacific Conf Ser 74:153-161

Schwartzman DW, McMenamin M, Volk T (1993) Did surface temperatures constrain microbial evolution? Bioscience 43:390-393

Schwartzman DW, Shore SN (1996) Biotically mediated surface cooling and habitability for complex life. In: Doyle LR (ed) Circumstellar Habitable Zones, p 421-443 Travis House Publishers, Menlo Park, California

Schwartzman DW, Volk T (1989) Biotic enhancement of weathering and the habitability of Earth. Nature 340:457-459

Schwartzman DW, Volk T (1991) Biotic enhancement of weathering and surface temperatures on Earth since the origin of life. Paleogeog Paleoclimat Paleoecol 90:357-371

Singh B, Gilkes RJ (1993) Weathering of spodumene to smectite in a lateritic environment. Clays Clay Minerals 41:624-630

Spyridakis DE, Chesters G, Wilde SA (1967) Kaolinisation of biotite as a result of coniferous and seedling growth. Soil Sci Soc Am Proc 31:203-210

Staudigel H, Chastain RA, Yayanos A, Bourcier W (1995) Biologically mediated dissolution of glass. Chem Geol 126:147-154

Stelzer R, Lehman H (1993) Recent developments in electron microscopical techniques for studying ion localization in plant cells. Plant and Soil 155/156:33-43

Stevens TO, McKinley JP (1995) Lithoautotrophic microbial ecosystems in deep basalt aquifers. Science 270:450-454

Stillings LL, Drever JI, Brantley SL, Sun Y, Oxburgh R (1996) Rates of feldspar dissolution at pH 3-7 with 0-8 mM oxalic acid. Chem Geol 132:79-90

Stoodley P, DeBeer D, Lewandowski Z (1994) Liquid flow in biofilm systems. Appl Environ Microbiol 60:2711-2716

Surdam RC, Boese SW, and Crossey LJ (1984) The chemistry of secondary porosity. In: McDonald DA and Surdam RC (eds) Clastic Diagenesis 37:127-149 Am Assoc Petrol Geol Memoirs, Tulsa, Oklahoma

Swoboda-Colberg NG, Drever JI (1992) Mineral dissolution rates: a comparison of laboratory and field studies. In: Kharaka YK, Maest AS (eds) Water-Rock Interaction-7, 1:115-118 AA Balkema, Rotterdam

Thorseth IH, Furnes H, Heldal M (1992) The importance of microbial activity in the alteration of natural basaltic glass. Geochim Cosmochim Acta 56:845-850

Thorseth IH, Furnes H, Tumyr O (1995a) Textural and chemical effects of bacterial activity on basaltic glass: an experimental approach. Chem Geol 119:139-160

Thorseth IH, Torsvik T, Furnes H, Muehlenbachs K (1995b) Microbes play an important role in the alteration of oceanic crust. Chem Geol 126:137-146

Thurman EM (1985) Organic Geochemistry of Natural Waters, Martinus Nijhoff/Dr W Junk Publishers, Dorderct, the Netherlands.

Tisdall JM (1994) Possible role of soil microorganisms in aggregation of soils. Plant Soil 159:115-121

Ullman WJ, Kirchman DL, Welch SA, Vandevivere P (1996) Laboratory evidence for the microbially mediated silicate mineral dissolution in nature. Chem Geol 132:11-17

Urrutia MM, Beveridge TJ (1993) Mechanism of silicate binding to the bacterial cell wall in Bacillus subtilis. J Bacteriol 175:1936-1945

Urrutia MM, Beveridge TJ (1994) Formation of fine-grained metal and silicate precipitates on a bacterial surface (Bacillus subtilis). Chem Geol 116:261-280

Urzi C, Lisi S, Criseo G, Pernice A (1991) Adhesion to and degradation of marble by a Micrococcus strain isolated from it. Geomicrobiology J 9:81-90

Vandevivere P, Kirchman DL (1993) Attachment stimulates exo-polysaccharide synthesis by a bacterium. Appl Environ Microbiol 59:3280-3286

Vandevivere P, Welch SA, Ullman WJ, Kirchman DL (1994) Enhanced dissolution of silicate minerals by bacteria at near-neutral pH. Microb Ecol 27:241-251

Varadachari C, Barman AK, Ghosh K (1994) Weathering of silicate minerals by organic acids II. Nature of residual products. Geoderma 61:251-268

Velbel MA (1986) Influence of surface area, surface characteristics, and solution composition on feldspar weathering rates. In: Davis JA, Hayes KF (eds) Geochemical Processes at Mineral Surfaces: Am Chem Soc Symp Ser 323:615-634

Velbel MA (1989) Weathering of hornblende to ferruginous products by a dissolution-reprecipitation mechanism: petrography and stoichiometry. Clays Clay Minerals 37:515-524

Velbel MA (1990) Influence of temperature and mineral surface characteristics on feldspar weathering rates in natural and artificial systems: a first approximation. Water Resources Res 26:3049-3053

Volk T (1987) Feedbacks between weathering and atmospheric CO_2 over the last 100 million years. Am J Sci 287:763-779

Walker CG, Hays PB, Kasting JF (1981) A negative feedback mechanism for the long-term stabilization of Earth's surface temperature. J Geophys Res 86:9776-9782

Walsh MM (1992) Microfossils and possible microfossils from the early Archean Onverwacht Group, Barberton Mountain Land, South Africa. PreCambrian Res 54:271-293

Watteau F, Berthelin J (1994) Microbial dissolution of iron and aluminum from soil minerals: efficiency and specificity of hydroxamate siderophores compared to aliphatic acids. J Soil Biol 30:1-9

Webley DM, Henderson MEK, Taylor IF (1963) The microbiology of rocks and weathered stones. J Soil Sci 14:102-112

Weed SB, Davey CB, Cook MG, (1969) Weathering of mica by fungi. Soil Sci Soc Am Proc 33:702-706

Welch SA (1991) The effect of organic acids on mineral dissolution rates and stoichiometry. MS Thesis, Univ of Delaware, Newark, Delaware

Welch SA, Ullman WJ (1992) Dissolution of feldspars in oxalic acid solutions. In: Kharaka YK Maest AS (eds) Water-Rock Interaction-7, 1:127-130 AA Balkema, Rotterdam

Welch SA, Ullman WJ (1993) The effect of organic acids on plagioclase dissolution rates and stoichiometry. Geochim Cosmochim Acta 57:2725-2736

Welch SA, Ullman WJ (1992) Microbially produced compounds and feldspar dissolution, V.M. Goldschmidt Conf, Reston, Virginia (May 1992)

Welch SA, Vandevivere P (1994) Effect of microbial and other naturally occurring polymers on mineral dissolution. Geomicrobiol. J 12:227-238

Welch, SA, Ullman WJ (1995) Effect of bacteria and organic acids on the apparent activation energy of feldspar dissolution at low temperatures, VM Goldschmidt Conf, Pennsylvania State Univ (May 1995)

Welch SA Ullman WJ (1996) Feldspar dissolution in acidic and organic solutions: Compositional and pH dependence of dissolution rate. Geochim Cosmochim Acta 60:2939-2948

Wierzchos J, Ascaso C (1994) Application of back-scattered electron imaging to the study of the lichen-rock interface. J Microscopy 175:54-59

Wierzchos J, Ascaso C (1996) Morphological and chemical features of bioweathered granitic biotite induced by lichen activity. Clays Clay Minerals 44:652-657

Williams ME, Rudolph ED (1974) The role of lichens and associated fungi in the chemical weathering of rock. Mycologia 66:648-660

Wilson MJ, Jones D (1984) The occurence and significance of manganese oxalate in *Pertusaria corallina* (Lichenes). Pedobiologia 26:373-379

Wilson MJ, Jones D, Russell JD (1980) Glushinskite, a naturally occuring magnesium oxalate. Mineral Mag 43:837-840

Wilson MJ, Jones D, McHardy WJ (1981) The weathering of serpentinite by *Lecanora atra*. Lichenologist 13:167-176

Wogelius RA, Walther JV (1991) Olivine dissolution at 25°C: Effects of pH, CO_2, and organic acids. Geochim Cosmochim Acta 55:943-954

Wolfaardt GM, Lawrence JR, Robarts RD, Caldwell SJ, Caldwell DE (1994) Multicellular organization in a degradative biofilm community. Appl Environ Microbiol 60:434-446

Zimmer C (1995) Triumph of the Archaea. Discover 16:30-31

Chapter 13

LONG-TERM EVOLUTION OF THE
BIOGEOCHEMICAL CARBON CYCLE

David J. Des Marais

Space Science Division
Ames Research Center
Moffett Field, California 94035 U.S.A.

INTRODUCTION

Both the habitability of Earth's environment and the origin and evolution of life have been shaped by physical and chemical interactions between the atmosphere, hydrosphere, geosphere and biosphere. These interactions can be examined for elements such as carbon (C) that have played central roles, both for life and the environment. The compounds of C are highly important, not only as organic matter, but also as atmospheric greenhouse gases, pH buffers in seawater, redox buffers virtually everywhere, and key magmatic constituents affecting plutonism and volcanism.

These multiple roles of C all interact with each other across a network of C reservoirs that are interconnected by an array of physical, chemical and biological processes. This network is termed the biogeochemical C cycle. As the word "cycle" implies, a major dynamic of the network is the movement of C around any of several cyclical pathways (Fig. 1). Although these pathways differ from each other regarding some of the reservoirs and processes involved, they share common ground in the hydrosphere and atmosphere. It is this common ground that unites the entire C cycle and allows even its most remote components to influence the surface environment and the biosphere.

This review gives a brief summary of the current state of the C cycle. Nonbiological factors which influenced the long-term evolution of the cycle are then discussed. Lastly, a scenario is presented for the long-term evolution of the C cycle and the biosphere.

THE BIOGEOCHEMICAL CARBON CYCLE TODAY

The C cycle can be envisioned as an integrated system of reservoirs and processes (Fig. 1). Principal C reservoirs are depicted as boxes and arranged to represent their true relative positions (i.e. atmospheric CO_2 reservoir at the top, mantle C reservoir at the bottom). The processes linking the reservoirs are represented by labelled arrows, and the arrays of boxes and arrows delineate the various subcycles. The timescales typically required for C to traverse each of the subcycles are indicated by the vertical bars at right. The actual boundaries between these subcycles should not be assumed to be precisely and sharply delineated in the natural environment; instead they define illustrative subdomains along the continuum of reservoirs and processes that collectively constitute the complete biogeochemical C cycle.

0275-0279/97/0035-0013$05.00

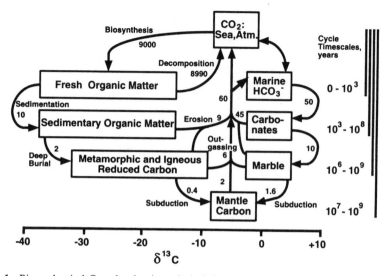

Figure 1. Biogeochemical C cycle, showing principal C reservoirs in the mantle, crust, oceans and atmosphere (boxes) and also the processes which unite these reservoirs (arrows). The range of each of these reservoir boxes along the horizontal axis gives a visual estimate of $\delta^{13}C$ values most typical of each reservoir. Numbers adjacent to the arrows give estimates of present-day fluxes, expressed in the units 10^{12} mol yr^{-1}. The vertical bars at right indicate the timeframes within which C typically completely traverses each of the four C subcycles (the HAB, SED, MET and MAN subcycles, see text). For example, C traverses the hydrosphere-atmosphere-biosphere (HAB) subcycle in 0 to 1000 years.

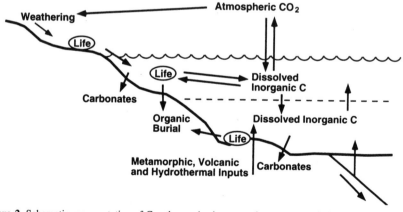

Figure 2. Schematic representation of C pathways in the atmosphere, ocean and shallow crust. The oval symbols depicting life represent the following three communities of primary producers: terrestrial photosynthetic, shallow water photosynthetic (both benthic and planktonic), and deep water and subsurface chemosynthetic (associated principally with hydrothermal activity and the alteration of basaltic crust).

The hydrosphere-atmosphere-biosphere (HAB) subcycle

The subcycle involving the hydrosphere, atmosphere and biosphere (the "HAB" subcycle, see Figs. 1 and 2) is distinguished by the small sizes of its reservoirs (Table 1), the very high rates of C exchange between them, and therefore the relatively short timescales required for cycling. Biota, included in the "fresh organic matter" reservoir in

Figure 1, dominate the exchange of C between CO_2 and organic matter. The fixation of CO_2 is achieved principally by photosynthetic organisms, and is shared almost equally by primary production in the marine (4000×10^{12} mol yr^{-1}; Martin et al. 1987) and terrestrial (5000×10^{12} mol yr^{-1}; Olson et al. 1985) environments. Because only about 10×10^{12} mol yr^{-1} of reduced C escapes into sediments (Berner and Canfield 1989), approximately 99.9% of the C fixed by the biosphere is decomposed relatively rapidly back to inorganic C. These figures testify to the high primary productivity achieved by oxygenic photosynthetic communities, which is sustained, in part, by highly efficient recycling of the critical nutrients N and P and trace metals.

The time required for C to traverse the HAB subcycle lies in the range of minutes to a thousand years. This range embraces, at short timescales of minutes to hours, the rapid cycling of C within microbial ecosystems (e.g. Des Marais 1995), and, at the thousand-year timescale, the present-day turnover rate of the global ocean (Broecker 1982).

Table 1. Reservoirs of carbon in the atmosphere, hydrosphere and geosphere.

Mass $\times 10^{18}$ moles

Reservoir	Reduced C	Oxidized C	Total C
Atmosphere	-	0.06[a]	0.06
Biosphere: plants and algae	0.13[b]	-	0.13
Hydrosphere	-	3.3[c]	3.3
Pelagic sediments	60[d]	1300[d]	1360
Continental margin sediments	>370[e]	>1000[e]	>1370
Sedimentary rocks	750[f]	3500[f]	4250
Crustal metamorphic and igneous rocks	100[g]	?	>100
Mantle			27000[h]

[a] Holland 1984.
[b] Mopper and Degens 1979, Olson et al. 1985.
[c] Holland 1984.
[d] Holser et al. 1988.
[e] Minimum inventories required for C isotopic mass balance, see text.
[f] Ronov 1980.
[g] Hunt 1972.
[h] Derived from estimates of mantle mass and C concentration, see text.

The sedimentary (SED) subcycle

Both the characteristics and the long-term stability of the HAB subcycle are strongly influenced by the sedimentary (SED) subcycle, which includes the reservoirs of sedimentary organic matter and carbonates (Fig. 1). For example, sedimentation limits global productivity by removing nutrients, phosphorus in particular (Holland 1984). Also, the balance between sedimentation of oxidized (carbonate, sulfate, Fe^{3+}) and reduced (organic C, sulfide, Fe^{2+}) species determines the abundances of O_2 and sulfate in the hydrosphere and atmosphere (Garrels and Perry 1974, Holland 1984). Weathering and erosion of igneous and sedimentary rocks release and transport nutrients to the oceans, thus affecting global biological productivity. Weathering rates can be enhanced substantially by biological activity (see Barker et al., this volume). The rates of weathering and the release of reduced species determine their rates of consumption of O_2. Thermal emanations of reduced species (H_2, sulfides, Fe^{2+}) also consume O_2. Thus the oxidation state of the surface environment is controlled by the balance between processes of biological productivity, organic decomposition, sedimentation, weathering and thermal emanations.

Rates of C sedimentation and weathering in the SED subcycle are considerably slower than those associated with C transformations in the HAB subcycle (Fig. 1). The rate at which C traverses the SED subcycle is determined principally by tectonic controls on rates of formation and destruction of sedimentary rocks; and the globally-averaged half-life of sedimentary rocks is about 200 million years (Derry et al. 1992). Organic C is buried principally in terrigenous deltaic-shelf sediments (8.7×10^{12} mol yr^{-1}) and in sediments beneath highly-productive open-ocean regions (0.8×10^{12} mol yr^{-1}); and global burial rates are about 10×10^{12} mol yr^{-1} (Berner and Canfield 1989). Although marine carbonate precipitation is presently dominated by calcareous plankton, the global net burial rate of carbonate (about 50×10^{12} mol yr^{-1}; Holser et al. 1988) is controlled ultimately by the riverine flux of Ca^{2+} and Mg^{2+} released by weathering, and by the global rate of reaction between seawater and submarine basalts (Garrels and Perry 1974). Sedimentary C reservoirs are much larger than the C reservoirs of the HAB subcycle (Table 1). Therefore, key biogeochemical properties of the oceans and atmosphere (in the HAB subcycle), including their nutrient inventories and oxidation states, are ultimately controlled by interactions with the SED subcycle. This control is exerted over timescales of typically tens of thousands to millions of years.

The metamorphic (MET) subcycle

A third (MET) subcycle (Fig. 1) is defined to include more deeply buried sedimentary and igneous rocks which experience metamorphism. This includes C that is contained in rocks entering subduction zones but that escapes injection into the mantle either because this C is degassed or because its host rock also escapes subduction (e.g. by lateral accretion into buoyant continental crust). The MET subcycle includes volumes of rock much larger than those in the SED subcycle (Lowe 1992), but the organic C reservoirs in the MET subcycle are considerably smaller (Hunt 1972), due both to losses during metamorphism of sedimentary rocks and to the typically much lower reduced C contents of igneous rocks. For example, hydrocarbon gases and CO_2 can be released by thermal decomposition of sedimentary organic matter (Hunt 1979). Carbonate C is also lost during metamorphism. Carbonates hosted in a range of rock compositions can yield CO_2 at elevated temperatures (Ferry 1991), for example:

Dolomite = Periclase + Calcite + CO_2

Tremolite + 11 Dolomite = 8 Forsterite + 13 Calcite + 9 CO_2 + H_2O

2 Talc + 3 Calcite = Dolomite + Tremolite + H_2O + CO_2

The time required for C to traverse the MET subcycle (millions to billions of years) is longer than it is in the SED subcycle (Fig. 1), due principally to the typically longer residence times of more deeply buried continental rocks.

The mantle-crust (MAN) subcycle

The deepest (MAN) subcycle includes the mantle C reservoir (Fig. 1) and the processes of subduction and mantle outgassing. Even though rocks that are subducted into the mantle become metamorphosed in the process, their associated C inventories which also reach the mantle become part of the MAN, not the MET, subcycle. The modern global rate of C outgassing from the mantle is approximately 2×10^{12} mol yr^{-1} (Des Marais 1985, Marty and Jambon 1987). The values shown in Figure 1 for subduction of carbonate and reduced C reflect the assumption that rates of C exchange between the mantle, crust and hydrosphere are in steady state, and that subduction does not discriminate between oxidized and reduced C species. Carbon requires on the order of 10's millions to billions of years to traverse the MAN subcycle (Fig 1). Assuming a C content of about 80 μg g^{-1} in the source region of mid-ocean ridge magmas (Pawley et al. 1992, Holloway, pers. comm.), the mantle C inventory is about 2.7×10^{22} mol (Table 1), which is much larger than the crustal C inventory. Such a large mantle inventory indicates that, over timescales of tens of millions to billions of years, any changes in the processes of mantle-crust exchange could have controlled both the abundance and the bulk oxidation state of the much smaller C reservoirs in the HAB and SED subcycles.

Biota that dwell at or below the sea floor (Fig. 2) obtain their energy principally from redox reactions involving reduced hydrothermal emanations (e.g. Jannasch and Wirsen 1979), and therefore these microbes occupy a "gray zone" between the HAB subcycle and the MAN subcycle. Accordingly, their primary productivity cannot exceed the flux of reduced species provided by thermal sources. The principal reduced species include H_2 (derived from water-rock reactions), reduced sulfur and Fe^{2+}. Today, this total flux, expressed as O_2 equivalents, is in the range $(0.2 \text{ to } 2.1) \times 10^{12}$ mol yr^{-1} (Elderfield and Schultz 1996).

THE CARBON ISOTOPE PERSPECTIVE

Additional insights into the biogeochemical C cycle are revealed by $\delta^{13}C$ values of carbonate (δ_{carb}) and reduced organic C (δ_{org}) in the C reservoirs. To the extent that sedimentary rocks have avoided deep burial and alteration, their δ_{carb} and δ_{org} values indicate the status of the C cycle at the time of their deposition. Ranges which represent the bulk of δ_{carb} and δ_{org} values for the various C reservoirs are depicted in Figure 1 (Deines 1980, Weber 1967).

Isotope discrimination

The $\delta^{13}C$ values of recently-deposited sedimentary C are created within the HAB subcycle and at its interface with the SED subcycle. These $\delta^{13}C$ values therefore can be attributed to those processes and reservoirs associated with erosion and outgassing of C, transport and chemical transformation of C within the hydrosphere and atmosphere, and sedimentation and burial of C. Organisms that assimilate compounds with single C atoms (e.g. CO_2 or CH_4) virtually always prefer ^{12}C over ^{13}C, thus forming organic matter having $^{13}C/^{12}C$ values lower than the C source. The net isotopic difference ($\delta_{carb} - \delta_{org} = \Delta_C$) reflects both (1) the metabolic pathways of CO_2 fixation and C metabolism by autotrophs, and (2) the pathways and mechanisms by which organic C is transformed or destroyed. The values δ_{carb} are determined by the $\delta^{13}C$ values of dissolved inorganic C (DIC) in seawater. Therefore, carbonates deposited in open marine environments have δ_{carb} values which are typically representative of the global marine DIC. Today, marine δ_{org} values are established principally by isotopic discrimination during biological fixation of DIC by photosynthetic microorganisms (Hayes et al. 1989). Although marine δ_{org} values can vary regionally, a globally-representative array of samples usually can give an accurate estimate of average global δ_{org}. Anaerobic organic diagenesis can substantially modify δ_{org} values (see discussion below) under conditions where methanogenesis and methylotrophy are quantitatively important pathways for C cycling (Hayes 1994).

Isotopic mass balance

Relative rates of burial of carbonate and organic C in sediments can be estimated by using δ_{carb} and δ_{org} values to construct an isotopic mass balance, as follows:

$$\delta_{in} = f_{carb} \delta_{carb} + f_{org} \delta_{org} \qquad (1)$$

where δ_{in} is the mean $\delta^{13}C$ value of C entering the HAB subcycle, and the equation's right side is the weighted-average δ value of freshly-buried C; f_{carb} and f_{org} are the fractions of C buried as carbonates and organic matter (f_{carb} = 1 - f_{org}), respectively. Rearranging Equation (1) and substituting for f_{carb} yields:

$$f_{org} = (\delta_{carb} - \delta_{in})/(\delta_{carb} - \delta_{org}) \tag{2}$$

Values of f_{org} are controlled by biological productivity, efficiency of organic decomposition, and physical processes controlling C burial. If there is a net reduction of inorganic C to organic C in the HAB subcycle, and if this redox change is transmitted to the sediments (that is, f_{org} increases), then a net oxidation of other redox-sensitive species (e.g. S, H_2O, Fe, Mn) must occur in order to maintain a redox balance. Thus, changes in δ_{carb} and δ_{org} reflect changes, not only in the formation and/or burial of carbonates and organic C, but also in the oxidation state of the oceans and atmosphere.

Isotopic mass balance offers a useful constraint upon estimates of crustal C reservoir sizes. For example, this is particularly helpful because the C inventories in continental margin sediments are very poorly constrained by direct measurements (Table 1; Holser et al. 1988). An estimate of these C inventories can be obtained by combining other observations with an isotope mass balance calculation. Direct abundance measurements are in agreement with mass balance estimates of the relative sizes of organic and inorganic C reservoirs in older Phanerozoic platform and syncline sedimentary rocks (Ronov 1980, Holser et al. 1988) and indicate that the average f_{org} value during the Phanerozoic was approximately 0.18. Incorporating estimates of modern f_{carb} and f_{org} values (+1 and -26, respectively; Deines 1980) into Equation (2), the present-day value for f_{org} lies near 0.21. This f_{org} value, together with the assumption that organic C concentrations in recent continental margin sediments must at least equal or exceed organic C contents in equivalent older rocks (Holser et al. 1988), allows a minimum carbonate C inventory to be estimated ($\sim 10^{21}$ mol) for continental margin sediments (Table 1).

NONBIOLOGICAL AGENTS OF LONG-TERM CHANGE

The sun

Most current models of stellar evolution predict that the sun was less luminous when it first entered the main sequence about 4.6 billion years ago (Newman and Rood 1977). As the sun burned H to He, its core became denser and thus hotter. This increased the rate of thermonuclear burning, which, in turn, increased the sun's luminosity over time. According to one estimate (Gough 1981), the sun was only ~70-71% as luminous 4.6 billion years ago as it is today (Fig. 3). This less luminous sun presents a paradox for earth's early climate (Sagan and Mullen 1972), because, if one assumes that parameters such as atmospheric composition and planetary albedo had been the same as today, then Earth's mean surface temperature would have been below freezing during the first 2 to 2.5 billion years of its history. However, the earliest known (3.8 billion-year-old) metasedimentary rocks indicate that liquid water was abundant (Schopf 1983).

The most direct solution to the faint early sun problem is to invoke a stronger greenhouse effect in the early atmosphere that was sustained by substantially higher atmospheric CO_2 levels (Owen et al. 1979). One-dimensional climate models have been used to estimate the CO_2 levels required (e.g. Kasting 1987). For example, if the global mean temperatures 4.5 and 2.5 billion years ago were equal to the modern global mean temperature, pure CO_2 atmospheres of 1 and 0.1 bar, respectively, would have been required. Therefore, CO_2 declined perhaps by a factor of 1000 or more, from 4.5 billion years ago to the present.

This large CO_2 decline required that the HAB and SED subcycles changed over time. These changes could have occurred as an expression of a self-regulating climate control

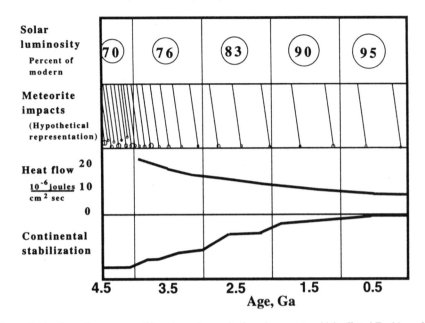

Figure 3. A schematic summary of long-term changes in those key agents which affected Earth's surface environment and the C cycle. The numbers in the circles are estimates of solar luminosity over time, expressed as a percentage of present-day luminosity (Newman and Rood 1977). Hypothetical representation of the distribution of meteorite impacts on Earth through time is based upon an extrapolation of the lunar impact chronology (Maher and Stevenson 1988). The curve estimating change in Earth's heat flow is based upon model calculations (Turcotte 1980). The curve depicting qualitatively the progress of "continental stabilization" is based upon reviews of continental evolution (Lowman 1989, Rogers 1996, Windley 1984).

system (Walker et al. 1981). For example, low solar luminosity would have favored low global temperatures, which in turn would have reduced rates of water evaporation, precipitation and hence low rates of chemical erosion of silicate rocks. Low erosion rates lead to low rates of CO_2 removal from the atmosphere, which would have allowed thermal CO_2 sources to increase atmospheric CO_2 levels. This would have raised surface temperatures until the rate of CO_2 removal by weathering achieved a balance with the thermal sources. As solar luminosity increased slowly over time, the CO_2 levels required to maintain the temperature at which erosional (CO_2 sink) and thermal (CO_2 source) processes balanced would have slowly declined. If the pH of the global ocean remained reasonably constant over time, then seawater HCO_3^- and $CO_3^=$ levels would have declined in close parallel with the atmospheric CO_2 decline. Such a substantial decline in seawater HCO_3^- concentrations (and, therefore, $CO_3^=$ concentrations) undoubtedly affected processes of carbonate precipitation (Grotzinger and Kasting 1993).

Impacts

The role of large impacts upon the C cycle merits more attention than it has received to date. The effect on the C cycle of the ancient (pre-3.8 billion year) heavy bombardment (Fig. 3) should have been substantial, given the sheer magnitude and frequency of those impacts and the likelihood that they affected the origin and survival of our biosphere (e.g. Maher and Stevenson 1988). Impacts on Mars have been credited with releasing crustal volatiles (CO_2, H_2O, etc.), which helped to enhance an ancient greenhouse (Newsom 1980, Brakenridge et al. 1985). However impacts also created dust clouds which, in the aftermath, might have weathered quickly, thus rapidly removing atmospheric CO_2 and perhaps triggering periodic profound cooling of the Hadean climate (N. Sleep, pers. comm.).

More recent impacts probably affected the C cycle, at least over shorter timescales. For example, the impact at the Cretaceous-Tertiary boundary probably released large quantities of CO_2 from the sedimentary carbonates and upper mantle materials that were within the Yucatan target zone (Toon 1997). Impacts of this size very likely affected the HAB subcycle over timescales up to thousands of years. Such impacts might have exerted more profound and permanent effects upon the biosphere, but interpretations of the specific details of these effects have been considerably more controversial.

Heat flow

Mantle outgassing. Over timescales of tens of millions to billions of years, the crustal C inventory was influenced by those processes which govern the exchange of C between Earth's surface and the deep crust and upper mantle. Rubey first proposed that plutonic gas emanations furnished the "excess" volatiles situated within Earth's crust, oceans and atmosphere (Rubey 1951). Mid-ocean ridge volcanism is quantitatively the most important source of mantle volatiles, and crustal production rates along the ridges have apparently varied by several percent over the past 100 Ma (Hays and Pitman 1973). The heat flow from Earth's interior was substantially greater during the earlier Precambrian (e.g. Lambert 1976). During the past 3.0 billion years, radioactive decay of the elements U, Th and ^{40}K has been the principal source of this heat, therefore their decay over time has caused global heat flow to decline (Fig. 3). Thermal fluxes of volatiles, including CO_2, can be scaled linearly to mid-ocean ridge spreading rates, and these rates vary with the square of heat flow (Sleep 1979). Thus, for example, heat flow 3.0 billion years ago has been estimated to have been 2.2 times its modern value (Turcotte 1980), therefore the mid-ocean ridge mantle CO_2 flux was perhaps approximately 5 times its present-day value.

Subduction. C is returned to the mantle via subduction at plate margins. Taking into account the evolution of temperature and pressure regimes of subducting slabs of oceanic crust, together with the stability of C species in such slabs, it was concluded that, for the modern Earth, a substantial fraction of the subducted C could escape dissociation and melting and be carried to considerable depths, possibly to be retained in the mantle for very long periods of time (Huang et al. 1980). However, 3 billion years ago, subducted C would, for a given pressure, have experienced considerably greater temperatures (McCulloch 1993). These greater temperatures enhanced the likelihood that subducted C reacted to form mobile phases which migrated upward and therefore escaped injection into the mantle (Des Marais 1985).

Even with a hotter earlier mantle, rates of C exchange between mantle and crust might have approached near-steady state (Figs. 4a and 4b) if, despite the hotter mantle, certain factors caused the subduction of C to become more effective. For example, high rates of volcanic outgassing probably sustained an atmosphere 3.8 billion years ago that was dominated by abundant CO_2, N_2 and H_2O, with lesser amounts of CO, H_2 and reduced sulfur gases (Kasting 1993a). Dissolved inorganic C levels in seawater would thus have been higher (also see discussion in "The sun" section, above). Accordingly, the removal of CO_2 by weathering of the sea floor (Staudigel and Hart 1983) should have been more extensive. Pervasive carbonation of Archean submarine basalts did indeed occur (Roberts 1987). Subducting slabs of these basalts would have been more extensively carbonated, perhaps contributing to higher C subduction rates.

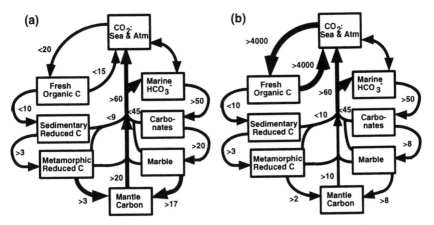

Figure 4. The biogeochemical C cycle at two stages in the early history of Earth. (a) C cycle before the advent of oxygenic photosynthesis, showing the low global primary productivity and the high rates of thermal emanation of C (see text). (b) C cycle during the late Archean, showing enhancement of rates of C cycling in the HAB subcycle due to the enhancement of primary productivity by oxygenic photosynthesis, and showing the declining, but still substantial thermal C flux (see text). Flux estimates are highly approximate, and are included principally to illustrate the directions of change over geologic time.

Metamorphism. In their classic model of the geochemical C cycle, Berner and coworkers recognized that, if the total C inventory in the HAB subcycle has been maintained over time at near-steady state, then net losses of C from the HAB subcycle during sedimentation and burial must have been balanced by the release of CO_2 during thermal metamorphism of sediments (Berner et al. 1983). For example, because carbonates are decarbonated during the subduction of sediments, CO_2 outgassing should covary with global mean spreading rates (which equal global mean subduction rates). Also, greater CO_2 outgassing from deep continental interiors should be correlated with greater worldwide tectonism which also should correlate with spreading rate (Berner et al. 1983). Therefore the mean global rate of CO_2 release from rock metamorphism should be proportional to global heat flow. Consequently, the long-term decline in global heat flow should have favored declining rates of transfer of CO_2 from the crustal sedimentary rocks to the HAB subcycle by thermal processes.

Continental architecture

Role of continents. Continents are important for the C cycle because subaerial weathering is a key CO_2 sink, rivers strongly affect seawater chemistry, and continents have been much more stable repositories of sedimentary C reservoirs than have ocean basins. Accordingly, any long-term changes in the areal extent, thickness, stability and subaerial exposure of continents would have contributed to the long-term evolution of the C cycle.

Ancient continents and the C cycle. An important recent debate within the earth sciences has been whether (1) modern-day continental crust represents nearly all of the crust that accumulated relatively slowly over Earth history, or (2) substantial amounts of crust formed very early and have subsequently experienced extensive recycling into the mantle (Bowring and Housh 1995). Theory (2) thus implies that modern crust is a remnant that has escaped extensive recycling.

If continental crust indeed formed only slowly over geologic time, then the earliest C cycle would have been substantially different. In an ancient world dominated by pervasive, volcanically active and unstable oceanic crust, both organic and carbonate C reservoirs would have resided in less-stable crustal sediments and therefore experienced shorter residence times. Rates of CO_2 removal by subaerial weathering would have been lower due to less land area. Consequently, atmospheric and oceanic C reservoirs would have been larger, perhaps by several orders of magnitude (see, e.g. Berner et al. 1983, Walker 1985).

The ancient C cycle would have been different even if continental crust had formed very early and then sustained extensive processing and recycling. Ancestral continental crust required a considerable interval of time to become stabilized by anatexis, metamorphism and underplating (Lowman 1989). Due to higher heat flow on early Earth, recycling of virtually all crust would have been more vigorous, and most, but not all (Boak and Dymek 1982), of the continental crust that persisted would have experienced higher geothermal gradients. Sedimentary organic C is particularly vulnerable to loss by thermal decomposition. These factors would have shortened the residence times of sedimentary C reservoirs. Higher heat flow sustained higher sea floor spreading rates (Sleep 1979), lower mean ages of ocean crust and, therefore, oceanic crust which was, on average, hotter and more buoyant. An overall greater buoyancy of oceanic crust favored shallower ocean basins, which, in turn, displaced more seawater onto the continents (Hays and Pitman 1973). Thus, even if the inventory of ancient continental crust had been similar to its modern inventory, land area on the ancient, Archean Earth would have been smaller.

COEVOLUTION OF THE C CYCLE AND EARLY LIFE

The Archean

The Archean rock record supports the view that the basic architecture of the C cycle, namely the nested HAB, SED, MET and MAN cycles (Figs. 1 and 4) was in place before 3.5 billion years ago. The major changes over time occurred mainly in the relative sizes of the C reservoirs and the C fluxes between them. The early state of the MAN and HAB subcycles will be discussed, then changes in the C cycle over time will be described as recorded in the sedimentary C reservoirs of the SED subcycle.

The MAN subcycle. The hotter early mantle must have influenced significantly the inventory of C in the crust, oceans and atmosphere. Higher rates of crustal production were accompanied by higher rates of mantle outgassing of C (Des Marais 1985; also, compare Fig. 4 with Fig. 1). A hotter mantle retained subducted C with greater difficulty (McCulloch 1993). These considerations are consistent with an early Earth in which the crustal C inventory actually might have exceeded the modern inventory (Des Marais 1985; Zhang and Zindler 1993), and C cycling between the mantle and crust was more vigorous than today (Figs. 4a and 4b).

Life's early role in the C cycle. The consequences of the biosynthesis of abundant organic C probably represent life's most substantial impact upon the C cycle. Before life began, H_2 gas from thermal sources had escaped to the atmosphere; some of this H_2 had even escaped to space. The earliest biosphere enhanced organic inventories by efficiently oxidizing reduced volcanic and hydrothermal species in order to convert CO_2 to organic matter. After oxygenic photosynthesis developed, the burial in sediments of photosynthetically-derived organic C allowed a stoichiometrically equivalent amount of O_2 and/or its oxidation products to accumulate slowly in the surface environment (e.g. Garrels and Perry 1974). Thus life greatly increased inventories of organic C in the HAB and SED subcycles, and it has simultaneously substantially increased the oxidation state of the hydrosphere and atmosphere.

For example, Figure 4a depicts a hypothetical C cycle before the biosphere invented oxygenic photosynthesis. Unable to split H_2O to obtain H for the reduction of CO_2 to organic C, the biosphere obtained reducing power for organic biosynthesis only by utilizing reduced species derived from volcanism, hydrothermal activity and weathering. Therefore global rates of primary productivity would have approximately equalled the rate of thermal emanations of reduced species. Today, the global flux of reduced species from the mid-ocean ridges is in the range 0.2 to 2×10^{12} mol yr^{-1} (Elderfield and Schultz 1996). This flux was greater in the distant past, due to higher heat flow. Therefore, for the pre-oxygenic-photosynthetic biosphere, it is assumed that global primary productivity scaled with the square of heat flow (the same scaling factor assumed for the mantle CO_2 flux, see "mantle outgassing" section, above), and thus was estimated to have been in the range (2 to 20) $\times 10^{12}$ mol yr^{-1} (Fig. 4).

Climate, atmospheric composition and weathering. Chemical weathering was very effective during the production of Archean sediments, consistent with high CO_2 levels (Walker 1985) and a warm climate (Lowe 1992). Biological activity might have played a major role in sustaining these high weathering rates (see Barker et al., this volume). The crust was tectonically and magmatically unstable and produced thick first cycle sediments in the greenstone belts. The weathering of a typical uplifted rock sequence produced coarse clastic sediments that became enriched in the most chemically-resistant components of the sequence, such as cherts and silicified komatiitic and dacitic tuffs (Nocita and Lowe 1990). These components were derived from silicified sedimentary units that constituted less than 20 percent of the original rock volume. Thus, despite the rapid uplift and transport of these rocks and their debris, their less chemically-resistant components were efficiently degraded. The highly effective weathering implied by these observations is consistent both with relatively warm, moist conditions and with elevated atmospheric CO_2 concentrations (Lowe 1994). Altered evaporites also occur in greenstone sequences between 3.5 and 3.2 billion years ago (e.g. Lowe and Knauth 1977, Buick and Dunlop 1990). Their formation in such tectonically unstable environments is consistent with high rates of evaporation that were promoted by elevated temperatures and dry conditions. The deposition of gypsum rather than anhydrite (Barley et al. 1979, Lowe 1983) indicates that temperatures very likely were below 58°C (Walker 1982).

Stability of the continental crust. Very little well-preserved sedimentary rock and associated C can be found among the relatively few provinces of lithosphere older than 3.0 billion years. The tectonically more active regime within the Archean marine basins favored rapid destruction by continental collisions, partial melting and mantle/crust exchange (Windley 1984). Most of Archean continental crust had yet to become stabilized by cratonization (Rogers 1996), therefore a greater fraction of continental sediments experienced relatively higher rates of instability and thermal alteration. The freshest sediments occur in the 3.5 to 3.2 billion-year-old Kapvaal Craton of South Africa and the Pilbara Block of Western Australia (Lowe 1992). These deposits are associated with episodes of greenstone activity and intrusive events which created stable microcontinents or cratons. These cratons later became the nuclei of full-sized modern continents.

Oxygenic photosynthesis. Figure 4b depicts the C cycle after the advent of oxygenic photosynthesis, which occurred sometime prior to 2.7 billion years ago (Buick 1992). Note the photosynthetically-enhanced global productivity and C recycling rates. Despite the absence of land plants, Archean global productivity was assumed to exceed somewhat the modern rates of marine productivity (Martin et al. 1987). This is because the remains of mid-Proterozoic terrestrial photosynthetic microbial communities have been reported in karst silcretes (e.g. Horodyski and Knauth 1994). Therefore productive Archean terrestrial communities seem possible.

The addition of oxygenic photosynthetic organic C to the HAB subcycle enhanced the net flux of organic C to sediments (compare Figs. 4a and 4b). This flux allowed a stoichiometrically-equivalent net flux of oxidized species, namely O_2 and its oxidation products (e.g. Fe^{3+} and $SO_4^=$), to enter the hydrosphere and atmosphere (Garrels and Perry 1974). However, at first, this novel source of oxidants was nearly totally consumed by reactions with volcanic and hydrothermal emanations.

The late Archean and early Proterozoic

Changes in the MAN subcycle. The evolution of Earth's mantle and crust during the Late Archean and Proterozoic substantially affected the C cycle. Following the inevitable decay of radioactive nuclides in the mantle, the heat flow from Earth's interior declined (Turcotte 1980). This decreased the rates of both sea floor hydrothermal circulation and volcanic outgassing of reduced species. The style of subduction also changed (McCulloch 1993). In the Early- to Mid-Archean, subducted slabs were dehydrated, sustained partial melting, and largely disaggregated in the upper 200 km of the mantle. Later, the reduced heat flow and lower temperatures permitted colder, stronger oceanic lithosphere to form. Subducting slabs thus sustained perhaps only partial dehydration and, together with volatiles such as CO_2 and H_2O, penetrated to depths exceeding 600 km (McCulloch 1993). An increased subduction efficiency of C probably lowered the crustal and atmospheric C inventory over long timescales, but the magnitude of this effect is presently unknown. It has been proposed (Kasting et al. 1993b) that the upper mantle was oxidized by the subduction of water, followed by the escape of reduced gases. A progressive oxidation of the upper mantle has not yet been demonstrated, but, if it had occurred, its effect upon the redox balance of the C cycle would have been substantial.

Tectonic effects on the SED and MET subcycles. The reworking of Archean continental crust by tectonism, igneous activity and metamorphism also had important consequences for the C cycle. Through a process termed 'internal differentiation' (Dewey and Windley 1981), preexisting crust may have become vertically zoned into granitic upper and granulitic lower parts. Also, a subcontinental lithosphere formed perhaps by the extraction of basaltic constituents from the mantle (Jordan 1988, Hoffman 1990), and contributed to a thickening and stabilization of continental crust. Thus, crustal evolution during the Late Archean and Early Proterozoic involved the modification, rearrangement and thickening (over- and underplating) of preexisting crust (Lowman 1989). In addition, lower thermal gradients and longer survival times of oceanic crust favored cooler, less buoyant crust and therefore ocean basins that, on average, became deeper. These trends probably favored the net emergence of continental land area in the interval 3.0 to 2.0 billion years ago, which led to increased subaerial weathering. This helps to explain the more abundant deposition of abundant clastic sediments observed in progressively younger sequences (Lowe 1992). Marine carbonates of this age record a substantial increase in $^{87}Sr/^{86}Sr$ values, indicating greater continental erosion and runoff (Mirota and Veizer 1994). New and extensive stable shallow water platforms became sites for the deposition and long-term preservation of carbonates (Grotzinger 1989) and organic C (Des Marais 1994).

Changes in the HAB and SED subcycles. Increased global continental erosion rates also would have accelerated the rate of decline of atmospheric CO_2 (Berner 1991, Walker 1990). Increased subaerial weathering would have enhanced the delivery of nutrients to coastal waters, enhancing biological productivity (Betts and Holland 1991). Greater productivity would have removed more CO_2 from surface seawater, but, given the still-higher-than-present oceanic and atmospheric inorganic C contents, the effect of this productivity on the atmosphere should have been minor. However, some authors (Lovelock and Whitfield 1982, Schwartzman and Volk 1991) have proposed that soil biota accelerated the weathering process on land, drawing CO_2 levels down even further. In any case, a declining atmospheric CO_2 inventory would have contributed to a late Archean decline in global temperatures. The first well-recorded glaciations occurred in the Late Archean (von Brunn and Gold 1993) and Early Proterozoic (Harland 1983). Perhaps these events represent the consequences of this declining atmospheric CO_2 inventory.

Patterns of carbonate deposition, as well as the presence/absence of gypsum/anhydrite in associated evaporites, indicate that seawater concentrations of HCO_3^-, $CO_3^=$, have declined, and $SO_4^=$ has increased since the late Archean (Grotzinger and Kasting 1993). Late Archean platform sequences include relatively abundant evidence of abiotic carbonate precipitation as tidal flat tufas and marine cements. Evaporite sequences often proceed directly from carbonate to halite deposition, thus excluding gypsum/anhydrite deposition. These observations are consistent with significantly lower seawater $SO_4^=$ concentrations and/or considerably greater HCO_3^- concentrations. In either case, the ratio of HCO_3^- to Ca^{2+} was significantly large to prevent deposition of gypsum/anhydrite in marine or marginal-marine environments. These observations are consistent with the view that inorganic C reservoirs within the HAB subcycle were much higher during the Archean and Paleoproterozoic (Walker 1985).

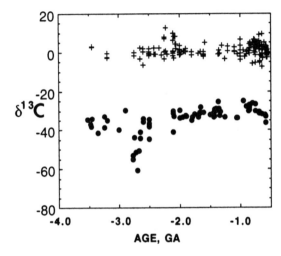

Figure 5. Plot of age versus δ_{carb} (+) and δ_{org} (●) for Archean and Proterozoic kerogens. Kerogen data are corrected for the effects of thermal alteration (Des Marais 1997). Between 2.2 to 2.0 billion years ago, note the high δ_{carb} values and the virtual disappearance thereafter of δ_{org} values more negative than -36. Other evidence indicates that atmospheric O_2 increased substantially at this time (see text).

Evolution of the Proterozoic biosphere and its environment

Carbon isotopic evidence. A survey of $\delta^{13}C$ values in Proterozoic sedimentary rocks (Fig. 5) reveals the antiquity of the isotopic dichotomy between carbonates and reduced C, which is a legacy of isotopic discrimination during organic biosynthesis (Schidlowski 1988). Several other robust features are seen in Figure 5. First, in rocks older than 2.1 billion years, δ_{org} values range widely from the mid -20 range to as low as   -65, with many values more negative than -35. In contrast, rocks younger than 2.0 billion years have δ_{org} values in the more positive, narrower range -25 to -35, with virtually no values more negative than -36. Substantial positive excursions in δ_{carb} are observed between 2.2 and 2.0 billion years ago and between 1.0 and 0.6 billion years ago (Fig. 5). Hayes proposed that the extremely negative δ_{org} values observed in ~2.8 to 2.6 billion-year-old kerogens were created by extensive methane recycling during organic diagenesis (Hayes 1983, Hayes 1994). A positive trend in δ_{org} values was identified between 2.5 and 0.6 billion years ago (Strauss et al. 1992a, Des Marais et al. 1992), which they attributed to progressively increasing relative rates of organic burial and preservation. A large positive δ_{carb} excursion between 2.2 and 2.0 billion years ago was attributed to an episode of enhanced organic burial (Baker and Fallick 1989, Karhu and Holland 1996). Knoll and coworkers reported several positive δ_{carb} excursions in Neoproterozoic (0.8 to 0.6 billion-year-old) sediments (Knoll et al. 1986); they also proposed that these excursions indicated global episodes of enhanced relative rates of organic sedimentation.

The above observations reconstruct a self-consistent view of the relationship between changing C isotopic abundances and an evolving global environment and biosphere (Des Marais 1997). The Archean atmosphere had only trace amounts of O_2, and its CO_2 inventory was perhaps two to three orders of magnitude greater than it is today (Kasting 1993a). With very high CO_2 availability, CO_2-fixing bacteria (e.g. cyanobacteria, anoxygenic photosynthetic bacteria and chemoautotrophs) can assimilate ^{12}C preferentially over ^{13}C to the maximum degree possible (Des Marais et al. 1989, Guy et al. 1993, Ruby et al. 1987). Given a supply of CO_2 with a δ value near -10 (assuming isotopic equilibrium between marine CO_2 and carbonates at ambient temperatures), many autotrophic prokaryotes, particularly the globally-abundant cyanobacteria, create δ_{org} values more negative than -30. Subsequent anaerobic processing of this organic C by fermenters, methanogens and methylotrophs (CH_4-utilizing bacteria) can create substantially lower δ_{org} values (Hayes 1994). Because methanogens are obligate anaerobes and methylotrophs are microaerophiles, a global environment having low but nonzero concentrations of atmospheric O_2 would have been optimal for sustaining a globally prominent methanogen-methylotroph cycle (Hayes 1994). Indeed, using a range of redox indicators, Holland estimates that atmospheric $[O_2]$ levels prior to 2.2 billion years ago were no more than 1 to 2% of the present atmospheric level (or PAL, Holland 1994). Thus, the numerous low (<-35) δ_{org} values in kerogens older than 2.2 billion years fulfill expectations based upon independent assessments of the environment and the biosphere.

Conditions were altered substantially during the interval 2.2 to 1.9 billion years ago. Atmospheric $[O_2]$ levels rose to more than 15% PAL (Holland 1994, Knoll and Holland 1995). Sedimentary $\delta_{sulfide}$ values became more variable and ranged to include substantially lower values (Cameron 1982). This evidence for enhanced isotopic discrimination during bacterial sulfate reduction to sulfide in shallow marine sediments indicated that $SO_4^=$ concentrations in seawater had increased significantly. Microfossils (Han and Runnegar 1992) and organic biomarker compounds (Summons and Walter 1990) appeared, indicating that O_2-requiring eukaryotes had become globally prominent. By virtually all accounts, the environment of the HAB subcycle had become substantially more oxidized.

The record of δ_{carb} and δ_{org} values provides further evidence that the Proterozoic environment became more oxidized between 2.2 and 1.9 billion years ago (Fig. 5). The large positive δ_{carb} excursion between 2.2 and 2.06 billion years ago (Baker and Fallick 1989, Karhu and Holland 1996) indicates, by isotopic mass balance considerations (Eqn. 1), that the relative rate of organic burial increased (that is, f_{org} increased; see Eqn. 2). Increased net organic burial led to increased $[O_2]$, $[SO_4^=]$ and sedimentary Fe^{3+} (Garrels and Perry 1974). These oxidants drove the once- pervasive methanogen/methylotroph cycle principally into restricted sedimentary enclaves. As the Pasteur Point (the $[O_2]$ level at which oxic respiration becomes viable) was exceeded, oxic respiration and bacterial sulfate reduction became, as they are today, globally-dominant pathways for the decomposition of organic C. Because these pathways cause minimal C isotopic discrimination (e.g. Blair et al. 1985, Kaplan and Rittenberg 1964), they created few opportunities for sedimentary organic C to assume δ_{org} values lower than -35 during diagenesis, which is consistent with the δ_{org} record after 2.1 billion years ago (Fig. 5). After 1.9 billion years ago, biological CO_2-assimilation probably became the dominant mechanism controlling the carbon isotopic depletion of sedimentary organic C, relative to coeval carbonate.

When populations eukaryotic photosynthetic microbes assumed global prominence, as evidenced by their appearance in the fossil record between 2.0 and 1.7 billion years ago (Han and Runnegar 1992, Summons and Walter 1990), they also began to influence global mean δ_{org} values substantially. When global δ_{carb} values were near 0, minimum δ_{org} values would therefore lie in the range -34 to -37, consistent with the record (Fig. 5). As described by models of climate (Kasting 1993a) and declining volcanic emanations (Des Marais 1985) environmental [CO_2] levels declined during the Proterozoic, gradually restricting the CO_2 available for organic biosynthesis. Declining CO_2 availability ultimately will suppress C isotopic discrimination during CO_2 uptake (e.g. Calder and Parker 1978, Des Marais et al. 1989, Des Marais et al. 1992). Perhaps the detectible decline in isotopic discrimination between 2.0 and 0.6 billion years ago (Δ_C declined from 33 to 28; Fig. 5) was a consequence of this long-term CO_2 decline. Phanerozoic Δ_C values are typically 26 (Hayes et al. 1989, Deines 1980).

Changing patterns of mineral deposition. The declining atmospheric CO_2 levels and the increasing O_2 levels were also heralded by the first appearance, 1.7 to 1.6 billion years ago, of massive bedded gypsum deposits (Grotzinger and Kasting 1993). This appearance is consistent with a decline in HCO_3^- and an increase in $SO_4^=$ concentrations in the global oceans. Abiotic carbonate precipitation as tidal flat tufas and marine cements, relatively abundant in Archean and early Proterozoic sequences, declined markedly in the mid-Proterozoic (Grotzinger and Kasting 1993). Such trends reflected fundamental changes in the abundance, style and biological control of mineral precipitation in the global oceans. Algal control of carbonate precipitation ultimately became a globally-important process (see de Vrind-de Jong and de Vrind, this volume).

Organic C concentrations in sedimentary rocks. Those sedimentary rocks which escaped moderate to high-grade metamorphism contain more than three-fourths of the crustal inventory of organic C (Hunt 1979), even though they constitute a relatively small volume of the crust. In their examination of Proterozoic kerogens, Strauss et al. (Strauss et al. 1992b) restricted their attention principally to rocks having organic C contents greater than 0.05 mg/g. Organic C contents less than 0.05 mg/g are frequently obscured by contamination from exogenous, younger sources.

A logarithmic plot of total organic C abundance versus sample age (Fig. 6) illustrates that more than 96 percent of the samples analyzed yielded values less than 30 mg/g and that the mean content was in the range 3 to 5 mg/g. Shales have generally higher organic C contents than carbonates and cherts, consistent with observations of Phanerozoic sediments (Ronov 1980). Organic C concentrations in shales and carbonates are also similar to those in Phanerozoic sediments, even when the greater degree of thermal destruction of the Proterozoic kerogens is taken into account (Strauss et al. 1992b).

Thus, no long-term trend in the concentration of organic C is apparent among sedimentary rocks representing the past 2.5 billion years. Consequently, the proposed long-term increase in the size of the crustal organic C reservoir could only have been accommodated by long-term changes in the global sedimentation rates and/or the preservation of rocks rich in organic C (Des Marais 1994).

Summary. The isotopic and fossil records are consistent with a progressively oxidizing environment that diminished anaerobic and microaerophilic populations of microbes and allowed O_2-dependent populations to assume global prominence. Evidence for increasing seawater $SO_4^=$ concentrations are also consistent with a more oxidizing

GEOMICROBIOLOGY

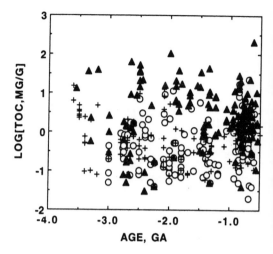

Figure 6. Logarithm of total organic carbon (TOC) concentrations in sedimentary rocks versus age. Data points depict lithologies as follows:

    ○   carbonates
    +   cherts
    ▲   shales

environment. Both the isotopic and other geochemical lines of evidence reveal that this oxidation was episodic in nature (Knoll 1985, Baker and Fallick 1989, Des Marais et al. 1992, Karhu and Holland 1996).

The long-term $[CO_2]$ decline was also eventually expressed, both in the progressive decline in C isotopic discrimination during the mid- and late-Proterozoic (Des Marais 1997), and in the decline in the abundance of massive abiotic carbonate precipitates (Grotzinger and Kasting 1993).

The C cycle, O_2 and the evolution of Eukarya

Because Eukarya (i.e. the Eukaryotes; e.g. algae, fungi, ultimately also plants, animals, etc.) offer a vivid example of biological evolution during the Proterozoic Eon, their potential relationship to changes in the C cycle deserves mention. The oldest known Eukarya are spirally-coiled, megascopic fossils in 2.1 billion-year-old shales (Han and Runnegar 1992). A major increase in the diversity of fossil algae and other unicellular Eukarya is observed in rocks approximately 1.0 billion years in age (Knoll and Walter 1992). The terminal Proterozoic diversification which led to plants and animals occurred between 0.53 and 0.59 billion years ago.

The evolutionary history of Eukarya might be linked to the C cycle through O_2 (e.g. Knoll and Holland 1995). O_2 is required by virtually all Eukarya having organelles (e.g. all algae, fungi, plants and animals have mitochondria; plants also have chloroplasts). To the extent that certain subgroups of Eukarya require specific minimum O_2 levels, specific groups perhaps appeared soon after their required O_2 levels were first attained. The earliest fossils of Eukarya appeared during the Paleoproterozoic "C isotope" event between 2.2 and 2.0 billion years ago (see above). The C isotope event at the end of the Proterozoic accompanied the earliest well-documented occurrence of multicellular life (Kaufman and Knoll 1995). The coincidence in timing between these C isotope events, the other evidence for in atmospheric O_2 increases, and the evolutionary events in Eukarya indicates that an important biogeochemical linkage might exist, however the evidence is not yet compelling (Knoll and Holland 1995).

The biosphere – C cycle connection: future work

Although the preceding discussion indicates that key interactions between the C cycle, atmosphere, hydrosphere and biosphere have occurred, a definitive proof of a "cause-and-effect" relationship between changes in the C cycle and biological evolution has not yet been fully achieved. What else is required? Paleoproterozoic paleontology must establish more convincingly the time of first appearance of O_2-requiring Eukarya (Knoll and Holland 1995). The causes of the truly remarkable positive excursions in δ_{carb} between 2.2 and 2.0 billion years ago and during the Neoproterozoic (Fig. 5) must be understood. Mechanistic relationships must be clarified which link the growth of the $SO_4^=$ and Fe^{3+} reservoirs to the coeval tectonic and geothermal processes. Evidence for changes in the reservoirs of O_2, $SO_4^=$, and Fe^{3+} during the Neoproterozoic should be better quantified. A more precise reconstruction of such biogeochemical phenomena promises us a deeper understanding, both of the C cycle and of our own origins.

ACKNOWLEDGMENTS

This review was supported by a grant from NASA's Exobiology Program. The author is indebted to J. Banfield for her patience and encouragement during the preparation of this manuscript.

REFERENCES

Baker AJ, Fallick AE (1989) Evidence from Lewisian limestones for isotopically heavy carbon in two-thousand-million-year-old-sea water. Nature 337:352-354

Barley ME, Dunlop JSR, Glover JE, Groves DI (1979) Sedimentary evidence for an Archaean shallow-water volcanic-sedimentary facies, eastern Pilbara Block, Western Australia. Earth Planet Sci Lett 43:74-84

Berner RA (1991) A model for atmospheric CO_2 over Phanerozoic time. Am J Sci 291:339-376

Berner RA, Canfield DE (1989) A new model for atmospheric oxygen over Phanerozoic time. Amer J Sci 289:333-361

Berner RA, Lasaga AC, Garrels RM (1983) The carbonate-silicate geochemical cycle and its effect on atmospheric carbon dioxide over the past 100 million years. Am J Sci 283:641-683

Betts JN, Holland HD (1991) The oxygen content of ocean bottom waters, the burial efficiency of organic carbon, and the regulation of atmospheric oxygen. Palaeogeogr, Palaeoclimatol, Palaeoecol 97:5-18

Blair N, Leu A, Munoz E, Olson J, Des Marais DJ (1985) Carbon isotopic fractionation in heterotrophic microbial metabolism. Appl Microbiol 50:996-1001

Boak JL, Dymek RF (1982) Metamorphism of the ca. 3800 Ma supracrustal rocks at Isua, West Greenland: Implications for early Archean crustal evolution. Earth Planet Sci Lett 59:155-176

Bowring SA, Housh T (1995) The Earth's early evolution. Science 269:1535-1540

Brakenridge GR, Newsom HE, Baker VR (1985) Ancient hot spring on Mars: origins and paleoenvironmental significance of small Martian valleys. Geology 13:859

Broecker WS (1982) Tracers in the Sea. Lamont-Doherty Geological Observatory, Palisades, New York

Buick R (1992) The antiquity of oxygenic photosynthesis: Evidence from stromatolites in sulphate-deficient Archaean lakes. Science 255:74-77

Buick R, Dunlop JSR (1990) Evaporitic sediments of early Archean age from the Warrawoona Group, North Pole, Western Australia. Sedimentol 37:247-277

Calder JA, Parker PL (1978) Geochemical implications of induced changes in C^{13} fractionation by blue-green algae. Geochim Cosmochim Acta 37:133-140

Cameron EM (1982) Sulphate and sulphate reduction in early Precambrian oceans. Nature 296:145-148

Deines P (1980) The isotopic composition of reduced organic carbon. In: Fritz P, Fontes JC (eds) Handbook of Environmental Isotope Geochemistry, p 329-406 Elsevier, Amsterdam

Derry LA, Kaufman AJ, Jacobsen SB (1992) Sedimentary cycling and environmental change in the Late Proterozoic: Evidence from stable and radiogenic isotopes. Geochim Cosmochim Acta 56:1317-1329

Des Marais DJ (1985) Carbon exchange between the mantle and crust and its effect upon the atmosphere: today compared to Archean time. In: Sundquist ET, Broecker WS (eds) The Carbon Cycle and Atmospheric CO_2: Natural Variations Archean to Present, p 602-611 American Geophysical Union, Washington, DC

Des Marais DJ (1994) Tectonic control of the crustal organic carbon reservoir during the Precambrian. Chem Geol 114:303-314

Des Marais DJ (1995) The biogeochemistry of subtidal marine hypersaline microbial mats, Guerrero Negro, Baja California Sur, Mexico. In: Jones JG (ed) Advances in Microbial Ecology, p 251-274 Plenum, New York

Des Marais DJ (1997) The isotopic evolution of the biogeochemical carbon cycle during the Proterozoic Eon. Organic Geochem (in press)

Des Marais DJ, Cohen Y, Nguyen H, Cheatham M, Cheatham T, Munoz E (1989) Carbon isotopic trends in the hypersaline ponds and microbial mats at Guerrero Negro, Baja California Sur, Mexico: Implications for Precambrian stromatolites. In: Cohen Y, Rosenberg E (eds) Microbial Mats: Physiological Ecology of Benthic Microbial Communities, p 191-205 Am Soc Microbiology, Washington, DC

Des Marais DJ, Strauss H, Summons RE, Hayes JM (1992) Carbon isotope evidence for the stepwise oxidation of the Proterozoic environment. Nature 359:605-609

Dewey JF, Windley BF (1981) Growth and differentiation of the continental crust. Phil Trans R Soc London A 301:189-206

Elderfield H, Schultz A (1996) Mid-ocean ridge hydrothermal fluxes and the chemical composition of the ocean. Ann Rev Earth Planetary Sci 24:191-224

Ferry JM (1991) Dehydration and decarbonation reactions as a record of fluid infiltration. In: Derrick DM (ed) Contact Metamorphism. Rev Mineral 26:351-394

Garrels RM, Perry EA, Jr. (1974) Cycling of carbon, sulfur, and oxygen through geologic time. In: Goldberg ED (ed) The Sea, p 303-336 John Wiley & Sons, New York

Gough DO (1981) Solar interior structure and luminosity variations. Solar Physics 74:21-34

Grotzinger JP (1989) Facies and evolution of Precambrian carbonate depositional systems: emergence of the modern platform archetype. In: Crevello PD, Wilson JL, Sarg JF, Read JF (eds) Controls on Carbonate Platform and Basin Development, p 79-106 Soc Econ Paleontol Mineral, Tulsa, Oklahoma

Grotzinger JP, Kasting JF (1993) New constraints on Precambrian ocean composition. J Geol 101:235-243

Guy RD, Fogel ML, Berry JA (1993) Photosynthetic fractionation of the stable isotopes of oxygen and carbon. Plant Physiol 101:37-47

Han TM, Runnegar B (1992) Megascopic eukaryotic algae from the 2.1-billion-year-old Negaunee Iron-Formation, Michigan. Science 257:232-235

Harland WB (1983) The Proterozoic glacial record. Mem Geol Soc Am 161:279-288

Hayes JM (1983) Geochemical evidence bearing on the origin of aerobiosis, a speculative hypothesis. In: Schopf JW (ed) Earth's Earliest Biosphere, p 291-301 Princeton Univ Press, Princeton, New Jersey

Hayes JM (1994) Global methanotrophy at the Archean-Proterozoic transition. In: Bengtson S (ed) Early Life on Earth. Nobel Symp 84, p 220-236 Columbia Univ Press, New York

Hayes JM, Popp BN, Takigiku R, Johnson MW (1989) An isotopic study of biogeochemical relationships between carbonates and organic carbon in the Greenhorn Formation. Geochim Cosmochim Acta 53:2961-2972

Hays JD, Pitman WC (1973) Lithospheric plate motion, sea level changes, and ecological consequences. Nature 246:18-22

Hoffman PF (1990) Geological constraints on the origin of the mantle root beneath the Canadian shield. Phil Trans R Soc London A 331:523-532

Holland HD (1984) The Chemical Evolution of the Atmosphere and Oceans. Princeton Univ Press, Princeton, New Jersey

Holland HD (1994) Early Proterozoic atmospheric change. In: Bengtson S (ed) Early Life on Earth, p 237-244 Columbia Univ Press, New York

Holser WT, Schidlowski M, Mackenzie FT, Maynard JB (1988) Geochemical cycles of carbon and sulfur. In: Gregor CB, Garrels RM, Mackenzie FT and Maynard JB (ed) Chemical Cycles in the Evolution of the Earth, p 105-173 John Wiley & Sons, New York

Horodyski RJ, Knauth LP (1994) Life on land in the Precambrian. Science 263:494-498

Huang W-L, Wyllie PJ, Nehru CE (1980) Subsolidus and liquidus phase relationships in the system CaO-SiO_2-CO_2 to 30 kbar with geological applications. Am Mineral 65:285-301

Hunt JM (1972) Distribution of carbon in crust of earth. Am Assoc Pet Geol Bull 56:2273-2277

Hunt JM (1979) Petroleum Geochemistry and Geology. W H Freeman, San Francisco

Jannasch HW, Wirsen CO (1979) Chemosynthetic primary production at East Pacific sea floor spreading centers. BioSci 29:592-598

Jordan TH (1988) Structure and formation of the continental lithosphere. J Petrology Special Lithosphere Issue, p 11-37

Kaplan IR, Rittenberg SC (1964) Carbon isotope fractionation during metabolism of lactate by *Desulfovibrio desulfuricans*. J Gen Microbiol 34:213-217

Karhu JA, Holland HD (1996) Carbon isotopes and the rise of atmospheric oxygen. Geology 24:867-870

Kasting JF (1987) Theoretical constraints on oxygen and carbon dioxide concentrations in the Precambrian atmosphere. Precamb Res 34:205-229

Kasting JF (1993a) Earth's early atmosphere. Science 259:920-926
Kasting JF, Eggler DH, Raeburn SP (1993b) Mantle redox evolution and the oxidation state of the Archean atmosphere. J Geol 101:245-257
Kaufman AJ, Knoll AH (1995) Neoproterozoic variations in the C-isotopic composition of seawater: stratigraphic and biogeochemical implications. Precamb Res 73:27-49
Knoll AH (1985) Exceptional preservation of photosynthetic organisms in silicified carbonates and silicified peats. Phil Trans R Soc Lond B B311:111-122
Knoll AH, Hayes JM, Kaufman AJ, Swett K, Lambert IB (1986) Secular variation in carbon isotope ratios from Upper Proterozoic successions of Svalbard and East Greenland. Nature 321:832-838
Knoll AH, Holland HD (1995) Oxygen and Proterozoic evolution: an update. In: Stanley S (ed) Effects of Past Global Change on Life, p 21-33 National Academy Press, Washington, DC
Knoll AH, Walter MR (1992) Latest Proterozoic stratigraphy and Earth history. Nature 356:673-678
Lambert RSJ (1976) Archean thermal regimes, crustal and upper mantle temperatures, and a progressive evolutionary model for the Earth. In: Windley BF (ed) The Early History of the Earth, p 363-373 John Wiley, New York
Lovelock JE, Whitfield M (1982) Lifespan of the biosphere. Nature 296:561-563
Lowe DR (1983) Restricted shallow-water sedimentation of 3.4 Byr-old stromatolitic and evaporitic strata of the Strelley Pool Chert, Pilbara Block, Western Australia. Precamb Res 19:239-283
Lowe DR (1992) Major events in the geological development of the Precambrian Earth. In: Schopf JW and Klein C (ed) The Proterozoic Biosphere: a multidisciplinary study, p 67-76 Cambridge Univ Press, New York
Lowe DR (1994) Early environments: constraints and opportunities for early evolution. In: Bengtson S (ed) Early Life on Earth. Nobel Symposium 84, p 24-35 Columbia Univ Press, New York
Lowe DR, Knauth LP (1977) Sedimentology of the Onverwacht Group (3.4 billion years), Transvaal, South Africa, and its bearing on the characteristics and evolution of the early Earth. J Geol 85:699-723
Lowman PDJ (1989) Comparative planetology and the origin of continental crust. Precamb Res 44:171-195.
Maher KA, Stevenson DJ (1988) Impact frustration of the origin of life. Nature 331:612-614
Martin JH, Knauer GA, Karl DM, Broenkow WW (1987) VERTEX: carbon cycling in the northeast Pacific. Deep Sea Res 34:267-285
Marty B, Jambon A (1987) C/^{3}He in volatile fluxes from the solid Earth: implicatons for carbon geodynamics. Earth Planet Sci Lett 83:16-26
McCulloch MT (1993) The role of subducted slabs in an evolving earth. Earth Planet Sci Lett 115:89-100.
Mirota MD, Veizer J (1994) Geochemistry of Precambrian carbonates. 6. Aphebian Albanel formations, Quebec, Canada. Geochim Cosmochim Acta 58:1735-1745
Mopper K, Degens ET (1979) Organic carbon in the ocean: nature and cycling. In: Bolin B (ed) The Global Carbon Cycle, p 293-316 Wiley, New York
Newman MJ, Rood RT (1977) Implications of solar evolution for the Earth's early atmosphere. Science 198:1035-1037
Newsom HE (1980) Hydrothermal alteration of impact melt sheets with implications for Mars. Icarus 44:207
Nocita BW, Lowe DR (1990) Fan-delta sequence in the Archean Fig Tree Group. Precamb Res 48:375-393.
Olson JS, Garrels RM, Berner RA, Armentano TV, Dyer MI, Taalon DH (1985) The natural carbon cycle. In: Trabalka JR (ed) Atmospheric Carbon Dioxide and the Global Carbon Cycle, p 175-213 US Department of Energy, Washington, DC
Owen T, Cess RD, Ramanathan V (1979) Early Earth: an enhanced carbon dioxide greenhouse to compensate for reduced solar luminosity. Nature 277:640-642
Pawley AR, Holloway JR, McMillan PF (1992) The effect of oxygen fugacity on the solubility of carbon-oxygen fluids in basaltic melt. Earth Planet Sci Lett 110:213-225
Roberts RG (1987) Ore deposit models No. 11. Archean lode gold deposits. Geosci Can 14:37-52
Rogers JJW (1996) A history of continents in the past three billion years. J Geol 104:91-107
Ronov AB (1980) Osadochnaja Oblochka Zemli (Sedimentary Layer of the Earth). Nauka, Moscow
Rubey WW (1951) Geologic history of sea water—an attempt to state the problem. Geol Soc Am Bull 62:1111-1148
Ruby EG, Jannasch HW, Deuser WG (1987) Fractionation of stable carbon isotopes during chemoautotrophic growth of sulfur-oxidizing bacteria. Appl Microbiol 53:1940-1943
Sagan C, Mullen G (1972) Earth and Mars: Evolution of atmospheres and surface temperatures. Science 177:52-56
Schidlowski M (1988) A 3,800-million-year isotopic record of life from carbon in sedimentary rocks. Nature 333:313-318
Schopf JW (1983) Earth's Earliest Biosphere. Princeton Univ Press, Princeton, New Jersey

Schwartzman DW, Volk T (1991) Biotic enhancement of weathering and surface temperatures on Earth since the origin of life. Global Planet Change 90:357-371

Sleep NH (1979) Thermal history and degassing of the Earth: some simple calculations. J Geol 87:671-686

Staudigel H, Hart SR (1983) Alteration of basaltic glass: Mechanism and significance for the oceanic crust-seawater budget. Geochim Cosmochim Acta 47:337-350

Strauss H, Des Marais DJ, Hayes JM, Summons RE (1992a) Proterozoic organic carbon - its preservation and isotopic record. In: Schidlowski M (ed) Proc Final Meeting of IGCP Project 157, Maria Laach, West Germany, p 203-211 Springer-Verlag, Berlin

Strauss H, Des Marais DJ, Summons RE, Hayes JM (1992b) Concentrations of organic carbon and maturities and elemental compositions of kerogens. In: Schopf JW, Klein C (ed) The Proterozoic Biosphere: a multidisciplinary study, p 95-100 Cambridge Univ Press, New York

Summons RE, Walter MR (1990) Molecular fossils and microfossils of prokaryotes and protists from Proterozoic sediments. Am J Sci 290-A:212-244

Toon OB (1997) Environmental perturbation caused by the impacts of asteroids and comets. Rev Geophys 35:41-78

Turcotte DL (1980) On the thermal evolution of the Earth. Earth Planet Sci Lett 48:53-58

von Brunn V, Gold DJC (1993) Diamictite in the Archaean Pongola Sequence of southern Africa. J Afr Earth Sci 16:367-374

Walker JCG (1982) Climatic factors on the Archean earth. Palaeogeogr, Palaeoclimatol, Palaeoecol 40:1-11

Walker JCG (1985) Carbon dioxide on the early Earth. Orig Life 16:117-127

Walker JCG (1990) Precambrian evolution of the climate system. Palaeogeogr, Palaeoclimatol, Palaeoecol 82:261-289

Walker JCG, Hays PB, Kasting JF (1981) A negative feedback mechanism for the long-term stabilization of Earth's surface temperature. J Geophys Res 86:9776-9782

Weber JN (1967) Possible changes in the isotopic composition of the oceanic and atmospheric carbon reservoir over geologic time. Geochim Cosmochim Acta 31:2343-2351

Windley BF (1984) The Evolving Continents. John Wiley & Sons, New York

Zhang Y, Zindler A (1993) Distribution and evolution of carbon and nitrogen in Earth. Earth Planet Sci Lett 117:331-345